D0782569

FOUNDATIONS
OF
FACTOR ANALYSIS
SECOND EDITION

Chapman & Hall/CRC

Statistics in the Social and Behavioral Sciences Series

Series Editors

A. Colin Cameron
University of California, Davis, USA

J. Scott Long
Indiana University, USA

Andrew Gelman
Columbia University, USA

Sophia Rabe-Hesketh
University of California, Berkeley, USA

Anders Skrondal
Norwegian Institute of Public Health, Norway

Aims and scope

Large and complex datasets are becoming prevalent in the social and behavioral sciences and statistical methods are crucial for the analysis and interpretation of such data. This series aims to capture new developments in statistical methodology with particular relevance to applications in the social and behavioral sciences. It seeks to promote appropriate use of statistical, econometric and psychometric methods in these applied sciences by publishing a broad range of reference works, textbooks and handbooks.

The scope of the series is wide, including applications of statistical methodology in sociology, psychology, economics, education, marketing research, political science, criminology, public policy, demography, survey methodology and official statistics. The titles included in the series are designed to appeal to applied statisticians, as well as students, researchers and practitioners from the above disciplines. The inclusion of real examples and case studies is therefore essential.

Published Titles

Analysis of Multivariate Social Science Data, Second Edition
David J. Bartholomew, Fiona Steele, Irini Moustaki, and Jane I. Galbraith

Bayesian Methods: A Social and Behavioral Sciences Approach, Second Edition
Jeff Gill

Foundations of Factor Analysis, Second Edition
Stanley A. Mulaik

Linear Causal Modeling with Structural Equations
Stanley A. Mulaik

Multiple Correspondence Analysis and Related Methods
Michael Greenacre and Jorg Blasius

Multivariable Modeling and Multivariate Analysis for the Behavioral Sciences
Brian S. Everitt

Statistical Test Theory for the Behavioral Sciences
Dato N. M. de Gruijter and Leo J. Th. van der Kamp

Chapman & Hall/CRC
Statistics in the Social and Behavioral Sciences Series

FOUNDATIONS OF FACTOR ANALYSIS

SECOND EDITION

STANLEY A. MULAIK

CRC Press
Taylor & Francis Group
Boca Raton London New York

CRC Press is an imprint of the
Taylor & Francis Group an informa business
A CHAPMAN & HALL BOOK

Chapman & Hall/CRC
Taylor & Francis Group
6000 Broken Sound Parkway NW, Suite 300
Boca Raton, FL 33487-2742

© 2010 by Taylor and Francis Group, LLC
Chapman & Hall/CRC is an imprint of Taylor & Francis Group, an Informa business

No claim to original U.S. Government works

Printed in the United States of America on acid-free paper
10 9 8 7 6 5 4 3 2 1

International Standard Book Number: 978-1-4200-9961-4 (Hardback)

This book contains information obtained from authentic and highly regarded sources. Reasonable efforts have been made to publish reliable data and information, but the author and publisher cannot assume responsibility for the validity of all materials or the consequences of their use. The authors and publishers have attempted to trace the copyright holders of all material reproduced in this publication and apologize to copyright holders if permission to publish in this form has not been obtained. If any copyright material has not been acknowledged please write and let us know so we may rectify in any future reprint.

Except as permitted under U.S. Copyright Law, no part of this book may be reprinted, reproduced, transmitted, or utilized in any form by any electronic, mechanical, or other means, now known or hereafter invented, including photocopying, microfilming, and recording, or in any information storage or retrieval system, without written permission from the publishers.

For permission to photocopy or use material electronically from this work, please access www.copyright.com (http://www.copyright.com/) or contact the Copyright Clearance Center, Inc. (CCC), 222 Rosewood Drive, Danvers, MA 01923, 978-750-8400. CCC is a not-for-profit organization that provides licenses and registration for a variety of users. For organizations that have been granted a photocopy license by the CCC, a separate system of payment has been arranged.

Trademark Notice: Product or corporate names may be trademarks or registered trademarks, and are used only for identification and explanation without intent to infringe.

Library of Congress Cataloging-in-Publication Data

Mulaik, Stanley A., 1935-
Foundations of factor analysis / Stanley A. Mulaik. -- 2nd ed.
p. cm.
Includes bibliographical references and index.
ISBN 978-1-4200-9961-4 (hardcover : alk. paper)
1. Factor analysis. I. Title.

BF39.M845 2010
519.5'354--dc22 2009032729

Visit the Taylor & Francis Web site at
http://www.taylorandfrancis.com

and the CRC Press Web site at
http://www.crcpress.com

Contents

Preface to the Second Edition

This is a book for those who want or need to get to the bottom of things. It is about the foundations of factor analysis. It is for those who are not content with accepting on faith the many equations and procedures that constitute factor analysis but want to know where these equations and procedures came from. They want to know the assumptions underlying these equations and procedures so that they can evaluate them for themselves and decide where and when they would be appropriate. They want to see how it was done, so they might know how to add modifications or produce new results.

The fact that a major aspect of factor analysis and structural equation modeling is mathematical means that getting to their foundations is going to require dealing with some mathematics. Now, compared to the mathematics needed to fully grasp modern physics, the mathematics of factor analysis and structural equation modeling is, I am happy to say, relatively easy to learn and not much beyond a sound course in algebra and certainly not beyond a course in differential calculus, which is often the first course in mathematics for science and engineering majors in a university. It is true that factor analysis relies heavily on concepts and techniques of linear algebra and matrix algebra. But these are topics that can be taught as a part of learning about factor analysis. Where differential calculus comes into the picture is in those situations where one seeks to maximize or minimize some algebraic expression. Taking derivatives of algebraic expressions is an analytic process, and these are algebraic in nature. While best learned in a course on calculus, one can still be shown the derivatives needed to solve a particular optimization problem. Given that the algebra of the derivation of the solution is shown step by step, a reader may still be able to follow the argument leading to the result. That, then, is the way this book has been written: I teach the mathematics needed as it is needed to understand the derivation of an equation or procedure in factor analysis and structural equation modeling.

This text may be used at the postgraduate level as a first-semester course in advanced correlational methods. It will find use in psychology, sociology, education, marketing, and organizational behavior departments, especially in their quantitative method programs. Other ancillary sciences may also find this book useful. It can also be used as a reference for explanations of various options in commercial computer programs for performing factor analysis and structural equation modeling.

There is a logical progression to the chapters in this text, reflecting the hierarchical structure to the mathematical concepts to be covered. First, in Chapter 2 one needs to learn the basic mathematics, principally linear algebra and matrix algebra and the elements of differential calculus. Then one needs to learn about composite variables and their means, variances, covariances, and correlation in terms of means, variances, and covariances among their component variables. Then one builds on that to deal with multiple and partial correlations, which are special forms of composite variables. This is accomplished in Chapter 3.

Differential calculus will first be encountered in Chapter 4 in demonstrating where the estimates of the regression weights come from, for this involves finding the weights of a linear combination of predictor variables that has either the maximum correlation with the criterion or the minimum expected difference with the criterion.

In Chapter 5 one uses the concepts of composite variables and multiple and partial regression to understand the basic properties of the multivariate normal distribution. Matrix algebra becomes essential to simplify notations that often involve hundreds of variables and their interrelations.

By this point, in Chapter 6 one is ready to deal with factor analysis, which is an extension of regression theory wherein predictor variables are now unmeasured, and hypothetical latent variables and dependent variables are observed variables. Here we describe the fundamental equation and fundamental theorem of factor analysis, and introduce the basic terminology of factor analysis. In Chapter 7, we consider how common factors may be extracted. We show that the methods used build on concepts of regression and partial correlation. We first look at a general algorithm for extracting factors proposed by Guttman. We then consider the diagonal and the centroid methods of factoring, both of which are of historical interest. Next we encounter eigenvectors and eigenvalues. Eigenvectors contain coefficients that are weights used to combine additively observed variables or their common parts into variables that have maximum variances (the eigenvalues), because they will be variables that account for most of the information among the original variables in any one dimension.

To perform a common-factor analysis one must first have initial estimates of the communalities and unique variances. Lower-bound estimates are frequently used. These bounds and their effect on the number of factors to retain are discussed in Chapter 8. The unique variances are either subtracted from the diagonal of the correlation matrix or the diagonal matrix of the reciprocals of their square roots are pre- and postmultiplied times the correlation matrix, and then the eigenvectors and eigenvalues of the resulting matrix are obtained. Different methods use different estimates of unique variances. The formulas for the eigenvectors and eigenvalues of a correlation matrix, say, are obtained by using differential calculus to solve the maximization problem of finding the weights of a linear combination of the variables that has the maximum variance under the constraint that the sum of the squares of the weights adds to unity. These will give rise to the factors, and the common factors will in turn be basis vectors of a common factor space, meaning that the observed variables are in turn linear combinations of the factors.

Maximum-likelihood-factor analysis, the equations for which were ultimately solved by Karl Jöreskog (1967), building on the work of precursor statisticians, also requires differential calculus to solve the maximization problem involved. Furthermore, the computational solution for the maximum-likelihood estimates of the model parameters cannot be solved directly by any algebraic analytic procedure. The solution has to be obtained numerically and iteratively. Jöreskog (1967) used a then new computer algorithm for nonlinear optimization, the Fletcher–Powell algorithm. We will explain how this works to obtain the maximum-likelihood solution for the exploratory factor-analysis model.

Chapter 9 examines several variants of the factor-analysis model: principal components, weighted principal components, image analysis, canonical factor analysis, descriptive factor analysis, and alpha factor analysis.

In Chapters 10 (simple structure and graphical rotation), 11 (orthogonal analytic rotation), and 12 (oblique analytic rotation) we consider factor rotation. Rotation of factors to simple structures will concern transformations of the common-factor variables into a new set of variables that have a simpler relationship with the observed variables. But there are numerous mathematical criteria of what constitutes a simple structure solution. All of these involve finding the solution for a transformation matrix for transforming the initial "unrotated solution" to one that maximizes or minimizes a mathematical expression constituting the criterion for simple structures. Thus, differential calculus is again involved in finding the algorithms for conducting a rotation of factors, and the solution is obtained numerically and iteratively using a computer.

Chapter 13 addresses whether or not it is possible to obtain scores on the latent common factors. It turns out that solutions for these scores are not unique, even though they are optimal. This is the factor-indeterminacy problem, and it concerns more than getting scores on the factors: there may be more than one interpretation for a common factor that fits the data equally well.

Chapter 14 deals with factorial invariance. What solutions for the factors will reveal the same factors even if we use different sets of observed variables? What coefficients are invariant in a factor-analytic model under restriction of range? Building on ideas from regression, the solution is effectively algebraic.

While much of the first 14 chapters is essentially unchanged from the first edition, developments that have taken place since 1972, when the first edition was published, have been updated and revised.

I have changed the notation to adopt a notation popularized by Karl Jöreskog for the common-factor model. I now write the model equation as $\mathbf{Y} = \mathbf{\Lambda X} + \mathbf{\Psi E}$ instead of $\mathbf{Z} = \mathbf{FX} + \mathbf{UV}$ and the equation of the fundamental theorem as $\mathbf{R}_{YY} = \mathbf{\Lambda \Phi}_{XX}\,\mathbf{\Lambda}' + \mathbf{\Psi}^2$ instead of $\mathbf{R}_{ZZ} = \mathbf{FC}_{XX}\mathbf{F}' + \mathbf{U}^2$.

I have added a new Chapter 5 on the multivariate normal distribution and its general properties along with the concept of maximum-likelihood estimation based on it. This will increase by one the chapter numbers for subsequent chapters corresponding to those in the first edition. However,

Chapter 12, on procrustean rotation in the first edition, has been dropped, and this subject has been briefly described in the new Chapter 12 on oblique rotation. Chapters 13 and 14 deal with factor scores and indeterminacy, and factorial invariance under restriction of range, respectively.

Other changes and additions are as follows. I am critical of several of the methods that are commonly used to determine the number of factors to retain because they are not based on sound statistical or mathematical theory. I have now directed some of these criticisms toward some methods in renumbered Chapter 8. However, since then I have also realized that, in most studies, a major problem with determining the number of factors concerns the presence of doublet variance uncorrelated with $n - 2$ observed variables and correlated between just the two of them. Doublet correlations contribute to the communalities, but lead to an overestimate of the number of overdetermined common factors. But the common-factor model totally ignores the possibility of doublets, while they are everywhere in our empirical studies. Both unique factor variance and doublet variance should be separated from the overdetermined common-factor variance. I have since rediscovered that a clever but heuristic solution for doing this, ignored by most factor-analytic texts, appeared in a paper I cited but did not understand sufficiently to describe in the first edition. This paper was by John Butler (1968), and he named his method "descriptive factor analysis." I now cover this more completely (along with a new method of my own I call "doublet factor analysis") in Chapter 9 as well as other methods of factor analysis. It provides an objective way to determine the number of overdetermined common factors to retain that those who use the eigenvalues greater than 1.00 rule of principal components will find more to their liking in the smaller number of factors it retains.

I show in Chapter 9 that Kaiser's formula (1963) for the principal axes factor structure matrix for an image analysis, $\Lambda = \mathbf{SA}_r\left[(\gamma_i - 1)^2/\gamma_i\right]_r^{1/2}$, is not the correct one to use, because this represents the "covariances" between the "image" components of the variables and the underlying "image factors." The proper factor structure matrix that represents the correlations between the unit variance "observed" variables and the unit variance image factors is none other than the weighted principal component solution: $\Lambda = \mathbf{SA}_r[\gamma_i]_r^{1/2}$.

In the 1990s, several new approaches to oblique rotation to simple structure were published, and these and earlier methods of rotation were in turn integrated by Robert Jennrich (2001, 2002, 2004, 2006) around a simple core computing algorithm, the "gradient projection algorithm," which seeks the transformation matrix for simultaneously transforming all the factors. I have therefore completely rewritten Chapter 12 on analytic oblique rotation on the basis of this new, simpler algorithm, and I show examples of its use.

In the 1970s, factor score indeterminacy was further developed by several authors, and in 1993 many of them published additional exchanges on the subject. A discussion of these developments has now been included in an expanded Chapter 13 on factor scores.

Factor analysis was also extended to confirmatory factor analysis and structural equation modeling by Jöreskog (1969, 1973, 1974, 1975). As a consequence, these later methodologies diverted researchers from pursuing exclusively exploratory studies to pursuing hypothesis-testing studies. Confirmatory factor analysis is now best treated separately in a text on structural equation modeling, as a special case of that method, but I have rewritten a chapter on confirmatory factor analysis for this edition. Exploratory factor analysis still remains a useful technique in many circumstances as an abductive, exploratory technique, justifying its study today, although its limitations are now better understood.

I wish to thank Robert Jennrich for his help in understanding the gradient projection algorithm. Jim Steiger was very helpful in clarifying Peter Schönemann's papers on factor indeterminacy, and I owe him a debt of gratitude.

I wish to thank all those who, over the past 30 years, have encouraged me to revise this text, specifically, Jim Steiger, Michael Browne, Abigail Panter, Ed Rigdon, Rod McDonald, Bob Cudeck, Ed Loveland, Randy Engle, Larry James, Susan Embretson, and Andy Smith. Without their encouragement, I might have abandoned the project. I owe Henry Kaiser, now deceased, an unrepayable debt for his unselfish help in steering me into a career in factor analysis and in getting me the postdoctoral fellowship at the University of North Carolina. This not only made the writing of the first edition of this book possible, but it also placed me in the intellectually nourishing environment of leading factor analysts, without which I could never have gained the required knowledge to write the first edition or its current sequel.

I also acknowledge the loving support my wife Jane has given me through all these years as I labored on this book into the wee hours of the night.

Stanley A. Mulaik

Preface to the First Edition

When I was nine years old, I dismantled the family alarm clock. With gears, springs, and screws—all novel mechanisms to me—scattered across the kitchen table, I realized that I had learned two things: how to take the clock apart and what was inside it. But this knowledge did not reveal the mysteries of the all-important third and fourth steps in the process: putting the clock back together again and understanding the theory of clocks in general. The disemboweled clock ended up in the trash that night. A common experience, perhaps, but nonetheless revealing.

There are some psychologists today who report a similar experience in connection with their use of factor analysis in the study of human abilities or personality. Very likely, when first introduced to factor analysis they had the impression that it would allow them to analyze human behavior into its components and thereby facilitate their activities of formulating structural theories about human behavior. But after conducting a half-dozen factor analyses they discovered that, in spite of the plethora of factors produced and the piles of computer output stacked up in the corners of their offices, they did not know very much more about the organization of human behavior than they knew before factor analyzing. Perhaps after factor analyzing they would admit to appreciating more fully the complexity of human behavior, but in terms of achieving a coherent, organized conception of human behavior, they would claim that factor analysis had failed to live up to their expectations.

With that kind of negative appraisal being commonly given to the technique of factor analysis, one might ask why the author insisted on writing a textbook about the subject. Actually, I think the case against factor analysis is not quite as grim as depicted above. Just as my youthful experience with the alarm clock yielded me fresh but limited knowledge about the works inside the clock, factor analysis has provided psychologists with at least fresh but limited knowledge about the components, although not necessarily the organization, of human psychological processes.

If some psychologists are disillusioned by factor analysis' failure to provide them with satisfactory explanations of human behavior, the fault probably lies not with the model of factor analysis itself but with the mindless application of it; many of the early proponents of the method encouraged this mindless application by their extravagant claims for the method's efficacy in discovering, as if by magic, underlying structures. Consequently, rather than use scientific intuition and already-available knowledge about the properties of variables under study to construct theories about the nature of relationships among the variables and formulating these theories as factor-analytic models to be tested against empirical data, many researchers have randomly picked variables representing a domain to be studied, intercorrelated the variables, and then factor-analyzed them expecting that the theoretically important variables of the domain would be revealed by the analysis. Even Thurstone (1947, pp. 55–56), who considered exploration of a new domain a legitimate application of factor-analytic methods, cautioned that exploration with factor analysis required carefully chosen variables and that results from using the method were only provisional in suggesting ideas for further research. Factor analysis is not a method for discovering full-blown structural theories about a domain. Experience has justified Thurstone's caution for, more often than one would like to admit, the factors obtained from purely exploratory studies have been difficult to integrate with other theory, if they have been interpretable at all. The more substantial contributions of factor analysis have been made when researchers postulated the existence of certain factors, carefully selected variables from the domain that would isolate their existence, and then proceeded to factor-analyze in such a way as to reveal these factors as clearly as possible. In other words, factor analysis has been more profitably used when the researcher knew what he or she was looking for.

Several recent developments, discussed in this book, have made it possible for researchers to use factor-analytic methods in a hypothesis-testing manner.

For example, in the context of traditional factor-analytic methodology, using procrustean transformations (discussed in Chapter 12), a researcher can rotate an arbitrary factor-pattern matrix to approximate a hypothetical factor-pattern matrix as much as possible. The researcher can then examine the goodness of fit of the rotated pattern matrix to the hypothetical pattern matrix to evaluate his or her hypothesis.

In another recent methodological development (discussed in Chapter 15) it is possible for the researcher to formulate a factor-analytic model for a set of variables to whatever degree of completeness he or she may desire, leaving the unspecified parameters of the model to be estimated in such a way as to optimize goodness of fit of the hypothetical model to the data, and then to test the overall model for goodness of fit against the data. The latter approach to factor analysis, known as confirmatory factor analysis or analysis of covariance structures, represents a radical departure from the traditional methods of performing factor analysis and may eventually become the predominant method of using the factor-analytic model.

The objective of this book, as its title *Foundations of Factor Analysis* suggests, is to provide the reader with the mathematical rationale for factor-analytic procedures. It is thus designed as a text for students of the behavioral or social sciences at the graduate level. The author assumes that the typical student who uses this text will have had an introductory course in calculus, so that he or she will be familiar with ordinary differentiation, partial differentiation, and the maximization and minimization of functions using calculus. There will practically be no reference to integral calculus, and many of the sections will be comprehensible with only a good grounding in matrix algebra, which is provided in Chapter 2. Many of the mathematical concepts required to understand a particular factor-analytic procedure are introduced along with the procedure.

The emphasis of this book is algebraic rather than statistical. The key concept is that (random) variables may be treated as vectors in a unitary vector space in which the scalar product of vectors is a defined operation. The empirical relationships between the variables are represented either by the distances or the cosines of the angles between the corresponding vectors. The factors of the factor-analytic model are basis vectors of the unitary vector space of observed variables. The task of a factor-analytic study is to find a set of basis vectors with optimal properties from which the vectors corresponding to the observed variables can be derived as linear combinations.

Chapter 1 provides an introduction to the role factor-analytic models can play in the formulation of structural theories, with a brief review of the history of factor analysis. Chapter 2 provides a mathematical review of concepts of algebra and calculus and an introduction to vector spaces and matrix algebra. Chapter 3 introduces the reader to properties of linear composite variables and the representation of these properties by using matrix algebra. Chapter 4 considers the problem of finding a particular linear combination of a set of random variables that is minimally distant from some external random variable in the vector space containing these variables. This discussion introduces the concepts of multiple correlation and partial correlation; the chapter ends with a discussion on how image theory clarifies the meaning of multiple correlation. Multiple- and partial-correlational methods are seen as essential, later on, to understanding methods of extracting factors.

Chapter 5 plunges into the theory of factor analysis proper with a discussion on the fundamental equations of common-factor analysis. Chapter 6 discusses methods of extracting factors; it begins with a discussion on a general algorithm for extracting factors and concludes with a discussion on methods for obtaining principal-axes factors by finding the eigenvectors and eigenvalues of a correlation matrix.

Chapter 7 considers the model of common-factor analysis in greater detail with a discussion on (1) the importance of overdetermining factors by the proper selection of variables, (2) the inequalities regarding the lower bounds to the communalities of variables, and (3) the fitting of the unrestricted common-factor model to a correlation matrix by least-squares and maximum-likelihood estimation.

Chapter 8 discusses factor-analytic models other than the common-factor model, such as component analysis, image analysis, image factor analysis, and alpha factor analysis. Chapter 9 introduces the reader to the topic of factor rotation to simple structure using graphical rotational methods. This discussion is followed by a discussion on analytic methods of orthogonal rotation in Chapter 10 and of analytic methods of oblique rotation in Chapter 11.

Chapters 12 and 13 consider methods of procrustean transformation of factors and the meaning of factorial indeterminacy in common-factor analysis in connection with the problem of estimating the common factors, respectively. Chapter 14 deals with the topic of factorial invariance of the common-factor model over sampling of different variables and over selection of different populations. The conclusion this chapter draws is that the factor-pattern matrix is the only invariant of a common-factor-analysis model under conditions of varied selection of a population.

Chapter 15 takes up the new developments of confirmatory factor analysis and analysis of covariance structures, which allow the researcher to test hypotheses using the model of common-factor analysis. Finally, Chapter 16 shows how various concepts from factor analysis can be applied to methods of multivariate analysis, with a discussion on multiple correlations, step-down regression onto a set of several variables, canonical correlations, and multiple discriminant analysis.

Some readers may miss discussions in this book of recent offshoots of factor-analytic theory such as 3-mode factor analysis, nonlinear factor analysis, and nonmetric factor analysis. At one point of planning, I felt such topics might be included, but as the book developed, I decided that if all topics pertinent to factor analysis were to be included, the book might never be finished. And so this book is confined to the more classic methods of factor analysis.

I am deeply indebted to the many researchers upon whose published works I have relied in developing the substance of this book. I have tried to give credit wherever it was due, but in the event of an oversight I claim no credit for any development made previously by another author.

I especially wish to acknowledge the influence of my onetime teacher, Dr. Calvin W. Taylor of the University of Utah, who first introduced me to the topic of factor analysis while I was studying for my PhD at the University of Utah, and who also later involved me in his factor-analytic research as a research associate. I further wish to acknowledge his role as the primary contributing cause in the chain of events that led to the writing of this book, for it was while substituting for him, at his request, as the instructor of his factor-analysis course at the University of Utah in 1965 and 1966 that I first conceived of writing this book. I also wish to express my gratitude to him for generously allowing me to use his complete set of issues of *Psychometrika* and other factor-analytic literature, which I relied upon extensively in the writing of the first half of this book.

I am also grateful to Dr. Henry F. Kaiser who, through correspondence with me during the early phases of my writing, widened my horizons and helped and encouraged me to gain a better understanding of factor analysis.

To Dr. Lyle V. Jones of the L. L. Thurstone Psychometric Laboratory, University of North Carolina, I wish to express my heartfelt appreciation for his continued encouragement and support while I was preparing the manuscript of this book. I am also grateful to him for reading earlier versions of the manuscript and for making useful editorial suggestions that I have tried to incorporate in the text. However, I accept full responsibility for the final form that this book has taken.

I am indebted to the University of Chicago Press for granting me permission to reprint Tables 10.10, 15.2, and 15.8 from Harry H. Harman's *Modern Factor Analysis*, first edition, 1960. I am also indebted to Chester W. Harris, managing editor of *Psychometrika*, and to the following authors for granting me permission to reprint tables taken from their articles that appeared in *Psychometrika*: Henry F. Kaiser, Karl G. Jöreskog, R. Darrell Bock and Rolf E. Bargmann, R. I. Jennrich, and P. F. Sampson. I am especially grateful to Lyle V. Jones and Joseph M. Wepman for granting me permission to reprint a table from their article, "Dimensions of language performance," which appeared in the *Journal of Speech and Hearing Research*, September, 1961.

I am also pleased to acknowledge the contribution of my colleagues, Dr. Elliot Cramer, Dr. John Mellinger, Dr. Norman Cliff, and Dr. Mark Appelbaum, who in conversations with me at one time or another helped me to gain increased insights into various points of factor-analytic theory, which were useful when writing this book. In acknowledging this contribution, however, I take full responsibility for all that appears in this book.

I am also grateful for the helpful criticism of earlier versions of the manuscript of this book, which were given to me by my students in my factor-analysis classes at the University of Utah and at the University of North Carolina.

I also wish to express my gratitude to the following secretaries who at one time or another struggled with the preparation of portions of this manuscript: Elaine Stewart, Judy Nelson, Ellen Levine, Margot Wasson, Judy Schenck, Jane Pierce, Judy Schoenberg, Betsy Schopler, and Bess Autry. Much of their help would not have been possible, however, without the support given to me by the L. L. Thurstone Psychometric Laboratory under the auspices of Public Health Service research grant No. M-10006 from the National Institute of Mental Health, and National Science Foundation Science Development grant No. GU 2059, for which I am ever grateful.

Stanley A. Mulaik

1

Introduction

1.1 Factor Analysis and Structural Theories

By a structural theory we shall mean a theory that regards a phenomenon as an aggregate of elemental components interrelated in a lawful way. An excellent example of a structural theory is the theory of chemical compounds: Chemical substances are lawful compositions of the atomic elements, with the laws governing the compositions based on the manner in which the electron orbits of different atoms interact when the atoms are combined in molecules.

Structural theories occur in other sciences as well. In linguistics, for example, structural descriptions of language analyze speech into phonemes or morphemes. The aim of structural linguistics is to formulate laws governing the combination of morphemes in a particular language. Biology has a structural theory, which takes, as its elemental components, the individual cells of the organism and organizes them into a hierarchy of tissues, organs, and systems. In the study of the inheritance of characters, modern geneticists regard the manifest characteristics of an organism (phenotype) as a function of the particular combination of genes (genotype) in the chromosomes of the cells of the organism.

Structural theories occur in psychology as well. At the most fundamental level a psychologist may regard behaviors as ordered aggregates of cellular responses of the organism. However, psychologists still have considerable difficulty in formulating detailed structural theories of behavior because many of the physical components necessary for such theories have not been identified and understood. But this does not make structural theories impossible in psychology. The history of other sciences shows that scientists can understand the abstract features of a structure long before they know the physical basis for this structure. For example, the history of chemistry indicates that chemists could formulate principles regarding the effects of mixing compounds in certain amounts long before the atomic and molecular aspects of matter were understood. Gregor Mendel stated the fundamental laws of inheritance before biologists had associated the chromosomes of the cell with inheritance. In psychology, Isaac Newton, in 1704, published a simple mathematical model of the visual effects of mixing different hues, but nearly a hundred years elapsed before Thomas Young postulated the existence of three types of color receptors in the retina to account for the relationships described in Newton's model. And only a half-century later did physiologist Helmholtz actually give a physiological basis to Young's theory. Other physiological theories subsequently followed. Much of psychological theory today still operates at the level of stating relationships among stimulus conditions and gross behavioral responses.

One of the most difficult problems of formulating a structural theory involves discovering the rules that govern the composition of the aggregates of components. The task is much easier if the scientist can show that the physical structure he is concerned with is isomorphic to a known mathematical structure. Then, he can use the many known theorems of the mathematical structure to make predictions about the properties of the physical structure. In this regard, George Miller (1964) suggests that psychologists have used the structure of euclidean space more than any other mathematical structure to represent structural relationships of psychological processes. He cites, for example, how Isaac Newton's (1704) model for representing the effects of mixing different hues involved taking the hues of the spectrum in their natural order and arranging them as points appropriately around the circumference of a circle. The effects of color mixtures could be determined by proportionally weighting the points of the hues in the mixture according to their contribution to the mixture and finding the center of gravity of the resulting points. The closer this center of gravity approached the center of the color circle, the more the resulting color would appear gray. In addition, Miller cites Schlosberg's (1954) representation of perceived emotional similarities among facial expressions by a two-dimensional graph with one dimension interpreted as pleasantness versus unpleasantness and the other as rejection versus attention, and Osgood's (1952) analysis of the components of meaning of words into three primary components: (1) evaluation, (2) power, and (3) activity.

Realizing that spatial representations have great power to suggest the existence of important psychological mechanisms, psychologists have developed techniques, such as metric and nonmetric factor analysis and metric and nonmetric multidimensional scaling, to create, systematically, spatial representations from empirical measurements. All four of these techniques represent objects of interest (e.g., psychological "variables" or stimulus "objects") as points in a multidimensional space. The points are so arranged with respect to one another in the space as to reflect relationships of similarity among the corresponding objects (variables) as given by empirical data on these objects.

Although a discussion of the full range of techniques using spatial representations of relationships found in data would be of considerable interest, we shall confine ourselves, in this book, to an in depth examination of the methods of factor analysis. The reason for this is that the methodology of factor analysis is historically much more fully developed than, say, that of multidimensional scaling; as a consequence, prescriptions for the ways of doing factor analysis are much more established than they are for these other techniques. Furthermore, factor analysis, as a technique, dovetails very nicely with such classic topics in statistics as correlation, regression, and multivariate analysis, which are also well developed. No doubt, as the gains in the development of multidimensional scaling, especially the nonmetric versions of it, become consolidated, there will be authors who will write textbooks about this area as well. In the meantime, the interested reader can consult Torgerson's (1958) textbook on metric multidimensional scaling for an account of that technique.

1.2 Brief History of Factor Analysis as a Linear Model

The history of factor analysis can be traced back into the latter half of the nineteenth century to the efforts of the British scientist Francis Galton (1869, 1889) and other scientists to discover the principles of the inheritance of manifest characters (Mulaik, 1985, 1987). Unlike Gregor Mendel (1866), who is today considered the founder of modern genetics, Galton did not try to discover these principles chiefly through breeding experiments using simple, discrete characters of organisms with short maturation cycles; rather, he concerned himself with human traits such as body height, physical strength, and intelligence, which today are not believed to be simple in their genetic determination. The general question asked by Galton was: To what extent are individual differences in these traits inherited and by what mechanism? To be able to answer this question, Galton had to have some way of quantifying the relationships of traits of parents to traits of offspring. Galton's solution to this problem was the method of regression. Galton noticed that, when he took the heights of sons and plotted them against the heights of their fathers, he obtained a scatter of points indicating an imperfect relationship. Nevertheless, taller fathers tended strongly to have on the average taller sons than shorter fathers. Initially, Galton believed that the average height of sons of fathers of a given height would be the same as the height of the fathers, but instead the average was closer to the average height of the population of sons as a whole. In other words, the average height of sons "regressed" toward the average height in the population and away from the more extreme height of their fathers. Galton believed this implied a principle of inheritance and labeled it "regression toward the mean," although today we regard the regression phenomenon as a statistical artifact associated with the linear-regression model. In addition, Galton discovered that he could fit a straight line, called the regression line, with positive slope very nicely through the average heights of sons whose fathers had a specified height. Upon consultation with the mathematician Karl Pearson, Galton learned that he could use a linear equation to relate the heights of fathers to heights of sons (cf. Pearson and Lee, 1903):

$$Y = a + bX + E \tag{1.1}$$

Here

- Y is the height of a son
- a the intercept with the Y-axis of the regression line passing through the averages of sons with fathers of fixed height
- b the slope of the regression line
- X the height of the father
- E an error of prediction

As a measure of the strength of relationship, Pearson used the ratio of the variance of the predicted variable, $\hat{Y} = a + bX$, to the variance of Y.

Pearson, who was an accomplished mathematician and an articulate writer, recognized that the mathematics underlying Galton's regression method had already been worked out nearly 70 years earlier by Gauss and other mathematicians in connection with determining the "true" orbits of planets from observations of these orbits containing error of observation. Subsequently, because the residual variate E appeared to be normally distributed in the prediction of height, Pearson identified the "error of observation" of Gauss's theory of least-squares estimation with the "error of prediction" of Equation 1.1 and treated the predicted component $\hat{Y} = a + bX$ as an estimate of an average value. This initially amounted to supposing that the average heights of sons would be given in terms of their fathers' heights by the equation $Y = a + bX$ (without the error term), if nature and environment did not somehow interfere haphazardly in modifying the predicted value. Although Pearson subsequently found the field of biometry on such an exploitation of Gauss's least-squares theory of error, modern geneticists now realize that heredity can also contribute to the E term in Equation 1.1 and that an uncritical application of least-squares theory in the study of the inheritance of characters can be grossly misleading.

Intrigued with the mathematical problems implicit in Galton's program to metricize biology, anthropology, and psychology, Pearson became Galton's junior colleague in this endeavor and contributed enormously as a mathematical innovator (Pearson, 1895). After his work on the mathematics of regression, Pearson concerned himself with finding an index for indicating the type and degree of relationship between metric variables (Pearson, 1909). This resulted in what we know today as the product-moment correlation coefficient, given by the formula

$$\rho_{XY} = \frac{E[(X - E(X))(Y - E(Y))]}{\sqrt{E[(X - E(X))^2]E[(Y - E(Y))^2]}} \tag{1.2}$$

where

- $E[\,]$ is the expected-value operator
- X and Y are two random variables

This index takes on values between −1 and +1, with 0 indicating no relationship. A deeper meaning for this coefficient will be given later when we consider that it represents the cosine of the angle between two vectors, each standing for a different variable.

In and of itself the product-moment correlation coefficient is descriptive only in that it shows the existence of a relationship between variables without showing the source of this relationship, which may be causal or coincidental in nature. When the researcher obtains a nonzero value for the product-moment correlation coefficient, he must supply an explanation for the relationship between the related variables. This usually involves finding that one variable is the cause of the other or that some third variable (and maybe others) is a common cause of them both. In any case, interpretations of correlations are greatly facilitated if the researcher already has a structural model on which to base his interpretations concerning the common component producing a correlation.

To illustrate, some of the early applications of the correlation coefficient in genetics led nowhere in terms of furthering the theory of inheritance because explanations given to nonzero correlation coefficients were frequently tautological, amounting to little more than saying that a relationship existed because such-and-such relationship-causing factor was present in the variables exhibiting the relationship. However, when Mendel's theory of inheritance (Mendel, 1865) was rediscovered during the last decade of the nineteenth century, researchers of hereditary relationships had available to them a structural mechanism for understanding how characters in parents were transmitted to offspring. Working with a trait that could be measured quantitatively, a geneticist could hypothesize a model of the behavior of the genes involved and from this model draw conclusions about, say, the correlation between relatives for the trait in the population of persons. Thus, product-moment correlation became not only an exploratory, descriptive index but an index useful in hypothesis testing. R. A. Fisher (1918) and Sewell Wright (1921) are credited with formulating the methodology for using correlation in testing Mendelian hypotheses.

With the development of the product-moment correlation coefficient, other related developments followed: In 1897, G. U. Yule published his classic paper on multiple and partial correlation. The idea of multiple correlation was this: Suppose one has p variables, $X_1, X_2, \ldots, X_p$, and wishes to find that linear combination

$$\hat{X}_1 = \beta_2 X_2 + \cdots + \beta_p X_p \qquad (1.3)$$

of the variables $X_2, \ldots, X_p$, which is maximally correlated with X_1. The problem is to find the weights $\beta_2, \ldots, \beta_p$ that make the linear combination $\hat{X}_1$ maximally correlated with X_1. After Yule's paper was published, multiple correlation became quite useful in prediction problems and turned out to be systematically related but not exactly equivalent to Gauss's solution for linear least-squares estimation of a variable, using information obtained on several independent variables observed at certain preselected values. In any case, with multiple correlation, the researcher can consider several components (on which he had direct measurements) as accounting for the variability in a given variable.

At this point, the stage was set for the development of factor analysis. By the time 1900 had arrived, researchers had obtained product-moment correlations on many variables such as physical measurements of organisms, intellectual-performance measures, and physical-performance measures. With variables showing relationships with many other variables, the need existed to formulate structural models to account for these relationships. In 1901, Pearson published a paper on lines and planes of closest fit to systems of points in space, which formed the basis for what we now call the principal-axes method of factoring. However, the first common-factor-analysis model is attributed to Spearman (1904). Spearman intercorrelated the test scores of 36 boys on topics such as classics, French, English, mathematics, discrimination of tones, and musical talent. Spearman had a theory, primarily attributed by him to Francis Galton and Herbert Spencer, that the abilities involved in taking each of these six tests were a general ability, common to all the tests, and a specific ability, specific to each test. Mathematically, this amounts to the equation

$$Y_j = a_j G + \psi_j \tag{1.4}$$

where

- Y_j is the jth manifest variable (e.g., test score in mathematics)
- a_j is a weight indicating the degree to which the latent general-ability variable G participates in Y_j
- ψ_j is an ability variable uncorrelated with G and specific to Y_j

Without loss of generality, one can assume that $E(Y_j) = E(\psi_j) = E(G) = 0$, for all j, implying that all variables have zero means. Then saying that ψ_j is specific to Y_j amounts to saying that ψ_j does not covary with another manifest variable Y_k, so that $E(Y_k\ \psi_j) = 0$, with the consequence that $E(\psi_j\ \psi_k) = 0$ (implying that different specific variables do not covary). Thus the covariances between different variables are due only to the general-ability variable, that is,

$$\begin{aligned} E(Y_j Y_k) &= E[(a_j G + \psi_j)(a_k G + \psi_k)] \\ &= E(a_j a_k G^2 + a_j G \psi_k + a_k G \psi_j + \psi_j \psi_k) \\ &= a_j a_k E(G^2) \end{aligned} \tag{1.5}$$

From covariance to correlation is a simple step. Assuming in Equation 1.5 that $E(G^2) = 1$ (the variance of G is equal to 1), we then can derive the correlation between Y_j and Y_k:

$$\rho_{jk} = a_j a_k \tag{1.6}$$

Spearman noticed that the pattern of correlation coefficients obtained among the six intellectual-test variables in his study was consistent with the model of a single common variable and several specific variables.

For the remainder of his life Spearman championed the doctrine of one general ability and many specific abilities, although evidence increasingly accumulated which disconfirmed such a simple model for all intellectual-performance variables. Other British psychologists contemporary with Spearman either disagreed with his interpretation of general ability or modified his "two-factor" (general and specific factors) theory to include "group factors," corresponding to ability variables not general to all intellectual variables but general to subgroups of them. The former case was exemplified by Godfrey H. Thomson (cf. Thomson, 1956), who asserted that the mind does not consist of a part which participates in all mental performance and a large number of particular parts which are specific to each performance. Rather, the mind is a relatively undifferentiated collection of many tiny parts. Any intellectual performance that we may consider involves only a "sample" of all these many tiny parts. Two performance measures are intercorrelated because they sample overlapping sets of these tiny parts. And Thomson was able to show how data consistent with Spearman's model were consistent with his sampling-theory model also.

Another problem with G is that one tends to define it mathematically as "the common factor common to all the variables in the set" rather than in terms of something external to the mathematics, for example "rule inferring ability." This is further exacerbated by the fact that to know what something is, you need to know what it is not. If all variables in a set have G in common, you have no instance for which it does not apply, and G easily becomes mathematical G by default. If there were other variables that were not due to G in the set, this would narrow the possibilities as to what G is in the world by indicating what it is not pertinent to. This problem has subtly bedeviled the theory of G throughout its history.

On the other hand, psychologists such as Cyril Burt and Philip E. Vernon took the view that in addition to a general ability (general intelligence), there were less general abilities such as verbal–numerical–educational ability and practical–mechanical–spatial–physical ability, and even less general abilities such as verbal comprehension, reading, spelling, vocabulary, drawing, handwriting, and mastery of various subjects (cf. Vernon, 1961). In other words, the mind was organized into a hierarchy of abilities running from the most general to the most specific. Their model of test scores can be represented mathematically much like the equation for multiple correlation as

$$Y_j = a_j G + b_{j1} G_1 + \cdots + b_{js} G_s + c_{j1} H_1 + \cdots + c_{jt} H_t + \psi_j$$

where

Y_j is a manifest intellectual-performance variable

$a_j, b_{j1}, \ldots, b_{js}, c_{j1}, \ldots, c_{jt}$ are the weights

G is the latent general-ability variable

$G_1, \ldots, G_s$ are the major group factors
$H_1, \ldots, H_t$ are the minor group factors
ψ_j is a specific-ability variable for the jth variable

Correlations between two observed variables, Y_j and Y_k, would depend upon having not only the general-ability variable in common but group-factor variables in common as well.

By the time all these developments in the theory of intellectual abilities had occurred, the 1930s had arrived, and the center of new developments in this theory (and indirectly of new developments in the methodology of common-factor analysis) had shifted to the United States where L. L. Thurstone at the University of Chicago developed his theory and method of multiple-factor analysis. By this time, the latent-ability variables had come to be called "factors" owing to a usage of Spearman (1927).

Thurstone differed from the British psychologists over the idea that there was a general-ability factor and that the mind was hierarchically organized. For him, there were major group factors but no general factor. These major group factors he termed the primary mental abilities. That he did not cater to the idea of a hierarchical organization for the primary mental abilities was most likely because of his commitment to a principle of parsimony; this caused him to search for factors which related to the observed variables in such a way that each factor pertained as much as possible to one nonoverlapping subset of the observed variables. Sets of common factors displaying this property, Thurstone said, had a "simple structure." To obtain an optimal simple structure, Thurstone had to consider common-factor variables that were intercorrelated. And in the case of factor analyses of intellectual-performance tests, Thurstone discovered that usually his common factors were all positively intercorrelated with one another. This fact was considerably reassuring to the British psychologists who believed that by relying on his simple-structure concept Thurstone had only hidden the existence of a general-ability factor, which they felt was evidenced by the correlations among his factors.

Perhaps one reason why Thurstone's simple-structure approach to factor analysis became so popular—not just in the United States but in recent years in England and other countries as well—was because simple-structure solutions could be defined in terms of more-or-less objective properties which computers could readily identify and the factors so obtained were easy to interpret. It seemed by the late 1950s when the first large-scale electronic computers were entering universities that all the drudgery could be taken out of factor-analytic computations and that the researcher could let the computer do most of his work for him. Little wonder, then, that not much thought was given to whether theoretically hierarchical solutions were preferable to simple-structure solutions, especially when hierarchical solutions did not seem to be blindly obtainable. And believing that factor analysis could automatically and blindly find the key latent variables in a domain, what researchers would want hierarchical solutions which might be more difficult to interpret than simple-structure solutions?

The 1950s and early 1960s might be described as the era of blind factor analysis. In this period, factor analysis was frequently applied agnostically, as regards structural theory, to all sorts of data, from personality-rating variables, Rorschach-test-scoring variables, physiological variables, semantic differential variables, and biographical-information variables (in psychology), to characteristics of mining districts (in mineralogy), characteristics of cities (in city planning), characteristics of arrowheads (in anthropology), characteristics of wasps (in zoology), variables in the stock market (in economics), and aromatic-activity variables (in chemistry), to name just a few applications. In all these applications the hope was that factor analysis could bring order and meaning to the many relationships between variables.

Whether blind factor analyses often succeeded in providing meaningful explanations for the relationships among variables is a debatable question. In the case of Rorschach-test-score variables (Cooley and Lohnes, 1962) there is little question that blind factor analysis failed to provide a manifestly meaningful account of the structure underlying the score variables. Again, factor analyses of personality trait-rating variables have not yielded factors universally regarded by psychologists as explanatory constructs of human behavior (cf. Mulaik, 1964; Mischel, 1968). Rather, the factors obtained in personality trait-rating studies represent confoundings of intrarater processes (semantic relationships among trait words) with intraratee processes (psychological and physiological relationships within the persons rated). In the case of factor-analytic studies of biographical inventory items, the chief benefit has been in terms of classifying inventory items into clusters of similar content, but as yet no theory as to life histories has emerged from such studies. Still, blind factor analyses have served classification purposes quite well in psychology and other fields, but these successes should not be interpreted as generally providing advances in structural theories as well.

In the first 60 years of the history of factor analysis, factor-analytic methodologists developed heuristic algebraic solutions and corresponding algorithms for performing factor analyses. Many of these methods were designed to facilitate the finding of approximate solutions using mechanical hand calculators. Harman (1960) credits Cyril Burt with formulating the centroid method, but Thurstone (1947) gave it its name and developed it more fully as an approximation to the computationally more challenging principal axes, the eigenvector–eigenvalue solution put forth by Hotelling (1933). Until the development of electronic computers, the centroid method was a simple and straightforward solution that highly approximated the principal axes solution. But in the 1960s, computers came on line as the government poured billions into the development of computers for decryption work and into the mathematics of nuclear physics in developing nuclear weapons. Out of the latter came fast computer algorithms for finding eigenvectors and eigenvalues. Subsequently, factor analysts discovered the computer, and the eigenvector and eigenvalue routines and began programming them to obtain principal axes solutions, which rapidly became the standard approach. Nevertheless most of the procedures initially used were still based on least-squares methods, for the statistically more sophisticated method of maximum-likelihood estimation was still both mathematically and computationally challenging.

Throughout the history of factor analysis there were statisticians who sought to develop a more rigorous statistical theory for factor analysis. In 1940, Lawley (1940) made a major breakthrough with the development of equations for the maximum-likelihood estimation of factor loadings (assuming multivariate normality for the variables), and he followed up this work with other papers (1942, 1943, 1949) that sketched a framework for statistical testing in factor analysis. The problem was, to use these methods you needed maximum-likelihood estimates of the factor loadings. Lawley's computational recommendations for finding solutions were not practical for more than a few variables. So, factor analysts continued to use the centroid method and to regard any factor loading less than .30 as "nonsignificant."

In the 1950s, Rao (1955) developed an iterative computer program for obtaining maximum-likelihood estimates, but this was later shown not to converge. Howe (1955) showed that the maximum-likelihood estimates of Lawley (1949) could be derived mathematically without making any distributional assumptions at all by simply seeking to minimize the determinant of the matrix of partial correlations among residual variables after partialling out common factors from the original variables. Brown (1961) noted that the same idea was put forth on intuitive grounds by Thurstone in 1953. Howe also provided a far more efficient Gauss–Seidel algorithm for computing the solution. Unfortunately, this was ignored or unknown. In the meantime, Harman and Jones (1966) presented their Gauss–Seidel minres method of least-squares estimation, which rapidly converged and yielded close approximations to the maximum-likelihood estimates.

The major breakthrough mathematically, statistically, and computationally in maximum-likelihood exploratory factor analysis, was made by Karl Jöreskog (1967), then a new PhD in mathematical statistics from the University of Uppsala in Sweden. He applied a then recently developed numerical algorithm of Fletcher and Powell (1963) to the maximum-likelihood estimation of the full set of parameters of the common-factor model. The algorithm was quite rapid in convergence. Jöreskog's algorithm has been the basis for maximum-likelihood estimation in most commercial computer programs ever since. However, the program was not always well integrated with other computing methods in some major commercial programs, so that the program reports principal components eigenvalues rather than those of the weighted reduced correlation matrix of the common-factor model provided by Jöreskog's method, which Jöreskog used in initially determining the number of factors to retain.

Recognizing that more emphasis should be placed on the testing of hypotheses in factor-analytic studies, factor analysts in the latter half of the 1960s began increasingly to concern themselves with the methodology of hypothesis testing in factor analysis. The first efforts in this regard, using what are known as procrustean transformations, trace their beginnings to a paper by Mosier (1939) that appeared in *Psychometrika* nearly two decades earlier. The techniques of procrustean transformations seek to transform (by a linear transformation) the obtained factor-pattern matrix (containing regression coefficients for the observed variables regressed onto the latent, underlying factors) to be as much as possible like a hypothetical factor-pattern matrix constructed according to some structural hypothesis pertaining to the variables studied. When the transformed factor-pattern matrix is obtained it is tested for its degree of fit to the hypothetical factor-pattern matrix. For example, Guilford (1967) used procrustean techniques to isolate factors predicted by his three-faceted model of the intellect. However, hypothesis testing with procrustean transformations have been displaced in favor of confirmatory factor analysis since the 1970s, because the latter is able to assess how well the model reproduces the sample covariance matrix.

Toward the end of the 1960s, Bock and Bargmann (1966) and Jöreskog (1969a) considered hypothesis testing from the point of view of fitting a hypothetical model to the data. In these approaches the researcher specifies, ahead of time, various parameters of the common-factor-analysis model relating manifest variables to hypothetical latent variables according to a structural theory pertaining to the manifest variables. The resulting model is then used to generate a hypothetical covariance matrix for the manifest variables that is tested for goodness of fit to a corresponding empirical covariance matrix (with unspecified parameters of the factor-analysis model adjusted to make the fit to the empirical covariance matrix as good as possible). These approaches to factor analysis have had the effect of encouraging researchers to have greater concern with substantive, structural theories before assembling collections of variables and implementing the factor-analytic methods.

We will treat confirmatory factor analysis in Chapter 15, although it is better treated as a special case of structural equation modeling, which would be best dealt in a separate book. The factor analysis we primarily treat in this book is exploratory factor analysis, which may be regarded as an "abductive," "hypothesis-generating" methodology rather than a "hypothesis-testing" methodology. With the development of structural equation, modeling researchers have come to see traditional factor analysis as a methodology to be used, among other methods, at the outset of a research program, to formulate hypotheses about latent variables and their relation to observed variables. Furthermore it is now regarded as just one of several approaches to formulating such hypotheses, although it has general applications any time one believes that a set of observed variables are dependent upon a set of latent common factors.

1.3 Example of Factor Analysis

At this point, to help the reader gain a more concrete appreciation of what is obtained in a factor analysis, it may help to consider a small factor-analytic study conducted by the author in connection with a research project designed to predict the reactions of soldiers to combat stress. The researchers had the theory that an individual soldier's reaction to combat stress would be a function of the degree to which he responded emotionally to the potential danger of a combat situation and the degree to which he nevertheless felt he could successfully cope with the situation. It was felt that, realistically, combat situations should arouse strong feelings of anxiety for the possible dangers involved. But ideally these feelings of anxiety should serve as internal stimuli for coping behaviors which would in turn provide the soldier with a sense of optimism in being able to deal with the situation. Soldiers who respond pessimistically to strong feelings of fear or anxiety were expected to have the greatest difficulties in managing the stress of combat. Soldiers who showed little appreciation of the dangers of combat were also expected to be unprepared for the strong anxiety they would likely feel in a real combat situation. They would have difficulties in managing the stress of combat, especially if they had past histories devoid of successful encounters with stressful situations.

To implement research on this theory, it was necessary to obtain measures of a soldier's emotional concern for the danger of a combat situation and of his degree of optimism in being able to cope with the situation. To obtain these measures, 14 seven-point adjectival rating scales were constructed, half of which were selected to measure the degree of emotional concern for threat, and half of which were selected to measure the degree of optimism in coping with the situation. However, when these adjectival scales were selected, the researchers were not completely certain to what extent these scales actually measured two distinct dimensions of the kind intended.

Thus, the researchers decided to conduct an experiment to isolate the common-meaning dimensions among these 14 scales. Two hundred and twenty-five soldiers in basic training were asked to rate the meaning of "firing my rifle in combat" using the 14 adjectival scales, with ratings being obtained from each soldier on five separate occasions over a period of 2 months. Intercorrelations among the 14 scales were then obtained by summing the cross products over the 225 soldiers and five occasions. (Intercorrelations were obtained in this way because the researchers felt that, although on any one occasion various soldiers might differ in their conceptions of "firing my rifle in combat" and on different occasions an individual soldier might have different conceptions, the major determinants of covariation among the adjectival scales would still be conventional-meaning dimensions common to the scales.) The matrix of intercorrelations, illustrated in Table 1.1, was then subjected to image factor analysis (cf. Jöreskog, 1962), which is a relatively

TABLE 1.1

Intercorrelations among 14 Scales

1	1.00														Frightening
2	.20	1.00													Useful
3	.65	−.26	1.0												Nerve-shaking
4	−.26	.74	−.32	1.00											Hopeful
5	.71	−.27	.70	−.32	1.00										Terrifying
6	−.25	.64	−.30	.68	−.31	1.00									Controllable
7	.64	−.30	.73	−.34	.74	−.34	1.00								Upsetting
8	−.40	.39	−.44	.40	−.47	.39	−.53	1.00							Painless
9	−.13	.24	−.11	.24	−.16	.26	−.17	.27	1.00						Exciting
10	−.45	.36	−.49	.42	−.53	.39	−.53	.58	.32	1.00					Nondepressing
11	.59	−.26	.63	−.31	.32	−.32	.69	−.45	−.15	−.51	1.00				Disturbing
12	−.30	.69	−.35	.75	−.36	.68	−.39	.44	.28	.48	−.38	1.00			Successful
13	−.36	.35	−.50	.43	−.42	.38	−.52	.45	.20	.49	−.45	.46	1.00		Settling (vs. unsettling)
14	−.35	.62	−.36	.65	−.38	.62	−.46	.50	.36	.51	−.45	.67	.49	1.00	Bearable

accurate but simple-to-compute approximation of common-factor analysis. Four orthogonal factors were retained, and the matrix of "loadings" associated with the "unrotated factors" is given in Table 1.2. The coefficients in this matrix are correlations of the observed variables with the common factors. However, the unrotated factors of Table 1.2 are not readily interpretable, and they do not in this form appear to correspond to the two expected-meaning dimensions used in selecting the 14 scales. At this point it was decided, after

TABLE 1.2

Unrotated Factors

		1	2	3	4
1	Frightening	.73	.35	.01	−.15
2	Useful	−.56	.60	−.10	.12
3	Nerve-shaking	.78	.31	−.05	−.12
4	Hopeful	−.62	.60	−.09	.13
5	Terrifying	.78	.34	.43	.00
6	Controllable	−.59	.52	−.06	.08
7	Upsetting	.84	.29	−.08	.00
8	Painless	−.65	.07	.03	−.33
9	Exciting	−.29	.21	−.02	−.35
10	Nondepressing	−.70	.04	.03	−.36
11	Disturbing	.70	.13	−.59	−.05
12	Successful	−.67	.54	−.04	.05
13	Settling	−.63	.09	.08	−.16
14	Bearable	−.69	.43	.04	−.12

some experimentation, to rotate only the first two factors and to retain the latter two unrotated factors as "difficult to interpret" factors. Rotation of the first two factors was done using Kaiser's normalized Varimax method (cf. Kaiser, 1958). The resulting rotated matrix is given in Table 1.3.

The meaning of "rotation" may not be clear to the reader. Therefore let us consider the plot in Figure 1.1 of the 14 variables, using for their coordinates the loadings on the variables on the first two unrotated factors. Here we see that the coordinate axes do not correspond to variables that would be clearly definable by their association with the variables. On the other hand, note the cluster of points in the upper right-hand quadrant (variables 1, 3, 5, and 7) and the cluster of points in the upper left-hand quadrant (variables 2, 4, 6, 12, and 14). It would seem that one could rotate the coordinate axes so as to have them pass near these clusters. As a matter of fact, this is what has been done to obtain the rotated coordinates in Table 1.3, which are plotted in Figure 1.2.

Rotated factor 1 appears almost exclusively associated with variables 1, 3, 5, 7, and 11, which were picked as measures of a fear response, whereas rotated factor 2 appears most closely associated with variables 2, 4, 6, 12, and 14, which were picked as measures of optimism regarding outcome. Although variables 8, 10, and 13 appear now to be consistent in their relationships to these two dimensions, they are not unambiguous measures of either factor. Variable 9 appears to be a poor measure of these two dimensions.

Some factor analysts at this point might prefer to relax the requirement that the obtained factors be orthogonal to one another. They would, in this case, most likely construct a unit-length vector collinear with variable 7 to

TABLE 1.3

Rotated Factors

		1	2
1	Frightening	.83	−.10
2	Useful	−.11	.85
3	Nerve-shaking	.84	−.16
4	Hopeful	−.17	.87
5	Terrifying	.86	−.14
6	Controllable	−.18	.80
7	Upsetting	.89	−.22
8	Painless	−.50	.44
9	Exciting	−.12	.34
10	Nondepressing	−.57	.43
11	Disturbing	.67	−.29
12	Successful	−.24	.85
13	Settling	−.48	.43
14	Bearable	−.33	.76

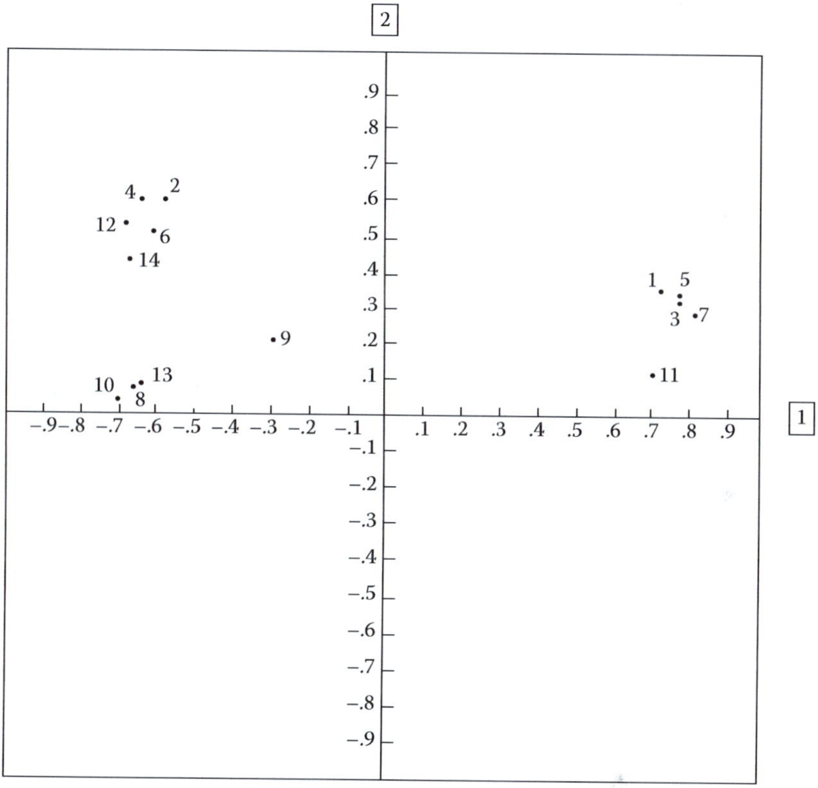

FIGURE 1.1
Plot of 14 variables on unrotated factors 1 and 2.

represent an "oblique" factor 1 and another unit-length vector collinear with variable 12 to represent an "oblique" factor 2. The resulting oblique factors would be negatively correlated with one another and would be interpreted as dimensions that are slightly negatively correlated. Such "oblique" factors are drawn in Figure 1.2 as arrows from the origin.

In conclusion, factor analysis has isolated two dimensions among the 14 scales which appear to correspond to dimensions expected to be present when the 14 scales were chosen for study. Factor analysis has also shown that some of the 14 scales (variables 8, 9, 10, and 13) are not unambiguous measures of the intended dimensions. These scales can be discarded in constructing a final set of scales for measuring the intended dimensions. Factor analysis has also revealed the presence of additional, unexpected dimensions among the scales. Although it is possible to hazard guesses as to the meaning of these additional dimensions (represented by factors 3 and 4), such guessing is not strongly recommended. There is considerable likelihood that the interpretation of these dimensions will be spurious. This is not

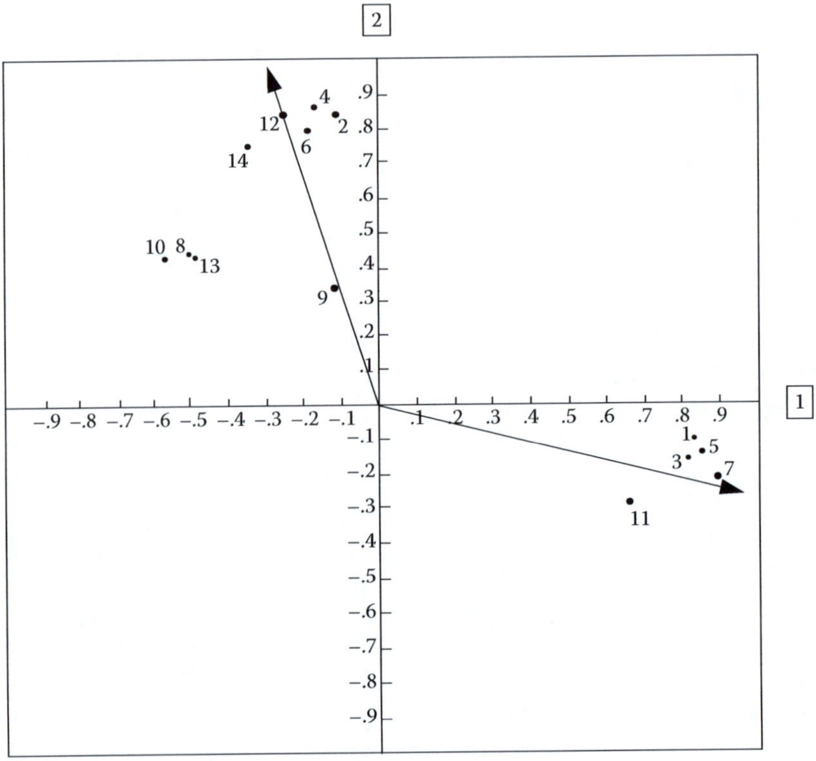

FIGURE 1.2
Plot of 14 variables on rotated factors 1 and 2.

to say that factor analysis cannot, at times, discover something unexpected but interpretable. It is just that in the present data the two additional dimensions are so poorly determined from the variables as to be interpretable only with a considerable risk of error.

This example of factor analysis represents the traditional, exploratory use of factor analysis where the researcher has some idea of what he will encounter but nevertheless allows the method freedom to find unexpected dimensions (or factors).

2

Mathematical Foundations for Factor Analysis

2.1 Introduction

Ideally, one begins a study of factor analysis with a mathematical background of up to a year of calculus. This is not to say that factor analysis requires an extensive knowledge of calculus, because calculus is used in only a few instances, such as in finding weights to assign to a set of independent variables to maximize or minimize the variance of a resulting linear combination. Or it will be used in finding a transformation matrix that minimizes, say, a criterion for simple structure in factor rotation. But having calculus in one's background provides sufficient exposure to working with mathematical concepts so that one will have overcome reacting to a mathematical subject such as factor analysis as though it were an esoteric subject comprehensible only to select initiates to its mysteries. One will have learned those subjects such as trigonometry, college algebra, and analytical geometry upon which factor analysis draws heavily.

In practice, however, the author recognizes that many students who now undertake a study of factor analysis come from the behavioral, social, and biological sciences, where mathematics is not greatly stressed. Consequently, in this chapter the author attempts to provide a brief introduction to those mathematical topics from modern algebra, trigonometry, analytic geometry, and calculus that will be necessary in the study of factor analysis. Moreover, the author also provides a background in operations with vectors and matrix algebra, which even students having just had first-year calculus will more than likely find them new. In this regard most students will find themselves on even ground if they had up to college algebra, because operations with vectors and matrix algebra are extensions of algebra using a new notation.

2.2 Scalar Algebra

By the term "scalar algebra," we refer to the ordinary algebra applicable to the real numbers with which the reader should already be familiar. The use of this term distinguishes ordinary algebra from operations with vectors and matrix algebra, which we will take up shortly.

2.2.1 Fundamental Laws of Scalar Algebra

The following laws govern the basic operations of scalar algebra such as addition, subtraction, multiplication, and division:

1. Closure law for addition: $a + b$ is a unique real number.
2. Commutative law for addition: $a + b = b + a$.
3. Associative law for addition: $(a + b) + c = a + (b + c)$.
4. Closure law for multiplication: ab is a unique real number.
5. Commutative law for multiplication: $ab = ba$.
6. Associative law for multiplication: $a(bc) = (ab)c$.
7. Identity law for addition: There exists a number 0 such that

$$a + 0 = 0 + a = a$$

8. Inverse law for addition: $a + (-a) = (-a) + a = 0$.
9. Identity law for multiplication: There exists a number 1 such that

$$a1 = 1a = a$$

10. Inverse law for multiplication: $a\frac{1}{a} = \frac{1}{a}a = 1$.
11. Distributive law: $a(b + c) = ab + ac$.

The above laws are sufficient for dealing with the real-number system. However, the special properties of zero should be pointed out:

$$\frac{0}{a} = 0$$

$$\frac{a}{0} \text{ is undefined}$$

$$\frac{0}{0} \text{ is indeterminate}$$

2.2.1.1 Rules of Signs

The rule of signs for multiplication is given as

$$a(-b) = -(ab) \quad \text{and} \quad (-a)(-b) = +(ab)$$

The rule of signs for division is given as

$$\frac{-a}{b} = \frac{a}{-b} \quad \text{and} \quad \frac{-a}{-b} = \frac{a}{b}$$

The rule of signs for removing parentheses is given as

$$-(a-b) = -a+b \quad \text{and} \quad -(a+b) = -a-b$$

2.2.1.2 Rules for Exponents

If n is a positive integer, then x^n will stand for

$$xx \cdots x \text{ with } n \text{ terms}$$

If $x^n = a$, then a, is known as the nth root of a. The following rules govern the use of exponents:

1. $x^a x^b = x^{a+b}$.
2. $(x^a)^b = x^{ab}$.
3. $(xy)^a = x^a y^a$.
4. $\left(\frac{x}{y}\right)^a = \frac{x^a}{y^a}$.
5. $\frac{x^a}{x^b} = x^{a-b}$.

2.2.1.3 Solving Simple Equations

Let x stand for an unknown quantity, and let a, b, c, and d stand for known quantities. Then, given the following equation

$$ax + b = cx + d$$

the unknown quantity x can be found by applying operations to both sides of the equation until only an x remains on one side of the equation and the known quantities on the other side. That is,

$$ax - cx + b = d \text{ (subtract cx from both sides)}$$

$$ax - cx = d - b \text{ (subtract b from both sides)}$$

$$(a-c)x = d - b \text{ (by reversing the distributive law)}$$

$$x = \frac{d-b}{a-c} \text{ (by dividing both sides by } a-c\text{)}$$

2.3 Vectors

It may be of some help to those who have not had much exposure to the concepts of modern algebra to learn that one of the essential aims of modern algebra is to classify mathematical systems—of which scalar algebra is an example—according to the abstract rules which govern these systems. To illustrate, a very simple mathematical system is a group. A group is a set of elements and a single operation for combining them (which in some cases is addition and in some others multiplication) behaving according to the following properties:

1. If a and b are elements of the group, then $a+b$ and $b+a$ are also elements of the group, although not necessarily the same element (closure).
2. If a, b, and c are elements of the group, then

$$(a+b)+c = a+(b+c) \text{ (associative law)}$$

3. There is an element 0 in the group such that for every a in the group

$$a+0 = 0+a = a \text{ (identity law)}$$

4. For each a in the group, there is an element $(-a)$ in the group such that

$$a+(-a) = (-a)+a = 0 \text{ (inverse law)}$$

One should realize that the nature of the elements in the group has no bearing upon the fact that the elements constitute a group. These elements may be integers, real numbers, vectors, matrices, or positions of an equilateral triangle; it makes no difference what they are, as long as under the operator (+) they behave according to the properties of a group. Obviously, a group is a far simpler system than the system which scalar algebra exemplifies. Only 4 laws govern the group, whereas 11 laws are needed to govern the system exemplified by scalar algebra, which, by the way, is known to mathematicians as a "field."

Among the various abstract mathematical systems, the system known as a "vector space" is the most important for factor analysis. (*Note*. One should not let the term "space" unduly influence one's concept of a vector space. A vector space need have no geometric connotations but may be treated entirely as an abstract system. Geometric representations of vectors are only particular examples of a vector space.) A vector space consists of two sets of mathematical objects—a set V of "vectors" and a set R of elements of a "field" (such as the scalars of scalar algebra)—together with two operations for combining them. The first operation is known as "addition of vectors" and has the following properties such that for every **u**, **v**, **w** in the set *V* of vectors:

1. $\mathbf{u} + \mathbf{v}$ is also a uniquely defined vector in *V*.
2. $\mathbf{u} + (\mathbf{v} + \mathbf{w}) = (\mathbf{u} + \mathbf{v}) + \mathbf{w}$.
3. $\mathbf{u} + \mathbf{v} = (\mathbf{v} + \mathbf{w})$.
4. There exists a vector **0** in *V* such that $\mathbf{u} + \mathbf{0} = \mathbf{0} + \mathbf{u} = \mathbf{u}$.
5. For each vector in *V* there exists a unique vector $-\mathbf{u}$ such that

$$\mathbf{u} + (-\mathbf{u}) = (-\mathbf{u}) + \mathbf{u} = \mathbf{0}$$

(One may observe that under addition of vectors the set *V* of vectors is a group.)

The second operation governs the combination of the elements of the field *R* of scalars with the elements of the set *V* of vectors and is known as "scalar multiplication." Scalar multiplication has the following properties such that for all elements *a*, *b* from the field *R* of scalars and all vectors **u**, **v** from the set *V* of vectors:

6. $a\mathbf{u}$ is a vector in *V*.
7. $a(\mathbf{u} + \mathbf{v}) = a\mathbf{u} + a\mathbf{v}$.
8. $(a + b)\mathbf{u} = a\mathbf{u} + b\mathbf{u}$.
9. $a(b\mathbf{u}) = ab(\mathbf{u})$.
10. $1\mathbf{u} = \mathbf{u}$; $0\mathbf{u} = \mathbf{0}$.

In introducing the idea of a vector space as an abstract mathematical system, we have so far deliberately avoided considering what the objects known as vectors might be. Our purpose in doing so has been to have the reader realize at the outset that a vector space is an abstract system that may be found in connection with various kinds of mathematical objects. For example, the vectors of a vector space may be identified with the elements of any field such as the set of real numbers. In such a case, the addition of vectors corresponds to the addition of elements in the field. In another example, the vectors of a vector space may be identified with *n*-tuples, which are ordered sets of real numbers. (In the upcoming discussion we will develop the properties of vectors more fully in connection with vectors represented by *n*-tuples.) Vectors may also be identified with the unidimensional random variables of mathematical statistics. This fact has important implications for multivariate statistics and factor analysis, and structural equation modeling, in particular, because it means one may use the vector concept to unify the treatment of variables in both finite and infinite populations. The reader should note at this point that the key idea in this book is that, in any linear analysis of variables of the behavioral, social, or biological sciences, the variables may be treated as if they are vectors in a linear vector space. The concrete representation of these variables may differ in various contexts (i.e., may be n-tuples or random variables), but they may always be considered vectors.

2.3.1 *n*-Tuples as Vectors

In a vector space of n-tuples, by a vector we shall mean a point in n-dimensional space designated by an ordered set of numbers known as an n-tuple which are the coordinates of the point. For example, (1,3,2) is a 3-tuple which represents a vector in three-dimensional space which is 1 unit from the origin (of the coordinate system) in the direction along the x-axis, 3 units from the origin in the direction of the y-axis, and 2 units from the origin in the direction of the z-axis. Note that in a set of coordinates the numbers appear in special positions to indicate the distance one must go along each of the reference axes to arrive at the point. In graphically portraying a vector, we shall use the convention of drawing an arrow from the origin of the coordinate system to the point. For example, the vector (1,3,2) is illustrated in Figure 2.1.

In a vector space of n-tuples we will be concerned with certain operations to be applied to the n-tuples which define the vectors. These operations will define addition, subtraction, and multiplication in the vector space of n-tuples. As a notational shorthand to allow us to forgo writing the coordinate numbers in full when we wish to express the equations in vector notation, we will designate individual vectors by lowercase boldface letters. For example, let **a** stand for the vector (1,3,2).

2.3.1.1 Equality of Vectors

Two vectors are equal if they have the same coordinates.

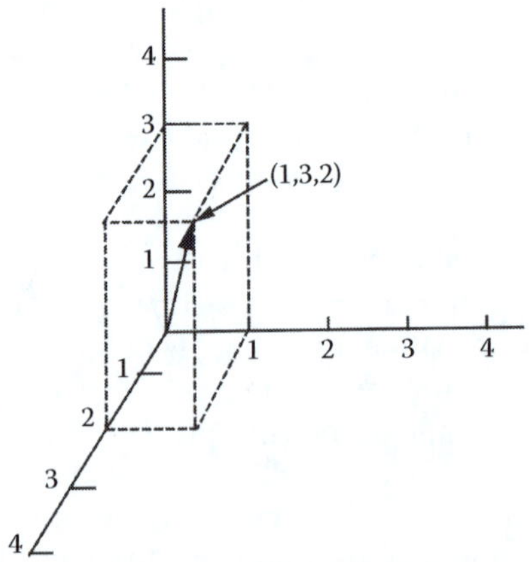

FIGURE 2.1
Graphical representation of vector (1,3,2) in three-dimensional space.

For example, if $\mathbf{a}=(1,2,4)$ and $\mathbf{b}=(1,2,4)$, then $\mathbf{a}=\mathbf{b}$. A necessary condition that two vectors are equal is that they have the same number of coordinates. For example, if

$$\mathbf{a}=(1,2,3,4) \quad \text{and} \quad \mathbf{b}=(1,2,3)$$

then $\mathbf{a} \neq \mathbf{b}$. In fact, when two vectors have different numbers of coordinates, they refer to a different order of n-tuples and cannot be compared or added.

2.3.2 Scalars and Vectors

Vectors compose a system complete in themselves. But they are of a different order from the algebra we normally deal with when using real numbers. Sometimes we introduce real numbers into the system of vectors. When we do this, we call the real numbers "scalars." In our notational scheme, we distinguish scalars from vectors by writing the scalars in italics and the vectors in lowercase boldface characters. Thus, $a \neq \mathbf{a}$.

In vector notation, we will most often consider vectors in the abstract. Then, rather than using actual numbers to stand for the coordinates of the vectors, we will use scalar quantities expressed algebraically. For example, let a general vector $\mathbf{a}$ in five-dimensional space be written as

$$\mathbf{a}=(a_1,a_2,a_3,a_4,a_5)$$

In this example a stands for a coordinate, each being distinguished from others by a different subscript. Whenever possible, algebraic expressions for the coordinates of vectors should take the same character as the character standing for the vector itself. This will not always be done, however.

2.3.3 Multiplying a Vector by a Scalar

Let $\mathbf{a}$ be a vector such that

$$\mathbf{a}=(a_1,a_2,\ldots,a_n)$$

and λ a scalar; then the operation

$$\lambda\mathbf{a}=\mathbf{c}$$

produces another vector $\mathbf{c}$ such that

$$\mathbf{c}=(\lambda a_1,\lambda a_2,\ldots,\lambda a_n)$$

In other words, multiplying a vector by a scalar produces another vector that has for components the components of the first vector each multiplied by the scalar. To cite a numerical example, let $\mathbf{a}=(1,3,4,5)$; then

$$2\mathbf{a}=(2\times1,\ 2\times3,\ 2\times4,\ 2\times5)=(2,6,8,10)=\mathbf{c}$$

2.3.4 Addition of Vectors

If $\mathbf{a}=(1,3,2)$ and $\mathbf{b}=(2,1,5)$, then their sum, denoted as $\mathbf{a}+\mathbf{b}$ is another vector $\mathbf{c}$ such that

$$\mathbf{c}=\mathbf{a}+\mathbf{b}=((1+2),(3+1),(2+5))=(3,4,7)$$

When we add two vectors together, we add their corresponding coordinates together to obtain a new vector.

The addition of vectors is found in physics in connection with the analysis of forces acting upon a body where vector addition leads to the "resultant" by the well-known "parallelogram law." This law states that if two vectors are added together, then lines drawn from the points of these vectors to the point of the vector produced by the addition will make a parallelogram with the original vectors. In Figure 2.2, we show the result of adding two vectors (1,3) and (3,2) together.

If more than two vectors are added together, the result is still another vector. In some factor-analytic procedures, this vector is known as a "centroid," because it tends to be at the center of the group of vectors added together.

2.3.5 Scalar Product of Vectors

In some vector spaces, in addition to the two operations of addition of vectors and scalar multiplication, a third operation known as the scalar product of two vectors (written $\mathbf{xy}$ for each pair of vectors $\mathbf{x}$ and $\mathbf{y}$) is defined which associates a scalar with each pair of vectors. This operation has the following abstract properties, given that $\mathbf{x}$, $\mathbf{y}$, and $\mathbf{z}$ are arbitrary vectors and a an arbitrary scalar:

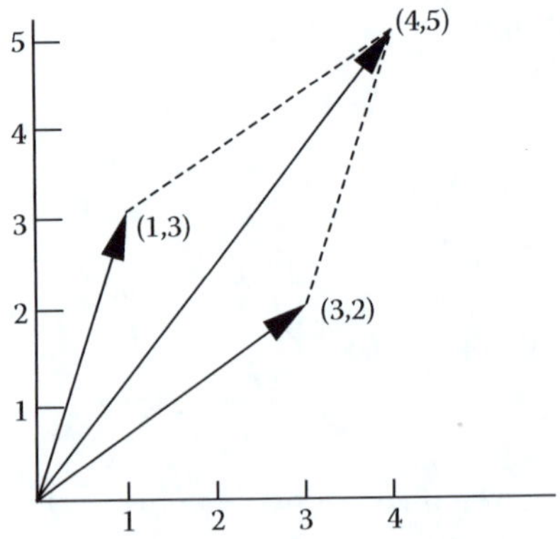

FIGURE 2.2
Graphical representation of sum of vectors (1,3) and (3,2) by vector (4,3) in two-dimensional space, illustrating the parallelogram law for addition of vectors.

11. $\mathbf{xy} = \mathbf{yx} =$ a scalar.
12. $\mathbf{x}(\mathbf{y} + \mathbf{z}) = \mathbf{xy} + \mathbf{xz}$.
13. $\mathbf{x}(a\mathbf{y}) = a(\mathbf{xy})$.
14. $\mathbf{xx} \geq 0$; $\mathbf{xx} = 0$ implies $\mathbf{x} = \mathbf{0}$.

When the scalar product is defined in a vector space, the vector space is known as a "unitary vector space." In a unitary vector space, it is possible to establish the length of a vector as well as the cosine of the angle between pairs of vectors. Factor analysis as well as other multivariate linear analyses is concerned exclusively with unitary vector spaces.

We will now consider the definition of the scalar product for a vector space of n-tuples. Let $\mathbf{a}$ be the vector $(a_1,a_2,\ldots,a_n)$ and $\mathbf{b}$ the vector $(b_1,b_2,\ldots,b_n)$; then the vector product of $\mathbf{a}$ and $\mathbf{b}$, written $\mathbf{ab}$, is the sum of the products of corresponding components of the vectors, that is,

$$\mathbf{ab} = a_1b_1 + a_2b_2 + \cdots + a_nb_n \tag{2.1}$$

To use a simple numerical example, let $\mathbf{a} = (1,2,5)$ and $\mathbf{b} = (3,3,4)$; then $\mathbf{ab} = 29$.

As a further note on notation, consider that an expression such as Equation 2.1, containing a series of terms to be added together which differ only in their subscripts, can be shortened by using the summational notation. For example, Equation 2.1 can be rewritten as

$$\mathbf{ab} = \sum_{i=1}^{n} a_ib_i \tag{2.2}$$

As an explanation of this notation, the expression a_ib_i on the right-hand side of the sigma sign stands for a general term in the series of terms, as in Equation 2.1, to be added. The subscript i in the term a_ib_i stands for a general subscript. The expression $\Sigma_{i=1}^{n}$ indicates that one must add together a series of subscripted terms. The expression "$i = 1$" underneath the sigma sign indicates which subscript is pertinent to the summation governed by this summation sign—in the present example the subscript i is pertinent—as well as the first value in the series of terms that the subscript will take. The expression n above the sigma sign indicates the highest value that the subscript will take. Thus we are to add together all terms in the series with the subscript i ranging in value from 1 to n.

2.3.6 Distance between Vectors

In high school geometry we learn from the Pythagorean theorem that the square on the hypotenuse of a right triangle is equal to the sum of the squares on the other two sides. This theorem forms the basis for finding the distance between two vector points. We shall define the distance between two vectors **a** and **b**, written

$$|\mathbf{a}-\mathbf{b}| = \left[\sum_{i=1}^{n}(a_i - b_i)^2\right]^{1/2}$$

where **a** and **b** are both vectors with n components. This means that we find the sum of squared differences between the corresponding components of the two vectors and then take the square root of that. In Figure 2.3, we have diagrammed the geometric equivalent of this formula.

$$|\mathbf{a}-\mathbf{b}| = \sqrt{(a_1 - b_1)^2 + (a_2 - b_2)^2} \tag{2.3}$$

2.3.7 Length of a Vector

Using Equation 2.3, we can find an equation for the length of a vector, that is, the distance of its point from the origin of the coordinate system. If we define the "zero vector" as

$$\mathbf{0} = (0, 0, \ldots, 0)$$

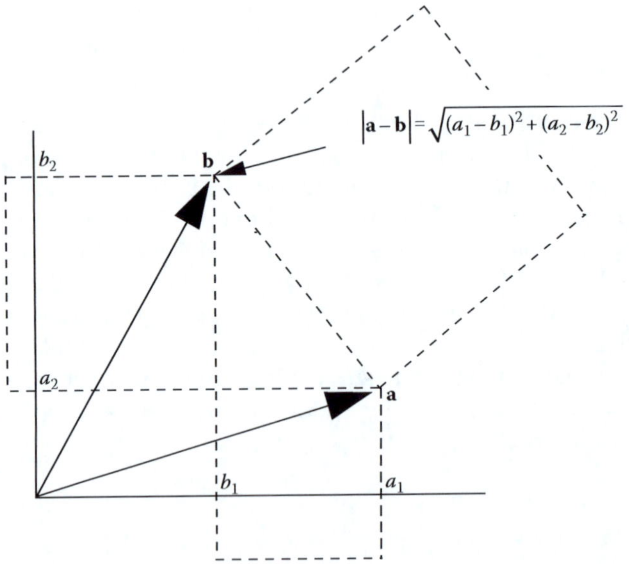

FIGURE 2.3
Graphical illustration of application of Pythagorean theorem to the determination of the distance between 2 two-dimensional vectors **a** and **b**.

Then

$$|\mathbf{a}| = \|\mathbf{a} - \mathbf{0}\| = \left[\sum_{i-1}^{n} (a_i - 0)^2\right]^{1/2}$$

$$= \left[\sum_{i=1}^{n} a_i^2\right]^{1/2} \qquad (2.4)$$

The length of a vector **a**, denoted $|\mathbf{a}|$, is the square root of the sum of the squares of its components. (*Note*. Do not confuse $|\mathbf{a}|$ with $|\mathbf{A}|$ which is a determinant.)

2.3.8 Another Definition for Scalar Multiplication

Another way of expressing scalar multiplication of vectors is given by the formula

$$\mathbf{ab} = |\mathbf{a}|\,|\mathbf{b}|\cos\theta \qquad (2.5)$$

where θ is the angle between the vectors. In other words, the scalar product of one vector times another vector is equivalent to the product of the lengths of the two vectors times the cosine of the angle between them.

2.3.9 Cosine of the Angle between Vectors

Since we have raised the concept of the cosine of the angle between vectors, we should consider the meaning of the cosine function as well as other important trigonometric functions.

In Figure 2.4, there is a right triangle where θ is the value of the angle between the base of the triangle and the hypotenuse. If we designate the length of the side opposite the angle θ as a, the length of the base as b, and the length of the hypotenuse as c, the ratio of these sides to one another gives the following trigonometric functions:

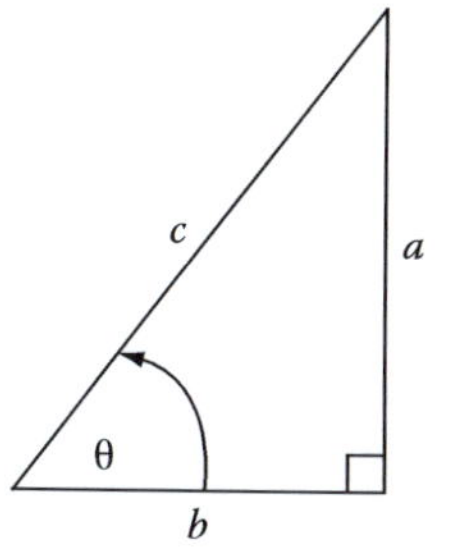

FIGURE 2.4
A right triangle with angle θ between base and hypotenuse.

$$\tan\theta = \frac{a}{b}$$

$$\sin\theta = \frac{a}{c}$$

$$\cos\theta = \frac{b}{c}$$

Reflection on the matter will show that these ratios are completely determined by the value of the angle θ in the right triangle and are independent of the size of the triangle.

Of particular interest to us here is what happens to the value of cos θ when θ is close to 0° and when θ is close to 90°. In Figure 2.5a, we illustrate a triangle in which the angle θ is close to 0°. Hence, the value of cos θ as the ratio of b to c is close to 1.00.

In Figure 2.5b, we illustrate a triangle in which the angle θ is close to 90°. In this case the length b is quite small relative to the length of c. Hence, the value of cos θ is close to 0.

In Figure 2.5c, we illustrate a triangle in which the angle is greater than 90°. In this case b is given a negative value, because the length of the base is measured from the origin of the angle in a direction opposite from the direction in which we normally measure the length of the base. By convention the length of the hypotenuse is always positive. Thus the cosine of an angle between 90° and 180° is a negative value. Moreover, as the angle θ approaches 180°, cos θ approaches −1.

The cosine function thus serves as a useful index of the relationship between vectors. When two vectors have a very small angle between them, the cosine of the angle between them is nearly 1. When the two vectors are 90° apart, nothing projects from one vector onto the other, and the cosine of the angle between them is 0. When the angle between them is between 90° and 180°, the cosine of the angle is negative, indicating that one vector points in the opposite direction, to some extent from the other.

We can use Equation 2.5 to find the cosine of the angle between two vectors in terms of their components. If we divide both sides of Equation 2.5 by the expression $|\mathbf{a}||\mathbf{b}|$, we have

$$\cos\theta = \frac{\mathbf{ab}}{|\mathbf{a}||\mathbf{b}|} \tag{2.6}$$

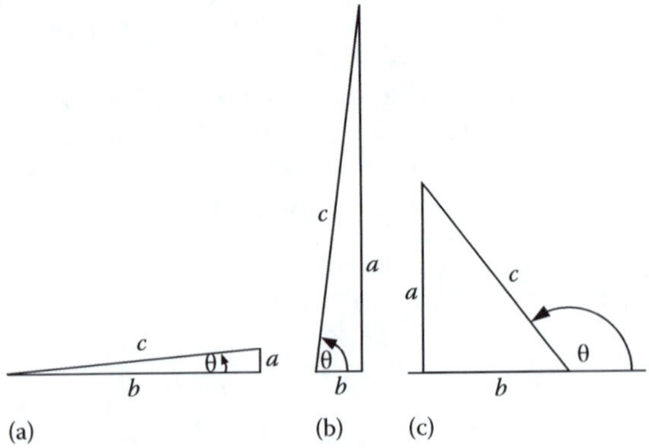

FIGURE 2.5
Illustrations of different right triangles.

But if we use Equations 2.2 and 2.4, Equation 2.6 can be rewritten as

$$\cos\theta = \frac{\sum_{i=1}^{n} a_i b_i}{\left[\left(\sum_{i=1}^{n} a_i^2\right)\left(\sum_{i=1}^{n} b_i^2\right)\right]^{1/2}} \tag{2.7}$$

Equation 2.7 has a direct bearing on the formula for the correlation coefficient. To illustrate, let us construct an N-component vector **x** such that the components of this vector are the respective deviation scores of N individuals on a variable denoted by X. (A deviation score, by the way, is defined as the difference between an actual score and the mean of the scores of a variable. In other words, if X_i is the ith subject's actual score on variable X, then the ith subject's deviation score x_i equals $X_i - \bar{X}$, where $\bar{X}$ is the mean of scores on X.) Similarly, let us construct an N-component vector Y such that the components of this vector are the respective deviation scores of the same N individuals on a variable Y. Then the correlation between variables X and Y is equivalent to the cosine of the angle between the vectors **x** and **y**, that is,

$$r_{XY} = \frac{\sum_{i=1}^{n} x_i y_i}{\left[\left(\sum_{i=1}^{n} x_i^2\right)\left(\sum_{i=1}^{n} y_i^2\right)\right]^{1/2}} \tag{2.8}$$

This formula is equivalent to one of the forms in which the sample correlation coefficient is given in most statistics texts.

2.3.10 Projection of a Vector onto Another Vector

In Figure 2.6 **x** and **y** are two vectors having an angle θ between them. The projection of **x** onto **y** is the vector **p** obtained by dropping a perpendicular from **x** to a line collinear with **y** and drawing a vector to that point. c is the length of **x** and b is the length of **p**. Because $\cos\theta = b/c$, $b = c\cos\theta$ is the length

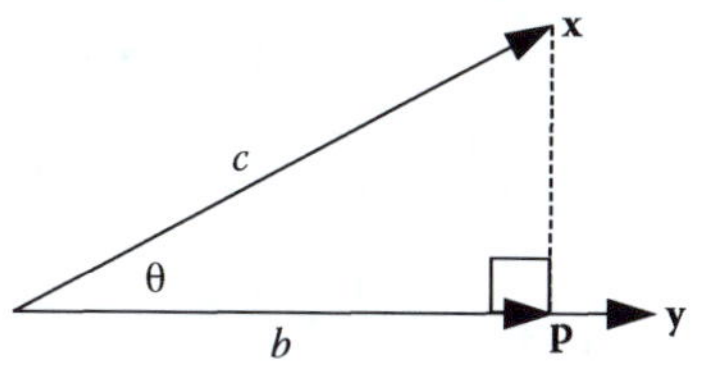

FIGURE 2.6
The vector **p** is the projection of the vector **x** onto the vector **y**. A line perpendicular to **y** is drawn from **x** to a line through **y** and **p** is the vector collinear with **y** up to that point.

of $\mathbf{p}$. Also $c = \sqrt{\mathbf{x}'\mathbf{x}}$ and $\cos\theta = \mathbf{x}'\mathbf{y}/(\sqrt{\mathbf{x}'\mathbf{x}}\sqrt{\mathbf{y}'\mathbf{y}})$ hence $b = \mathbf{x}'\mathbf{y}/\sqrt{\mathbf{y}'\mathbf{y}}$. The proportion of length of $\mathbf{y}$ due to length of $\mathbf{p}$ is $b/\sqrt{\mathbf{y}'\mathbf{y}} = \mathbf{x}'\mathbf{y}/\sqrt{\mathbf{y}'\mathbf{y}}$. Projection of $\mathbf{x}$ on $\mathbf{y}$ is thus $\mathbf{p} = (\mathbf{x}'\mathbf{y}/\sqrt{\mathbf{y}'\mathbf{y}})\mathbf{y}$. The concept of projection is central to least squares regression of one variable onto another. Here $\mathbf{p}$ is the predictable part of $\mathbf{y}$ due to $\mathbf{x}$.

2.3.11 Types of Special Vectors

The "null vector" $\mathbf{0}$ is the origin of the coordinate system in a space and is defined by

$$\mathbf{0} = (0, 0, \ldots, 0)$$

The "sum vector" $\mathbf{1} = (1, 1, \ldots, 1)$ is used when we wish to sum the components of another vector $\mathbf{a}$ as

$$\sum_{i=1}^{n} a_i = \mathbf{1a}$$

Note that $1\mathbf{a} \neq \mathbf{1a}$.

We shall say that two vectors are orthogonal if their scalar product is 0. That is, if $\mathbf{a}$ and $\mathbf{b}$ are n-component vectors (neither of which is the null vector), then $\mathbf{a}$ is orthogonal to $\mathbf{b}$ if

$$\mathbf{ab} = \sum_{i=1}^{n} a_i b_i = 0$$

It follows then, from Equation 2.7, that any two vectors having for their scalar product the value 0 must have the cosine of the angle between them equal to 0. Consequently, the angle between them must be 90°.

Example. If $\mathbf{a} = (1, 3, 5)$ and $\mathbf{b} = (1, 3, -2)$ then

$$\mathbf{ab} = 1 + 9 - 10 = 0$$

A "normalized" vector is a vector having a length of 1. Any given vector (other than the null vector) can be transformed into a normalized vector by dividing each of its components by the value of its length. In other words, if $\mathbf{v}$ is a vector, $\mathbf{v}$ can be transformed into a corresponding normalized vector $\mathbf{u}$ such that

$$\mathbf{u} = \frac{1}{|\mathbf{v}|}\mathbf{v} \quad \text{where } |\mathbf{u}| = 1$$

The "unit vector" is frequently used as a basis vector in a set of basis vectors from which all other vectors occupying the space can be derived as linear combinations. The ith unit vector in a space of n dimensions is denoted by $\mathbf{e}_i$, where

$$\mathbf{e}_1 = (1,0,0,\ldots,0)$$

$$\mathbf{e}_2 = (0,1,0,\ldots,0)$$

$$\mathbf{e}_3 = (0,0,1,\ldots,0)$$

$$\cdots\cdots\cdots\cdots\cdots\cdots$$

$$\mathbf{e}_n = (0,0,0,\ldots,1)$$

Note that any two different unit vectors are orthogonal to one another, that is,

$$\mathbf{e}_j\mathbf{e}_k = 0 \quad j \neq k$$

2.3.12 Linear Combinations

We shall say that a vector is a "linear combination" of some other vectors, if it can be obtained as the sum of scalar multiples of the other vectors. That is to say, let $\mathbf{v}_1, \mathbf{v}_2, \mathbf{v}_3, \ldots, \mathbf{v}_n$ be vectors. Then the vector $\mathbf{a}$ is a linear combination of these vectors if

$$\mathbf{a} = w_1\mathbf{v}_1 + w_2\mathbf{v}_2 + w_3\mathbf{v}_3 + \cdots + w_n\mathbf{v}_n \tag{2.9}$$

where $w_1, w_2, w_3, \ldots, w_n$ are scalars (numbers, not vectors). For example, if

$$\mathbf{a} = (1,1), \quad \mathbf{v}_1 = (2,3), \quad \text{and} \quad \mathbf{v}_2 = (3,5)$$

it can be shown that

$$\mathbf{a} = 2\mathbf{v}_1 + (-1)\mathbf{v}_2 = (4,6) - (3,5) = (1,1)$$

We can show that any vector in an n-dimensional space is a linear combination of the n unit vectors in that space. To illustrate, suppose we have a vector

$$\mathbf{a} = (a_1, a_2)$$

and two other vectors $\mathbf{a}_1 = (a_1, 0)$ and $\mathbf{a}_2 = (0, a_2)$, then

$$\mathbf{a} = \mathbf{a}_1 + \mathbf{a}_2 = (a_1, 0) + (0, a_2)$$

But $\mathbf{a}_1 = a_1(1,0)$ and $\mathbf{a}_2 = a_2(0,1)$

Hence

$$\mathbf{a} = (a_1, a_2) = a_1(1,0) + a_2(0,1) = a_1\mathbf{e}_1 + a_2\mathbf{e}_2$$

2.3.13 Linear Independence

We shall say that a set of n vectors is linearly independent if no vector in that set can be derived as a linear combination from the rest of the vectors in the set. Conversely, if any vector in a set of n vectors is a linear combination of some of the other vectors in the set, then the set is "linearly dependent."

The mathematical definition of linear dependence states that a set of vectors $\mathbf{v}_1, \mathbf{v}_2, \mathbf{v}_3, \ldots, \mathbf{v}_n$ is linearly dependent if some scalars $w_1, w_2, w_3, \ldots, w_n$ can be found (not all zero) such that

$$w_1\mathbf{v}_1 + w_2\mathbf{v}_2 + \cdots + w_n\mathbf{v}_n = 0 \tag{2.10}$$

If Equation 2.10 is true, we can derive any vector $\mathbf{v}_k$ in the set of vectors $\mathbf{v}_1, \mathbf{v}_2, \mathbf{v}_3, \ldots, \mathbf{v}_n$ as a linear combination of the rest. For proof, subtract the kth term $w_k\,\mathbf{v}_k$ from both sides of Equation 2.10. Then,

$$w_1\mathbf{v}_1 + w_2\mathbf{v}_2 + \cdots + w_{k-1}\mathbf{v}_{k-1} + w_{k+1}\mathbf{v}_{k+1} + \cdots + w_n\mathbf{v}_n = -w_k\mathbf{v}_k$$

or

$$-\frac{w_1}{w_k}\mathbf{v}_1 - \frac{w_2}{w_k}\mathbf{v}_2 - \cdots - \frac{w_n}{w_k}\mathbf{v}_n = \mathbf{v}_k$$

The significance of a linearly independent set is that it is a collection of vectors which contains no redundant information.

2.3.14 Basis Vectors

In a space of n dimensions, a linearly independent set of vectors occupying that space can contain at most n vectors. Such linearly independent sets of vectors, from which all other vectors in the space can be derived as linear combinations, are known as sets of "basis vectors." There is an unlimited number of different sets of basis vectors that may be obtained for any n-dimensional space.

2.4 Matrix Algebra

Matrix algebra is, in some ways, a logical extension of n-tuple vector algebra. Whereas n-tuple vector algebra deals with the operations for manipulating individual vectors, matrix algebra deals with the operations for manipulating whole collections of n-tuple vectors simultaneously.

2.4.1 Definition of a Matrix

A matrix is defined as a rectangular array of numbers arranged into rows and columns. The following is the manner in which a matrix expressed algebraically is written in full:

$$\begin{bmatrix} a_{11} & a_{12} & a_{13} & a_{14} & a_{15} \\ a_{21} & a_{22} & a_{23} & a_{24} & a_{25} \\ a_{31} & a_{32} & a_{33} & a_{34} & a_{35} \\ a_{41} & a_{42} & a_{43} & a_{44} & a_{45} \end{bmatrix}$$

This array represents a 4 × 5 matrix (4 rows by 5 columns). As a rule of notation, brackets [], parentheses (), or double lines || || are used to enclose the rectangular array of numbers to designate it as a matrix. Remember that a matrix has no numerical value. Rather, it can be thought of as a collection of either row vectors or column vectors of numbers.

Because matrices are two-dimensional arrays of numbers, a system of double subscripts must be used to identify the elements in the matrix. According to convention among mathematicians, the first subscript (reading from left to right) designates the number of the row in which the element appears. The second subscript designates the column in which the element appears. Hence a_{23} designates an element in the second row and third column; a_{ij} refers to a general element appear in the *i*th row and *j*th column.

The following arrays of numbers are matrices:

$$\begin{bmatrix} 1 & 2 \\ 3 & 1 \end{bmatrix} \quad \begin{bmatrix} 3 & 4 & 5 & 4 & 3 \\ 3 & 6 & 8 & 4 & 2 \\ 7 & 12 & 8 & 4 & 2 \end{bmatrix} \quad (4,5,6) \quad \begin{bmatrix} 2 \\ 5 \\ 7 \end{bmatrix}$$

Other conventions which we will follow are these:

Matrices will be denoted by uppercase boldface letters (**A**, **B**, etc.) and elements by lowercase italicized letters (a_{ij}, b_{kj}, etc.). Usually the elements of a matrix take the same lowercase italicized alphabetical letter as the uppercase letter standing for the matrix, but not always.

If we refer to a matrix as **A**, this does not tell us how many rows or columns the matrix has. We must learn this from other sources than the notation before we start performing operations on the matrix. Usually this is indicated at the outset of a discussion of a matrix in the following manner: **A** is an $m \times n$ matrix.

Another short way of denoting a matrix is with double lines and a single typical element, or with brackets and a single typical element:

$$\mathbf{A} = \|a_{ij}\| \qquad \mathbf{A} = [a_{ij}]$$

2.4.2 Matrix Operations

Just as with vectors, matrices also have operations of addition, subtraction, equality, multiplication by a scalar, and matrix multiplication.

2.4.2.1 Equality

Two matrices **A** and **B** are said to be equal, written **A** = **B**, if they have identical corresponding elements. That is **A** = **B**, only if $a_{ij} = b_{ij}$ for every i and j. Both **A** and **B** must have the same dimensions. If **A** is not equal to **B**, we write **A** ≠ **B**.

2.4.2.2 Multiplication by a Scalar

Given a scalar v and a matrix **A**, the product v**A** is defined as

$$v\mathbf{A} = \begin{bmatrix} va_{11} & \cdots & va_{1n} \\ va_{21} & \cdots & va_{2n} \\ \cdots & \cdots & \cdots \\ va_{m1} & \cdots & va_{mn} \end{bmatrix}$$

As with vectors, multiplying a scalar by a matrix implies multiplying the scalar by every element of the matrix. The order of multiplication does not matter so that

$$v\mathbf{A} = \mathbf{A}v$$

Examples. If

$$v = 3 \text{ and } \mathbf{A} = \begin{bmatrix} 2 & 3 \\ 1 & 0 \end{bmatrix} \text{ then } v\mathbf{A} = \begin{bmatrix} 6 & 9 \\ 3 & 0 \end{bmatrix}$$

If

$$v = -2 \text{ and } \mathbf{A} = \begin{bmatrix} 1 & 3 & 2 \\ -2 & 1 & 3 \end{bmatrix} \text{ then } v\mathbf{A} = \begin{bmatrix} -2 & -6 & -4 \\ 4 & -2 & -6 \end{bmatrix}$$

2.4.2.3 Addition

The sum of two matrices **A** and **B**, both having the same number of rows and columns, is a matrix **C** whose elements are the sums of the corresponding elements in **A** and **B**. The sum is written

$$\mathbf{A} + \mathbf{B} = \mathbf{C}$$

and for any element c_{ij} of **C**

$$c_{ij} = a_{ij} + b_{ij}$$

Note. If the two matrices **A** and **B** do not have the same number of rows and columns, then addition is undefined and cannot be carried out for them.

Example. Let

$$\mathbf{A} = \begin{bmatrix} 3 & 2 & 1 \\ -1 & 3 & 4 \end{bmatrix} \qquad \mathbf{B} = \begin{bmatrix} 4 & -1 & 2 \\ 7 & 1 & 3 \end{bmatrix}$$

$$\mathbf{A} + \mathbf{B} = \begin{bmatrix} 7 & 1 & 3 \\ 6 & 4 & 7 \end{bmatrix}$$

2.4.2.4 Subtraction

Subtraction is analogous to the addition of matrices. In the case of the matrices **A** and **B** just defined

$$\mathbf{A} - \mathbf{B} = \begin{bmatrix} -1 & 3 & -1 \\ -8 & 2 & 1 \end{bmatrix}$$

2.4.2.5 Matrix Multiplication

If **A** is an $n \times m$ matrix and **B** an $m \times p$ matrix, then the matrix multiplication of **A** times **B** produces another matrix **C** having n rows and p columns. This is written as

$$\mathbf{AB} = \mathbf{C} \tag{2.11}$$

Perhaps the simplest way to understand how matrix multiplication works is to assume that the premultiplying matrix **A** consists of n row vectors and the postmultiplying matrix **B** consists of p column vectors. Then the matrix **C** is a rectangular array containing the scalar products of each row vector in **A** multiplied times each column vector in **B**. Each resulting scalar product is placed in the corresponding row of **A** and the corresponding column of **B** in **C**.

To illustrate, let us write Equation 2.11 in full, arranging the elements of **A** to represent n row vectors and the elements of $\boldsymbol{B}$ to represent p column vectors:

$$\begin{bmatrix} (a_{11} & a_{12} & \cdots & a_{1m}) \\ (a_{21} & a_{22} & \cdots & a_{2m}) \\ \cdots & \cdots & \cdots & \cdots \\ (a_{n1} & a_{n2} & \cdots & a_{nm}) \end{bmatrix} \left[\begin{bmatrix} b_{11} \\ b_{21} \\ \vdots \\ b_{m1} \end{bmatrix} \begin{bmatrix} b_{12} \\ b_{22} \\ \vdots \\ b_{m2} \end{bmatrix} \begin{matrix} \cdots \\ \cdots \\ \cdots \\ \cdots \end{matrix} \begin{bmatrix} b_{1p} \\ b_{2p} \\ \vdots \\ b_{mp} \end{bmatrix} \right]$$

$$= \begin{bmatrix} c_{11} & c_{12} & \cdots & c_{1p} \\ c_{21} & c_{22} & \cdots & c_{2p} \\ \cdots & \cdots & \cdots & \cdots \\ c_{n1} & c_{n2} & \cdots & c_{np} \end{bmatrix}$$

To find the matrix **C**, the number of elements in the rows of **A** must equal the number of elements in the columns of **B**, to allow us to obtain the scalar products of the row vectors of **A** times the column vectors of **B**. In other words, the number of columns of **A** must equal the number of rows of **B**. **C** will have as many columns as **B**.

We begin with the first row vector of **A** and find the scalar product of this row vector times each of the column vectors of **B**. The resulting scalar products are placed in the first row and corresponding columns of **C**. Next, we take the second row of **A** and multiply it times each of the column vectors of **B**. Again the scalar products of these multiplications are placed in the second row of C in the position corresponding to the respective column of **B**. The process continues with multiplying a successive row vector of **A** times each of the column vectors of **B** and placing the resulting scalar product values in the respective row and columns of **C**. In general, if $\mathbf{a}_i'$ is the ith row vector in **A** and $\mathbf{b}_j$ is the jth column vector of **B**, then

$$c_{ij} = \mathbf{a}_i'\mathbf{b}_j = \sum_{k=1}^{m} a_{ik}b_{kj}$$

where c_{ij} is the element in the ith row and jth column of **C**. Defined in terms of the operation of the scalar product of vectors, matrix multiplication requires that the number of elements in the rows of the premultiplying matrix equal the number of elements in the columns of the postmultiplying matrix. This is necessary so that the components of the row and column vectors, respectively, will match in performing the scalar product of vectors.

Examples

1. Let

$$\mathbf{A} = \begin{bmatrix} 1 & 3 & 2 & 1 \\ -1 & 2 & 3 & 5 \end{bmatrix} \quad \text{and} \quad \mathbf{B} = \begin{bmatrix} 1 & 3 & 4 \\ 2 & 1 & 2 \\ 3 & 4 & 5 \\ -2 & 1 & 2 \end{bmatrix}$$

then

$$\mathbf{AB} = \mathbf{C} = \begin{bmatrix} 11 & 15 & 22 \\ 2 & 16 & 25 \end{bmatrix}$$

2. Let

$$\mathbf{G} = \begin{bmatrix} 1 & 3 & 5 \\ 2 & 1 & -3 \end{bmatrix} \quad \text{and} \quad \mathbf{H} = \begin{bmatrix} 1 & 2 \\ 3 & 1 \\ -2 & 5 \end{bmatrix}$$

then

$$\mathbf{GH} = \begin{bmatrix} 0 & 30 \\ 11 & -10 \end{bmatrix}$$

3. Let

$$\mathbf{Q} = \begin{bmatrix} 1 \\ 5 \\ 7 \end{bmatrix} \quad \text{and} \quad \mathbf{R} = (3 \quad 4 \quad 2) \quad \text{then} \quad \mathbf{QR} = \begin{bmatrix} 3 & 4 & 2 \\ 15 & 20 & 10 \\ 21 & 28 & 14 \end{bmatrix}$$

Unlike in ordinary algebra, it does not always follow (in fact, it rarely follows) that $\mathbf{AB} = \mathbf{BA}$ in matrix algebra. This is because by reversing which matrix is the first matrix on the left-hand side, one may upset the match of row elements to column elements, in which case the necessary scalar product of vectors could not be found, leaving matrix multiplication undefined for the pair of matrices. This is seen in example (1) where **BA** is not defined although **AB** is. Or the resulting matrix may be different in size if one reverses the order of multiplication (and multiplication is defined) which is seen in example (2) where **GH** is a 2×2 matrix but **HG** is a 3×3 matrix, and in example (3) where **QR** is a 3×3 matrix while **RQ** is a 1×1 matrix (or a single element).

However, the following expressions do hold:

$$(\mathbf{AB})\mathbf{C} = \mathbf{A}(\mathbf{BC}) = \mathbf{ABC} \text{ (associative law)}$$

$$\mathbf{A}(\mathbf{B} + \mathbf{C}) = \mathbf{AB} + \mathbf{AC} \text{ (distributive law)}$$

Note. In manipulating matrix equations, students sometimes forget that the order of multiplication is important when they apply the distributive law. For example,

$$\mathbf{A}(\mathbf{B} + \mathbf{C}) \neq \mathbf{BA} + \mathbf{CA}$$

or

$$(\mathbf{B} + \mathbf{C})\mathbf{A} \neq \mathbf{AB} + \mathbf{AC}$$

2.4.3 Identity Matrix

The identity matrix is a special matrix that plays the role in matrix algebra which the number 1 plays in ordinary scalar algebra. The identity matrix is denoted by the symbol **I** or $\mathbf{I}_n$, where n is the order (size) of the identity matrix in question. It is a square matrix having as many rows as columns and with 1s running down its main diagonal from the upper left-hand side corner to the lower right-hand side corner. Zeros are found in every other position off the diagonal. That is,

$$\mathbf{I} = \begin{bmatrix} 1 & 0 & 0 & 0 & \cdots & 0 \\ 0 & 1 & 0 & 0 & \cdots & 0 \\ 0 & 0 & 1 & 0 & \cdots & 0 \\ 0 & 0 & 0 & 1 & \cdots & 0 \\ \vdots & \vdots & \vdots & \vdots & \ddots & \vdots \\ 0 & 0 & 0 & 0 & \cdots & 1 \end{bmatrix}$$

Identity matrices come in different sizes: The second-order identity matrix is

$$\mathbf{I} = \mathbf{I}_2 = \begin{bmatrix} 1 & 0 \\ 0 & 1 \end{bmatrix}$$

The identity matrix of order 3 is

$$\mathbf{I} = \mathbf{I}_3 = \begin{bmatrix} 1 & 0 & 0 \\ 0 & 1 & 0 \\ 0 & 0 & 1 \end{bmatrix}$$

If $\mathbf{A}$ is a square matrix of order n and $\mathbf{I}_n$ the identity matrix of order n, then

$$\mathbf{I}_n\mathbf{A} = \mathbf{A}\mathbf{I}_n$$

However, if $\mathbf{A}$ is an $n \times m$ matrix, then, although $\mathbf{A}\mathbf{I}_m$ is defined, $\mathbf{I}_m\mathbf{A}$ is not defined. However, $\mathbf{I}_n\mathbf{A}$ is defined. The result of these multiplications of $\mathbf{A}$ by the identity matrix yields

$$\mathbf{I}_n\mathbf{A} = \mathbf{A}\mathbf{I}_m = \mathbf{A}$$

2.4.4 Scalar Matrix

Sometimes we wish to multiply every element in a matrix by a scalar number v. This can be accomplished by a square scalar matrix $\mathbf{S}$ such that

$$\mathbf{S} = v\mathbf{I}$$

Then if $\mathbf{A}$ is a square matrix which commutes with $\mathbf{S}$

$$\mathbf{SA} = v\mathbf{A}$$

2.4.5 Diagonal Matrix

A matrix with different values for its main diagonal elements but with zeros as the values of the off-diagonal elements is known as a "diagonal matrix." **D** is often (but not always) used to designate a diagonal matrix.

Examples

$$\mathbf{D} = \begin{bmatrix} 3 & 0 & 0 \\ 0 & 2 & 0 \\ 0 & 0 & 1 \end{bmatrix} \qquad \mathbf{U} = \begin{bmatrix} .3 & 0 \\ 0 & .1 \end{bmatrix}$$

Diagonal matrices have some interesting and useful properties. Suppose **D** is a diagonal matrix as just shown above and

$$\mathbf{A} = \begin{bmatrix} 4 & 1 & 3 & 2 \\ 5 & 2 & 6 & 7 \\ 1 & 3 & 8 & 4 \end{bmatrix} \qquad \mathbf{B} = \begin{bmatrix} -2 & 5 & 1 \\ 3 & 7 & 3 \\ 4 & 1 & 6 \\ 5 & -1 & 4 \end{bmatrix}$$

Then

$$\mathbf{DA} = \begin{bmatrix} 3 & 0 & 0 \\ 0 & 2 & 0 \\ 0 & 0 & 1 \end{bmatrix} \begin{bmatrix} 4 & 1 & 3 & 2 \\ 5 & 2 & 6 & 7 \\ 1 & 3 & 8 & 4 \end{bmatrix} = \begin{bmatrix} 12 & 3 & 9 & 6 \\ 10 & 4 & 12 & 14 \\ 1 & 3 & 8 & 4 \end{bmatrix}$$

while

$$\mathbf{BD} = \begin{bmatrix} -2 & 5 & 1 \\ 3 & 7 & 3 \\ 4 & 1 & 6 \\ 5 & -1 & 4 \end{bmatrix} \begin{bmatrix} 3 & 0 & 0 \\ 0 & 2 & 0 \\ 0 & 0 & 1 \end{bmatrix} = \begin{bmatrix} -6 & 10 & 1 \\ 9 & 14 & 3 \\ 12 & 2 & 6 \\ 15 & -2 & 4 \end{bmatrix}$$

Note that when **D** is the premultiplying matrix, as in **DA**, its diagonal elements multiply respectively each of the elements in the corresponding row of postmultiplying **A**. When **D** is the postmultiplying matrix, as in **BD**, the diagonal elements of **D** are multiplied respectively times the elements of the corresponding columns of the premultiplying **B**. Knowing these properties can save you from multiplying by a diagonal matrix in the usual, more complex way. By multiplying the above expressions in the usual way, you will get the same results.

2.4.6 Upper and Lower Triangular Matrices

An $n \times n$ square matrix some of whose elements on the principal diagonal and above the principal diagonal are nonzero, while all elements below the principal diagonal are 0 is known as an "upper triangular matrix."

An $n \times n$ square matrix whose nonzero elements are on the principal diagonal and below the principal diagonal, while those above are 0 is a "lower triangular matrix."

$$\begin{bmatrix} a_{11} & a_{12} & a_{13} & \cdots & a_{1n} \\ 0 & a_{22} & a_{23} & \cdots & a_{2n} \\ 0 & 0 & a_{33} & \cdots & a_{3n} \\ 0 & 0 & 0 & \ddots & \vdots \\ 0 & 0 & 0 & 0 & a_{nn} \end{bmatrix} \text{ is an upper triangular matrix}$$

$$\begin{bmatrix} a_{11} & 0 & 0 & 0 & 0 \\ a_{21} & a_{22} & 0 & 0 & 0 \\ a_{31} & a_{32} & a_{33} & 0 & 0 \\ \vdots & \vdots & \vdots & \ddots & 0 \\ a_{n1} & a_{n2} & a_{n3} & \cdots & a_{nn} \end{bmatrix} \text{ is a lower triangular matrix}$$

2.4.7 Null Matrix

A matrix of whatever dimensions having only 0s in its cells is known as a "null matrix." It is denoted as **0**. A null matrix has the following algebraic properties provided that it commutes with the matrices in question:

$$\mathbf{A} + \mathbf{0} = \mathbf{A} \quad \mathbf{A} - \mathbf{A} = \mathbf{0} \quad \mathbf{A0} = \mathbf{0}$$

However, if $\mathbf{AB} = \mathbf{0}$, this does not necessarily imply that either **A** or **B** is a null matrix. It is easy to find nonnull matrices **A, B** whose product $\mathbf{AB} = \mathbf{0}$. This is an instance where matrix algebra is not like ordinary algebra.

2.4.8 Transpose Matrix

The transpose of a matrix **A** is another matrix designated **A′** which is formed from **A** by interchanging rows and columns of **A** so that row i becomes column i and column j becomes row j of **A′**. That is, if

$$\mathbf{A} = \|a_{ij}\|, \quad \text{then} \quad \mathbf{A}' = \|a_{ji}\|$$

Examples

$$\mathbf{A} = \begin{bmatrix} 1 & 3 & 2 \\ 2 & 0 & 1 \end{bmatrix} \qquad \mathbf{A}' = \begin{bmatrix} 1 & 2 \\ 3 & 0 \\ 2 & 1 \end{bmatrix}$$

$$\mathbf{B} = \begin{bmatrix} 1 & 3 \\ 2 & 4 \end{bmatrix} \qquad \mathbf{B}' = \begin{bmatrix} 1 & 2 \\ 3 & 4 \end{bmatrix}$$

A way of visualizing transposing a matrix is to imagine that it is flipped over, using as a hinge the first diagonal elements of the original matrix going from left to right and down. For example, the elements 1 in the first row and 0 in the second row of **A** constitute the first diagonal of **A**. The elements 1 and 4 in the first and second row are the principal diagonal of **B**.

The transpose matrix $(\mathbf{AB})'$ equals $\mathbf{B'A'}$, reversing the order of multiplication and then transposing the individual matrices. The pattern holds for any number of multiplied matrices

$$(\mathbf{ABCDE \cdots XYZ}) = (\mathbf{Z'Y'X' \cdots E'D'C'B'A'})$$

But there is no effect on summing matrices, except to transpose them individually:

$$(\mathbf{A}+\mathbf{B}+\mathbf{C})' = \mathbf{A}'+\mathbf{B}'+\mathbf{C}'$$

Note that

$$\mathbf{I}' = \mathbf{I} \quad \text{and} \quad (\mathbf{A}')' = \mathbf{A}$$

2.4.9 Symmetric Matrices

A symmetric matrix is a square matrix for which its transpose is equal to itself. It is a matrix in which any element a_{ij} equals the corresponding element a_{ji}. That is, let **R** be a square symmetric matrix; then

$$\mathbf{R}' = \mathbf{R}$$

All covariance and intercorrelation matrices for a single set of variables are symmetric matrices.

Example

$$\mathbf{R} = \begin{bmatrix} 1.00 & -0.21 & 0.32 \\ -0.21 & 1.00 & 0.43 \\ 0.32 & 0.43 & 1.00 \end{bmatrix}$$

2.4.10 Matrix Inverse

In scalar algebra, if $ab = 1$, then b is a special element with respect to a, that is, $1/a$, such that multiplying a by b yields 1. Is there an analogous relationship among matrices? If such a matrix exists for a given matrix **A**, we shall call it the "matrix inverse" of **A**. It will be denoted by $\mathbf{A}^{-1}$.

The following necessary (but not sufficient) condition must be met for the matrix **A** to have an inverse: The matrix must be square to have an inverse. However, not every square matrix has an inverse.

When the inverse matrix exists for a matrix, the following algebraic relationships hold:

$$(\mathbf{AB})^{-1} = \mathbf{B}^{-1}\mathbf{A}^{-1}$$

$$(\mathbf{A}^{-1})^{-1} = \mathbf{A}$$

$$(\mathbf{A}')^{-1} = (\mathbf{A}^{-1})'$$

$$(\mathbf{A}^{-1})'\mathbf{A}' = \mathbf{A}'(\mathbf{A}^{-1})' = \mathbf{I}$$

Methods for finding the inverse of any given square matrix are too complicated to be given here. However, for matrices of up to order 3, a method based upon determinants can be used to find the matrix inverse, if it exists. Before going into this method, we should therefore, take up the topic of determinants.

2.4.11 Orthogonal Matrices

A square matrix **P** is said to be orthogonal if

$$\mathbf{PP}' = \mathbf{P}'\mathbf{P} = \mathbf{I}$$

Looking at this on the level of vector multiplication we see that different column (or row) vectors of the matrix are orthogonal to one another, and the length of each column (row) vector is 1. It also follows that the inverse of an orthogonal matrix is its transpose.

Example

$$\mathbf{P} = \begin{bmatrix} 1/\sqrt{3} & 1/\sqrt{6} & -1/\sqrt{2} \\ 1/\sqrt{3} & 1/\sqrt{6} & 1/\sqrt{2} \\ 1/\sqrt{3} & -2/\sqrt{6} & 0 \end{bmatrix} \qquad \mathbf{P}' = \begin{bmatrix} 1/\sqrt{3} & 1/\sqrt{3} & 1/\sqrt{3} \\ 1/\sqrt{6} & 1/\sqrt{6} & -2/\sqrt{6} \\ -1/\sqrt{2} & 1/\sqrt{2} & 0 \end{bmatrix}$$

$$\mathbf{PP}' = \begin{bmatrix} 1 & 0 & 0 \\ 0 & 1 & 0 \\ 0 & 0 & 1 \end{bmatrix}$$

Orthogonal matrices will be important to exploratory factor analysis, because maximum likelihood and other solutions obtain eigenvector matrices which are orthogonal matrices.

Sometimes an orthogonal matrix is called an "orthonormal" matrix because its column or row vectors have unit lengths and are mutually orthogonal.

2.4.12 Trace of a Matrix

Let

$$A = \begin{bmatrix} a_{11} & a_{12} & a_{13} & a_{14} & a_{15} \\ a_{21} & a_{22} & a_{23} & a_{24} & a_{25} \\ a_{31} & a_{32} & a_{33} & a_{34} & a_{35} \\ a_{41} & a_{42} & a_{43} & a_{44} & a_{45} \\ a_{51} & a_{52} & a_{53} & a_{54} & a_{55} \end{bmatrix}$$

Then

$$\text{tr}\,\mathbf{A} = a_{11} + a_{22} + a_{33} + a_{44} + a_{55}$$

is known as the trace of **A**. The trace of any square matrix is the sum of its diagonal elements. The result is a scalar. In general,

$$\text{tr}\,\mathbf{A} = \sum_{i=1}^{n} a_{ii}$$

2.4.13 Invariance of Traces under Cyclic Permutations

Suppose the $n \times n$ square matrix **A** is itself the product of other matrices that are not all square. Given

$$\mathbf{A}_{n\times n} = \mathbf{F}_{n\times r}\mathbf{C}_{r\times r}\mathbf{F}'_{r\times n}$$

then

$$\text{tr}\,\mathbf{A}_{n\times n} = \text{tr}\,\mathbf{F}_{n\times r}\mathbf{C}_{r\times r}\mathbf{F}'_{r\times n} = \text{tr}\,\mathbf{F}'_{r\times n}\mathbf{F}_{n\times r}\mathbf{C}_{r\times r} = \text{tr}\,\mathbf{C}_{r\times r}\mathbf{F}'_{r\times n}\mathbf{F}_{n\times r}$$

Although the products of the matrices in the equality above are all square matrices, they are not of the same dimension. The first product produces an $n\times n$ matrix, the second and third produce $|\mathbf{A}|$ matrices. Nevertheless each of their traces will be the same. Note that the square matrices are produced by cycling the right most matrix of a matrix product around to the left-most position. This produces a different ordering or permutation of the component matrices, hence cyclic permutations. In general, if **ABCDE** is square,

$$\text{tr}(\mathbf{ABCDE}) = \text{tr}(\mathbf{EABCD}) = \text{tr}(\mathbf{DEABC}) = \text{tr}(\mathbf{CDEAB}) = \text{tr}(\mathbf{BCDEA})$$

Traces are often used in defining in matrix notation a least-squares criterion to be optimized in least-squares estimation. For example, given **S** we seek a matrix

$\Sigma = \Lambda\Phi\Lambda' + \Psi^2$ by varying the elements of the matrices Λ, Φ, Ψ^2 on the right-hand side, such that $\mathbf{E} = \mathbf{S} - \Sigma = \mathbf{S} - (\Lambda\Phi\Lambda' + \Psi^2)$ and $\mathrm{tr}\mathbf{E}'\mathbf{E}$ is a minimum. **E** represents the element-by-element difference between the matrix **S** and the matrix Σ. The diagonal elements of $\mathbf{E}'\mathbf{E}$ contain the column sum of squares of the elements of **E**, which are corresponding differences between elements of **S** and Σ. Then by summing the column sum of squares, $\mathrm{tr}\mathbf{E}'\mathbf{E}$ contains the sum of all squared differences between elements of **S** and Σ. And the values for elements of Λ, Φ, Ψ^2 which make $\mathrm{tr}\mathbf{E}'\mathbf{E}$ a minimum are least squares estimates of them.

2.5 Determinants

For any square matrix **A** there exists a number uniquely determined by the elements of **A** known as the determinant of **A**. This number is designated $|\mathbf{A}|$ but is also written as

$$|\mathbf{A}| = \begin{vmatrix} a_{11} & a_{12} & \cdots & a_{1n} \\ a_{21} & a_{22} & \cdots & a_{2n} \\ \vdots & \vdots & \ddots & \vdots \\ a_{n1} & a_{n2} & \cdots & a_{nn} \end{vmatrix}$$

that is, by enclosing the matrix elements within vertical straight lines.

For a 1×1 matrix $\mathbf{A} = a$, the determinant of **A** is simply a. For a 2×2 matrix **A**, the determinant of **A** is

$$|\mathbf{A}| = \begin{vmatrix} a_{11} & a_{12} \\ a_{21} & a_{22} \end{vmatrix} = a_{11}a_{22} - a_{12}a_{21}$$

For a 3×3 matrix **A**, the determinant of **A** is written

$$|\mathbf{A}| = \begin{vmatrix} a_{11} & a_{12} & a_{13} \\ a_{21} & a_{22} & a_{23} \\ a_{31} & a_{32} & a_{33} \end{vmatrix} = a_{11}a_{22}a_{33} + a_{12}a_{23}a_{31} + a_{13}a_{21}a_{32} - a_{13}a_{22}a_{31} - a_{11}a_{23}a_{32} - a_{12}a_{21}a_{33}$$

One computational device for finding determinants of 2×2 and 3×3 matrices is to copy all but the last column of the determinant to the right-hand side of the determinant. Then draw arrows through the elements as shown. Then for each arrow, multiply together the terms through which the arrow passes. Multiply the product by +1 if the arrow points down, and multiply the product by −1 if the arrow points up. Add the products together, each multiplied by 1 or −1, respectively.

Note: This does not work with 4×4 matrices or larger.

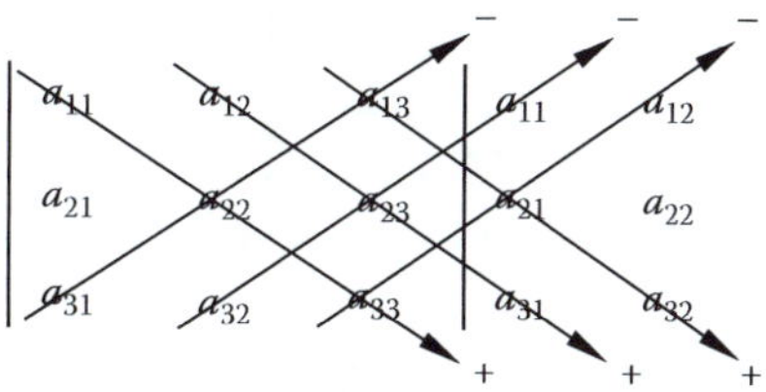

Unfortunately in factor analysis and structural equation modeling the determinants are of matrices that are much larger. Usually they are determinants of the covariance or correlation matrix among a large number of variables subjected to analysis. So, we need a more general definition of a determinant to include large matrices.

The determinant of an nth order $n \times n$ matrix $\mathbf{A}$ is defined by mathematicians as

$$|\mathbf{A}| = \sum (-1)^t a_{1i} a_{2h} a_{3r} \cdots a_{ns} \tag{2.12}$$

where the sum is taken across all $n!$ permutations of the right-hand side subscripts, and t is the number of inversions or interchanges of pairs of adjacent elements in the product needed to bring the right-hand side subscripts into ascending numerical order. In effect, the different product terms in the sum are obtained by selecting each element to enter the term from a different row and column of the matrix. (One should recognize that this does not denote the absolute value of a number because the symbol between the vertical lines is a matrix and not a scalar.)

By an interchange is meant exchanging the position of two adjacent numbers in a permutation of the integers 1, 2, 3,…,n, so that if one number is smaller than the other, it is placed before the larger number on the left in the resulting new permutation. For example, suppose $n = 6$, and we have the permutation

$$314265$$

of the first six integers. Then an interchange would be

$$\mathbf{31}4265 \rightarrow \mathbf{13}4265$$

where the pair 3 1 are interchanged to become 1 3 because 1 comes before 3. Another interchange might then be to interchange 4 2 to 2 4, so we would have

$$13\mathbf{42}65 \rightarrow 13\mathbf{24}65.$$

And even further, we would interchange 3 2 to become 2 3:

$$132465 \rightarrow 123465$$

And finally, we could interchange 6 5 to 5 6:

$$123465 \rightarrow 123456$$

Now, in the present example we made four interchanges to bring the original permutation into ascending order. The quantity t in the formula for the determinant is the number of interchanges of the elements of the permutation needed to bring its right-hand side subscripts into ascending order. If t is odd, then (–1)t is equal to –1 and if t is even, then (–1)t is equal to 1.

Although considerable mathematical research was conducted on determinants in the nineteenth and early twentieth centuries, their importance today is somewhat lessened. Many of the uses for them, in the theory of inverses, areas of parallelepipeds, in the concept of rank of a matrix, in the theory of multiple correlation have been replaced by other concepts that allow one to achieve much the same thing as with determinants. However, they do play an important role in multivariate statistics, insofar as the determinant of a covariance matrix is found in the formula for the multivariate normal distribution, and the concept of a generalized variance is based on the determinant of the variance–covariance matrix of a set of variables. The concept of the rank of a matrix, of nonsingular and singular matrices and tests for linear independence of variables also derive from the theory of determinants, and these are important concepts in factor analysis and structural equation modeling.

Nevertheless, in some cases the use of determinants leads to the simplest way to derive some mathematical results. Computationally, the formula in the definition of a determinant is not a very efficient way to obtain the value of a determinant. Alternative methods for finding the value of determinants usually involve transforming the matrix into an upper or lower triangular matrix or into a diagonal matrix, and then multiplying the diagonal elements of the resulting matrix together to yield the determinant. In factor analysis, the correlation matrix is transformed into a diagonal matrix of eigenvalues and then the eigenvalues are multiplied together to get the determinant of the correlation matrix.

Examples

$$\begin{vmatrix} 1 & 2 \\ 3 & 4 \end{vmatrix} = -2 \qquad \begin{vmatrix} 3 & 2 & 1 \\ 1 & 2 & 2 \\ 3 & 1 & 3 \end{vmatrix} = 13$$

2.5.1 Minors of a Matrix

The determinant of an rth-order square submatrix of a square matrix $\mathbf{A}$ is known as a minor of the matrix $\mathbf{A}$. In the above examples, the individual elements of the two matrices are first-order minors, whereas in the case of the 3×3 matrix

$$\begin{vmatrix} 3 & 2 \\ 1 & 2 \end{vmatrix} \quad \begin{vmatrix} 1 & 2 \\ 3 & 3 \end{vmatrix} \quad \begin{vmatrix} 2 & 2 \\ 1 & 3 \end{vmatrix}$$

are second-order minors.

2.5.2 Rank of a Matrix

The rank of a matrix is the order of the highest-order nonzero determinant obtainable from square submatrices of elements of the matrix obtained by deleting rows and/or columns of the matrix. For example, the matrix

$$\begin{bmatrix} 2 & 3 & 5 \\ 1 & 1 & 6 \\ 3 & 5 & 6 \\ 3 & 1 & 7 \end{bmatrix}$$

has at most a rank of 3 because the largest square submatrix that can be formed from this matrix by deleting rows and/or columns is a 3×3 matrix, and no determinants of larger order may be formed. It will have a rank of 3 if there exists a 3×3 submatrix of this matrix with a determinant not equal to 0. If no 3×3 submatrix has a determinant not equal to 0, then the rank may be 2, and will be determined to be 2 if at least one 2×2 nonzero determinant can be formed by deleting rows and columns of the original matrix. And it will have a rank of 1 if at least one nonzero element exists in the matrix. The matrix will have rank 0 if all elements in the matrix are 0, for then all 1×1 determinants will be 0.

The rank of a matrix indicates the maximum number of linearly independent rows (columns) of the matrix.

Thurstone (1947) used the concept of "minimum rank" of a correlation matrix whose diagonal elements had been replaced with "communalities" as the number of common factors to retain.

2.5.3 Cofactors of a Matrix

In a matrix **A**, we will designate A_{ij} to be the "cofactor" of the element a_{ij}, found by deleting the *i*th row and *j*th column from the matrix **A**, taking the determinant of the remaining submatrix of elements, and giving it the sign of $(-1)^{i+j}$.

Example

$$\mathbf{A} = \begin{bmatrix} a_{11} & a_{12} & a_{13} \\ a_{21} & a_{22} & a_{23} \\ a_{31} & a_{32} & a_{33} \end{bmatrix} \qquad A_{12} = -\begin{vmatrix} a_{21} & a_{23} \\ a_{31} & a_{33} \end{vmatrix}$$

2.5.4 Expanding a Determinant by Cofactors

Let **A** be an $n \times n$ matrix. The determinant of **A** can be found by taking any row or column of the matrix and summing the products of the elements of that row or column, each multiplied by its corresponding cofactor. For example, if we choose the second row of the matrix **A** to find the determinant of **A**, we have

$$|\mathbf{A}| = a_{21}A_{21} + a_{22}A_{22} + \cdots + a_{2n}A_{2n}$$

Example

$$\mathbf{A} = \begin{bmatrix} 1 & 3 & 5 \\ 4 & 2 & 3 \\ 1 & 4 & 2 \end{bmatrix}$$

$$|\mathbf{A}| = -4\begin{vmatrix} 3 & 5 \\ 4 & 2 \end{vmatrix} + 2\begin{vmatrix} 1 & 5 \\ 1 & 2 \end{vmatrix} - 3\begin{vmatrix} 1 & 3 \\ 1 & 4 \end{vmatrix}$$

$$= -4(6-20) + 2(2-5) - 3(4-3) = 47$$

2.5.5 Adjoint Matrix

If **A** is an $n \times n$ matrix, the transpose of a matrix obtained from **A** by replacing its elements by their corresponding cofactors is known as the "adjoint matrix" for **A** and is designated by $\mathbf{A}^+$. For example, if

$$\mathbf{A} = \begin{bmatrix} a_{11} & a_{12} & a_{13} \\ a_{21} & a_{22} & a_{23} \\ a_{31} & a_{32} & a_{33} \end{bmatrix} \qquad \mathbf{A}^+ = \begin{bmatrix} A_{11} & A_{21} & A_{31} \\ A_{12} & A_{22} & A_{32} \\ A_{13} & A_{23} & A_{33} \end{bmatrix}$$

The inverse of a matrix **A** can be obtained by multiplying its adjoint matrix $\mathbf{A}^+$ by the reciprocal of the determinant of **A**, that is,

$$\mathbf{A}^{-1} = \frac{1}{|\mathbf{A}|}\mathbf{A}^+$$

Example

$$\mathbf{A} = \begin{bmatrix} 1 & 3 & 5 \\ 4 & 2 & 3 \\ 1 & 4 & 2 \end{bmatrix} \qquad \mathbf{A}^{-1} = \frac{1}{47}\begin{bmatrix} -8 & 14 & 1 \\ -5 & -3 & 17 \\ 14 & -1 & -10 \end{bmatrix}$$

For proof that the right-hand side matrix is the inverse of the left-hand side matrix, multiply the matrices together to see if the result is an identity matrix.

2.5.6 Important Properties of Determinants

1. The determinant of the transpose matrix equals the determinant of the original matrix, i.e., $|\mathbf{A}'| = |\mathbf{A}|$.
2. Any theorem about $|\mathbf{A}|$ that is true for rows (columns) of a matrix **A** is also true for columns (rows).
3. If any two rows (columns of the matrix **A** are "interchanged," the determinant of the resulting matrix equals $-|\mathbf{A}|$.
4. Let **B** be a matrix formed from the matrix **A** by multiplying one of its rows (columns) by k, then $|\mathbf{B}| = k|\mathbf{A}|$.
5. Adding k times one row (column) of **A** to another row (column) of **A** produces a matrix whose determinant is still $|\mathbf{A}|$.
6. If two rows (columns) of a matrix **A** are identical, then $|\mathbf{A}| = 0$.
7. If any column of a matrix **A** is a linear combination of other columns of **A**, then $|\mathbf{A}| = 0$.
8. If any row (column) of **A** consists of only 0s, then $|\mathbf{A}| = 0$.
9. If the determinant of a square matrix **A** is 0, we say the matrix is "singular," otherwise we say that the matrix is "nonsingular." Singular matrices have no inverses.
10. If **A** and **B** are $n \times n$ square matrices, then $|\mathbf{AB}| = |\mathbf{A}|\,|\mathbf{B}|$. This result extends to the determinant of the product of any number of $n \times n$ matrices.
11. The determinant of a diagonal matrix equals the product of its diagonal elements, i.e., $|\mathbf{D}| = \prod_{i=1}^{n} d_{ii}$.
12. The determinant of an upper (lower) triangular matrix equals the product of the elements in its principal diagonal $|\Delta| = \prod_{i=1}^{n} \delta_{ii}$.
13. The determinant of an orthogonal matrix **P** equals ±1.
14. $|\mathbf{A}^{-1}| = \dfrac{1}{|A|}$.
15. $|\mathbf{I}| = 1$.
16. $|\mathbf{0}| = 0$.

2.5.7 Simultaneous Linear Equations

Consider the following system of simultaneous linear equations in which there are n equations and n unknowns:

$$a_{11}x_1 + a_{12}x_2 + \cdots + a_{1n}x_n = b_1$$

$$a_{21}x_1 + a_{22}x_2 + \cdots + a_{2n}x_n = b_2$$

$$\cdots\cdots\cdots\cdots\cdots\cdots\cdots\cdots\cdots$$

$$a_{n1}x_1 + a_{n2}x_2 + \cdots + a_{nn}x_n = b_n$$

This system can be expressed in the matrix form as

$$\begin{bmatrix} a_{11} & a_{12} & \cdots & a_{1n} \\ a_{21} & a_{22} & \cdots & a_{2n} \\ \vdots & \vdots & \ddots & \vdots \\ a_{n1} & a_{n2} & \cdots & a_{nn} \end{bmatrix} \begin{bmatrix} x_1 \\ x_2 \\ \vdots \\ x_n \end{bmatrix} = \begin{bmatrix} b_1 \\ b_2 \\ \vdots \\ b_n \end{bmatrix}$$

or in the equation form as

$$\mathbf{Ax} = \mathbf{b} \tag{2.13}$$

where

A is the square matrix of known coefficients
x is the n-component column vector of unknown values $x_1, x_2, \ldots, x_n$
b is the n-component column vector of known quantities

The task is to solve for the unknown quantities, $x_1, x_2, \ldots, x_n$, in terms of known quantities in **A** and **b**.

Although there is a way to solve for the unknown quantities using determinants, known as "Cramer's rule," we will instead consider how we might do this using matrix algebra. This is more practical. Suppose we multiply both sides of Equation 2.13 by $\mathbf{A}^{-1}$, we obtain

$$\mathbf{A}^{-1}\mathbf{Ax} = \mathbf{A}^{-1}\mathbf{b}$$

$$\mathbf{x} = \mathbf{A}^{-1}\mathbf{b} \tag{2.14}$$

In other words, the column vector can be solved for by premultiplying both sides of the equation by the inverse matrix of **A**.

If **A** is singular, that is, has a 0 determinant, then no solution is possible, because there will be no inverse matrix.

2.6 Treatment of Variables as Vectors

We indicated in Section 2.3 that the key idea of this book is that variables may be treated as vectors. In this section, we intend to show how this may be done.

There are several ways to define a variable. A common definition is that a variable is a quantity that may take any one of a set of possible values. Another definition is that a variable is a property on which individuals of a population differ. For the purposes of this book, however, we will use the following definition, with qualification: "A variable is a functional relation that associates members of a first set (population) with members of a second set of ordered sets of real numbers in such a way that no member of the first set (population) is associated with more than one ordered set of real numbers at a time." This definition is quite general for it includes both the familiar case of the unidimensional variable (involving a single quantifiable property or attribute) and the case of the multidimensional variable (involving simultaneously several quantifiable properties or attributes). Because most theoretical work, however, deals with relations among individual attributes, we shall hereafter usually mean by the term "variable" a unidimensional variable that as a functional relation associates each member of a population with only one real number at a time.

There are several ways to classify variables. For example, statisticians frequently classify variables according to whether they are discrete or continuous. A discrete variable can take on at most a countably infinite number of values, whereas a continuous variable can take on an uncountably infinite number of values. (A set is said to have a countably infinite number of members if it is infinite and each member of the set can be put in a one-to-one correspondence with members of the set of integers.) Mathematical operations on discrete variables involve only algebra, whereas on continuous variables they involve calculus. Discrete variables may be represented by N-tuples but continuous variables must be represented in other ways. But a major concern is not whether variables are discrete or continuous but whether they involve finite or infinite populations.

2.6.1 Variables in Finite Populations

Variables in finite populations are treated as vectors in a unitary vector space of N-tuples. For example, suppose we have a finite population of five individuals whose scores on a variable are, respectively 3, 1, 1, 4, 2. Let us represent these scores by a 5-tuple (3,1,1,4,2). Each coordinate of the 5-tuple corresponds to an individual member of the population and takes on the values of the variable for that individual.

When more than one variable is defined for the members of a finite population, we represent the variables by N-tuples arranged in the form of $N \times 1$

column vectors (with N the number of individuals in the population). When the variables are dealt with collectively (as in the case of finding linear combinations of the variables), we transpose the column vectors to row vectors and arrange them as the rows of an $n \times N$ matrix (with n the number of variables). For example, let $\mathbf{x}_1, \mathbf{x}_2, \ldots, \mathbf{x}_n$ be n N-tuple column vectors standing for n variables $X_1, X_2, \ldots, X_n$ defined in a finite population of N individuals. By $\mathbf{X}$ we shall mean the $n \times N$ matrix, whose row vectors are the transposed variable vectors $\mathbf{x}_1, \mathbf{x}_2, \ldots, \mathbf{x}_n$.

$$\mathbf{X} = \begin{bmatrix} \mathbf{x}_1' \\ \mathbf{x}_2' \\ \vdots \\ \mathbf{x}_n' \end{bmatrix} = \begin{bmatrix} x_{11} & x_{12} & \cdots & x_{1N} \\ x_{21} & x_{22} & \cdots & x_{2N} \\ \cdots & \cdots & \cdots & \cdots \\ x_{n1} & x_{n2} & \cdots & x_{nN} \end{bmatrix} \tag{2.15}$$

Consider now the linear combination

$$\mathbf{y}' = \mathbf{a}'\mathbf{X} = (a_1, a_2, \ldots, a_n) \begin{bmatrix} \mathbf{x}_1' \\ \mathbf{x}_2' \\ \vdots \\ \mathbf{x}_n' \end{bmatrix} = (y_1, y_2, \ldots, y_N) \tag{2.16}$$

An important matrix in factor analysis and structural equation modeling would be the matrix of variances and covariances among these variables:

$$\mathbf{S}_{XX} = \frac{1}{N}\left[\mathbf{XX}' - \frac{1}{N}\mathbf{X11'X}'\right] \tag{2.17}$$

Here $\mathbf{1}$ is the column sum vector consisting of all 1s. $\mathbf{X}$ is defined in Equation 2.15. $\mathbf{XX}'$ is a square symmetric $n \times n$ matrix of sums of squares and cross products among the variables, summed across individuals.

$$\mathbf{XX}' = \begin{bmatrix} \sum_{i=1}^{N} X_{1i}^2 & \sum_{i=1}^{N} X_{1i}X_{2i} & \cdots & \sum_{i=1}^{N} X_{1i}X_{ni} \\ \sum_{i=1}^{N} X_{2i}X_{1i} & \sum_{i=1}^{N} X_{2i}^2 & \cdots & \sum_{i=1}^{N} X_{2i}X_{ni} \\ \vdots & \vdots & \ddots & \vdots \\ \sum_{i=1}^{N} X_{ni}X_{1i} & \sum_{i=1}^{N} X_{ni}X_{2i} & & \sum_{i=1}^{N} X_{ni}^2 \end{bmatrix}$$

$\mathbf{X1}$ is a $n \times 1$ column vector of sums of scores in each respective row of $\mathbf{X}$. The column of variable means would be given by

$$\bar{\mathbf{X}} = \frac{1}{N}\mathbf{X1} = \frac{1}{N}\begin{bmatrix} x_{11} & x_{12} & \cdots & x_{1N} \\ x_{21} & x_{22} & \cdots & x_{2N} \\ \cdots & \cdots & \cdots & \cdots \\ x_{n1} & x_{n2} & \cdots & x_{nN} \end{bmatrix}\begin{bmatrix} 1 \\ 1 \\ \vdots \\ 1 \end{bmatrix} = \frac{1}{N}\begin{bmatrix} \sum_{i=1}^{N} X_{1i} \\ \sum_{i=1}^{N} X_{2i} \\ \vdots \\ \sum_{i=1}^{N} X_{ni} \end{bmatrix} = \begin{bmatrix} \bar{X}_1 \\ \bar{X}_2 \\ \vdots \\ \bar{X}_n \end{bmatrix}$$

Another important matrix derivable from the matrix $\mathbf{X}$ is the symmetric $n \times n$ matrix, $\mathbf{R}_{XX}$, containing the intercorrelation coefficients for pairs of variables in $\mathbf{X}$. To obtain $\mathbf{R}_{XX}$, let $\mathbf{D}_X^2 = [\text{diag}\,\mathbf{S}_{XX}]$. Here [diag $\mathbf{S}_{XX}$] means to extract the diagonal elements of $\mathbf{S}_{XX}$ and place them in the principal diagonal of a diagonal matrix with zero off-diagonal elements. Then

$$\mathbf{R}_{XX} = \mathbf{D}_X^{-1}\mathbf{S}_{XX}\mathbf{D}_X^{-1} \tag{2.18}$$

The matrix $\mathbf{D}_X^2$ has as its diagonal elements the variances of the respective variables in $\mathbf{X}$.

2.6.2 Variables in Infinite Populations

Variables defined for a countably infinite population may be represented by N-tuples with infinitely many coordinates; however, not all operations defined on N-tuples with a finite number of coordinates apply to N-tuples with infinitely many coordinates. In particular, the definition given for the scalar product of N-tuples of finite order does not work on "N-tuples" of infinite order, because in the case of N-tuples of infinite order sums of infinitely many product terms are involved, and these normally do not converge to finite scalar values. Because of this difficulty and because of the additional difficulty that variables in uncountably infinite populations are not expressible as N-tuples, other notations have been sought to represent variables in infinite populations.

Mathematical statisticians have developed a useful notation for dealing with variables defined on infinite populations. They let a capital letter stand for a variable and use a corresponding lowercase letter to stand for a particular value of the variable. For example, X is a variable, and x is some particular value of the variable X. Mathematical statisticians have also adopted the convention of combining variables in this notation as if variables were elements of a scalar algebra. For example, U, V, W, X, Y, and Z are variables defined on a population, then the expression $U = V/W + XY - Z^2$ is a permissible expression and means that if u, v, w, x, y, and z are the respective real values of these variables for any arbitrary member of the population, then $u = v/w + xy - z^2$. Thus, there is a one-to-one correspondence between expressions involving variables and expressions involving scalars, which reveals the fact that variables form a field under the operations of addition and multiplication. An immediate consequence of this fact is that variables may serve as vectors in a vector space, with the operation of addition of vectors in the vector space corresponding to the operation of addition of variables in the field of variables.

However, a vector space of variables is not a unitary vector space—which we require for multivariate linear analysis—unless we can find a way to define for the vector space an operation having the properties of the scalar product of vectors. The simple product XY of two variables X and Y does not satisfy the requirements of a scalar product because the result is a variable and not a scalar. However, we might consider the possibility that the simple product XY is part of, but not all, the solution to the requirements for a scalar product. What we need to look for is an operation that takes a random variable and produces from it a unique number known as its scalar product. Since XY is a random variable representing a unique combination of the variables X and Y, applying this operation to XY will produce a unique scalar corresponding to this pair.

The operation we have in mind is the "expected value operator," denoted by $E()$, that obtains the mean of a random variable. Applied to a discrete variable (a variable that takes on only a finite number of distinct values) this operator resembles in many ways the operation for finding the mean of a variable for a finite population (or sample) by the method of grouping of individuals by scores. In the latter method, for each value on the discrete variable X one obtains the proportion $p(x_i)=f_i/N$ of the individuals in the finite population having value x_i on the discrete variable, where f_i is the number of individuals having the value x_i and N is the number of individuals in the population (or sample). Then the mean is given as $E(X)=\sum_{i=1}^{m} x_i p(x_i)$ with m the number of possible values of X in the population.

Regardless, in the general discrete case with countably infinite populations, for each value x_i mathematical statisticians assume that there exists a nonnegative number $p(x_i)$, which, roughly speaking, represents the proportion of the total population with value x_i of the variable X. Then the "expected value" of X denoted $E(X)$ is given as

$$E(X)=\sum x_i p(x_i) \quad \text{for all } p(x_i)>0 \tag{2.19}$$

In the case of an uncountably infinite population, it is not possible to obtain the proportions of the total population by a process of counting. As a matter of fact, if a value x_i on a variable X has only a finite or countably infinite individuals in an uncountably infinite population associated with it, then it still has, paradoxically a "zero" proportion of the total population associated with it. We will not attempt to explain the basis for this paradox since it is based on one of the subtle features of measure theory, which is beyond the scope of this book (cf. Singh, 1959).

The expected value of a continuous random variable (a variable which can take on any of the values in a continuum of values on any interval or intervals of the real numbers) has to be defined differently. Given the continuous variable X, it is postulated that there exists a function $f(x)$ known as the "density function" of X that allows one to determine the proportion $P(X|a < x \leq b)$ of the total population having values x in any continuous interval $a < x \leq b$ over which the variable is defined. Since this is a density function, the total area under the curve $f(x)$ equals unity. If one refers to the graph of the function $f(x)$ in Figure 2.7, the proportion $P(X|a < x \leq b)$ corresponds to the area under the curve of the function $f(x)$ in the interval $a < x \leq b$. This area may be obtained by integral calculus from the integral equation

$$P(X|a < x \leq b) = \int_a^b f(x)dx$$

The expected value of the continuous variable X is given as

$$E(X) = \int_{-\infty}^{+\infty} xf(x)dx \tag{2.20}$$

Returning now to the problem of defining the scalar product, **xy**, of two vectors, **x** and **y**, representing two random variables X and Y defined on an infinite population, it would be possible to define

$$xy = E(XY) \quad \text{and} \quad |\mathbf{x}| = \sqrt{E(X^2)}$$

If X and Y are both discrete variables, then

$$E(XY) = E(XY) = \sum x_i y_i p(x_i, y_i) \quad \text{over all } p(x_i \cdot y_i) > 0$$

where $p(x_i,y_i) > 0$ is the probability in the population of having simultaneously the values x_i and y_i on the variables X and Y, respectively. If X and Y are continuous random variables, then,

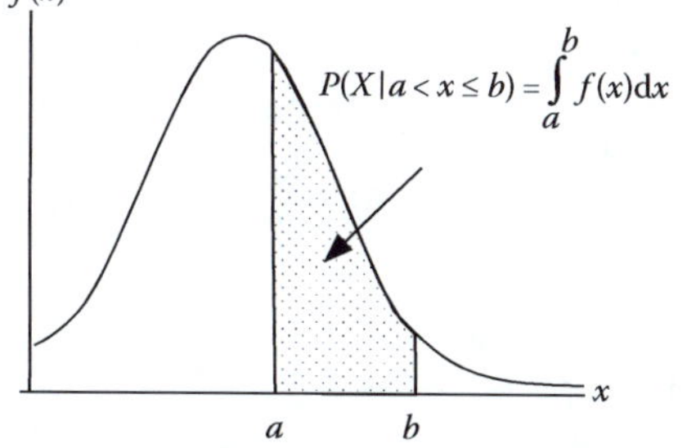

FIGURE 2.7
The probability that the random variable X takes on values greater than a or less than or equal to b is given by the area under the density function, $f(x)$, between a and b on the x-axis.

$$E(XY) = \int_{-\infty}^{+\infty} \int_{-\infty}^{+\infty} xy\, f(x,y)\,dx\,dy$$

where $f(x,y)$ is the joint density function of X and Y.

But $E(XY)$ and $\sqrt{E(XX)} = \sqrt{E(X^2)}$ are not pivotal concepts in multivariate statistics. They do become pivotal if we consider

$$\mathbf{xy} = \text{cov}(X,Y) = E(XY) - E(X)E(Y) = E[(X - E(X))(Y - E(Y))]$$

where cov(X, Y) denotes the covariance between variables X and Y. Here X and Y have their means subtracted from them before obtaining the products. The resulting transformed variables $X^* = X - E(X)$ and $Y^* = Y - E(Y)$ now have zero means, because $E(X^*) = E[X - E(X)] = E(X) - E(X) = 0$ and the same for $E(Y^*)$. Then $\text{cov}(X^*,Y^*) = E(X^*Y^*) - E(X^*)E(Y^*) = E(X^*Y^*)$. This suggests that without loss of generality in those cases involving linear algebra, we may assume that all variables under consideration have zero means. In much of this book, we will make this assumption (there will be exceptions and these will be noted). Consequently, we will assume that for variables having zero means

$$\mathbf{xy} = E(XY) = \text{cov}(X,Y)$$

and

$$|\mathbf{x}| = \sqrt{\mathbf{xx}} = \sqrt{E(XX)} = \sqrt{E(X^2)} = \sqrt{\text{var}(X)} \tag{2.21}$$

where

$|\mathbf{x}|$ denotes the length of the vector $\mathbf{x}$ corresponding to the variable X
var(X) denotes the variance of the variable X

So, the length of a vector corresponding to a random variable is the standard deviation of the variable.

The point of all this discussion of how to define a scalar product between vectors representing random variables is to establish that most multivariate statistics implicitly assumes that the linear algebra of unitary vector spaces applies to its topics. While the notation may not explicitly reveal the vectors involved, they are there in the random variables under operations of addition and scalar multiplication and the expected value of the product of two random variables.

2.6.3 Random Vectors of Random Variables

Mathematical statisticians do not readily acknowledge that the random variables they pack into what they call "random vectors" are themselves vectors. Nor do they explicitly indicate that much of multivariate statistics with random vectors of random variables is an application of linear algebra. But when they use their random-vector notation to obtain covariance matrices, they are using the scalar product of vectors, applied simultaneously to many vectors, each corresponding to a random variable. We will not dwell excessively on this fact beyond this point, since we will conform to convention and work with the random-vector notation. But, it pays to recognize that the underlying structure of multivariate statistics is linear algebra.

As we may have already indicated, in mathematical statistics a random variable is a random real-valued quantity whose values depend on the outcomes of an experiment governed by chance. A single random variable is represented by an italicized uppercase letter, with particular values of the variable represented by subscripted lowercase versions of the same letter. Several random variables are treated collectively as the "coordinates" of what is known as a "random vector." For example, if $X_1, X_2, \ldots, X_n$ are n random variables with zero means, then the random vector $\mathbf{X}$ is defined as the column vector with coordinates the random variables $X_1, X_2, \ldots, X_n$, that is

$$\mathbf{X} = \begin{bmatrix} X_1 \\ X_2 \\ \vdots \\ X_n \end{bmatrix}$$

Although this looks like an ordinary n-tuple, it is not because the elements of an ordinary n-tuple are constants, and here they are random variables. Assume now that the random variables have zero means. We may now wish to obtain a linear combination of these random variables $Y = a_1X_1 + a_2X_2 + \cdots + a_nX_n$ with the a's constant weights. This linear combination may be written in matrix notation as $Y = \mathbf{a}'\mathbf{X}$. The result Y, however, is not a constant but also a random variable. Now, consider the $n \times n$ matrix

$$\mathbf{\Sigma}_{XX} = E(\mathbf{XX}')$$

An arbitrary element, σ_{jk}, of the matrix, $\mathbf{\Sigma}_{XX}$, has its value $\sigma_{jk} = E(X_jX_k)$ since the expected value of any matrix (in this case the expected value of the matrix $\mathbf{XX}'$) is the matrix of expected values of the corresponding elements of the original matrix. The matrix $\mathbf{XX}'$ is itself a random matrix of random variables for its elements:

$$\mathbf{XX}' = \begin{bmatrix} X_1 \\ X_2 \\ \vdots \\ X_n \end{bmatrix} \begin{pmatrix} X_1 & X_2 & \cdots & X_n \end{pmatrix} = \begin{bmatrix} X_1X_1 & X_1X_2 & \cdots & X_1X_n \\ X_2X_1 & X_2X_2 & \cdots & X_2X_n \\ \vdots & \vdots & \ddots & \vdots \\ X_nX_1 & X_nX_2 & \cdots & X_nX_n \end{bmatrix}$$

Hence

$$E(\mathbf{XX}') = \begin{bmatrix} E(X_1X_1) & E(X_1X_2) & \cdots & E(X_1X_n) \\ E(X_2X_1) & E(X_2X_2) & \cdots & E(X_2X_n) \\ \vdots & \vdots & \ddots & \vdots \\ E(X_nX_1) & E(X_nX_2) & \cdots & E(X_nX_n) \end{bmatrix} = \begin{bmatrix} \sigma_1^2 & \sigma_{12} & \cdots & \sigma_{1n} \\ \sigma_{21} & \sigma_2^2 & \cdots & \sigma_{2n} \\ \vdots & \vdots & \ddots & \vdots \\ \sigma_{n1} & \sigma_{n2} & \cdots & \sigma_n^2 \end{bmatrix}$$

Again, if $\mathbf{X}$ is an $n \times 1$ random vector of random variables, $X_1, X_2, \ldots, X_n$, and $\mathbf{Y}$ is a $p \times 1$ vector of random variables, $Y_1, Y_2, \ldots, Y_p$, then

$$\mathbf{\Sigma}_{XY} = E(\mathbf{XY}')$$

is an $n \times p$ covariance matrix of cross covariances between the two sets of variables with a typical element, $\sigma_{gh} = E(X_g Y_h)$. In effect, covariance matrices are matrices of scalar products of vectors.

2.7 Maxima and Minima of Functions

There are numerous occasions scattered throughout multivariate statistics, including multiple regression, factor analysis, and structural equation modeling, where we must find the value or (values) of an independent variable (or variables), which will maximize (or minimize) some dependent variable. For example, in multiple regression we seek values of weights to assign to predictor variables in a regression equation that minimize the average value of the squared differences between the predictive composite scores and the actual scores on the criterion. Or in principal components analysis we seek the weights of a linear combination (under the restraint that the sum of squares of the weights add up to 1) of a set of variables that will have the maximum variance over all possible such linear combinations. Or in the Varimax method of rotation, we seek values of the elements of the orthogonal transformation matrix that will rotate the factors so that the sum of the variances of the squared loadings on the respective factors will be a maximum. In maximum-likelihood factor analysis, we seek values for the factor loadings, and unique variances that will maximize the joint likelihood function. In structural equation modeling, using maximum-likelihood estimation, we seek values of the free model parameters that will minimize the likelihood fit function.

In this section, we expect the student fresh to the topic of calculus to obtain nothing more than an intuitive understanding of what is involved in the solution of maximization and minimization problems by calculus. To obtain a more thorough background in this subject, the student should take a course in elementary calculus. Those sections in this book involving obtaining the solution to the maximization or minimization of some function will be marked with an asterisk in front of the section heading and may be passed over by those who are mathematically challenged. Much of this book does not require calculus.

2.7.1 Slope as the Indicator of a Maximum or Minimum

Consider the graphical representation of the function $y=f(x)$ in Figure 2.8. The function portrayed in this figure is not designed to represent any particular function other than one having a minimum and maximum value over the range of the values of x depicted.

Now, consider the effects of drawing straight lines tangent to the surface of the curve at those points on the curve corresponding to the values of x_1, x_2, x_3, and x_4. We see, going from left to right along the x-axis, that the slope of the line tangent at the point corresponding to x_1 is downward, indicating the function is descending at that point. At x_2 the tangent line is horizontal, and the function is at a local minimum. At x_3 the tangent line is sloping upward, indicating a rise in the function at that point. And at x_4 the tangent line is again horizontal, but this time at a maximum for the function. We see at this point that lines tangent to minima or maxima are horizontal. On the basis of this simple observation, calculus builds the methodology for finding points that correspond to the maxima or minima or functions.

2.7.2 Index for Slope

Let us now consider quantifying what we mean by the slope of a straight line drawn tangent to a point on the curve of some function. Consider the graph of the function $y=f(x)$ in Figure 2.9. Let the point R be a point on the curve with coordinates $(x,f(x))$. Draw a line tangent to the curve at point R. Next, consider a line drawn from the point R to another point S on the curve a short distance away from the point R in the direction of increasing x.

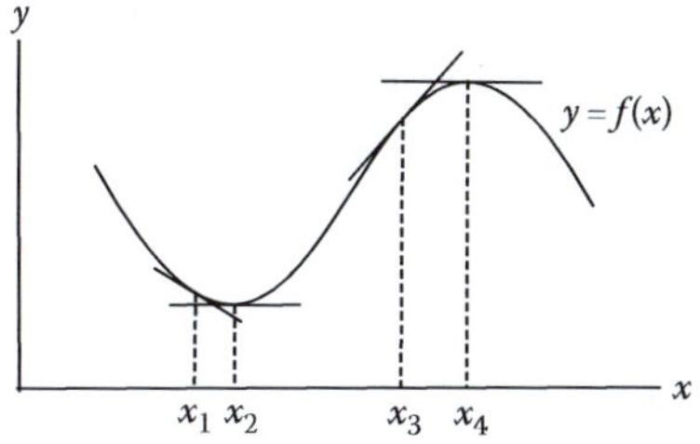

FIGURE 2.8
Graph of function $y=f(x)$ showing slopes of curve of function corresponding to different values of x.

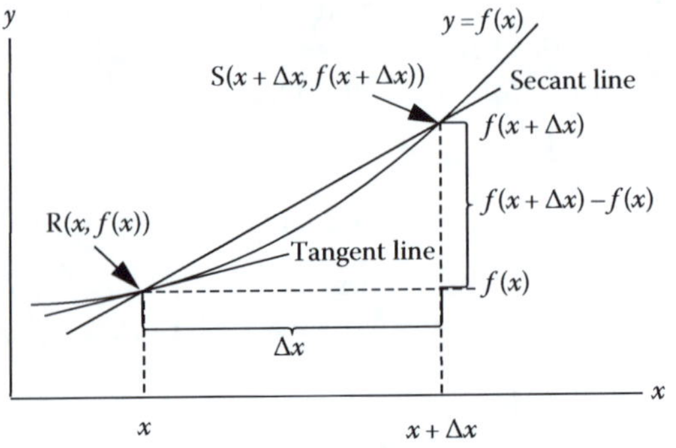

FIGURE 2.9
Graph of function $y=f(x)$ that shows how a line tangent to a curve at point $R(x,f(x))$ may be approximated by a secant line drawn between points $R(x, f(x))$ and $S(x+\Delta x, f(x+\Delta x))$. The approximation improves as Δx approaches 0.

The coordinates of the point S are $(x+\Delta x, f(x+\Delta x))$, where Δx is some small increment added to x. Intuitively, we can see that if we make Δx smaller and smaller, the point S will come closer and closer to the point R. Also the slope of the line passing through R and S will come closer and closer to the slope of the line tangent to the curve at R. Using this fact, let us first define the slope of the line passing through R and S as

$$\text{Slope(RS)} = \frac{f(x+\Delta x) - f(x)}{\Delta x}$$

We see that this is simply the ratio of the side opposite to the side adjacent in a right triangle. Next we will define the slope of the line tangent to the curve at point R, written as dy/dx, to be

$$\frac{dy}{dx} = \lim_{\Delta x \to 0} \frac{f(x+\Delta x) - f(x)}{\Delta x} \tag{2.22}$$

Although to those not familiar with taking the limit of an expression, Equation 2.22 may appear to involve dividing by 0, such is not the case. Generally, we can simplify the right-hand side expression so that the denominator, Δx, cancels out with quantities in the numerator before we take the limit when Δx approaches 0.

2.7.3 Derivative of a Function

An important point to realize about Equation 2.22 is that it also represents a function of the variable x. In this case, Equation 2.22 is a function relating the values of the slope of a line drawn tangent to the curve of the function $y=f(x)$ to the values of x corresponding to the points to which the line is

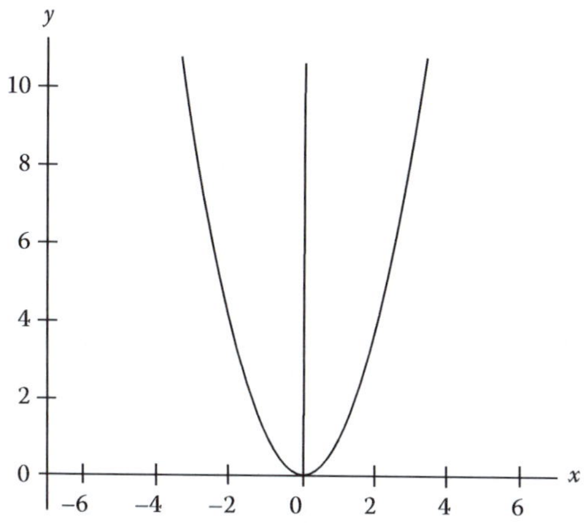

FIGURE 2.10
Graph of function $y = x^2$ which is minimized when $x = 0$.

drawn tangent. The slope function corresponding to the function $y = f(x)$ is known as the "derivative" of the function $y = f(x)$ with respect to x.

For example, consider the function $y = x^2$ shown in Figure 2.10. Let us look for the derivative of this function, using the definition in Equation 2.22:

$$\frac{dy}{dx} = \lim_{\Delta x \to 0} \frac{(x + \Delta x)^2 - x^2}{\Delta x}$$

$$= \lim_{\Delta x \to 0} \frac{x^2 + 2x\Delta x + \Delta x^2 - x^2}{\Delta x} = \frac{2x\Delta x + \Delta x^2}{\Delta x}$$

$$= \lim_{\Delta x \to 0} 2x + \Delta x = 2x$$

Using the derivative, we can find the value of the slope of a line drawn tangent to any point on the curve $y = x^2$ corresponding to a given value of x. Consider, for example, the slope of a line at the point on the curve where x equals 1. Substituting the value of 1 for x in the derivative $dy/dx = 2x$, we find that the slope is equal to 2 at this point. Or when x is 3 we find that the line drawn tangent to a corresponding point on the curve has a slope of 6.

But in the equation $y = x^2$ at what value of x is y a minimum? This can be found by setting the derivative equal to 0 and solving for x.

$$2x = 0 \quad \text{then } x = 0$$

When x is 0, the slope is 0, indicating that the tangent line drawn to the curve at the point where x equals 0 is horizontal. If we examine our graph for $y = x^2$, we will see that this corresponds to the minimum point on the curve.

2.7.4 Derivative of a Constant

Consider the function $y = k$, where k is some constant. Then using Equation 2.22, we have $f(x + \Delta x) = k$ and $f(x) = k$, so that,

$$\frac{dy}{dx} = \lim_{\Delta x \to 0} \frac{k - k}{\Delta x} = \lim_{\Delta x \to 0} 0 = 0$$

Again we do not have to divide by 0, because $(k - k) = 0$ and $0/\Delta x = 0$. Geometrically, the curve of $y = k$ is a horizontal straight line at altitude k. At any point on the line, the slope is 0.

2.7.5 Derivative of Other Functions

The derivatives for different functions are normally different themselves. As a consequence, one must determine for each function what its derivative is. This might seem to involve a bit of labor, especially if we had to apply the definition in Equation 2.22 to complicated functions. But fortunately this is not necessary, because most functions can be analyzed into parts and the derivatives of these parts taken separately and then added together. Moreover, it is possible to memorize tables of derivatives of common elementary functions which have already been derived using Equation 2.22. Some common derivatives are given in Table 2.1.

As an example of the use of Table 2.1, that consider the problem of finding values of x that will either maximize or minimize the function

$$y = \frac{x^3}{3} - \frac{x^2}{2} - 2x + 1$$

By rule 8 in Table 2.1, we see that the derivative of a function which is the sum of some other functions is equal to the sum of the derivatives of the other functions.

Hence

$$\frac{dy}{dx} = x^2 - x - 2$$

To find the points of x which correspond to maximum or minimum values of y, we need to set the derivative of y equal to 0, and then solve for x, that is, solve

$$x^2 - x - 2 = 0$$

The solution to this is obtained by factoring so that we have

$$(x + 1)(x - 2) = 0$$

TABLE 2.1

Derivatives of Common Functions

	Function	Derivative
1	$y = k$	$\frac{dy}{dx} = 0$
2	$y = x$	$\frac{dy}{dx} = 1$
3	$y = kx$	$\frac{dy}{dx} = k$
4	$y = ax^n$	$\frac{dy}{dx} = anx^{n-1}$
5	$y = e^x$	$\frac{dy}{dx} = e^x$
6	$y = a^x$	$\frac{dy}{dx} = a^x \ln a$
7	$y = \ln x$	$\frac{dy}{dx} = \frac{1}{x}$
If $u = u(x)$ and $v = v(x)$ are functions of x, then for		
8	$y = u + v$	$\frac{dy}{dx} = \frac{du}{dx} + \frac{dv}{dx}$
9	$y = uv$	$\frac{dy}{dx} = u\frac{dv}{dx} + v\frac{du}{dx}$
10	$y = \frac{u}{v}$	$\frac{dy}{dx} = \frac{v(du/dx) - u(dv/dx)}{v^2}$
11	$y = \sin x$	$\frac{dy}{dx} = \cos x$
12	$y = \cos x$	$\frac{dy}{dx} = -\sin x$
13	$y = \tan x$	$\frac{dy}{dx} = \sec^2 x$

We see that solutions that will make the left-hand side of the equation equal to 0 are $x = -1$ and $x = 2$. In other words, at $x = -1$ and $x = +2$, y is either a maximum or a minimum.

When a function is known to have only a maximum or a minimum, there is no question of whether one has found either a maximum or a minimum when he or she finds an x that makes the derivative of the function equal to 0. However, in those cases where both maximum and minimum solutions exist, ambiguity exists.

Several methods may be used to resolve this ambiguity: (1) By direct computation, one can compute values of the function $y = f(x)$ for values in the vicinity of x_m, where x_m is a value of x at which the derivative of the function equals 0. If x_m leads to a value for y that is less than values of y in the vicinity of x_m, then x_m is where the minimum is located. If the value of y at x_m is greater than values of y in the vicinity of x_m, then x_m is likely a maximum. (2) A more certain method to use is based upon finding the "second derivative" of $y=f(x)$. The second derivative is simply the derivative of the function that is the derivative of $y=f(x)$. One then finds the value of the second derivative at when $x=x_m$. If the second derivative is positive, then x_m corresponds to a minimum y. If the second derivative is negative, then x_m corresponds to a maximum y.

To illustrate the use of the second derivative, consider the problem thus cited in which

$$\frac{dy}{dx} = x^2 - x - 2$$

The second derivative is

$$\frac{d^2y}{dx^2} = 2x - 1$$

where d^2y/dx^2 is the symbol for the second derivative. Substituting –1 for x into the equation for the second derivative, we have

$$2(-1) - 1 = -3$$

Since the value of the second derivative when $x=-1$ is negative, y must be a maximum when $x=-1$. Substitute $x=2$ into the equation for the second derivative:

$$2(2) - 1 = 3$$

Since the value of the second derivative is positive when $x=2$, y must be a minimum when $x=2$.

A word of caution is necessary regarding using the zero slope as a sufficient indication of a maximum or minimum for a function. Some functions will have zero slope where there is neither a maximum or a minimum; for example, the function $y=x^3$ has a zero derivative when x equals 0, but the function is at neither a maximum nor a minimum at that point. So, zero derivative of the function at some point is only a necessary but not a sufficient condition that there exists a maximum or a minimum at that point. A sufficient condition for a maximum or a minimum is that the second derivative does not equal 0 at the point where the function has zero slope.

2.7.6 Partial Differentiation

In multivariate techniques, such as factor analysis, we are concerned with functions not of a single independent variable but of many independent variables. Thus we may be interested in knowing how to treat a change in the function in connection with changes in the independent variables. This may be done by allowing only one variable at a time to change while holding the other variables constant and by observing the degree of change in the value of the function. In this connection, it is possible to obtain the derivative of the function in the direction of one of the independent variables, treating all other independent variables as constants. Such a directional derivative is known as a "partial derivative." For example, suppose we wish to know the rate of change of the function $y = 3x^2 + 12xz + 4z^2$ when we change x while holding z constant. We can obtain the partial derivative of y with respect to x as

$$\frac{\partial y}{\partial x} = 6x + 12z$$

Here we obtained the derivative in the usual way, but we treated the other independent variable, z, as a constant for the differentiation. On the other hand, the partial derivative of y with respect to z is

$$\frac{\partial y}{\partial z} = 12x + 8z$$

Again, we now treated x as if it were a constant while taking the derivatives of the terms with respect to z. So, there is very little new to learn to obtain partial derivatives.

2.7.7 Maxima and Minima of Functions of Several Variables

In Figure 2.11, we illustrate a function $z = f(x, y)$ of two independent variables x and y having a relative maximum. Consider that at a maximum or

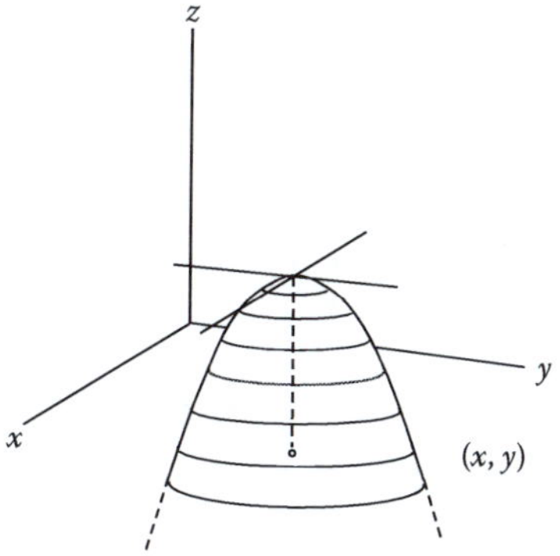

FIGURE 2.11
Graph of a function of two variables illustrating point (x, y) at which function $z = f(x, y)$ is a maximum. Note that tangents to surface of function in directions of x- and y-axes at point (x, y) are horizontal, indicating zero slopes in those directions.

a minimum for the function, there must be some point (x, y) on the plane surface defined by the x- and y-axes where the slopes of the function in the directions of the x- and y-axes are, respectively, 0. In other words, a necessary condition for a maximum or a minimum is that at some point (x, y).

$$\frac{\partial z}{\partial x} = 0$$

$$\frac{\partial z}{\partial y} = 0$$

To establish the sufficient conditions for a maximum or a minimum, define the "Jacobian matrix" of second derivatives of z evaluated at (x,y) as

$$\mathbf{J} = \begin{bmatrix} \frac{\partial^2 z}{\partial x^2} & \frac{\partial^2 z}{\partial x \partial y} \\ \frac{\partial^2 z}{\partial y \partial x} & \frac{\partial^2 z}{\partial y^2} \end{bmatrix}$$

A sufficient condition that there exists a maximum or a minimum at this point (x,y), where $\partial z/\partial x = 0$ and $\partial z/\partial y = 0$ is that the determinant $|\mathbf{J}| \neq 0$. If $|\mathbf{J}| < 0$, then z is a relative maximum at (x,y); on the other hand, if $|\mathbf{J}| > 0$, then z is a relative minimum at (x,y).

For example, investigate the relative maxima or minima of the function $f(x, y) = 2x^2 + y^2 + 4x + 6y + 2xy$. Taking the partial derivatives of $f(x,y)$ with respect to x and y and setting them equal to 0, we obtain

$$\frac{\partial f}{\partial x} = 4x + 4 + 2y = 0$$

$$\frac{\partial f}{\partial y} = 2y + 6 + 2x = 0$$

or

$$4x + 2y = -4$$

$$2x + 2y = -6$$

Solving this system of simultaneous liner equations for x and y, we obtain $x = 1$ and $y = -4$ as the point where $f(x,y)$ is at a possible extremum. The second derivatives of $f(x,y)$ are

$$\frac{\partial^2 f}{\partial x^2} = 4 \qquad \frac{\partial^2 f}{\partial x \partial y} = 2$$

$$\frac{\partial^2 f}{\partial y \partial x} = 2 \qquad \frac{\partial^2 f}{\partial y^2} = 2$$

Hence the Jacobian matrix is

$$\mathbf{J} = \begin{bmatrix} 4 & 2 \\ 2 & 2 \end{bmatrix}$$

whose determinant is $|\mathbf{J}| = 4$. Hence, with the determinant of the Jacobian matrix positive, an extremum must exist and it is a minimum.

2.7.8 Constrained Maxima and Minima

Up to now we have considered maximizing (or minimizing) a function of the type $z = f(x, y)$, where x and y are independent of one another. We shall now consider maximizing (minimizing) a function of the $z = f(x, y)$ where x and y must maintain certain dependent relationships between them as defined by one or more equations of the type $g(x, y) = 0$. The most general and powerful method for finding solutions for x and y that maximize $f(x, y)$ under constraints is the method of Lagrangian multipliers. Assuming there is one equation of constraint $g(x, y) = 0$, we first form the function

$$F(x, y) = f(x, y) + \theta g(x, y)$$

where θ is an unknown multiplier (to be determined) multiplied times the equation of constraint. Then we form the equations

$$\frac{\partial F}{\partial x} = \frac{\partial f}{\partial x} + \theta \frac{\partial g}{\partial x} = 0$$

$$\frac{\partial F}{\partial y} = \frac{\partial f}{\partial y} + \theta \frac{\partial g}{\partial y} = 0$$

$$\frac{\partial F}{\partial \theta} = g(x, y) = 0$$

which we then solve for x, y, and θ.

For example, maximize the function $f(x, y) = 10 - 2x^2 - y^2$ subject to the constraint $x + y - 1 = 0$. First, form the equation

$$F = 10 - 2x^2 - y^2 + \theta(x + y - 1)$$

Then

$$\frac{\partial F}{\partial x} = -4x + \theta = 0$$

$$\frac{\partial F}{\partial y} = -2y + \theta = 0$$

$$\frac{\partial F}{\partial g} = x + y - 1 = 0$$

Solving this system of equations for x, y, and θ, we obtain $x = 1/3$, $y = 2/3$, and $\theta = 4/3$. In other words, $f(x, y)$ is a maximum at $x = 1/3$ and $y = 2/3$ when x and y are constrained so that $x + y - 1 = 0$.

The method of Lagrangian multipliers just described may be generalized to functions of more than two independent variables and problems with more than one equation of constraint. In general, given a function $f(x_1, x_2, \ldots, x_n)$ to maximize subject to the constraints defined by the equations $g_1(x_1, x_2, \ldots, x_n) = 0, \ldots, g_m(x_1, x_2, \ldots, x_n) = 0$, $m \leq n$, form the function

$$F(x_1, x_2, \ldots, x_n) + \theta_1 g_1(x_1, x_2, \ldots, x_n) + \cdots + \theta_m g_m(x_1, x_2, \ldots, x_n)$$

Then, obtain the equations

$$\frac{\partial F}{\partial x_1} = \frac{\partial F}{\partial x_2} = \cdots = \frac{\partial F}{\partial x_n} = 0$$

which may be combined with the m equations of constraint to form a system of $n + m$ equations in $n + m$ unknowns which may then be solved. In factor analysis and structural equation modeling, we will find that the solutions for such systems of equations require iterative, numerical methods instead of direct algebraic manipulation.

3

Composite Variables and Linear Transformations

3.1 Introduction

We begin this section with definitions of univariate means, variances, then bivariate covariances, and correlations, all in scalar algebra. In Section 3.2, we consider their multivariate counterparts using matrix algebra. In Section 3.3, we discuss the properties of linear composite variables, because these play a basic role in all multivariate statistics as well as factor analysis and structural equation modeling.

3.1.1 Means and Variances of Variables

As indicated in Chapter 2, random variables are denoted by uppercase roman letters, while their corresponding lowercase forms denote specific realizations or observations of these random variables. For example, X and Y are random variables, while x and y are specific values of these random variables as might be observed in some realization of them, as, for example, in a subject's test scores on two test variables denoted by X and Y.

By $E(X)$, we shall mean the expected value or population mean of the random variable X. By $\bar{X} = (1/N)\sum_{i=1}^{N} x_i$ we shall denote the sample mean of the random variable in a sample of N observations from the population of X.

By var(X), we shall mean the population variance of the random variable as given by

$$\text{var}(X) \equiv E(X^2) - [E(X)]^2 = E[(X - E(X))^2]$$

We will also use the notation, when referring to the variances of a variance–covariance matrix, $\sigma^2(X) = \text{var}(X)$.

The variance, roughly speaking, indicates the spread or dispersion of the variable's scores around the mean in the distribution. It is specifically the expected squared deviation from the mean.

The unbiased sample estimate of the population variance in a sample of size N is given by the formula

$$s^2(x) = \frac{1}{(N-1)}\left[\sum_{i=1}^{N} x_i^2 - \frac{\left(\sum_{i=1}^{N} x_i\right)^2}{N}\right]$$

The population covariance between two variables, *X* and *Y*, is defined by

$$\text{cov}(X,Y) \equiv E(XY) - E(X)E(Y) = E[(X - E(X))(Y - E(Y))]$$

Often, we will also use the notation $\sigma(X_i, X_j)$ to denote a covariance between two variables in a variance–covariance matrix.

The covariance is an index of association between two variables. It represents the expected product of the deviations of each variable from its mean. If the covariance is zero, the two variables are not linearly associated. (It is possible to have zero covariances with variables that are nonlinearly related. On the other hand, if two variables are statistically independent, their covariance is zero.) If the covariance is nonzero, this does not tell us how strong the association is between the variables. That will depend on knowing the variances of the variables as well, from which we can then find the correlation coefficient between them, which is a normalized index of association. The magnitude of the covariance depends on the units of measurement for the variables in question.

The variance of a variable is equivalent to the covariance of the variable with itself:

$$\text{var}(X) = \text{cov}(X,X) = E(XX) - E(X)E(X) = E(X^2) - (E(X))^2$$

The sample covariance between scores *x* and *y* in a sample of size *N* is given by

$$s(x,y) = \frac{1}{(N-1)}\left[\sum_{i=1}^{N} x_i y_i - \frac{\left(\sum_{i=1}^{N} x_i\right)\left(\sum_{i=1}^{N} y_i\right)}{N}\right]$$

The population correlation between two variables is given by

$$\rho(X,Y) \equiv \frac{\text{cov}(X,Y)}{\sqrt{\text{var}(X)}\sqrt{\text{var}(Y)}}$$

The correlation coefficient (like the cosine function) ranges between −1 and +1, with a 0 correlation indicating no linear association between the variables. A correlation of −1 indicates a perfect inverse linear relationship between two variables. High scores on one variable are matched by low scores on the other, and vice-versa. A correlation of +1 is a perfect direct linear relationship between two variables. High scores are matched with high scores and low scores with low scores. It is more common, however, to find empirical correlations between variables that are between the extremes of –1 and +1. These indicate weaker forms of linear statistical association between the pair of variables.

3.1.1.1 Correlation and Causation

The existence of a nonzero covariance or correlation between two variables in and of itself indicates nothing about the nature of causation between the two variables. If we indicate by an arrow from one variable to another that the first variable is the cause of the other, then the causal situation between two variables could be that $X \rightarrow Y$ (X is a cause of Y), or $X \leftarrow Y$ (X is caused by Y), or there exists some other variable Z such that $X \leftarrow Z \rightarrow Y$ (Z is a common cause of both X and Y, while neither is a direct cause of the other), or simply a coincidence.

On the other hand, if $X \rightarrow Y$ then X and Y will be correlated. So, the direction of inference is from causation to correlation (covariance), but not the other way around. In fact, structural equation modeling is based on the idea that a complex model of causal relations between multiple variables can determine a specific kind of pattern of correlations (covariances) among the variables. If the observed correlations (covariances) have the predicted pattern based on the model, then this gives support to the model.

A simple case would be a model with three variables in it $W \rightarrow X \rightarrow Y$. Suppose W is a manipulated variable that causes variation in X when W varies. If there is a causal connection between W and Y, mediated by X, there will be a correlation between W and Y, as well as between X and Y. If, on the other hand, the real situation is $W \rightarrow X \leftarrow Y$, W cannot be a cause of Y and will be uncorrelated with Y, even though X and Y are correlated. And it does not matter if there is a third variable, Z, not correlated with W, but a common cause of X and Y. W will still be correlated with Y only if X mediates the effect of W on Y, and is therefore a direct cause of Y. And that lends credence to the assertion that X is a cause of Y. W can have no indirect effect on Y via Z, otherwise this would mean W and Z are correlated, which they are not. The assumption that W is uncorrelated with any Z that is a common cause of X and Y is the crucial assumption here, the plausibility for which is strong if W is manipulated, say, at random, breaking any dependency between W and other causes of Y. Of course, it will also depend on there not being an unmeasured mediator cause between W and Y, and careful experimental technique can frequently ensure this. Well, this gives just a flavor of what interpretations we can give to correlations. Correlation does not imply causation, but causation implies correlation.

The sample correlation coefficient. The sample correlation coefficient is given by Pearson's product moment correlation coefficient:

$$r_{xy} = \frac{s(x,y)}{\sqrt{s^2(x)}\sqrt{s^2(y)}} = \frac{\sum_{i=1}^{N} x_i y_i - \left(\sum_{i=1}^{N} x_i\right)\left(\sum_{i=1}^{N} y_i\right)/N}{\sqrt{\sum_{i=1}^{N} x_i^2 - \left(\sum_{i=1}^{N} x_i\right)^2/N}\sqrt{\sum_{i=1}^{N} y_i^2 - \left(\sum_{i=1}^{N} y_i\right)^2/N}}$$

3.2 Composite Variables

You are probably already familiar with the idea of a composite variable from your experience with test scores. Most tests consist of more than one question, and the scores obtained on these tests are usually the sums of the scores obtained on each question. In this case, the different questions can be thought of as separate variables, and the total score on the test can be thought of as a composite variable dependent upon these separate variables. We say that a total test score is a composite variable because it is composed of several component variables, the individual-question scores. It is dependent upon the test question scores because it cannot exist without them but must be derived from them.

Actually, we must be more restrictive about what we mean by a composite variable than simply saying that it is a variable composed of several component variables. Consider the random variable $U = XYZ$. This variable is dependent upon three other random variables, X, Y, and Z, which could be said to be the components of U. This, however, is an example of a "nonlinear" composite variable, whereas this book is about "linear" composite variables of the type

$$Y = a_1 X_1 + \cdots + a_n X_n \tag{3.1}$$

where

Y is a linear composite variable

X_i's are component random variables

a_i's are numerical constants that serve as "weights" indicating by how much the scores on the components are multiplied, respectively, before entering the composite

In subsequent chapters, we will find that the fundamental equation of factor analysis is of the type of Equation 3.1. What this means is that factor analysis considers observed variables to be linear composites of some other variables. However, factor analysis works backward from the way we normally work with composite variables where we first begin with some component variables and then add them together (often with weights) to form the composite. Factor analysis begins with observed variables that are presumed to be linear composites of common factors or causes and works backward to find some likely component variables from which the observed variables were derived. In structural equation modeling with latent variables, the observed variables are assumed to be linear composites of the latent variables (which are specified, although no direct measures are obtained on them). According to the causal model, covariances or correlations among the observed variables are determined by the nature of the composite variables' relations to the latent variables.

So, before we get into these models proper, we need first to cover the basic properties of composite variables. Knowledge of these properties has wide application beyond factor analysis and structural equation modeling, so students need to learn these properties well.

3.3 Unweighted Composite Variables

The simplest composite variables to consider are those that do not weight the component variables differently. In effect each variable has the same weight, a, and without loss of generality, we may assume that the weight is 1. Thus suppose that we have the simple, unweighted composite:

$$X = X_1 + X_2 + X_3 + X_4 \tag{3.2}$$

3.3.1 Mean of an Unweighted Composite

The mean of the composite variable X is given by

$$E(X) = E(X_1 + X_2 + X_3 + X_4) = E(X_1) + E(X_2) + E(X_3) + E(X_4) \tag{3.3}$$

which is simple enough. In general, if $X = X_1 + \cdots + X_n$, then

$$E(X) = E(X_1) + \cdots + E(X_n)$$

The mean of a composite variable is the sum of the means of the components.

3.3.2 Variance of an Unweighted Composite

The variance of the composite is a bit more complicated. Drawing upon our definition for variance and applying it to the variable X, we get

$$\text{var}(X) = E(X^2) - [E(X)]^2$$

which becomes, after substituting Equation 3.2 for X and Equation 3.3 for $E(X)$,

$$\text{var}(X) = E[(X_1 + X_2 + X_3 + X_4)^2] - [E(X_1) + E(X_2) + E(X_3) + E(X_4)]^2 \tag{3.4}$$

In high school or college algebra, you may have learned to square a composite expression like this by setting up your expression like this:

$$\begin{array}{ll} (X_1+X_2+X_3+X_4) & E(X_1)+E(X_2)+E(X_3)+E(X_4) \\ \times(X_1+X_2+X_3+X_4) & \times E(X_1)+E(X_2)+E(X_3)+E(X_4) \end{array}$$

Then you would multiply each term in the bottom row times each term in the upper row, and then sum the resulting product terms. But that was pretty messy, if you will recall. There is an alternative way to set up these expressions, and that is with square tables that look like matrices:

$$X^2 \Rightarrow \begin{array}{c|cccc} & X_1 & X_2 & X_3 & X_4 \\ \hline X_1 & X_1^2 & X_1X_2 & X_1X_3 & X_1X_4 \\ X_2 & X_2X_1 & X_2^2 & X_2X_3 & X_2X_4 \\ X_3 & X_3X_1 & X_3X_2 & X_3^2 & X_3X_4 \\ X_4 & X_4X_1 & X_4X_2 & X_4X_3 & X_4^2 \end{array}$$

$$[E(X)]^2 \Rightarrow \begin{array}{c|cccc} & E(X_1) & E(X_2) & E(X_3) & E(X_4) \\ \hline E(X_1) & E(X_1)E(X_1) & E(X_1)E(X_2) & E(X_1)E(X_3) & E(X_1)E(X_4) \\ E(X_2) & E(X_2)E(X_1) & E(X_2)E(X_2) & E(X_2)E(X_3) & E(X_2)E(X_4) \\ E(X_3) & E(X_3)E(X_1) & E(X_3)E(X_2) & E(X_3)E(X_3) & E(X_3)E(X_4) \\ E(X_4) & E(X_4)E(X_1) & E(X_4)E(X_2) & E(X_4)E(X_3) & E(X_4)E(X_4) \end{array}$$

Focusing now on the expression, $E[(X_1+X_2+X_3+X_4)^2]$, in Equation 3.4, we see that we would have to get the expected value of each expression of products in the first table above and then sum these together. Similarly for the second expression $[E(X_1)+E(X_2)+E(X_3)+E(X_4)]^2$ in Equation 3.4, we would simply have to sum the elements in the resulting table. But instead of summing all the elements at this point, let us combine both boxes into a matrix expression to obtain

$$\begin{bmatrix} E(X_1^2) & E(X_1X_2) & E(X_1X_3) & E(X_1X_4) \\ E(X_2X_1) & E(X_2^2) & E(X_2X_3) & E(X_2X_4) \\ E(X_3X_1) & E(X_3X_2) & E(X_3^2) & E(X_3X_4) \\ E(X_4X_1) & E(X_4X_2) & E(X_4X_3) & E(X_4^2) \end{bmatrix}$$

$$-\begin{bmatrix} E(X_1)E(X_1) & E(X_1)E(X_2) & E(X_1)E(X_2) & E(X_1)E(X_4) \\ E(X_2)E(X_1) & E(X_2)E(X_2) & E(X_2)E(X_3) & E(X_2)E(X_4) \\ E(X_3)E(X_1) & E(X_3)E(X_2) & E(X_3)E(X_3) & E(X_3)E(X_4) \\ E(X_4)E(X_1) & E(X_4)E(X_2) & E(X_4)E(X_3) & E(X_4)E(X_4) \end{bmatrix}$$

We could sum the terms in each matrix above and then compute their difference. But we could also subtract the right-hand side matrix from the left-hand side matrix and get $\|E(X_i\ X_j) - E(X_i)\ E(X_j)\| = \|\text{cov}(X_i, X_j)\|$. We see that this is a variance–covariance matrix of covariances for each pair of variables, including the variables covaried with themselves.

$$\begin{bmatrix} \sigma(X_1^2) & \sigma(X_1,X_2) & \sigma(X_1,X_3) & \sigma(X_1,X_4) \\ \sigma(X_2,X_1) & \sigma(X_2^2) & \sigma(X_2,X_3) & \sigma(X_2,X_4) \\ \sigma(X_3,X_1) & \sigma(X_3,X_2) & \sigma(X_3^2) & \sigma(X_3,X_4) \\ \sigma(X_4,X_1) & \sigma(X_4,X_2) & \sigma(X_4,X_3) & \sigma(X_4^2) \end{bmatrix}$$

The variance–covariance matrix is more meaningful, as we will show later on with matrix expressions for composite variables. But for now, we see that to obtain the variance of the unweighted composite X, all we need to do is to add together all the elements of the variance–covariance matrix:

$$\text{var}(X) = \sum_{i=1}^{4}\sum_{j=1}^{4} \sigma(X_i, X_j)$$

In general, for composites with n components

$$\text{var}(X) = \sum_{i=1}^{n}\sum_{j=1}^{n} \sigma(X_i, X_j) \tag{3.5}$$

In other words, the variance of an unweighted composite is the sum of all the variances on the principal diagonal and all the covariances among its component variables off the diagonal of the variance–covariance matrix. This means that the variance of the composite does not depend simply on the variances of the individual components, but on the nature of their covariation.

Sometimes, it is convenient to write the equation in Equation 3.5 as

$$\text{var}(X) = \sum_{i=1}^{n} \sigma^2(X_i) + \sum_{i=1}^{n}\sum_{j=1}^{n} \sigma(X_i, X_j) \quad i \neq j$$

The provision that $i \neq j$ means "do not add the terms when $i = j$," which refers to the diagonal elements of the variance–covariance matrix. Since the variances are on the principal diagonal, we do not want to add them in twice, and this provision prevents that.

But there is another way to look at the variance of a composite, which yields some important insights. Consider the formula for the correlation coefficient

$$\rho(X,Y) \equiv \frac{\text{cov}(X,Y)}{\sqrt{\text{var}(X)}\sqrt{\text{var}(Y)}}$$

We see now that we can solve for cov(*X*,*Y*) in terms of the other expressions and write

$$\text{cov}(X,Y) = \rho(X,Y)\sqrt{\text{var}(X)}\sqrt{\text{var}(Y)}$$

We are going to focus on covariances, correlations, and variances, among component variables, and, to make the notation more compact, we will write σ_{gh} for cov(X_g, X_h), ρ_{gh} for $\rho(X_g, X_h)$, and σ_g^2 for var(X_g). Then,

$$\sigma^2(X) = \sum_{i=1}^{n} \sigma_i^2 + \sum_{i=1}^{n}\sum_{j=1}^{n} \sigma_{ij} \quad i \neq j$$

Now, we know that the mean of a series of scores is given by the formula

$$\bar{X} = \frac{1}{n}\sum_{i=1}^{n} x_i$$

from which we can derive an expression for the summation as

$$\sum_{i=1}^{n} x_i = n\bar{X}$$

We can similarly write

$$\sum_{i=1}^{n} \sigma_i^2 = n\bar{\sigma}_i^2 \quad \text{and} \quad \sum_{i=1}^{n}\sum_{\substack{j=1 \\ i\neq j}}^{n} \sigma_{ij} = n(n-1)\bar{\sigma}_{ij}$$

Hence

$$\sigma^2(X) = n\bar{\sigma}_i^2 + n(n-1)\bar{\sigma}_{ij} \tag{3.6}$$

Now we can consider what happens to the total variance of the composite under variation in the average covariance among the component variables. Suppose the average covariance among the components is zero, then the total variance in Equation 3.6 is due only to the sum of the variances of the components. So, when the component variables are mutually independent, they will have zero covariances among them, with the consequence that the variance of the composite is due only to the sum of the variances of the components. But if the average covariance between the items is positive, then the total variance will be greater than *n* times the sum of the average variance. On the other hand, if the average covariance is negative, the total variance will be less than *n* times the average variance.

3.3.3 Covariance and Correlation between Two Composites

Let X_1, X_2, and X_3 be random variables that are the components of a composite variable $X = X_1 + X_2 + X_3$, and let Y_1, Y_2, Y_3, and Y_4 be another set of random variables that are the components of a composite $Y = Y_1 + Y_2 + Y_3 + Y_4$. We seek to find the covariance between X and Y.

The definition of covariance is

$$\sigma(X, Y) = E(XY) - E(X)E(Y)$$

Substituting the definition for X and Y above in terms of their components, $\sigma(X, Y) = E[(X_1 + X_2 + X_3)(Y_1 + Y_2 + Y_3 + Y_4)] - E(X_1 + X_2 + X_3)E(Y_1 + Y_2 + Y_3 + Y_4)$, leads to the two tables

	Y_1	Y_2	Y_3	Y_4
X_1	X_1Y_1	X_1Y_2	X_1Y_3	X_1Y_4
X_2	X_2Y_1	X_2Y_2	X_2Y_3	X_2Y_4
X_3	X_3Y_1	X_3Y_2	X_3Y_3	X_3Y_4

	$E(Y_1)$	$E(Y_2)$	$E(Y_3)$	$E(Y_4)$
$E(X_1)$	$E(X_1)E(Y_1)$	$E(X_1)E(Y_2)$	$E(X_1)E(Y_3)$	$E(X_1)E(Y_4)$
$E(X_2)$	$E(X_2)E(Y_1)$	$E(X_2)E(Y_2)$	$E(X_2)E(Y_3)$	$E(X_2)E(Y_4)$
$E(X_3)$	$E(X_3)E(Y_1)$	$E(X_3)E(Y_2)$	$E(X_3)E(Y_3)$	$E(X_3)E(Y_4)$

Again, focusing on the left-hand side expression, we see that we would have to get the expected value of each expression of products in the box on the left-hand side above and then sum these together, and do the same for the box on the right-hand side. But let us just get the expected values of each element in the boxes without summing them together. Doing this and combining both boxes we get

$E(X_1Y_1)$	$E(X_1Y_2)$	$E(X_1Y_3)$	$E(X_1Y_4)$
$E(X_2Y_1)$	$E(X_2Y_2)$	$E(X_2Y_3)$	$E(X_2Y_4)$
$E(X_3Y_1)$	$E(X_3Y_2)$	$E(X_3Y_3)$	$E(X_3Y_4)$

−

$E(X_1)E(Y_1)$	$E(X_1)E(Y_2)$	$E(X_1)E(Y_3)$	$E(X_1)E(Y_4)$
$E(X_2)E(Y_1)$	$E(X_2)E(Y_2)$	$E(X_2)E(Y_3)$	$E(X_2)E(Y_4)$
$E(X_3)E(Y_1)$	$E(X_3)E(Y_2)$	$E(X_3)E(Y_3)$	$E(X_3)E(Y_4)$

Now, instead of first summing all elements in each box and then subtracting the sums, we instead subtract corresponding elements of the right-hand side table from the elements of the left-hand side table, then we get a table of typical elements $E(X_iY_j) - E(X_i)E(X_j)$, with each equal the covariance $\sigma(X_i, Y_j)$ between the respective pairs of component variables. In other words, we have the matrix table

$$\begin{bmatrix} \sigma(X_1,Y_1) & \sigma(X_1,Y_2) & \sigma(X_1,Y_3) & \sigma(X_1,Y_4) \\ \sigma(X_2,Y_1) & \sigma(X_2,Y_2) & \sigma(X_2,Y_3) & \sigma(X_2,Y_4) \\ \sigma(X_3,Y_1) & \sigma(X_3,Y_2) & \sigma(X_3,Y_3) & \sigma(X_3,Y_4) \end{bmatrix}$$

whose elements are the covariances between the component variables of the first set in X with the component variables in the second set of Y. The covariance between the two components, X and Y, is then simply the sum of the elements of this matrix:

$$\sigma(X,Y) = \sum_{i=1}^{3}\sum_{j=1}^{4}\sigma(X_i,Y_j)$$

In general, for composites of p variables in X and q variables in Y

$$\sigma(X,Y) = \sum_{i=1}^{p}\sum_{j=1}^{q}\sigma(X_i,Y_j) \tag{3.7}$$

The covariance between two composites is the sum of the elements of the covariance matrix between them.

3.3.4 Correlation of an Unweighted Composite with a Single Variable

Let $X = X_1 + \cdots + X_k$ be a composite of k component variables and let Y be a single variable. The correlation between Y and the composite X is then given by

$$\rho(Y,X) = \frac{\operatorname{cov}(Y,X)}{\sqrt{\sigma^2(Y)}\sqrt{\sigma^2(X)}} = \frac{\sum_{j=1}^{k}\operatorname{cov}(Y,X_j)}{\sqrt{\sigma^2(Y)}\sqrt{\sum_{i=1}^{k}\sum_{j=1}^{k}\sigma(X_i,X_j)}}$$

Let us now write each covariance in terms of the correlation between the respective variables multiplied by their standard deviations.

$$\rho(Y,X) = \frac{\sum_{j=1}^{k}\rho_{yj}\sigma_y\sigma_j}{\sigma_y\sqrt{\sum_{i=1}^{k}\sum_{j=1}^{k}\rho_{ij}\sigma_i\sigma_j}} = \frac{\sigma_y\sum_{j=1}^{k}\rho_{yj}\sigma_j}{\sigma_y\sqrt{\sum_{i=1}^{k}\sum_{j=1}^{k}\rho_{ij}\sigma_i\sigma_j}} = \frac{\sum_{j=1}^{k}\rho_{yj}\sigma_j}{\sqrt{\sum_{i=1}^{k}\sum_{j=1}^{k}\rho_{ij}\sigma_i\sigma_j}} \tag{3.8}$$

We see that in the numerator a common term, σ_y, could be factored out of the summation and then cancelled with the σ_y in the denominator.

Let us now make a further simplification that does not lose generality. We will assume that all variables have unit variances (e.g., they are standard score variables).

Then we get

$$\rho(Y,X) = \frac{\sum_{j=1}^{k} \rho_{yj}}{\sqrt{\sum_{i=1}^{k}\sum_{j=1}^{k} \rho_{ij}}} = \frac{\sum_{j=1}^{k} \rho_{yj}}{\sqrt{k + \sum_{\substack{i=1 \\ i \neq j}}^{k}\sum_{j=1}^{k} \rho_{ij}}} = \frac{k\bar{\rho}_{yj}}{\sqrt{k + k(k-1)\bar{\rho}_{ij}}} \tag{3.9}$$

Let us now divide both the numerator and denominator of Equation 3.9 by $1/k$. In the denominator, $1/k$ is taken inside the square root as $1/k^2$ and then multiplied by each term, allowing us to cancel a k in each term, leaving

$$\rho(X,Y) = \frac{\bar{\rho}_{yj}}{\sqrt{1/k + [(k-1)/k]\bar{\rho}_{ij}}} \tag{3.10}$$

Now, we are in a position to see some interesting results. First, suppose we take the limit as we increase the number of components k without bound

$$\lim_{k \to \infty} \frac{\bar{\rho}_{yj}}{\sqrt{1/k + [(k-1)/k]\bar{\rho}_{ij}}} = \frac{\bar{\rho}_{yj}}{\sqrt{\bar{\rho}_{ij}}} \tag{3.11}$$

By increasing the number of components in the composite, without bound (while maintaining constant the average correlations in the numerator and denominator), Equation 3.10 converges to a theoretical limit in Equation 3.11. This limit tells us the correlation we theoretically could expect to obtain between a composite and an external variable if we increase the number of components indefinitely while maintaining the same average correlation of the components with the external variable Y and the same average correlation among the components of the composite.

To illustrate, suppose we have a 10-item test in which the average intercorrelation among the components items is $\bar{\rho}_{ij} = .50$ and the average correlation of the components with the external criterion Y is .20. According to Equation 3.10, the correlation between the 10-item composite and Y is .27. Can we improve the validity by indefinitely increasing the number of items, while maintaining the average intercorrelations among the components and them with the criterion? From Equation 13.11, we see that in this case the correlation would only increase to a tiny bit over .28. There would be little incentive then to add more component items under these conditions.

In the above example, the average interitem correlation $\bar{\rho}_{ij}$ is greater than the average item–criterion correlation $\bar{\rho}_{yj}$. Let us now consider a different example where the average item–criterion correlation is greater than the average interitem correlation $\bar{\rho}_{ij}$. Suppose the average item–criterion correlation $\bar{\rho}_{yj}$ is .20, but the average interitem correlation is only .04. For a 10-item test of this kind the correlation between the total test score and the criterion would be .54 according to Equation 3.10. But if we make this test indefinitely long while maintaining the same average intercorrelations, the correlation of the total test score with the criterion would be perfect, i.e., equal to 1.00!

We can now summarize three principles concerning the correlation between a composite variable and another variable: (1) As long as $\bar{\rho}_{yj}$ and $\bar{\rho}_{ij}$ remain unchanged, adding additional components to the composite will increase the correlation between the composite variable and the external criterion. (2) There is an upper limit to the size of the correlation coefficient between the composite and the external criterion that can be obtained by increasing the number of components in the composite. If $\bar{\rho}_{ij} > \bar{\rho}_{yj}$, then the greater the difference between them, the less would be accomplished by adding items to the composite as long as $\bar{\rho}_{yj}$ and $\bar{\rho}_{ij}$ remain unchanged. (3) If $\bar{\rho}_{ij} < \bar{\rho}_{yj}$, a substantial improvement in the prediction can be obtained by adding additional components to the composite as long as $\bar{\rho}_{yj}$ and $\bar{\rho}_{ij}$ remain unchanged.

We are also in a position to establish bounds to $\bar{\rho}_{ij}$ given $\bar{\rho}_{yj}$, if we hope to obtain a perfect prediction of the criterion variable from the composite by increasing the number of components. Take Equation 3.10, set $\rho_{yx} = 1$, and solve for $\bar{\rho}_{ij}$. In this case, we find that $\bar{\rho}_{ij} = \bar{\rho}_{yj}^2$. Thus if the average correlation among the components within a composite is $\bar{\rho}_{yj}^2$, then increasing the number of components indefinitely would yield perfect prediction.

But more important, we must consider the following: If $\rho_{ij} < \bar{\rho}_{yj}^2$ then perfect prediction could be obtained with a finite number of components. In fact, the actual number of components that we would need could be determined by setting Equation 3.12 to 1 and solving for k, which would give us

$$k = \frac{1 - \bar{\rho}_{ij}}{\bar{\rho}_{yj}^2 - \bar{\rho}_{ij}} \tag{3.12}$$

For example, suppose $\bar{\rho}_{yj} = .30$ and $\bar{\rho}_{ij} = .06$; then the number of items k, that we would need to obtain perfect prediction according to Equation 3.12 would be approximately 31. However, if $\bar{\rho}_{ij}$ is greater than $\bar{\rho}_{yj}^2$, then k will be a negative number, which makes no sense, and thereby suggests that perfect prediction could never be obtained.

3.3.5 Correlation between Two Unweighted Composites

Let X and Y be two composite variables such that

$$X = X_1 + \cdots + X_k; \qquad E(X_1) = \cdots E(X_k) = 0 \tag{3.13}$$

$$Y = Y_1 + \cdots + Y_m; \qquad E(Y_1) = \cdots E(Y_m) = 0 \tag{3.14}$$

Assume that the two composites have no components in common and that $k \neq m$. What is the correlation between X and Y?

First, let us set up the problem as simply finding the correlation between two variables X and Y:

$$\rho(X,Y) = \frac{\sigma(X,Y)}{\sigma(X)\sigma(Y)} \tag{3.15}$$

Then, from Equation 3.9, we have

$$\rho(X,Y) = \frac{\sum_{i=1}^{k}\sum_{g=1}^{m}\sigma(X_i,Y_g)}{\sigma(X)\sigma(Y)} = \frac{\sum_{i=1}^{k}\sum_{g=1}^{m}\rho_{ig}\sigma_i\sigma_g}{\sigma(X)\sigma(Y)} \tag{3.16}$$

But, we can assume without loss of generality that all component variables have variances of unity. Then, after simplification analogous to that leading to Equation 3.10, we have

$$\rho(X,Y) = \frac{\bar{\rho}_{ig}}{\sqrt{1/k + [(k-1)/k]\bar{\rho}_{ij}}\sqrt{1/m + [(m-1)/m]\bar{\rho}_{gh}}} \tag{3.17}$$

Again as with the case of Equation 3.10, we can see that the correlation between two composites as expressed in Equation 3.17 is a function of the number of components in the composites as well as their interitem correlations. As a matter of fact the correlation between two composites is "attenuated" from what it could be if one increased the number of components in each composite, without bound, keeping the average interitem correlations constant. Therefore, let us see what the effect would be to increasing the number of items in each composite without bound while keeping the average interitem correlations the same. The value of $1/k$ and $1/m$ will approach zero, while $(k-1)/k$ and $(m-1)/m$ both approach 1.00, so that we will have

$$\lim_{k\to\infty}\lim_{m\to\infty}\rho(X,Y) = \frac{\bar{\rho}_{ig}}{\sqrt{\bar{\rho}_{ij}}\sqrt{\bar{\rho}_{gh}}} \tag{3.18}$$

Equations 3.17 and 3.18 yield the following conclusions: (1) The higher the average cross correlation between the components of different composites, the higher is the correlation between the two composites. (2) The lower the average correlation among the components of the same composite, that is, the lower $\bar{\rho}_{ij}$ and $\bar{\rho}_{gh}$, other things being equal, the higher is the correlation between the composites. (3) As the number of components in the two composites increases (provided that the average cross correlation and average interitem correlations remain constant), the higher will be the correlation between the composites. If the average cross correlation between the components of the two composites equals the average within-composite correlation among the components of both composites, then increasing the number of components in both composites without bound will increase the correlation between the two composites to perfect correlations of 1.00. (4) In general, one strives to achieve the following inequality in practice in order to maximize the correlation between two composites:

$$\bar{\rho}_{ig} \geq \bar{\rho}_{ij} \geq \bar{\rho}_{gh} \quad (3.19)$$

At first it may seem paradoxical; however, two composite variables may be highly correlated although the average cross correlation between the components of the two composites is low. This will be possible if the average within-composite correlation is also low, especially if lower than the average cross correlation.

Consider, the following example: Snow (1966) constructed several 25-item subtests for a vocabulary test, using the same words but with different item formats. For one of the subtests, he found that the average interitem correlation was .13. For another subtest he found that the average interitem correlation was .08. The average cross correlation between the items of the two subtests was .10. But the correlation between the total scores obtained on each subtest was .77!

This may seem surprising, but such a result is perfectly consistent with Equation 3.18.

The meaning of such a result is not that the two tests have a single thing in common, knowledge of vocabulary, but that they have several things in common. This illustrates a danger in using composite scores to study the complexity of a phenomenon. The composite scores will obscure any potential complexity that exists within the composites; as a result one may have the tendency in theoretical discussion to treat the composite as a homogeneous entity. For example, we might wonder whether the high correlations found between the total scores on the various tests of general intelligence represent, as is sometimes believed, a single dimension common to them all as much as they represent a hodgepodge of dimensions common to them, which as yet we know little about.

Equation 3.18 also indicates how we might assemble a composite of predictor variables to maximally predict a complex composite criterion. Above all, we should collect predictor variables that have only low correlations with one another. After this requirement is met, it may not matter very much if the correlations of the components of the predictor composite correlate also at a low level with the components of the criterion composite. The only consideration is that the cross correlations between the components of the predictor and criterion composites are not substantially lower than their within-composite correlations.

3.3.6 Summary Concerning Unweighted Composites

By inserting the appropriate expressions for composite scores into the typical formulas for the variances of and correlations between the variables, we can, with some algebraic manipulation, derive corresponding formulas for composite variables expressed in terms of their components. These formulas in general indicate that correlations between composite variables and other variables will increase as the average correlation between their components decreases, as the number of components in the composite increases, and as the correlations of the components with other variables increase.

3.4 Differentially Weighted Composites

Suppose the composite is a "weighted composite," i.e.,

$$X = w_1X_1 + w_2X_2 + w_3X_3 + w_4X_4$$

where the w_i's are multiplier weights of different values for multiplying each respective component variable before adding it to the composite.

Then

$$E(X) = w_1E(X_1) + w_2E(X_2) = w_3E(X_3) + w_4E(X_4) \tag{3.20}$$

and

$$\text{var}(X) = \sum_{i=1}^{n}\sum_{j=1}^{n} w_iw_j\sigma_{ij} = \sum_{i=1}^{n} w_i^2\sigma_i^2 + \sum_{i=1}^{n}\sum_{j=1}^{n} w_iw_j\sigma_{ij} \quad i \neq j \tag{3.21}$$

3.4.1 Correlation between a Differentially Weighted Composite and Another Variable

Suppose Y is an external variable, and $X = w_1X_1 + \cdots + w_kX_k$, a weighted composite. Then

$$\rho(Y, X) = \frac{\sum_{i=1}^{k} w_i\sigma_{yi}}{\sigma_y\sigma_x} = \frac{\sum_{i=1}^{k} w_i\,\rho_{yi}\sigma_y\sigma_i}{\sigma_y\sigma_x} = \frac{\sum_{i=1}^{k} w_i\,\rho_{yi}\sigma_i}{\sigma_x} \tag{3.22}$$

In other words, the correlation between the composite and the external variable is equal to the sum of the products of weights times item–criterion correlations times item standard deviations divided by the standard deviation of the composite.

3.4.2 Correlation between Two Differentially Weighted Composites

Let X and Y be two differentially weighted composite variables based on two distinct sets of component variables such that

$$X = w_1X_1 + \cdots + w_kX_k; \qquad E(X) = w_1E(X_1) = \cdots = w_kE(X_k) = 0$$
$$Y = v_1Y_1 + \cdots + v_mY_m; \qquad E(Y) = v_1E(Y_1) = \cdots = v_mE(Y_m) = 0 \tag{3.23}$$

Then

$$\rho(X,Y) = \frac{\sum_{i=1}^{k}\sum_{g=1}^{m} w_i v_g \sigma_{ig}}{\sigma_x \sigma_y} \tag{3.24}$$

In other words, the correlation between two composites is equal to the weighted sum of the cross covariances between items of the two sets divided by the product of the composite standard deviations.

3.5 Matrix Equations

3.5.1 Random Vectors, Mean Vectors, Variance–Covariance Matrices, and Correlation Matrices

Let

$$\mathbf{X} = \begin{bmatrix} X_1 \\ X_2 \\ \vdots \\ X_p \end{bmatrix} \tag{3.25}$$

be a $p \times 1$ column vector of random variables, known as a "random vector." We need now to know how to take the expected value of a matrix of random variables. What we do is take the expected value of each random variable in the matrix and replace that variable with its expected value. Because **X** is a random vector, that is, a $p \times 1$ matrix, its mean vector, $\boldsymbol{\mu}(\mathbf{X})$, is given by

$$\boldsymbol{\mu}(\mathbf{X}) = E(\mathbf{X}) = \begin{bmatrix} E(X_1) \\ E(X_2) \\ \vdots \\ E(X_p) \end{bmatrix} = \begin{bmatrix} \mu_1 \\ \mu_2 \\ \vdots \\ \mu_p \end{bmatrix}$$

The variance–covariance matrix, Σ_{XX}, for **X** is given by

$$\text{var}(\mathbf{X}) = \mathbf{\Sigma}_{XX} = E(\mathbf{XX}') - E(\mathbf{X})E(\mathbf{X}')$$

$$= E\left(\begin{bmatrix} X_1 \\ X_2 \\ \vdots \\ X_p \end{bmatrix} [X_1 \quad X_2 \quad \cdots \quad X_p]\right) - E\begin{bmatrix} X_1 \\ X_2 \\ \vdots \\ X_p \end{bmatrix} E[X_1 \quad X_2 \quad \cdots \quad X_p]$$

$$= E\begin{bmatrix} X_1^2 & X_1X_2 & \cdots & X_1X_p \\ X_2X_1 & X_2^2 & \cdots & X_2X_1 \\ \vdots & \vdots & \ddots & \vdots \\ X_2X_1 & X_2X_1 & \cdots & X_p^2 \end{bmatrix} - \begin{bmatrix} E(X_1) \\ E(X_2) \\ \vdots \\ E(X_p) \end{bmatrix} [E(X_1) \quad E(X_2) \quad \cdots \quad E(X_p)]$$

$$= \begin{bmatrix} E(X_1^2) & E(X_1X_2) & \cdots & E(X_1X_p) \\ E(X_2X_1) & E(X_2^2) & \cdots & E(X_2X_p) \\ \vdots & \vdots & \ddots & \vdots \\ E(X_pX_1) & E(X_pX_1) & \cdots & E(X_p^2) \end{bmatrix}$$

$$- \begin{bmatrix} [E(X_1)]^2 & E(X_1)E(X_2) & \cdots & E(X_1)E(X_p) \\ E(X_2)E(X_1) & [E(X_2)]^2 & \cdots & E(X_2)E(X_p) \\ \vdots & \vdots & \ddots & \vdots \\ E(X_p)E(X_1) & E(X_p)(X_2) & \cdots & [E(X_p)]^2 \end{bmatrix}$$

$$= \begin{bmatrix} \sigma_{11} & \sigma_{12} & \cdots & \sigma_{1p} \\ \sigma_{21} & \sigma_{22} & \cdots & \sigma_{2p} \\ \vdots & \vdots & \ddots & \vdots \\ \sigma_{p1} & \sigma_{p2} & \cdots & \sigma_{pp} \end{bmatrix} \tag{3.26}$$

Let

$$\mathbf{D}_X^2 = [\text{diag}\,\mathbf{\Sigma}_{XX}] = \begin{bmatrix} \sigma_{11} & 0 & \cdots & 0 \\ 0 & \sigma_{22} & \cdots & 0 \\ \vdots & \vdots & \ddots & \vdots \\ 0 & 0 & \cdots & \sigma_{pp} \end{bmatrix}. \tag{3.27}$$

The operation, [diag()], creates a diagonal matrix by extracting the elements in the principal diagonal of the matrix in parentheses and inserting them in the principal diagonal of a diagonal matrix. Then the matrix

$$\mathbf{D}_X = \begin{bmatrix} \sqrt{\sigma_{11}} & 0 & \cdots & 0 \\ 0 & \sqrt{\sigma_{22}} & \cdots & 0 \\ \vdots & \vdots & \ddots & \vdots \\ 0 & 0 & \cdots & \sqrt{\sigma_{pp}} \end{bmatrix} \quad \text{and} \quad \mathbf{D}_X^{-1} = \begin{bmatrix} \frac{1}{\sqrt{\sigma_{11}}} & 0 & \cdots & 0 \\ 0 & \frac{1}{\sqrt{\sigma_{22}}} & \cdots & 0 \\ \vdots & \vdots & \ddots & \vdots \\ 0 & 0 & \cdots & \frac{1}{\sqrt{\sigma_{pp}}} \end{bmatrix}$$

The population correlation matrix for the variables in **X** is given as follows:

$$\mathbf{P} = \mathbf{D}_X^{-1}\boldsymbol{\Sigma}_{XX}\mathbf{D}_X^{-1}$$

$$= \begin{bmatrix} \frac{1}{\sqrt{\sigma_{11}}} & 0 & \cdots & 0 \\ 0 & \frac{1}{\sqrt{\sigma_{22}}} & \cdots & 0 \\ \vdots & \vdots & \ddots & \vdots \\ 0 & 0 & \cdots & \frac{1}{\sqrt{\sigma_{pp}}} \end{bmatrix} \begin{bmatrix} \sigma_{11} & \sigma_{12} & \cdots & \sigma_{1p} \\ \sigma_{21} & \sigma_{22} & \cdots & \sigma_{2p} \\ \vdots & \vdots & \ddots & \vdots \\ \sigma_{p1} & \sigma_{p2} & \cdots & \sigma_{pp} \end{bmatrix} \begin{bmatrix} \frac{1}{\sqrt{\sigma_{11}}} & 0 & \cdots & 0 \\ 0 & \frac{1}{\sqrt{\sigma_{22}}} & \cdots & 0 \\ \vdots & \vdots & \ddots & \vdots \\ 0 & 0 & \cdots & \frac{1}{\sqrt{\sigma_{pp}}} \end{bmatrix}$$

$$= \left\| \frac{\sigma_{ij}}{\sqrt{\sigma_{ii}}\sqrt{\sigma_{jj}}} \right\| = \begin{bmatrix} 1 & \rho_{12} & \cdots & \rho_{1p} \\ \rho_{21} & 1 & \cdots & \rho_{2p} \\ \vdots & \vdots & \ddots & \vdots \\ \rho_{p1} & \rho_{p2} & \cdots & 1 \end{bmatrix} \tag{3.28}$$

3.5.2 Sample Equations

A realization of the p variables in **X** is given by the $p \times 1$ observation vector

$$\mathbf{x} = \begin{bmatrix} x_1 \\ x_2 \\ \vdots \\ x_p \end{bmatrix} \tag{3.29}$$

where the lowercase letter p stands for a specific realized value of the corresponding variable in **X**. This may represent, for example, the score of a subject on p test variables. When one has a sample of N observations, $\mathbf{x}_1$, $\mathbf{x}_2$, $\mathbf{x}_3,\ldots,\mathbf{x}_N$, these may be made the columns of a $p \times N$ data matrix:

$$\mathbf{X} = \begin{bmatrix} x_{11} & x_{12} & x_{13} & \cdots & x_{1N} \\ x_{21} & x_{22} & x_{23} & \cdots & x_{2N} \\ \vdots & \vdots & \vdots & \cdots & \vdots \\ x_{p1} & x_{p2} & x_{p3} & \cdots & x_{pN} \end{bmatrix} \tag{3.30}$$

Then the sample mean vector is given by

$$\bar{\mathbf{x}} = \frac{1}{N}\sum_{i=1}^{N}\mathbf{x}_i \tag{3.31}$$

We may rewrite this also in matrix terms using the $N \times 1$ sum vector to perform the summations

$$\bar{\mathbf{x}} = \begin{bmatrix} \bar{x}_1 \\ \bar{x}_2 \\ \vdots \\ \bar{x}_p \end{bmatrix} = \frac{1}{N}\begin{bmatrix} x_{11} & x_{12} & x_{13} & \cdots & x_{1N} \\ x_{21} & x_{22} & x_{23} & \cdots & x_{2N} \\ \vdots & \vdots & \vdots & \cdots & \vdots \\ x_{p1} & x_{p2} & x_{p3} & \cdots & x_{pN} \end{bmatrix}\begin{bmatrix} 1 \\ 1 \\ 1 \\ \vdots \\ 1 \end{bmatrix}$$

or simply

$$\bar{\mathbf{x}} = \frac{1}{N}\mathbf{X1} \tag{3.32}$$

The $p \times p$ sample variance–covariance matrix is given as

$$\mathbf{S} = \frac{1}{(N-1)}\left(\sum_{i=1}^{N}\mathbf{x}_i\mathbf{x}_i' - N\bar{\mathbf{x}}\bar{\mathbf{x}}'\right) \tag{3.33}$$

or, rewriting this in terms of the data matrix, **X**, and the $N \times 1$ sum vector, as

$$\mathbf{S} = \frac{1}{(N-1)}\left(\mathbf{XX}' - \frac{1}{N}\mathbf{X11}'\mathbf{X}'\right) = \frac{1}{(N-1)}\mathbf{X}\left(\mathbf{I}_N - \frac{1}{N}\mathbf{11}'\right)\mathbf{X}' \tag{3.34}$$

where $\mathbf{I}_N$ is an $N \times N$ identity matrix while $\mathbf{11}'$ is a $N \times N$ matrix all of whose elements are 1s.

The equations given here using the sum vector and data matrix are useful only in mathematical equations. Algorithms for obtaining the sample mean vector and the sample variance–covariance matrix by computers usually read in an observation vector at a time and accumulate the sum of squares and cross products of the scores in a square $p \times p$ matrix table as well as the sum of the scores in a $p \times 1$ vector table. Then, after all the summing is done, one computes the mean vector by dividing each of the elements in the vector of variable score sums by N and computes the variance–covariance matrix as in Equation 3.33, where $\sum_{i=1}^{N}\mathbf{x}_i\mathbf{x}_i'$ corresponds to the $p \times p$ matrix of the accumulated sum of squares and cross products.

The sample correlation matrix is derived from the sample variance–covariance matrix as follows

$$\mathbf{R} = \mathbf{D}_X^{-1}\mathbf{S}_{XX}\mathbf{D}_X^{-1} \tag{3.35}$$

where $\mathbf{S}_{XX}$ is the sample variance–covariance matrix for the variables in $\mathbf{X}$, and

$$\mathbf{D}_X^2 = [\text{diag } \mathbf{S}_{XX}]$$

3.5.3 Composite Variables in Matrix Equations

An unweighted composite variable is defined as the simple sum of the variables in $\mathbf{X}$ and is given as

$$X = \mathbf{1}'\mathbf{X}$$

where $\mathbf{1}$ is the $p \times 1$ sum vector. Then

$$\begin{aligned}
\sigma^2(X) &= E(X^2) - [E(X)]^2 \\
&= E(\mathbf{1}'\mathbf{X}\mathbf{X}'\mathbf{1}) - E(\mathbf{1}'\mathbf{X})E(\mathbf{X}'\mathbf{1}) \\
&= \mathbf{1}'E(\mathbf{X}\mathbf{X}')\mathbf{1} - \mathbf{1}'E(\mathbf{X})E(\mathbf{X}')\mathbf{1} \\
&= \mathbf{1}'[E(\mathbf{X}\mathbf{X}') - E(\mathbf{X})E(\mathbf{X}')]\mathbf{1} \\
&= \mathbf{1}'\boldsymbol{\Sigma}\mathbf{1}
\end{aligned} \tag{3.36}$$

Note that $\mathbf{1}'\boldsymbol{\Sigma}\mathbf{1}$ is but a way of summing all the elements of $\boldsymbol{\Sigma}$.

A weighted composite variable defined as $X = \sum_{i=1}^{n} w_i X_i = \mathbf{w}'\mathbf{X}$ has variance

$$\begin{aligned}
\sigma^2(X) &= E(X^2) - [E(X)]^2 \\
&= E(\mathbf{w}'\mathbf{X}\mathbf{X}'\mathbf{w}) - E(\mathbf{w}'\mathbf{X})E(\mathbf{X}'\mathbf{w}) \\
&= \mathbf{w}'E(\mathbf{X}\mathbf{X}')\mathbf{w} - \mathbf{w}'E(\mathbf{X})E(\mathbf{X}')\mathbf{w} \\
&= \mathbf{w}'[E(\mathbf{X}\mathbf{X}') - E(\mathbf{X})E(\mathbf{X}')]\,\mathbf{w} \\
&= \mathbf{w}'\boldsymbol{\Sigma}\mathbf{w}
\end{aligned}$$

The reader should "burn" this final result in his or her brain as an expression for the variance of a linear combination or composite of random variables. It occurs throughout multivariate statistics in various forms. Mathematically, it is said to be a quadratic form.

We will now consider the covariance between two composite variables. Let $\mathbf{X} = [X_1\ X_2 \ldots X_n]'$ and $\mathbf{Y} = [Y_1\ Y_2 \ldots Y_m]$. Define now $X = \mathbf{w}'\mathbf{X}$ and $Y = \mathbf{v}'\mathbf{Y}$. Then

$$\sigma(X,Y) = E(\mathbf{w}'\mathbf{X}\mathbf{Y}'\mathbf{v}) - E(\mathbf{w}'\mathbf{X})E(\mathbf{Y}'\mathbf{v}) = \mathbf{w}'[E(\mathbf{X}\mathbf{Y}') - E(\mathbf{X})E(\mathbf{X}')]\mathbf{v}$$
$$= \mathbf{w}'\mathbf{\Sigma}_{XY}\mathbf{v}$$

where $\mathbf{\Sigma}_{XY}$ is the cross covariance matrix between **X** and **Y**.

The correlation between the variables X and Y is given by

$$\rho(X,Y) = \frac{\sigma(X,Y)}{\sigma(X)\sigma(Y)} = \frac{\mathbf{w}'\mathbf{\Sigma}_{XY}\mathbf{v}}{\sqrt{\mathbf{w}'\mathbf{\Sigma}_{XX}\mathbf{w}}\sqrt{\mathbf{v}'\mathbf{\Sigma}_{YY}\mathbf{v}}}.$$

To have a better intuitive grasp of the concept of the correlation between a composite variable and an external variable, consider the case where an external variable represented by the vector **y** is to be correlated with a composite variable $\hat{\mathbf{y}}$, which is a linear combination of two variables represented by $\mathbf{x}_1$ and $\mathbf{x}_2$ and is the projection of **y** onto the space spanned by $\mathbf{x}_1$ and $\mathbf{x}_2$. Assuming the vectors $\mathbf{x}_1$ and $\mathbf{x}_2$ represent a linearly independent set of vectors, we may treat these vectors as a set of basis vectors in the two-dimensional space spanned by them. Since any linear combination of the basis vectors in a space produces a vector that also lies in the same space spanned by them, it must be true that $\mathbf{x}_c$ is a vector to be found in the plane spanned by the vectors $\mathbf{x}_1$ and $\mathbf{x}_2$. On the other hand, the vector **y** need not and, most likely, will not be in the plane defined by $\mathbf{x}_1$ and $\mathbf{x}_2$. The cosine of the angle between the vector **y** and the vector $\mathbf{x}_c$ corresponds to the correlation coefficient between variables Y and X_c. Figure 3.1 is an illustration of the relationships among the vectors $\mathbf{x}_1$, $\mathbf{x}_2$, $\mathbf{x}_c$, and **y**. Note that $\mathbf{x}_c$ is in the plane defined by $\mathbf{x}_1$ and $\mathbf{x}_2$.

3.5.4 Linear Transformations

We have already seen how a variable Y may be defined as a linear combination of some other variables, X_1, X_2, X_3, and X_4.

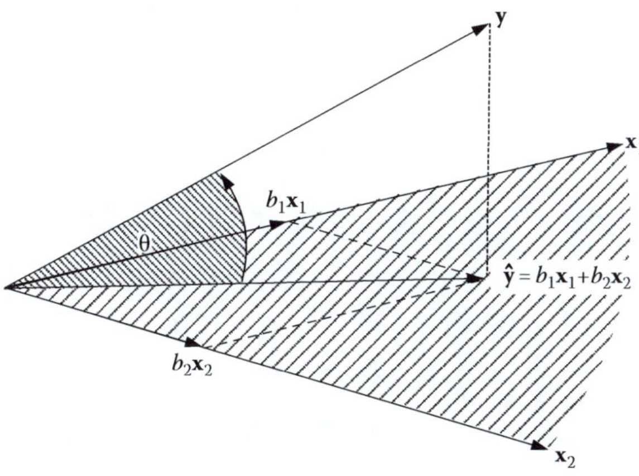

FIGURE 3.1
Vector $\hat{\mathbf{y}}$ is a linear combination of vectors $\mathbf{x}_1$ and $\mathbf{x}_2$. Cosine of angle θ between vector **y** and $\hat{\mathbf{y}}$ corresponds to correlation between them.

$$Y = [a_1 \quad a_2 \quad a_3 \quad a_4]\begin{bmatrix} X_1 \\ X_2 \\ X_3 \\ X_4 \end{bmatrix}$$

$$Y = \mathbf{a}'\mathbf{X}$$

Suppose more than one composite variable is defined in this way using different weight vectors, $\mathbf{a}_i$, applied to the same random vector $\mathbf{X}$, $Y_i = \mathbf{a}_i'\mathbf{X}$. If we wish to describe collectively these linear combinations of the X variables, we may arrange the weight vectors as rows in some matrix $\mathbf{A}$ and write a matrix equation to obtain the Y variables from the X variables. Here, we will illustrate with three Y variables:

$$\begin{bmatrix} Y_1 \\ Y_2 \\ Y_3 \end{bmatrix} = \begin{bmatrix} a_{11} & a_{12} & a_{13} & a_{14} \\ a_{21} & a_{22} & a_{23} & a_{24} \\ a_{31} & a_{32} & a_{33} & a_{34} \end{bmatrix}\begin{bmatrix} X_1 \\ X_2 \\ X_3 \\ X_4 \end{bmatrix}$$

$$\mathbf{Y} = \mathbf{AX}$$

Here, we say that the variables in $\mathbf{Y}$ are a "linear transformation" of the variables in $\mathbf{X}$. For our purposes in multivariate statistics, what will be of interest frequently is the variance–covariance matrix of the initial set of variables. For example, let $\mathbf{X}$ denote a $p \times 1$ random vector with a variance–covariance matrix

$$\boldsymbol{\Sigma} = \text{var}(\mathbf{X}) = E(\mathbf{XX}') - E(\mathbf{X})E(\mathbf{X}')$$

Define now the $q \times 1$ random vector

$$\mathbf{Y} = \mathbf{AX}$$

where $\mathbf{A}$ is a $q \times p$ "transformation" matrix. The variance–covariance matrix for the variables in $\mathbf{Y}$ is given by

$$\begin{aligned} \text{var}(\mathbf{Y}) &= E(\mathbf{YY}') - E(\mathbf{Y})E(\mathbf{Y}') \\ &= E(\mathbf{AXX}'\mathbf{A}') - E(\mathbf{AX})E(\mathbf{X}'\mathbf{A}') \\ &= \mathbf{A}E(\mathbf{XX}')\mathbf{A}' - \mathbf{A}E(\mathbf{X})E(\mathbf{X}')\mathbf{A}' = \mathbf{A}[E(\mathbf{XX}') - E(\mathbf{X})E(\mathbf{X}')]\mathbf{A}' \\ &= \mathbf{A\Sigma A}' \end{aligned}$$

Given that the means of the variables in **X** are given by $E(\mathbf{X})$, we may write

$$E(\mathbf{Y}) = E(\mathbf{AX}) = \mathbf{A}E(\mathbf{X})$$

for the means of the variables in **Y** as a linear transformation of the variables in **X**.

3.5.5 Some Special, Useful Linear Transformations

Adding subsets of a set of variables. Suppose I am given six variables and I want to select out from them the four variables, X_1, X_3, X_4, and X_6, and add them together to form a composite variable Y. I can do that easily enough with the following transformation

$$Y = \begin{bmatrix} 1 & 0 & 1 & 1 & 0 & 1 \end{bmatrix} \begin{bmatrix} X_1 \\ X_2 \\ X_3 \\ X_4 \\ X_5 \\ X_6 \end{bmatrix} \quad \text{or} \quad Y = \mathbf{w}'\mathbf{X}$$

Note how the 1s in the **w′** matrix combine with the corresponding variables in **X** while the 0s in **w′** combine with variables that are not to enter into the linear combination. The variance for Y is then given by

$$\sigma^2(Y) = \begin{bmatrix} 1 & 0 & 1 & 1 & 0 & 1 \end{bmatrix} \begin{bmatrix} \sigma_{11} & \sigma_{12} & \sigma_{13} & \sigma_{14} & \sigma_{15} & \sigma_{16} \\ \sigma_{21} & \sigma_{22} & \sigma_{23} & \sigma_{24} & \sigma_{25} & \sigma_{26} \\ \sigma_{31} & \sigma_{32} & \sigma_{33} & \sigma_{34} & \sigma_{35} & \sigma_{36} \\ \sigma_{41} & \sigma_{42} & \sigma_{43} & \sigma_{44} & \sigma_{45} & \sigma_{46} \\ \sigma_{51} & \sigma_{52} & \sigma_{53} & \sigma_{54} & \sigma_{55} & \sigma_{56} \\ \sigma_{61} & \sigma_{62} & \sigma_{63} & \sigma_{64} & \sigma_{65} & \sigma_{66} \end{bmatrix} \begin{bmatrix} 1 \\ 0 \\ 1 \\ 1 \\ 0 \\ 1 \end{bmatrix} = \mathbf{w}'\boldsymbol{\Sigma}\mathbf{w}$$

Similarly instead of unit weights, we might want a differentially weighted linear composite of these same variables:

$$Y = \begin{bmatrix} a_1 & 0 & a_3 & a_4 & 0 & a_5 \end{bmatrix} \begin{bmatrix} X_1 \\ X_2 \\ X_3 \\ X_4 \\ X_5 \\ X_6 \end{bmatrix} = \mathbf{a}'\mathbf{X}$$

Selecting and rearranging variables. Suppose we have a set of variables and wish to select a subset of them and rearrange them in a different order (e.g., we may want to put all the criteria together and all the predictors together).

$$\mathbf{Y} = \begin{bmatrix} X_3 \\ X_1 \\ X_2 \\ X_6 \end{bmatrix} = \begin{bmatrix} 0 & 0 & 1 & 0 & 0 & 0 \\ 1 & 0 & 0 & 0 & 0 & 0 \\ 0 & 1 & 0 & 0 & 0 & 0 \\ 0 & 0 & 0 & 0 & 0 & 1 \end{bmatrix} \begin{bmatrix} X_1 \\ X_2 \\ X_3 \\ X_4 \\ X_5 \\ X_6 \end{bmatrix} = \mathbf{LX}$$

What you see here is a "selection matrix." Each row of the selection matrix corresponds to a variable in the new random vector. You put a 1 in a column of the row to indicate which variable in the **X** vector to be "selected" to be the variable corresponding to that row. So, we see that the first variable in the **Y** vector is X_3 because multiplying the first row vector times the column vector **X** matches a nonzero element only with X_3.

Now the variance–covariance matrix of the new **Y** vector is simply

$$\mathbf{\Sigma}_{YY} = \mathbf{L\Sigma}_{XX}\mathbf{L}'$$

and the mean vector for the variables in **Y** is given by $\boldsymbol{\mu}_Y = \mathbf{L}\boldsymbol{\mu}_X$.

We can use the above results to select and extract in a certain order subsets of variables in a covariance matrix and then to obtain a resulting rearranged covariance matrix by pre- and postmultiplying the selection matrix times the original covariance matrix.

4

Multiple and Partial Correlations

4.1 Multiple Regression and Correlation

Throughout the remainder of this book, we will be dealing with differentially weighted composite variables having the distinguishing characteristic that they satisfy, uniquely, certain conditions to be imposed upon the weighting of the component variables. For example, in this section we will be concerned with the problem of finding that particular set of weights which when applied to a given set of component variables produces a linear composite variable such that the expected squared difference between its scores and the scores of some external variable is a minimum. To find such weights for a set of predictor variables is the problem of "multiple regression."

Later in this book, we will deal with the problem of finding a set of weights satisfying the condition that their sum of squares is equal to 1 while producing a linear composite having maximum variance from a given set of component variables. This is the problem of finding principal components. Exploratory common factor analysis will seek to find weights for a linear composite variable composed of the common parts of a set of variables that has maximum variance under the constraint that the sum of the squares of the weights adds up to 1.00. In other words, weights may be found for a set of component variables that produce composite variables satisfying certain criteria.

4.1.1 Minimizing the Expected Squared Difference between a Composite Variable and an External Variable

We will first consider a simple case with a single external variable. Let $\mathbf{X}$ be a $k \times 1$ random vector of k predictor variables and Y a criterion random variable. We will assume that $E(\mathbf{X}) = \mathbf{0}$ and $E(Y) = 0$. We seek to find the composite

$$\hat{Y} = b_1 X_1 + \cdots + b_k X_k \tag{4.1}$$

such that

$$L = E[(Y - \hat{Y})^2] = E\left[\left(Y - \sum_{j=1}^{k} b_j X_j\right)^2\right] \tag{4.2}$$

is a minimum. The problem is to find the weights b_j that make $\hat{Y}$ satisfy this criterion. This may be done by finding the respective partial derivatives of L with respect to each of the coefficients $b_1, \ldots, b_k$, setting these partial derivatives each equal to zero, and solving the resulting system of simultaneous linear equations for the coefficients $b_1, \ldots, b_k$.

Before finding the partial derivatives of L with respect to each of the coefficients $b_1, \ldots, b_k$, we first carry the expectation operator in Equation 4.2 inside the brackets after squaring the term in parentheses.

$$L = E\left[\left(Y^2 - 2\sum_{j=1}^{k} b_j X_j Y + \sum_{j=1}^{k}\sum_{h=1}^{k} b_j b_h X_j X_h\right)\right]$$

$$= E(Y^2) - 2\sum_{j=1}^{k} b_j E(X_j Y) + \sum_{j=1}^{k}\sum_{h=1}^{k} b_j b_h E(X_j X_h) \tag{4.3}$$

Taking the partial derivative of L with respect to any one of the coefficients, say, b_g, will result in an equation of the same form, so that we need not show each of the partial derivatives after one has been shown. We will now find the partial derivative of L with respect to b_g:

$$\frac{\partial L}{\partial b_g} = -2E(X_g Y) + 2\sum_{j=1}^{k} b_j E(X_j X_g) \tag{4.4}$$

If we set this partial derivative equal to zero, we may then move the first term on the right-hand side of Equation 4.4 over to the left-hand side of the equation, which yields, after dividing both sides by 2,

$$E(X_g Y) = \sum_{j=1}^{k} b_j E(X_j X_g) \tag{4.5}$$

There will be k equations of the type in Equation 4.5, and we may write them in the following way:

$$E(X_1 Y) = b_1 E(X_1^2) + b_2 E(X_1 X_2) + \cdots + b_k E(X_1 X_k)$$

$$E(X_2 Y) = b_1 E(X_2 X_1) + b_2 E(X_2^2) + \cdots + b_k E(X_2 X_k)$$

$$\cdots\cdots\cdots\cdots\cdots\cdots\cdots\cdots\cdots\cdots\cdots\cdots$$

$$E(X_k Y) = b_1 E(X_k X_1) + b_2 E(X_k X_2) + \cdots + b_k E(X_k^2) \tag{4.6}$$

Consider now that the system of equations in Equation 4.6 may be represented in matrix terms. Let the values on the left-hand side of Equation 4.6 be the elements of a $k \times 1$ vector of covariances $\mathbf{\Sigma}_{XY}$. On the other hand, let the expected values of the cross-product terms be arranged in a symmetric $k \times k$ covariance matrix $\mathbf{\Sigma}_{XX}$. Finally, treat the coefficients $b_1, \ldots, b_k$ as elements of a $k \times 1$ vector **b**. Then Equation 4.6 may be expressed by the matrix equation

$$\mathbf{\Sigma}_{XY} = \mathbf{\Sigma}_{XX}\mathbf{b} \tag{4.7}$$

Now, our task is to solve for **b**. To do this, premultiply both sides of Equation 4.7 by $\mathbf{\Sigma}_{XX}^{-1}$:

$$\mathbf{\Sigma}_{XX}^{-1}\mathbf{\Sigma}_{XY} = \mathbf{b} \tag{4.8}$$

This is a fundamental equation in regression theory. Memorize it.

*4.1.2 Deriving the Regression Weight Matrix for Multivariate Multiple Regression

Let **X** be a $k \times 1$ column random vector of k predictor random variables. Let **Y** be a $m \times 1$ random vector of m criterion variables. Without loss of generality, assume all variables have zero means. We will define the following covariance matrices

$$\mathbf{\Sigma}_{XX} = E(\mathbf{XX}') \quad \mathbf{\Sigma}_{XY} = E(\mathbf{XY}') \quad \mathbf{\Sigma}_{YY} = E(\mathbf{YY}') \tag{4.9}$$

Consider the model equation

$$\mathbf{Y}_{m\times 1} = \mathbf{B}'_{m\times k}\mathbf{X}_{k\times 1} + \mathbf{E}_{m\times 1} \tag{4.10}$$

Let $\hat{\mathbf{Y}} = \mathbf{B}'\mathbf{X}$, then

$$\mathbf{E} = (\mathbf{Y} - \hat{\mathbf{Y}}) \tag{4.11}$$

We seek to find that linear transformation of the **X** variables

$$\begin{bmatrix} \hat{Y}_1 \\ \hat{Y}_2 \\ \vdots \\ \hat{Y}_m \end{bmatrix} = \begin{bmatrix} b_{11} & b_{12} & \cdots & b_{1k} \\ b_{21} & b_{22} & \cdots & b_{2k} \\ \cdots & \cdots & \cdots & \cdots \\ b_{m1} & b_{m2} & \cdots & b_{mk} \end{bmatrix} \begin{bmatrix} X_1 \\ X_2 \\ \vdots \\ X_k \end{bmatrix}$$

$$\hat{\mathbf{Y}} = \mathbf{B}'\mathbf{X} \tag{4.12}$$

such that

$$L = \mathrm{tr}E\left[(\mathbf{Y}-\hat{\mathbf{Y}})(\mathbf{Y}-\hat{\mathbf{Y}})'\right] = \mathrm{tr}(\mathbf{EE}') \tag{4.13}$$

is a minimum.

We will need to obtain partial derivatives of L with respect to the elements of $\mathbf{B}$. Before we attempt to do so, we need to first substitute Equation 4.12 into Equation 4.13, expand, and then take the expectation of each term.

$$\begin{aligned} L &= \mathrm{tr}E\left[(\mathbf{Y}-\mathbf{B}'\mathbf{X})(\mathbf{Y}-\mathbf{B}'\mathbf{X})'\right] = \mathrm{tr}E\left[(\mathbf{Y}-\mathbf{B}'\mathbf{X})(\mathbf{Y}'-\mathbf{X}'\mathbf{B})\right] \\ &= \mathrm{tr}E[\mathbf{YY}' - \mathbf{YX}'\mathbf{B} - \mathbf{B}'\mathbf{XY} + \mathbf{B}'\mathbf{XX}'\mathbf{B}] \\ &= \mathrm{tr}\left[E(\mathbf{YY}') - E(\mathbf{YX}')\mathbf{B} - \mathbf{B}'E(\mathbf{XY}') + \mathbf{B}'E(\mathbf{XX}')\mathbf{B}\right] \\ &= \mathrm{tr}[\mathbf{\Sigma}_{YY} - \mathbf{\Sigma}_{YX}\mathbf{B} - \mathbf{B}'\mathbf{\Sigma}_{XY} + \mathbf{B}'\mathbf{\Sigma}_{XX}\mathbf{B}] \\ &= \mathrm{tr}[\mathbf{\Sigma}_{YY}] - \mathrm{tr}[\mathbf{\Sigma}_{YX}\mathbf{B}] - \mathrm{tr}[\mathbf{B}'\mathbf{\Sigma}_{XY}] + \mathrm{tr}[\mathbf{B}'\mathbf{\Sigma}_{XX}\mathbf{B}] \end{aligned}$$

But $\mathrm{tr}[\mathbf{\Sigma}_{YX}\mathbf{B}] = \mathrm{tr}[\mathbf{B}'\mathbf{\Sigma}_{XY}]$ because the first expression is the transpose of the second and the diagonal elements are the same. So, we may rewrite the last line as

$$L = \mathrm{tr}[g\mathbf{\Sigma}_{YY}] - 2\mathrm{tr}[\mathbf{\Sigma}_{YX}\mathbf{B}] + \mathrm{tr}[\mathbf{B}'\mathbf{\Sigma}_{XX}\mathbf{B}] \tag{4.14}$$

Schönemann (1965) derived the following useful expressions for the matrix derivatives of traces of matrices:

$$\frac{\partial\,\mathrm{tr}\,\mathbf{A}}{\partial \mathbf{X}} = \mathbf{0} \tag{4.15}$$

$$\frac{\partial\,\mathrm{tr}\,\mathbf{AX}}{\partial \mathbf{X}} = \mathbf{A}' \tag{4.16}$$

$$\frac{\partial\,\mathrm{tr}\,\mathbf{X}'\mathbf{AX}}{\partial \mathbf{X}} = (\mathbf{A}+\mathbf{A}')\mathbf{X} \tag{4.17}$$

$$\frac{\partial f}{\partial \mathbf{X}'} = \left(\frac{\partial f}{\partial \mathbf{X}}\right)' \quad \text{(transposition of independent variable)} \tag{4.18}$$

$$\frac{\partial\,\mathrm{tr}\,\mathbf{UVW}}{\partial \mathbf{X}} = \frac{\partial\,\mathrm{tr}\,\mathbf{WUV}}{\partial \mathbf{X}} = \frac{\partial\,\mathrm{tr}\,\mathbf{VWU}}{\partial \mathbf{X}} \quad \text{(invariance of cyclic permutations)} \tag{4.19}$$

$$\frac{\partial\,\mathrm{tr}\,\mathbf{Y}'}{\partial \mathbf{X}} = \frac{\partial\,\mathrm{tr}\,\mathbf{Y}}{\partial \mathbf{X}} \quad \text{(invariance under transposition of dependent variable)} \tag{4.20}$$

$$\frac{\partial \operatorname{tr} \mathbf{UV}}{\partial \mathbf{X}} = \frac{\partial \operatorname{tr} \mathbf{U}_c\mathbf{V}}{\partial \mathbf{X}} + \frac{\partial \operatorname{tr} \mathbf{UV}_c}{\partial \mathbf{X}} \quad \text{(product rule)} \tag{4.21}$$

where

U, **V**, **W**, and **Y** are matrix functions of **X**

$\mathbf{U}_c$ and $\mathbf{V}_c$ mean that these matrices are to be treated as constants in taking the partial derivative

Taking now the partial derivative of L with respect to **B** in Equation 4.14 leads to taking the partial derivative of each term on the right-hand side in Equation 4.14:

$$\frac{\partial L}{\partial \mathbf{B}} = \frac{\partial \operatorname{tr} \mathbf{\Sigma}_{YY}}{\partial \mathbf{B}} - \frac{2\partial \operatorname{tr}(\mathbf{\Sigma}_{YX}\mathbf{B})}{\partial \mathbf{B}} + \frac{\partial \operatorname{tr}(\mathbf{B}'\mathbf{\Sigma}_{XX}\mathbf{B})}{\partial \mathbf{B}} \tag{4.22}$$

The partial derivative of the first term on the right-hand side of the equation is **0**, from Equation 4.15. The partial derivative of the middle term on the right-hand side is $-2\mathbf{\Sigma}_{XY}$, from Equation 4.16. The partial derivative of the right-most term on the right-hand side is $2\mathbf{\Sigma}_{XX}\mathbf{B}$, from Equation 4.17 and the fact that $\mathbf{\Sigma}_{XX}$ is symmetric, implying its transpose equals the original matrix. Hence

$$\frac{\partial L}{\partial \mathbf{B}} = -2\mathbf{\Sigma}_{XY} + 2\mathbf{\Sigma}_{XX}\mathbf{B} \tag{4.23}$$

Setting this equation equal to the null matrix **0**, we have

$$-2\mathbf{\Sigma}_{XY} + 2\mathbf{\Sigma}_{XX}\mathbf{B} = \mathbf{0}$$

or

$$2\mathbf{\Sigma}_{XX}\mathbf{B} = 2\mathbf{\Sigma}_{XY}$$

Dividing both sides by 2, we then have

$$\mathbf{\Sigma}_{XX}\mathbf{B} = \mathbf{\Sigma}_{XY}$$

4.1.3 Matrix Equations for Multivariate Multiple Regression

Finally, we obtain **B** by multiplying both sides by $\mathbf{\Sigma}_{XX}^{-1}$ to obtain

$$\mathbf{B} = \mathbf{\Sigma}_{XX}^{-1}\mathbf{\Sigma}_{XY} \tag{4.24}$$

Again, memorize this equation!

Now let us substitute Equation 4.24 for **B** in Equation 4.12, to obtain $\hat{\mathbf{Y}} = \mathbf{\Sigma}_{YX}\mathbf{\Sigma}_{XX}^{-1}\mathbf{X}$, the predicted values for the Y variables based on the values of the X variables in **X**. The variance–covariance matrix, $\operatorname{var}(\hat{\mathbf{Y}}) = \mathbf{\Sigma}_{\hat{Y}\hat{Y}}$, is given by

$$\boldsymbol{\Sigma}_{\hat{Y}\hat{Y}} = E(\boldsymbol{\Sigma}_{YX}\boldsymbol{\Sigma}_{XX}^{-1}\mathbf{X}\mathbf{X}'\boldsymbol{\Sigma}_{XX}^{-1}\boldsymbol{\Sigma}_{XY}) = \boldsymbol{\Sigma}_{YX}\boldsymbol{\Sigma}_{XX}^{-1}E(\mathbf{X}\mathbf{X}')\boldsymbol{\Sigma}_{XX}^{-1}\boldsymbol{\Sigma}_{XY}$$
$$= \boldsymbol{\Sigma}_{YX}\boldsymbol{\Sigma}_{XX}^{-1}\boldsymbol{\Sigma}_{XX}\boldsymbol{\Sigma}_{XX}^{-1}\boldsymbol{\Sigma}_{XY} = \boldsymbol{\Sigma}_{YX}\boldsymbol{\Sigma}_{XX}^{-1}\boldsymbol{\Sigma}_{XY} \tag{4.25}$$

As an aside, we need to note the following identities:

$$\boldsymbol{\Sigma}_{\hat{Y}\hat{Y}} = \boldsymbol{\Sigma}_{YX}\boldsymbol{\Sigma}_{XX}^{-1}\boldsymbol{\Sigma}_{XY} = \mathbf{B}'\boldsymbol{\Sigma}_{XY} = \boldsymbol{\Sigma}_{YX}\mathbf{B} = \mathbf{B}'\boldsymbol{\Sigma}_{XX}\mathbf{B} \tag{4.26}$$

4.1.4 Squared Multiple Correlations

Let $\mathbf{D}_Y^2 = [\text{diag}\,\boldsymbol{\Sigma}_{YY}]$ be a diagonal matrix whose diagonal is the principal diagonal of $\boldsymbol{\Sigma}_{YY}$, the variance–covariance of the original criterion variables. Similarly, let $\mathbf{D}_{\hat{Y}}^2 = [\text{diag}\,\boldsymbol{\Sigma}_{\hat{Y}\hat{Y}}] = \text{diag}\left[\boldsymbol{\Sigma}_{YX}\boldsymbol{\Sigma}_{XX}^{-1}\boldsymbol{\Sigma}_{XY}\right]$. This contains, in its principal diagonal, the predicted variances of the predicted variables. Now, we define the diagonal matrix of squared multiple correlations for predicting each of the variables in **Y** from the variables in **X** as

$$\mathbf{P} = \mathbf{D}_Y^{-1}\mathbf{D}_{\hat{Y}}^2\mathbf{D}_Y^{-1} = \left\| \frac{\sigma^2(\hat{Y}_j)}{\sigma^2(Y_j)} \right\| \tag{4.27}$$

Each diagonal element of Equation 4.27 represents a ratio of the variance of the predicted variable divided by the actual variance of the same variable. Hence, a squared multiple correlation represents the proportion of total variance of the Y variable that is "due to" or predicted by the predictor variables in **X**.

If the random vector **X** has a multivariate normal distribution with the parameters, population mean vector, $\boldsymbol{\mu}$, and population covariance matrix, $\boldsymbol{\Sigma}_{XX}$, it is possible to test whether the population multiple-correlation coefficient, $\bar{R}$, is equal to zero or not, given a sample multiple correlation R computed from N randomly selected observations.

$$F = \frac{R^2}{1-R^2}\frac{N-k}{k-1}$$

This ratio is distributed as the F-distribution with $k-1$ and $N-k$ degrees of freedom when the population, $\bar{R}$, is equal to zero.

Morrison (1967, p. 104) points out that multiple-correlation coefficients sampled from a population with a multiple correlation, $\bar{R}=0$, are biased estimators of the population multiple-correlation coefficient. In particular, in this case,

$$E(R^2) = \frac{k-1}{N-1}$$

Thus, if the number of predictors, $k-1$, is large relative to N, the number of observations in a sample, the expected value of the sample multiple correlation will approach 1 in value, even when the multiple correlation in the population is zero. For this reason one should keep the number of predictor variables small, relative to the number of observations, when estimating multiple-correlation coefficients.

4.1.5 Correlations between Actual and Predicted Criteria

Let $\text{cov}(\mathbf{Y}, \hat{\mathbf{Y}}) = E(\mathbf{Y}\hat{\mathbf{Y}}') = \boldsymbol{\Sigma}_{Y\hat{Y}}$ be the matrix of covariances between the actual criterion variables in $\mathbf{Y}$ and the predicted values of these criterion variables based on the variables in $\mathbf{X}$. By making the substitution, $\hat{\mathbf{Y}} = \boldsymbol{\Sigma}_{YX}\boldsymbol{\Sigma}_{XX}^{-1}\mathbf{X}$, we obtain

$$E(\mathbf{Y}\hat{\mathbf{Y}}') = E(\mathbf{Y}\mathbf{X}'\boldsymbol{\Sigma}_{XX}^{-1}\boldsymbol{\Sigma}_{XY}) = \boldsymbol{\Sigma}_{YX}\boldsymbol{\Sigma}_{XX}^{-1}\boldsymbol{\Sigma}_{XY} \tag{4.28}$$

The population covariance matrix between the actual and the predicted scores of the criterion variables is the same as the covariance matrix among the predicted parts of the criterion variables:

$$\text{cov}(\mathbf{Y}, \hat{\mathbf{Y}}) = \text{cov}(\hat{\mathbf{Y}}, \hat{\mathbf{Y}}) \tag{4.29}$$

Now, the correlation matrix between the variables in $\mathbf{Y}$ and those in $\hat{\mathbf{Y}}$ is obtained as follows. Let $\mathbf{D}_Y^2 = [\text{diag}\,\boldsymbol{\Sigma}_{YY}]$ and $\mathbf{D}_{\hat{Y}}^2 = [\text{diag}\,\boldsymbol{\Sigma}_{\hat{Y}\hat{Y}}]$. These diagonal matrices contain the respective variances of the variables in the principal diagonals of $\boldsymbol{\Sigma}_{YY}$ and $\boldsymbol{\Sigma}_{\hat{Y}\hat{Y}}$. Then the matrix of correlations between the variables in $\mathbf{Y}$ and those in $\hat{\mathbf{Y}}$ is given by

$$\mathbf{R}_{Y\hat{Y}} = \mathbf{D}_Y^{-1}\boldsymbol{\Sigma}_{YX}\boldsymbol{\Sigma}_{XX}^{-1}\boldsymbol{\Sigma}_{XY}\mathbf{D}_{\hat{Y}}^{-1} \tag{4.30}$$

The principal diagonal of $\mathbf{R}_{Y\hat{Y}}$ contains the correlations between corresponding actual and predicted values of the same variable. So, let us extract the principal diagonal of $\mathbf{R}_{Y\hat{Y}}$ and place it in a diagonal matrix:

$$\begin{aligned}\text{diag}\,\mathbf{R}_{Y\hat{Y}} &= \text{diag}[\mathbf{D}_Y^{-1}\boldsymbol{\Sigma}_{YX}\boldsymbol{\Sigma}_{XX}^{-1}\boldsymbol{\Sigma}_{XY}\mathbf{D}_{\hat{Y}}^{-1}] = \mathbf{D}_Y^{-1}\,\text{diag}[\boldsymbol{\Sigma}_{YX}\boldsymbol{\Sigma}_{XX}^{-1}\boldsymbol{\Sigma}_{XY}]\mathbf{D}_{\hat{Y}}^{-1}\\ &= \mathbf{D}_Y^{-1}\mathbf{D}_{\hat{Y}}^2\mathbf{D}_{\hat{Y}}^{-1} = \mathbf{D}_Y^{-1}\mathbf{D}_{\hat{Y}}\end{aligned} \tag{4.31}$$

But a typical diagonal element of Equation 4.31 is $\dfrac{\sigma(\hat{Y}_j)}{\sigma(Y_j)}$, that is

$$\text{diag}\,\mathbf{R}_{Y\hat{Y}} = \left\|\frac{\sigma(\hat{Y})}{\sigma(Y)}\right\| \tag{4.32}$$

The diagonal elements of this matrix are the respective multiple correlations between the actual criterion variables and their predicted counterparts based on the variables in $\mathbf{X}$. A multiple-correlation coefficient is a correlation between a criterion variable and the composite of predicted values of that variable based on a set of predictor variables.

We also now see why

$$\mathbf{P} = \mathbf{D}_Y^{-1}\mathbf{D}_{\hat{Y}}^2\mathbf{D}_Y^{-1} = \left\| \frac{\sigma^2(\hat{Y}_j)}{\sigma^2(Y_j)} \right\|$$

in Equation 4.27 is called the diagonal matrix of squared multiple correlations between the actual and predicted values of the criterion variables based on the variables in **X**.

Under the assumptions that all of the variables jointly have a multivariate normal distribution, hypotheses about an individual population multiple-correlation coefficient can be tested using Fisher's Z-transformation of the sample multiple-correlation coefficient R and the value of the population multiple correlation under the null hypothesis R_0. We must obtain

$$Z_R = \tanh^{-1} R = \frac{1}{2}\log\left(\frac{1+R}{1-R}\right)$$

$$Z_{R_0} = \tanh^{-1} R_0 = \frac{1}{2}\log\left(\frac{1+R_0}{1-R_0}\right)$$

Then

$$z = \frac{Z_R - Z_{R_0}}{1/\sqrt{N-3}} \tag{4.33}$$

is approximately asymptotically normally distributed with a mean of zero and standard deviation of unity.

4.2 Partial Correlations

Partial correlation is based on the principles of multiple correlation. Partial correlation seeks to determine the correlations among a set of dependent variables after the components in these variables predictable from another set of independent variables have been removed. To illustrate, let us define the following random vectors and matrices:

$\mathbf{X}_1$ is a $q \times 1$ random vector of dependent variables. Assume

$$E(\mathbf{X}_1) = \mathbf{0}$$

$\mathbf{X}_2$ is a $p \times 1$ random vector of independent variables. Assume

$$E(\mathbf{X}_2) = \mathbf{0}$$

$$\mathbf{\Sigma}_{11} = E(\mathbf{X}_1\mathbf{X}_1')$$

$$\boldsymbol{\Sigma}_{22} = E(\mathbf{X}_2\mathbf{X}_2')$$

$$\boldsymbol{\Sigma}_{12} = E(\mathbf{X}_1\mathbf{X}_2')$$

$$\boldsymbol{\Sigma}_{21} = \boldsymbol{\Sigma}_{12}'$$

$\mathbf{B}$ is a $p \times q$ matrix of regression coefficients for predicting variables in $\mathbf{X}_1$ from the variables in $\mathbf{X}_2$, that is, $\mathbf{B} = \boldsymbol{\Sigma}_{22}^{-1}\boldsymbol{\Sigma}_{21}$

$\mathbf{X}_{1\cdot 2} = (\mathbf{X}_1 - \hat{\mathbf{X}}_1) = (\mathbf{X}_1 - \mathbf{B}'\mathbf{X}_2)$ is the random vector of residual variables in $\mathbf{X}_1$ with the components predicted from the variables in $\mathbf{X}_2$ removed

Let us now find the covariance matrix among the residual variables in $\mathbf{X}_{1\cdot 2}$.

$$\boldsymbol{\Sigma}_{11\cdot 2} = E(\mathbf{X}_{1\cdot 2}\mathbf{X}_{1\cdot 2}') \tag{4.34}$$

By substituting the expression defined above for $\mathbf{X}_{1\cdot 2}$ into Equation 4.33 and expanding, we can find an expression for the partial covariance matrix in terms of $\mathbf{X}_1$, $\mathbf{X}_2$, and $\mathbf{B}$.

$$\begin{aligned}
\boldsymbol{\Sigma}_{11\cdot 2} &= E[(\mathbf{X}_1 - \mathbf{B}'\mathbf{X}_2)(\mathbf{X}_1 - \mathbf{B}'\mathbf{X}_2)'] = E[(\mathbf{X}_1 - \mathbf{B}'\mathbf{X}_2)(\mathbf{X}_1' - \mathbf{X}_2'\mathbf{B})] \\
&= E(\mathbf{X}_1\mathbf{X}_1' - \mathbf{X}_1\mathbf{X}_2'\mathbf{B} - \mathbf{B}'\mathbf{X}_2\mathbf{X}_1 + \mathbf{B}\mathbf{X}_2\mathbf{X}_2'\mathbf{B}) \\
&= \boldsymbol{\Sigma}_{11} - \boldsymbol{\Sigma}_{12}\mathbf{B} - \mathbf{B}'\boldsymbol{\Sigma}_{21} + \mathbf{B}'\boldsymbol{\Sigma}_{22}\mathbf{B} \\
&= \boldsymbol{\Sigma}_{11} - \boldsymbol{\Sigma}_{12}\boldsymbol{\Sigma}_{22}^{-1}\boldsymbol{\Sigma}_{21} - \boldsymbol{\Sigma}_{12}\boldsymbol{\Sigma}_{22}^{-1}\boldsymbol{\Sigma}_{21} + \boldsymbol{\Sigma}_{12}\boldsymbol{\Sigma}_{22}^{-1}\boldsymbol{\Sigma}_{21} \\
&= \boldsymbol{\Sigma}_{11} - \boldsymbol{\Sigma}_{12}\boldsymbol{\Sigma}_{22}^{-1}\boldsymbol{\Sigma}_{21}
\end{aligned} \tag{4.35}$$

Now, to find the partial-correlation matrix, $\mathbf{R}_{11\cdot 2}$, first define

$$\mathbf{D}_{1\cdot 2}^2 = [\text{diag}\,\boldsymbol{\Sigma}_{11\cdot 2}]$$

which is a diagonal matrix containing the partial variances of the variables in $\mathbf{X}_{1\cdot 2}$. Then,

$$\mathbf{R}_{11\cdot 2} = \mathbf{D}_{1\cdot 2}^{-1}\boldsymbol{\Sigma}_{11\cdot 2}\mathbf{D}_{1\cdot 2}^{-1} \tag{4.36}$$

Equation 4.35 divides the partial variances and covariances by the partial standard deviations in a way analogous to the procedure for finding correlations from a variance–covariance matrix.

Individual sample partial-correlation coefficients may be obtained by simply substituting the sample variance–covariance matrices and diagonal matrices obtained from them for the population variance–covariance and diagonal matrices. Then individual partial correlations can be tested for the significance of their deviation from a hypothetical-population partial-correlation coefficient by transforming the sample partial-correlation coefficient, $r_{ij\cdot 1,\ldots,p}$, and the hypothesized value, $\rho_{ij\cdot 1,\ldots,p}$, by Fisher's Z-transformations:

$$Z_r = \tanh^{-1} r = \frac{1}{2}\log\left(\frac{1+r}{1-r}\right)$$

$$Z_\rho = \tanh^{-1}\rho = \frac{1}{2}\log\left(\frac{1+\rho}{1-\rho}\right)$$

Under the assumption of multivariate normality for all of the variables jointly, $Z_r - Z_\rho$ is asymptotically (as sample size increases without bound) distributed normally with a mean of zero and a variance of approximately $1/(N-p-3)$. Then the statistic

$$z = \frac{Z_r - Z_\rho}{1/\sqrt{N-p-3}} \tag{4.37}$$

is approximately normally distributed with a mean of zero and a standard deviation of unity (Muirhead, 1982), where N is the sample size, and p the number of predictors or variables conditioned on.

4.3 Determinantal Formulas

Up to now we have developed formulas for multiple and partial correlation in terms of matrix algebra. However, these formulas can also be expressed in terms of determinants. Although determinantal formulas for multiple correlation do not lend themselves to efficient computing routines, they do provide a powerful means for deriving some matrix equations that cannot be easily derived otherwise.

Let $\mathbf{R}$ be an $n \times n$ matrix of correlation coefficients among n variables:

$$\mathbf{R} = \begin{bmatrix} 1 & \rho_{12} & \rho_{13} & \cdots & \rho_{1n} \\ \rho_{21} & 1 & \rho_{23} & \cdots & \rho_{2n} \\ \rho_{31} & \rho_{32} & 1 & \cdots & \rho_{3n} \\ \vdots & \vdots & \vdots & \ddots & \vdots \\ \rho_{n1} & \rho_{n2} & \rho_{n3} & \cdots & 1 \end{bmatrix} \tag{4.38}$$

Let the first variable be a criterion variable. Let variables 2, 3, …, n be predictor variables. We wish to find a weight vector, $\boldsymbol{\beta}$, of β weights for predicting variable 1 from the $n-1$ other variables.

Let us partition the correlation matrix, $\mathbf{R}$, as follows:

$$\mathbf{R} = \begin{bmatrix} 1 & \mathbf{R}_{12} \\ \mathbf{R}_{21} & \mathbf{R}_{22} \end{bmatrix} \tag{4.39}$$

The number 1 is a scalar; $\mathbf{R}_{12}$ is a row vector containing the correlations in the first row of $\mathbf{R}$ between variable 1 and variable 2 through n. $\mathbf{R}_{22}$ is the submatrix of **R** containing the $n - 1$ other variables. With this partition, we identify some important submatrices of the matrix **R**.

Although the vector $\boldsymbol{\beta}$ may be obtained using the formula $\boldsymbol{\beta} = \mathbf{R}_{22}^{-1}\mathbf{R}_{21}$, we may also find the elements of $\boldsymbol{\beta}$ by the following determinantal formulas (the proof of which we shall not give here):

$$\beta_{12} = \frac{-R_{12}}{R_{11}}$$

$$\beta_{13} = \frac{-R_{13}}{R_{11}}$$

$$\cdots\cdots\cdots\cdots\cdots\cdots$$

$$\beta_{1n} = \frac{-R_{1n}}{R_{11}} \tag{4.40}$$

Here β_{1j} denotes the weight to assign to the normalized jth variable in predicting variable 1. $R_{12}, R_{13}, \ldots, R_{1j}, \ldots, R_{1n}$ are cofactors of the elements $\rho_{12}, \rho_{13}, \ldots, \rho_{1j}, \ldots, \rho_{1n}$, and ρ_{11}, respectively, of **R**.

4.3.1 Multiple-Correlation Coefficient

The square of the multiple-correlation coefficient for predicting variable 1 from the $n - 1$ predictor variables is

$$R_{1\cdot 23\cdots n}^2 = \boldsymbol{\beta}\mathbf{R}_{21}$$

However, we may also obtain the squared multiple correlation by the formula:

$$R_{1\cdot 23\cdots n}^2 = 1 - \frac{|\mathbf{R}|}{R_{11}} \tag{4.41}$$

where

$|\mathbf{R}|$ is the determinant of the matrix **R**

R_{11} is the cofactor of ρ_{11}

Evidently, the ratio $|\mathbf{R}|/R_{11}$ is the error of estimate in predicting variable 1, that is, the proportion of the variance in variable 1 that cannot be predicted from the $n - 1$ other variables in **R**. Furthermore,

$$R_{1\cdot 23\cdots n} = \sqrt{1 - \frac{|\mathbf{R}|}{R_{11}}} \tag{4.42}$$

is the multiple-correlation coefficient, and

$$\sigma_{1\cdot23\cdots n} = \sqrt{\frac{|\mathbf{R}|}{R_{11}}} \tag{4.43}$$

is the partial standard deviation of variable 1 after components due to variables 2,…, n have been partialled from variable 1.

In general, predicting the jth variable in a correlation matrix, **R**, from the $n-1$ remaining variables in **R**, we have

$$R_{j\cdot12\cdots(j)\cdots n} = \sqrt{1-\frac{|\mathbf{R}|}{R_{jj}}} \tag{4.44}$$

$$\sigma_{j\cdot12\cdots(j)\cdots n} = \sqrt{\frac{|\mathbf{R}|}{R_{jj}}} \tag{4.45}$$

4.3.2 Formulas for Partial Correlations

The partial correlation between variables j and k in **R** with the $n-2$ other variables in **R** partialled out or "held constant" is given by the equation

$$\rho_{jk\cdot1\cdots n} = \frac{-R_{jk}}{\sqrt{R_{jj}R_{kk}}} \tag{4.46}$$

where

R_{jk} is the cofactor of element ρ_{jk}

R_{jj} and R_{kk} are the cofactors of elements ρ_{jj} and ρ_{kk} of **R**

4.4 Multiple Correlation in Terms of Partial Correlation

Guttman (1953) in developing his theory of image analysis provided a very useful model for explaining what happens in multiple correlation. Here we will briefly outline what image analysis is about so that we can apply it to the problem of multiple correlation.

Consider that within any set of variables each variable can be thought of as consisting of two parts: (1) a part that is predictable by multiple correlation from the rest of the variables in the set and (2) a part that is not predictable from the rest of the variables in the set. For the sake of illustration, let us define $\mathbf{X}_2$ to be an $n \times 1$ random vector, the coordinates of which are normalized random variables with means of zero and standard deviations of 1.00. Now we will define an $n \times 1$ random vector, $\mathbf{X}_{p2}$, such that

$$\mathbf{X}_{p2} = \mathbf{W}'\mathbf{X}_2 \tag{4.47}$$

where **W** is the $n \times n$ multiple-correlation weight matrix for predicting each variable in $\mathbf{X}_2$ from every other variable in $\mathbf{X}_2$. In other words, if we take a coordinate of $\mathbf{X}_{p2}$ it will represent the predicted component of the corresponding random variable in $\mathbf{X}_2$ based on the $(n-1)$ other variables in $\mathbf{X}_2$.

Now, let us define a random vector $\mathbf{X}_{u2}$ such that

$$\mathbf{X}_{u2} = \mathbf{X}_2 - \mathbf{X}_{p2} = (\mathbf{X}_2 - \mathbf{W}'\mathbf{X}_2) \tag{4.48}$$

$\mathbf{X}_{u2}$ corresponds to the vector of "unpredicted parts" of the variables in $\mathbf{X}_2$.

From the above definition it follows, by moving **P** to the left-hand side of Equation 4.48 and rearranging, that

$$\mathbf{X}_2 = \mathbf{X}_{p2} + \mathbf{X}_{u2} \tag{4.49}$$

This is the fundamental equation of Guttman's image theory, which states that a variable consists of a predictable component and an unpredictable component with respect to the other variables in the set of variables of which it is a member.

4.4.1 Matrix of Image Regression Weights

Before we can go any further, we need to obtain the matrix **W** in Equation 4.47. This we can do by using the determinantal formulas developed in Section 4.3. In constructing the weight matrix, **W**, we will consider that a variable being predicted from the $n-1$ other variables will receive a weight of zero in the prediction equation involving the full set of n variables. This is because only the other variables must determine the predictable part of a predictable variable. Thus, we can use Equation 4.40 to construct a weight matrix based upon the cofactors of the matrix $\mathbf{R}_{22}$ of correlation coefficients among the variables in $\mathbf{X}_2$. For example, in a problem of four variables we have in terms of cofactors of $\mathbf{R}_{22}$:

$$\mathbf{W} = \begin{bmatrix} 0 & \dfrac{-R_{12}}{R_{22}} & \dfrac{-R_{13}}{R_{22}} & \dfrac{-R_{14}}{R_{22}} \\ \dfrac{-R_{21}}{R_{22}} & 0 & \dfrac{-R_{21}}{R_{22}} & \dfrac{-R_{24}}{R_{22}} \\ \dfrac{-R_{31}}{R_{22}} & \dfrac{-R_{32}}{R_{22}} & 0 & \dfrac{-R_{34}}{R_{22}} \\ \dfrac{-R_{41}}{R_{22}} & \dfrac{-R_{42}}{R_{22}} & \dfrac{-R_{43}}{R_{22}} & 0 \end{bmatrix} \tag{4.50}$$

However, to determine the matrix **W** in actual practice, using cofactors would be quite inefficient. It would be better if we could obtain a matrix equation that would generate the above matrix. Kaiser (1963), in an article discussing image analysis, provides the following matrix equation for **W**:

$$\mathbf{W} = (\mathbf{I} - \mathbf{R}_{22}^{-1}\mathbf{S}^2) \tag{4.51}$$

$$\mathbf{S}^2 = [\text{diag } \mathbf{R}_{22}^{-1}]^{-1} \tag{4.52}$$

The expression in Equation 4.52 means that $\mathbf{S}^2$ is a diagonal matrix containing in its principal diagonal the reciprocals of the principal diagonal elements of the inverse matrix of $\mathbf{R}_{22}$. (Do not confuse $\mathbf{S}^2$ with the sample variance–covariance matrix.) We will now show how Kaiser's matrix equation yields the matrix in Equation 4.50.

One way in which the inverse of a matrix can be obtained is by dividing the determinant of the matrix into each element of the corresponding adjoint matrix. The adjoint matrix is the transpose of the matrix containing the cofactors in place of the corresponding elements of the original matrix. In our four-variable problem the diagonal elements of the inverse matrix $\mathbf{R}_{22}^{-1}$ will be

$$\rho^{11} = \frac{R_{11}}{R} \quad \rho^{22} = \frac{R_{22}}{R} \quad \rho^{33} = \frac{R_{33}}{R} \quad \rho^{44} = \frac{R_{44}}{R} \tag{4.53}$$

where

the superscripts indicate elements of the inverse matrix $\mathbf{R}_{22}^{-1}$
R is the determinant of $\mathbf{R}_{22}$
R_{11}, R_{22}, R_{33}, and R_{44} are cofactors of the elements of $\mathbf{R}_{22}$

Consequently, in terms of these cofactor expressions, Kaiser's equation becomes

$$\mathbf{W} = \begin{bmatrix} 1 & & & \\ & 1 & & \\ & & 1 & \\ & & & 1 \end{bmatrix} - \begin{bmatrix} \frac{R_{11}}{R} & \frac{R_{12}}{R} & \frac{R_{13}}{R} & \frac{R_{14}}{R} \\ \frac{R_{21}}{R} & \frac{R_{22}}{R} & \frac{R_{23}}{R} & \frac{R_{24}}{R} \\ \frac{R_{31}}{R} & \frac{R_{32}}{R} & \frac{R_{33}}{R} & \frac{R_{34}}{R} \\ \frac{R_{41}}{R} & \frac{R_{42}}{R} & \frac{R_{43}}{R} & \frac{R_{44}}{R} \end{bmatrix} \begin{bmatrix} \frac{R}{R_{11}} & & & \\ & \frac{R}{R_{22}} & & \\ & & \frac{R}{R_{33}} & \\ & & & \frac{R}{R_{44}} \end{bmatrix} \tag{4.54}$$

The matrix on the left-hand side is the identity matrix $\mathbf{I}$. The matrix in the center is the inverse matrix $\mathbf{R}_{22}^{-1}$. The matrix on the right-hand side is $\mathbf{S}^2$. The effect of postmultiplying the matrix $\mathbf{R}_{22}^{-1}$ by the diagonal matrix $\mathbf{S}^2$ is to multiply the first-column elements of the inverse matrix by the first diagonal element of $\mathbf{S}^2$, the second-column elements of the inverse matrix by the second diagonal element of $\mathbf{S}^2$, and so on. The result will be that the diagonal elements of the inverse matrix will become 1s. The R's in the denominator of each element of the inverse matrix will be cancelled by the R's in the numerator of the diagonal elements of $\mathbf{S}^2$. The elements of the columns of the matrix will be divided respectively by the cofactors R_{11}, R_{22}, R_{33}, and R_{44}. When we subtract the transformed inverse matrix from the identity matrix, the diagonal unities will be cancelled to zeros, and the off-diagonal elements will be given minus signs. The result will be the weight matrix, $\mathbf{W}$, in Equation 4.50.

To gain more insight into this weight matrix, consider that Kaiser's equation in Equation 4.50 can be rewritten as

$$\mathbf{W} = \mathbf{R}_{22}^{-1}(\mathbf{R}_{22} - \mathbf{S}^2) \tag{4.55}$$

Looking closely at this equation, we may recognize it as the form of the general equation for finding a multiple-regression weight matrix presented earlier in this chapter in Equation 4.8 The matrix $(\mathbf{R}_{22} - \mathbf{S}^2)$ represents the covariances between the original variables and their predictable parts. The matrix $\mathbf{S}^2$ represents the errors of estimate of each variable with respect to the $n - 1$ other variables as predictors of it. We could have obtained Equation 4.55 had we known another way, besides the use of determinants to find the covariances between the original variables and their predictable parts. But no such other way was available. In practice, we do not use determinants to find the matrix $\mathbf{W}$; rather, we invert the matrix $\mathbf{R}_{22}$, find its principal diagonal elements, and then use Kaiser's Equation 4.51.

4.4.2 Meaning of Multiple Correlation

We are now ready to look deeper into the meaning of the β weights in multiple correlation. Let us define $\mathbf{X}_1$ as a $m \times 1$ random vector whose coordinates are m normalized random variables that do not belong to the set of variables composing $\mathbf{X}_2$. Assume $E(\mathbf{X}_1) = \mathbf{0}$. If we wished to predict the variables in $\mathbf{X}_1$ from the variables in $\mathbf{X}_2$, we would have to obtain a β-weight matrix using Equation 4.9. Here we will redefine it as

$$\boldsymbol{\beta} = \mathbf{R}_{22}^{-1}\mathbf{R}_{21} \tag{4.56}$$

However, let us set the $\boldsymbol{\beta}$ matrix aside for a moment and consider the question: What are the correlations between the scores of $\mathbf{X}_1$ and $\mathbf{X}_{u2}$? We shall see that the answer to this question has a bearing on the theory of multiple correlation. In effect, we will show that the correlations between $\mathbf{X}_1$ and $\mathbf{X}_{u2}$ are partial correlations between $\mathbf{X}_1$ and $\mathbf{X}_2$ based on partialling from $\mathbf{X}_2$ the variance in each variable that covaries with other variables in $\mathbf{X}_2$. These partial correlations have an intimate connection with the β-weight matrix.

Let the matrix of covariance between the variables in $\mathbf{X}_1$ and the variables in $\mathbf{U}$ be designated as $\mathbf{\Sigma}_{1U}$. In terms of the basic matrices, $\mathbf{X}_1$ and $\mathbf{X}_{u2}$, $\mathbf{\Sigma}_{1u2}$ is found as follows:

$$\begin{aligned}\boldsymbol{\Sigma}_{1u2} &= E(\mathbf{X}_1\mathbf{X}_{u2}) = E[\mathbf{X}_1(\mathbf{X}_2' - \mathbf{X}_2'\mathbf{W})] \\ &= E(\mathbf{X}_1\mathbf{X}_2' - \mathbf{X}_1\mathbf{X}_2'\mathbf{W}) \\ &= \mathbf{R}_{12} - \mathbf{R}_{12}\mathbf{W}\end{aligned}$$

The use of correlation matrices on the right-hand side of the equation is possible because the original scores are in the standard score form. However, $\boldsymbol{\Sigma}_{1u2}$ is still a covariance matrix because $\mathbf{X}_{u2}$ does not contain normalized variables. If we substitute Equation 4.51 into the above equation and expand the result

$$\boldsymbol{\Sigma}_{1u2} = \mathbf{R}_{12} - \mathbf{R}_{12}(\mathbf{I} - \mathbf{R}_{22}^{-1}\mathbf{S}^2) = \mathbf{R}_{12}\mathbf{R}_{22}^{-1}\mathbf{S}^2$$

or by substituting Equation 4.55 into this expression we have

$$\boldsymbol{\Sigma}_{1U} = \boldsymbol{\beta}'\mathbf{S}^2 \tag{4.57}$$

To find the correlation matrix between $\mathbf{X}_1$ and $\mathbf{X}_{u2}$, we have to find the expression for the variances of the variables in $\mathbf{X}_{u2}$. To do this, we first need to find the variance–covariance matrix $\boldsymbol{\Sigma}_{uu2}$ from the matrix $\mathbf{X}_{u2}$. This is possible from

$$\boldsymbol{\Sigma}_{UU} = E(\mathbf{UU}') = E[(\mathbf{X}_2 - \mathbf{W}'\mathbf{X}_2)(\mathbf{X}_2 - \mathbf{W}'\mathbf{X}_2)']$$

which, after substitution of Equation 4.51 for $\mathbf{W}$ and factoring, becomes

$$\boldsymbol{\Sigma}_{uu2} = \mathbf{S}^2\mathbf{R}_{22}^{-1}\mathbf{S}^2 \tag{4.58}$$

where $\mathbf{S}^2$ is defined in Equation 4.52.

Now, the diagonal elements of $\boldsymbol{\Sigma}_{uu2}$ will contain the variances of the variables in $\mathbf{U}$. But what are the diagonal elements of $\boldsymbol{\Sigma}_{uu2}$? Consider that the diagonal elements of $\boldsymbol{\Sigma}_{uu2}$ will equal $\mathbf{S}^2[\text{diag } \mathbf{R}_{22}^{-1}]\mathbf{S}^2$. But by definition $\mathbf{S}^2 = [\text{diag } \mathbf{R}_{22}^{-1}]^{-1}$. Hence $\mathbf{S}^{-2} = [\text{diag } \mathbf{R}_{22}^{-1}]$. Consequently $[\text{diag } \boldsymbol{\Sigma}_{uu2}] = \mathbf{S}^2$.

To find the correlation matrix $\mathbf{R}_{1u2}$ from $\boldsymbol{\Sigma}_{1u2}$, we postmultiply $\boldsymbol{\Sigma}_{1u2}$ by the matrix $\mathbf{S}^{-1}$, which contains the reciprocals of the standard deviations of variables in $\mathbf{X}_{u2}$:

$$\mathbf{R}_{1u2} = \boldsymbol{\Sigma}_{1u2}\mathbf{S}^{-1} = \boldsymbol{\beta}'\mathbf{S}^2\mathbf{S}^{-1} = \boldsymbol{\beta}'\mathbf{S} \tag{4.59}$$

Solving for $\boldsymbol{\beta}$ we have

$$\boldsymbol{\beta} = \mathbf{S}^{-1}\mathbf{R}_{u2,1} \tag{4.60}$$

where a given element β_{jq} in the matrix $\boldsymbol{\beta}$ can be expressed as

$$\beta_{jq} = \frac{\rho_{j,q\cdot 1\cdots n}}{\sigma_{j\cdot 1\cdots n}} \tag{4.61}$$

which is the partial correlation between the predictor j in $\mathbf{X}_2$ and the criterion q in $\mathbf{X}_1$ after the variance due to the other variables in the predictor set $\mathbf{X}_2$ have been partialled from variable j, divided by the partial standard deviation of the unpredicted component in variable j. In effect, the partial-correlation coefficient can be seen as the weight to assign to variable j in order to predict variable q, whereas the partial standard deviation scales the magnitude of variable j to a larger value depending upon the relative smallness of the unpredicted standard deviation.

We now see that multiple correlation depends upon the unique parts of the predictor variables and not upon their parts correlated with other predictors. Moreover, we see that there must be some correlation between the unique parts of the predictor variables and the criterion for the multiple-correlation weights to be greater than or less than zero.

4.4.3 Yule's Equation for the Error of Estimate

Working at the level of scalar algebra, Yule (1907) developed a subscript notation for multiple- and partial-correlation equations that is currently widely used. For example

$$\rho_{12\cdot 34}$$

represents the correlation between variables 1 and 2 after variables 3 and 4 have been partialled from them. On the other hand

$$\sigma_{1\cdot 34}$$

represents the standard deviation of variable 1 after variables 3 and 4 have been partialled from it.

$$\beta_{12\cdot 34\cdots p}$$

represents the partial-regression coefficient for predicting variable 1 from variable 2 after variables 3 to p have been partialled from variables 1 and 2. In the case of β weights the order of the subscripts to the left-hand side of the period is important. The first subscript on the left-hand side of the period represents the variable predicted; the second on the left-hand side represents the predictor variable. Thus

$$\beta_{12\cdot 34\cdots p} \neq \beta_{21\cdot 34\cdots p}$$

because the second expression represents the partial-regression coefficient for predicting variable 2 from variable 1 after variables 3 to p have been partialled from them.

Using this notation, Yule developed the following equation (cf. Yule, 1907) for the error of estimate (the variance in the predicted variable that cannot be predicted) for predicting variable 1 from variables 2 to p in a multiple-correlation problem

$$1 - R_{1\cdot 23\cdots p}^2 = \frac{\sigma_{1\cdot 23\cdots p}^2}{\sigma_1^2} = (1-\rho_{12}^2)(1-\rho_{13\cdot 2}^2)\cdots(1-\rho_{1p\cdot 23\cdots(p-1)}^2)$$

where

$R_{1\cdot 23\cdots p}^2$ is the square of the multiple-correlation coefficient for predicting variable 1 from variables 2 to p

$\sigma_{1\cdot 23\cdots p}^2$ is the variance in variable 1 that cannot be predicted by variables 2 to p

σ_1^2 is the original variance of variable 1

$\rho_{13\cdot 2}$ is the partial correlation between variable 1 and variable 3 after variable 2 has been partialled from both of them

Yule's equation can begin with any particular predictor variable and proceed in any order to partial preceding variables from succeeding variables, e.g.,

$$\frac{\sigma_{1\cdot 23\cdots p}^2}{\sigma_1^2} = (1-\rho_{13}^2)(1-\rho_{14\cdot 3}^2)\cdots(1-\rho_{1p\cdot 23\cdots(p-1)}^2) \quad (4.62)$$

As a consequence of Yule's equation, we can establish two important theorems: (1) A multiple-correlation coefficient is never less than the absolute value of any (in particular, the highest) correlation between a predictor and the predicted variable. (2) The addition of a variable to the set of predictor variables will increase or leave the same the multiple-correlation coefficient between the set of predictors and the predicted variable. Whichever happens depends upon whether or not the additional predictor variable contains variance correlated with the predicted variable that is uncorrelated with the other predictors.

4.4.4 Conclusions

A multiple-correlation coefficient depends upon the correlations between the unique parts of the predictor variables and the criterion. One can see that as the average correlation among the predictors decreases while the average correlation between the predictors and the criterion increases, then the average partial correlation between the predictors' unique parts and the criterion should increase. In other words, there will be, on an average, less covariance to be partialled out of the covariance between predictors and the criterion. As a result, we should increase the multiple-correlation coefficient. This corresponds with a principle developed in connection with simply weighted composite variables.

However, we must also realize that often much of the unique variance in a variable will be due to an error of measurement. Since an error will not correlate with anything (by the assumptions of classical reliability theory), its presence diminished the chances of the unique variance of an additional predictor variable covarying with the criterion. This leads us to the principle that "increasing the number of variables in a prediction equation is not guaranteed to give a substantial increase in the multiple-correlation coefficient." This is because to some extent the additional variable will contain nothing new correlated with the criterion not found in previous predictors. The only way to overcome this difficulty is to add variables to the predictor set that have substantial correlations with the criterion but low correlations with other predictors. In practice, we find this difficult to do.

When we increase the number of predictor variables without correspondingly increasing our sample of subjects, we augment the difficulties of sampling error. We get more spuriously large partial correlations between the predictors and the criterion, and as a result the multiple-correlation coefficient will be inflated. This relates to the bias of sample multiple-correlation coefficients mentioned previously in Section 4.1. Since partial-correlation coefficients are based on multiple correlation, we can also determine that sampling errors will affect partial-correlation coefficients as the number of predictor variables increases relative to the number of subjects. For these reasons, we must be very careful in working with large numbers of predictors in multiple-correlation and partial-correlation problems.

5

Multivariate Normal Distribution

5.1 Introduction

A generalization of the univariate normal distribution to more than one variable is the multivariate normal distribution. Much of the basic statistical theory underlying statistical inference in factor analysis and structural equation modeling is based on the multivariate normal distribution. Some of the reason for this is that it is often an excellent approximation to many real-world distributions, while at the same time mathematically tractable. It also lends itself well to linear functional relations among variables because linear transformations of multivariate, normally distributed, variables are variables that also have multivariate normal distributions. And sampling distributions of many multivariate statistics, including those of factor analysis, behave approximately as multivariate normal distributions because of the effect described by the central limit theorem.

5.2 Univariate Normal Density Function

Let us begin with something we should be familiar with from basic statistics, the univariate normal density function. This distribution, one should remember, is based on two parameters, the mean, μ, and the variance, σ^2, of the distribution. This distribution is a continuous distribution with a probability density function given by

$$f(y) = \frac{1}{\sqrt{2\pi\sigma^2}} e^{-(y-\mu)^2/2\sigma^2} \tag{5.1}$$

where "e" is the transcendental constant, and $-\infty < y < \infty$. The quantity $1/\sqrt{2\pi\sigma^2}$ is the normalizing constant, used to ensure that the area under the curve is unity, which is an essential property of a density function. The "bell curve" of a normal distribution with a mean of 5 and variance of 4 is shown in Figure 5.1.

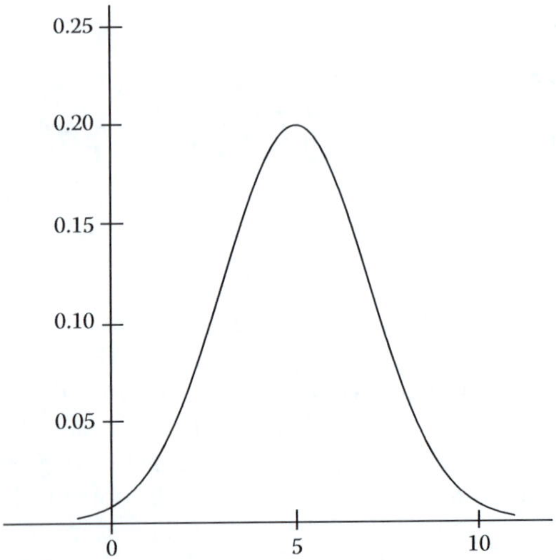

FIGURE 5.1
Curve of the normal distribution for a variable with a mean of 5 and variance of 4.

But another way of writing the formula for the normal distribution will make it look more like the multivariate normal density function that we are about to show.

$$f(y) = \frac{1}{\sqrt{2\pi\sigma^2}} e^{-(y-\mu)(\sigma^2)^{-1}(y-\mu)}.$$

5.3 Multivariate Normal Distribution

An $n \times 1$ random vector $\mathbf{Y}$ of n variables is said to have a multivariate normal distribution with a mean vector of $\boldsymbol{\mu}$ and variance–covariance matrix, $\boldsymbol{\Sigma}$, if its density function is given by

$$f(\mathbf{y}) = \frac{1}{(2\pi)^{n/2}|\Sigma|^{1/2}} e^{-(1/2)(\mathbf{y}-\boldsymbol{\mu})'\Sigma^{-1}(\mathbf{y}-\boldsymbol{\mu})} \tag{5.2}$$

The quantity $|\boldsymbol{\Sigma}|$ is known as the population generalized variance of the variables in **Y**. Since this is a determinant of the variance–covariance matrix, a small value for the determinant near zero means the **y**'s are all close to the mean vector, $\boldsymbol{\mu}$. It also suggests near multicollinearity of the variables (Rencher, 2002), with the consequence that prediction with these variables will be unstable because the standard errors of the β weights of regression will be large.

Notation for briefly describing a multivariate notation is $N_p(\boldsymbol{\mu}, \boldsymbol{\Sigma})$. N denotes a normal distribution, and the subscript p denotes the number of variables in the mean vector and variance–covariance matrix. $\boldsymbol{\mu}$ denotes the $p \times 1$ mean vector, and $\boldsymbol{\Sigma}$ the $p \times p$ variance–covariance matrix, which are the parameters of this distribution.

5.3.1 Bivariate Normal Distribution

If the number of variables is $n = 2$, then the multivariate normal distribution specializes to the bivariate normal distribution (Figure 5.2).

The volume under the curve for $-\infty < x_1 < \infty$ and $-\infty < x_2 < \infty$ is unity. The coefficient $1/((2\pi)^{n/2}\,|\boldsymbol{\Sigma}|^{1/2})$ is again the normalizing constant that adjusts the exponential function to have a unit volume under the curve.

Of special interest in the bivariate case is a contour plot of the bivariate normal distribution (Figure 5.3).

A contour plot is based on the idea that through points of equal elevation or density on the bivariate density function, one can draw lines around the "hill" formed by the density function. These lines are then projected down onto a floor below, as if one were looking down on the "hill" from above and seeing the elliptical lines projected on a plane at the base. The ellipses toward the middle correspond in this case to higher elevations or densities on the curve. As correlation between the two variables become higher and higher, the ellipses become narrower and narrower around their long axis. As the correlation between the two variables becomes closer and closer to zero, the ellipses approaches circles in shape. A contour plot gives information about the shape of the distribution.

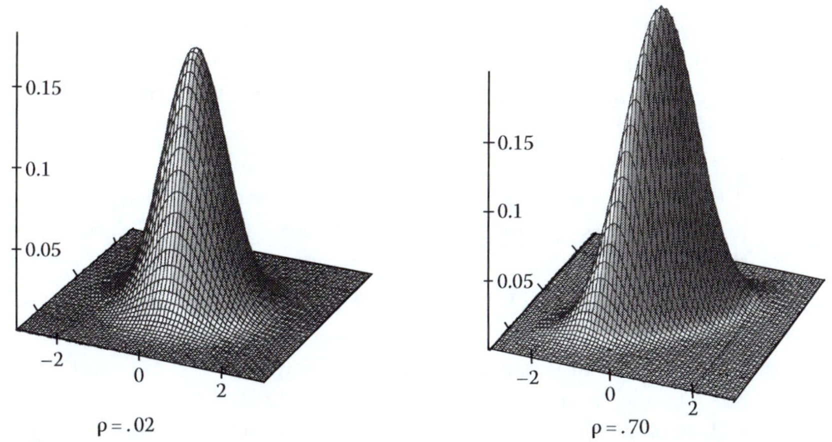

FIGURE 5.2
Graphs of density functions for the bivariate normal distribution for two variables with mean vector $\boldsymbol{\mu}' = [1, 2]$, variances of 1.00, and $\rho_{12} = .02$, and $\rho_{12} = .70$, respectively.

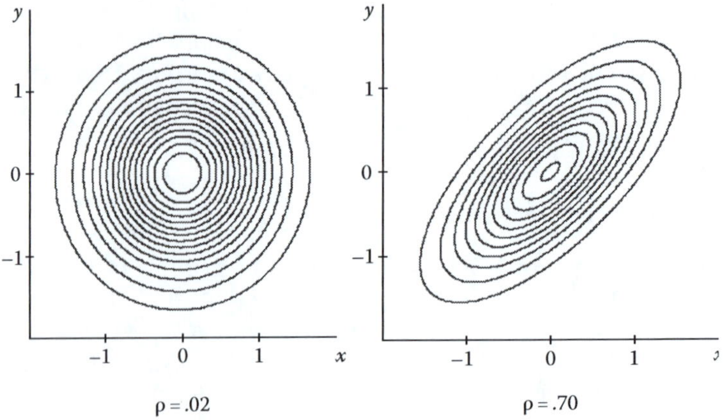

FIGURE 5.3
Contour plots of the bivariate normal distributions shown in Figure 5.2. Elliptical contour lines represent loci of points with equal density.

5.3.2 Properties of the Multivariate Normal Distribution

The following are properties of the multivariate normal distribution.

Let **X** be a $p \times 1$ random vector distributed $N_p(\boldsymbol{\mu}, \boldsymbol{\Sigma})$:

1. The contours of equal density for the $N_p(\boldsymbol{\mu}, \boldsymbol{\Sigma})$ multivariate normal distribution are ellipsoids defined by vector values **x** of **X** such that $(\mathbf{x} - \boldsymbol{\mu})'\boldsymbol{\Sigma}^{-1}(\mathbf{x} - \boldsymbol{\mu}) = c^2$ (Johnson and Wichern, 1998). The solid ellipsoid of **x** values satisfying $(\mathbf{x} - \boldsymbol{\mu})'\boldsymbol{\Sigma}^{-1}(\mathbf{x} - \boldsymbol{\mu}) \leq \chi^2_p(\alpha)$ has probability of $1 - \alpha$. This forms the basis for statistical tests using the chi-square distribution with the multivariate normal distribution. The ellipsoids are centered on the mean vector $\boldsymbol{\mu}$.
2. Linear combinations of the components of **X** are normally distributed (Johnson and Wichern, 1998; Rencher, 2002).
   a. Let **a** be a $p \times 1$ vector of constants. Define $Y = \mathbf{a}'\mathbf{X}$. Then Y is distributed $N(\mathbf{a}'\boldsymbol{\mu}, \mathbf{a}'\boldsymbol{\Sigma}\mathbf{a})$. The properties of the mean and variance of this composite are already established in Chapter 3. The proof that the composite variable has a normal distribution is beyond the level of this text but may be found in Anderson (1984).
   b. Let **A** be a $p \times q$ matrix of rank q whose columns are vectors $\mathbf{a}_j$ and $q \leq p$. Then $\mathbf{Y} = \mathbf{A}'\mathbf{X}$ is distributed $N_q(\mathbf{A}'\boldsymbol{\mu}, \mathbf{A}'\boldsymbol{\Sigma}\mathbf{A})$. The mean vector $\mathbf{A}'\boldsymbol{\mu}$ and variance–covariance matrix $\mathbf{A}'\boldsymbol{\Sigma}\mathbf{A}$ are based on the properties of linear transformations described in Chapter 3. The proof of multivariate normality is beyond the level of this text.
3. Any subset of variables in **X** has a multivariate normal distribution, and each individual X_i in **X**, has a unit normal distribution.

To illustrate, let us partition **X** into two sets of variables such that

$$\mathbf{X} = \begin{bmatrix} \mathbf{X}_1 \\ \mathbf{X}_2 \end{bmatrix} \quad \boldsymbol{\mu} = \begin{bmatrix} \boldsymbol{\mu}_1 \\ \boldsymbol{\mu}_2 \end{bmatrix} \quad \text{and} \quad \boldsymbol{\Sigma} = \begin{bmatrix} \boldsymbol{\Sigma}_{11} & \boldsymbol{\Sigma}_{12} \\ \boldsymbol{\Sigma}_{21} & \boldsymbol{\Sigma}_{22} \end{bmatrix}$$

Assume $\mathbf{X}_1$ is $r \times 1$ and $\mathbf{X}_2$ is $(p-r) \times 1$. Then according to Rencher (2002), if **X** is distributed $N_p(\boldsymbol{\mu}, \boldsymbol{\Sigma})$, then $\mathbf{X}_1$ is distributed $N_r(\boldsymbol{\mu}_1, \boldsymbol{\Sigma}_{11})$.

Since any individual variable X_i is a subset of **X**, then it must be the case that X_i has a univariate normal distribution with mean μ_ι and variance σ_ι^2.

However, Rencher (2002) notes that the converse is not true. If the density of each x_i for each variable X_i in **X** is univariate normal, it does not follow that **X** has a joint multivariate normal distribution.

In the following, we assume **X** is partitioned as in paragraph 3 above, but has $q + p$ variables. ($\mathbf{X}_1$ may also represent some q additional variables, such as criterion variables and $\mathbf{X}_2$ some p predictor variables or variables to condition on.) Then,

$$\begin{bmatrix} \mathbf{X}_1 \\ \mathbf{X}_2 \end{bmatrix} \text{ is } N_{q+p}\left[\begin{pmatrix} \boldsymbol{\mu}_1 \\ \boldsymbol{\mu}_2 \end{pmatrix}, \begin{pmatrix} \boldsymbol{\Sigma}_{11} & \boldsymbol{\Sigma}_{12} \\ \boldsymbol{\Sigma}_{21} & \boldsymbol{\Sigma}_{22} \end{pmatrix}\right] \tag{5.3}$$

Following Rencher (2002) we can now establish the following properties:

4. Independence
   a. The subvectors $\mathbf{X}_1$ and $\mathbf{X}_2$ are independent if $\boldsymbol{\Sigma}_{12} = \boldsymbol{\Sigma}_{21} = \mathbf{0}$.
   b. Any two individual variables X_i and X_j are independent if $\sigma_{ij} = 0$.
5. Conditional distribution

   If $\mathbf{X}_1$ and $\mathbf{X}_2$ are not independent then $\boldsymbol{\Sigma}_{12} \neq \mathbf{0}$ and the conditional distribution of $\mathbf{X}_1$ given $\mathbf{X}_2 = \mathbf{x}_2$, denoted by $f(\mathbf{x}_1|\mathbf{x}_2)$, is multivariate normal with

$$E(\mathbf{x}_1|\mathbf{x}_2) = \boldsymbol{\mu}_1 - \boldsymbol{\Sigma}_{12}\boldsymbol{\Sigma}_{22}^{-1}(\mathbf{x}_2 - \boldsymbol{\mu}_2) \tag{5.4}$$

$$\text{cov}(\mathbf{x}_1|\mathbf{x}_2) = \boldsymbol{\Sigma}_{11} - \boldsymbol{\Sigma}_{12}\boldsymbol{\Sigma}_{22}^{-1}\boldsymbol{\Sigma}_{21} \tag{5.5}$$

6. Distribution of sum of two random but independent subvectors

   If $p = q$ and $\mathbf{X}_1$ and $\mathbf{X}_2$ are independent, i.e., $\boldsymbol{\Sigma}_{12} = \boldsymbol{\Sigma}_{21} = \mathbf{0}$, then,

$$\mathbf{X}_1 + \mathbf{X}_2 \text{ is } N_p(\boldsymbol{\mu}_1 + \boldsymbol{\mu}_2, \boldsymbol{\Sigma}_{11} + \boldsymbol{\Sigma}_{22}) \tag{5.6}$$

$$\mathbf{X}_1 - \mathbf{X}_2 \text{ is } N_p(\boldsymbol{\mu}_1 - \boldsymbol{\mu}_2, \boldsymbol{\Sigma}_{11} - \boldsymbol{\Sigma}_{22}) \tag{5.7}$$

7. Standard normal variables

   Given, $\mathbf{X}$ is $p \times 1$ and distributed $N_p(\boldsymbol{\mu}, \boldsymbol{\Sigma})$, assume that there exists a $p \times p$ is a matrix $\mathbf{T}$ that is a factoring of $\boldsymbol{\Sigma}$ such that $\boldsymbol{\Sigma} = \mathbf{T}'\mathbf{T}$, then we may construct a $p \times 1$ random vector $\mathbf{Z}$ such that

$$\mathbf{Z} = (\mathbf{T}')^{-1}(\mathbf{X} - \boldsymbol{\mu}) \tag{5.8}$$

   and $\mathbf{Z}$ is distributed $N_p(\mathbf{0}, \mathbf{I})$. This follows from the fact that $\mathbf{Z}$ is a linear transformation of $\mathbf{X}$ and

$$E(\mathbf{Z}) = E\left[(\mathbf{T}')^{-1}(\mathbf{X} - \boldsymbol{\mu})\right] = (\mathbf{T}')^{-1}\left[E(\mathbf{X}) - \boldsymbol{\mu}\right] = (\mathbf{T}')^{-1}(\boldsymbol{\mu} - \boldsymbol{\mu}) = \mathbf{0} \tag{5.9}$$

   and

$$\begin{aligned}
\text{var}(\mathbf{Z}) &= E[\mathbf{Z}\mathbf{Z}'] = E\left[(\mathbf{T}')^{-1}(\mathbf{X} - \boldsymbol{\mu})(\mathbf{X} - \boldsymbol{\mu})'(\mathbf{T})^{-1}\right] \\
&= E\left[(\mathbf{T}')^{-1}(\mathbf{X}\mathbf{X}' - \mathbf{X}\boldsymbol{\mu}' - \boldsymbol{\mu}\mathbf{X}' + \boldsymbol{\mu}\boldsymbol{\mu}')(\mathbf{T})^{-1}\right] \\
&= (\mathbf{T}')^{-1}E(\mathbf{X}\mathbf{X}' - \boldsymbol{\mu}\boldsymbol{\mu}')(\mathbf{T})^{-1} = (\mathbf{T}')^{-1}\boldsymbol{\Sigma}(\mathbf{T})^{-1} \\
&= (\mathbf{T}')^{-1}(\mathbf{T}'\mathbf{T})(\mathbf{T})^{-1} = \mathbf{I}
\end{aligned} \tag{5.10}$$

   In later chapters we will discuss several methods by which a Gramian square symmetric matrix may be "factored" into a matrix $\mathbf{T}$ times its transpose.

8. The chi-square distribution

   Mathematical statisticians define a chi-square random variable as the sum of p independent squared standard normal variates, i.e., $\chi_p^2 = Z_1^2 + Z_2^2 + \cdots + Z_p^2$, where each Z_i is distributed $N(0, 1)$. We will now show that given $\mathbf{Z}$ defined as in Equation 5.8, $\mathbf{Z}'\mathbf{Z}$ is distributed as chi-square with p degrees of freedom. This follows because $\mathbf{Z}'\mathbf{Z}$ is a scalar representing the sum of the squares of the elements of $\mathbf{Z}$. Since $\mathbf{Z}'\mathbf{Z}$ is a scalar we may write this as

$$\mathbf{Z}'\mathbf{Z} = (\mathbf{X} - \boldsymbol{\mu})'(\mathbf{T})^{-1}(\mathbf{T}')^{-1}(\mathbf{X} - \boldsymbol{\mu}) = (\mathbf{X} - \boldsymbol{\mu})'\Sigma^{-1}(\mathbf{X} - \boldsymbol{\mu}) = \chi_p^2$$

   because $\boldsymbol{\Sigma} = \mathbf{T}'\mathbf{T}$ hence $\boldsymbol{\Sigma}^{-1} = (\mathbf{T}'\mathbf{T})^{-1} = (\mathbf{T})^{-1}(\mathbf{T}')^{-1}$.

*5.4 Maximum-Likelihood Estimation

5.4.1 Notion of Likelihood

R.A. Fisher invented the idea of likelihood to represent the "as if" probability with which an observed value of a random variable will occur given that the population parameters of a known or presumed distribution function for the variable are certain values. Likelihood is a concept to be used in counterfactual reasoning. To illustrate, suppose we know that a random variable X is distributed normally but with unknown mean and variance. Suppose we observe the value $x = 2$. Now, we can discuss counterfactually the probability with which this observed value will occur under different hypothesized values for the population mean and variance. For example, given the formula for the normal density function

$$f(x) = \frac{1}{\sqrt{2\pi\sigma^2}} e^{-(x-\mu)^2/2\sigma^2} \tag{5.11}$$

if we hypothesize $\mu = 1$ and $\sigma^2 = 2$, then the probability density associated with the observed value of $x = 2$ is

$$\ell(x = 2; \mu = 1, \sigma^2 = 2) = \frac{1}{\sqrt{2\pi}\sqrt{2}} e^{-(2-1)^2/2(2)} = .2197.$$

But this is only an as if probability density under the assumption that the distribution is normal with $\mu = 1$ and $\sigma^2 = 2$. So, we designate this as if probability density as a likelihood to distinguish it from the actual probability with which the value of 2 occurs, which may be unknown. The likelihood of an observation given hypothesized values for the parameters is an indication of the support for the hypothesis given the observation. The smaller this value, the less support for the hypothesis.

5.4.2 Sample Likelihood

If a random variable X is distributed with distribution function $f(x; \boldsymbol{\theta})$, where $\boldsymbol{\theta} = [\theta_1, \ldots, \theta_k]$ are parameters of the distribution function, then associated with a sample of independently distributed observed values $x_1, \ldots, x_N$ of X is an expression for the likelihood of the sample, $L(x_1, \ldots, x_N; \boldsymbol{\theta}) = f(x_1; \boldsymbol{\theta})\, f(x_2; \boldsymbol{\theta}) \ldots f(x_N; \boldsymbol{\theta})$, the product of the likelihoods of the individual observations in the sample. This represents the "joint likelihood" with which the sample would occur if the distribution function had a certain form and its parameters were equal to specified values for $\boldsymbol{\theta}$, and the observations were independently distributed of one another.

5.4.3 Maximum-Likelihood Estimates

Maximum-likelihood estimation is a method for estimating parameters popularized by R.A. Fisher. The method seeks either to find formulas for computing the parameter estimates or algorithms that iteratively converge on the optimal values for the parameters. For example, given a sample of values of a random variable X, $x_1, \ldots, x_N$, one obtains estimates of the unknown parameters of the probability distribution $f(x; \boldsymbol{\theta})$ by solving for those values of the unknown parameters that yield a maximum value for the sample likelihood over the range of values that $\boldsymbol{\theta}$ can potentially take. In other words, the maximum-likelihood estimate $\hat{\boldsymbol{\theta}}$ of the parameters in $\boldsymbol{\theta}$ are those values that would make the observed values $x_1,\ldots,x_N$ in the sample occur with the greatest joint likelihood. We designate the value of the maximum likelihood as max $L(\boldsymbol{\theta})$. Methods for finding maximum-likelihood estimates involve methods for finding maxima and minima of functions with respect to certain variable parameters. These usually resort to the use of differential calculus.

Example: Maximum-likelihood estimates of the mean and variance of a univariate normal distribution.

If X is normally distributed $N(x; \mu, \sigma^2)$ its density function is given by

$$f(x;\mu,\sigma^2)=\frac{1}{\sqrt{2\pi\sigma^2}}e^{-(x-\mu)^2/2\sigma^2}$$

If a random sample of N independent observations is obtained from a population with this distribution, then the sample values $x_1,\ldots,x_N$ have a joint likelihood (given values for the parameters of the distribution) of

$$L(x_1,\ldots,x_N;\mu,\sigma^2)=\left(\frac{1}{\sqrt{2\pi\sigma^2}}\right)^N\prod_{k=1}^{N}e^{-(x_k-\mu)^2/2\sigma^2} \tag{5.12}$$

We will seek solutions for the mean μ and variance σ^2 that yield the greatest joint likelihood by treating this as a problem of maximizing the likelihood function as a function of the mean and variance parameters as variables and using partial derivatives to locate where this function has a maximum. This will occur when the partial derivatives of the function with respect to the mean and variance parameters are simultaneously zero. The partial derivatives represent the slopes of lines tangent to the surface of the likelihood function curve at the respective values for the mean and variance in the directions of the respective parameter variables. A slope of zero corresponds to a maximum or a minimum. In this case it will correspond to a maximum.

Because taking the derivative of expressions involving products leads to complexities, mathematical statisticians routinely seek not the maximum of this function but rather of another function that is a monotonic transformation of it and that is easier to differentiate, i.e., the natural logarithm of the joint likelihood function (Figure 5.4). Whatever values for the parameters that maximize the natural logarithm of the joint likelihood function will also maximize the joint likelihood function itself. Recall the following properties of natural logarithms: (a) $\ln(ab)=\ln a+\ln b$, (b) $\ln(a/b)=\ln a-\ln b$, (c) $\ln(x^r)=r\ln x$, and (d) $\ln(e^x)=x$. The natural logarithm of the joint likelihood is then:

$$\ln L(x_1,\ldots,x_N|\mu,\sigma^2)=N\ln\left(\frac{1}{\sqrt{2\pi\sigma^2}}\right)+\sum_{k=1}^{N}\ln e^{-(x_k-\mu)^2/2\sigma^2}$$

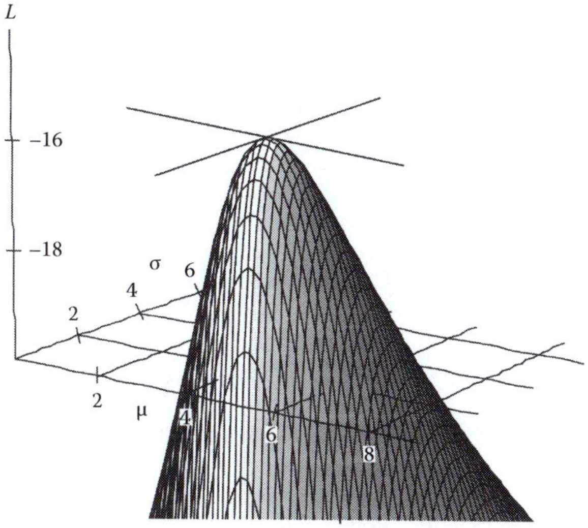

FIGURE 5.4
Joint likelihood function for mean and standard deviation based on a sample of eight observations. Tangent lines in direction of μ and σ at the maximum represent directional derivatives of L with respect to μ and σ at the maximum.

which may be further written as

$$\ln L(x_1,\ldots,x_N|\mu,\sigma^2) = -\frac{N}{2}\ln(2\pi\sigma^2) - \sum_{k=1}^{N}(x_k-\mu)^2/2\sigma^2 \qquad (5.13)$$

Now, a necessary (but not sufficient) condition that a function has a maximum (or minimum) is that the derivative exists at a point where the value of the derivative with respect to each of the parameters, respectively, equals zero, for this will be a local maximum or minimum. Because the partial derivative represents the slope of the function in the direction of one of the parameter variables, a zero value for the derivative indicates a zero slope. A line drawn tangent to the function at the maximum or minimum point in the direction of the parameter will be horizontal, indicating a local extreme value.

The present function has two parameters, so to find the maximum of the log joint likelihood function, we need to find the derivatives of this function with respect to the mean parameter μ and the variance parameter σ^2. (*Note:* Regard the variance not as the square of something else, but as a parameter in its own right. The symbol for the square is an integral part of the symbol for the variance.)

Thus

$$\frac{\partial \ln L}{\partial \mu} = -\frac{\partial \frac{N}{2}\ln(2\pi\sigma^2)}{\partial \mu} - \frac{\partial \sum_{k=1}^{N}(x_k-\mu)^2/2\sigma^2}{\partial \mu}$$

which, because the first term is zero, since no expression in μ, appears in it, we may further simplify as

$$\frac{\partial \ln L}{\partial \mu} = -\frac{1}{2\sigma^2}\sum_{k=1}^{N} 2(x_k - \mu)\frac{\partial(-\mu)}{\partial \mu}$$

or

$$\frac{\partial \ln L}{\partial \mu} = \frac{1}{\sigma^2}\sum_{k=1}^{N}(x_i - \mu)$$

Setting this equal to zero we obtain

$$\frac{1}{\sigma^2}\sum_{k=1}^{N}(x_i - \mu) = 0$$

which we may seek to solve for μ. First, let us multiply both sides by σ^2 to obtain

$$\sum_{k=1}^{N}(x_i - \mu) = 0$$

Next, carry the summation inside the parentheses

$$\sum_{k=1}^{N} x_k - N\mu = 0$$

or

$$\sum_{k=1}^{N} x_k = N\mu$$

Divide both sides by N to solve for μ:

$$\hat{\mu} = \frac{1}{N}\sum_{k=1}^{N} x_k \tag{5.14}$$

So, we see that the formula for the maximum-likelihood estimate of the mean of a normal distribution is the usual familiar formula for the sample mean.

Again take the partial derivative of the log joint likelihood function with respect to the variance parameter:

$$\frac{\partial \ln L}{\partial \sigma^2} = -\frac{\partial \frac{N}{2}\ln(2\pi\sigma^2)}{\partial \sigma^2} - \frac{\partial \sum_{k=1}^{N}(x_k - \mu)^2/2\sigma^2}{\partial \sigma^2}$$

or

$$\frac{\partial \ln L}{\partial \sigma^2} = -\frac{N}{2}\frac{2\pi}{2\pi\sigma^2} - \frac{-2\sum_{k=1}^{N}(x_k - \mu)^2}{4(\sigma^2)^2}$$

or

$$\frac{\partial \ln L}{\partial \sigma^2} = -\frac{N}{2\sigma^2} + \frac{\sum_{k=1}^{N}(x_k - \mu)^2}{2(\sigma^2)^2} \tag{5.15}$$

Now set this equal to zero:

$$-\frac{N}{2\sigma^2} + \frac{\sum_{k=1}^{N}(x_k - \mu)^2}{2(\sigma^2)^2} = 0$$

We now can solve for σ^2. Bring the left-hand side expression on the left over to the right-hand side of the equation:

$$\frac{\sum_{k=1}^{N}(x_k - \mu)^2}{2(\sigma^2)^2} = \frac{N}{2\sigma^2}$$

Now, multiply both sides by $2(\sigma^2)^2$

$$\sum_{k=1}^{N}(x_k - \mu)^2 = \frac{N2(\sigma^2)^2}{2\sigma^2}$$

or, simplifying further, we obtain

$$\sum_{k=1}^{N}(x_k - \mu)^2 = N\sigma^2$$

Divide both sides by N:

$$\sigma^2 = \frac{\sum_{k=1}^{N}(x_k - \mu)^2}{N}$$

But since μ is unknown and must be estimated, we replace it with the value of the maximum-likelihood estimate of the mean, previously derived:

$$\hat{\sigma}^2 = \frac{\sum_{k=1}^{N}(x_k - \bar{x})^2}{N} \quad \text{where } \bar{x} = \frac{1}{N}\sum_{k=1}^{N} x_k \tag{5.16}$$

Note that this is not the unbiased estimator of the population variance, so in this case the maximum-likelihood estimator is not an unbiased estimator. But if we divide instead by $(N-1)$, we get the unbiased estimator, and this is more frequently used. The maximum-likelihood estimator of the variance is the average squared deviation from the sample mean.

Maximum-likelihood estimators always require prespecifying the form of the distribution function whose parameters are to be estimated. In this case, we presume that the distribution was normal. When the population distribution function is unknown, we must fall back on other estimators that do not require making distributional assumptions, such as least-squares and generalized least-squares. But maximum-likelihood estimators almost universally have better properties, are always sufficient estimators, consistent, and even often unbiased as well as "robust," meaning they do not greatly differ from the "true" population parameters, and this often justifies using maximum-likelihood estimation when the distribution of the data is not greatly different from the theorized distribution. Maximum-likelihood estimation also lends itself to generalized likelihood ratio tests, like the chi-square test.

5.4.4 Multivariate Case

Let $\mathbf{Y}$ be an $n \times 1$ random vector distributed $N_n(\boldsymbol{\mu}, \boldsymbol{\Sigma})$. Given a random sample of independent observations of $\mathbf{Y}$, $\mathbf{y}_1,\ldots,\mathbf{y}_N$, the joint likelihood of the observations under the multivariate normal distribution is

$$L(\mathbf{y}_1,\ldots,\mathbf{y}_N;\boldsymbol{\mu},\boldsymbol{\Sigma}) = \prod_{k=1}^{N} f(\mathbf{y}_k;\boldsymbol{\mu},\boldsymbol{\Sigma})$$

$$= \prod_{k=1}^{N} \frac{1}{(\sqrt{2\pi})^n |\boldsymbol{\Sigma}|^{1/2}} e^{-(\mathbf{y}_k-\boldsymbol{\mu})'\boldsymbol{\Sigma}^{-1}(\mathbf{y}_k-\boldsymbol{\mu})/2}$$

We may reduce this further in a manner analogous to the univariate normal case:

$$L(\mathbf{y}_1,\ldots,\mathbf{y}_N;\boldsymbol{\mu},\boldsymbol{\Sigma}) = \left(\frac{1}{(\sqrt{2\pi})^n |\boldsymbol{\Sigma}|^{1/2}}\right)^N \prod_{k=1}^{N} e^{-\frac{1}{2}(\mathbf{y}_k-\boldsymbol{\mu})\boldsymbol{\Sigma}^{-1}(\mathbf{y}_k-\boldsymbol{\mu})}$$

But now, let us transform this monotonically by taking the natural logarithm of the joint likelihood function. Abbreviate the natural logarithm of the joint likelihood as $\ln L$. Then,

$$\ln L = -\frac{1}{2}Nn\ln 2\pi - \frac{1}{2}N\ln|\boldsymbol{\Sigma}| - \frac{1}{2}\sum_{k=1}^{N}(\mathbf{y}_k-\boldsymbol{\mu})'\boldsymbol{\Sigma}^{-1}(\mathbf{y}_k-\boldsymbol{\mu}) \tag{5.17}$$

Let us abbreviate the expression on the left-hand side for the natural logarithm of the joint likelihood to simply ln *L*. Then,

$$\frac{\partial \ln L}{\partial \boldsymbol{\mu}} = -\frac{1}{2}\frac{\partial \sum_{k=1}^{N} (\mathbf{y}_k - \boldsymbol{\mu})'\boldsymbol{\Sigma}^{-1}(\mathbf{y}_k - \boldsymbol{\mu})}{\partial \boldsymbol{\mu}}$$

We may rewrite this as

$$\frac{\partial \ln L}{\partial \boldsymbol{\mu}} = -\frac{1}{2}\sum_{k=1}^{N} \frac{\partial\, \mathrm{tr}\left[(\mathbf{y}_k - \boldsymbol{\mu})'\boldsymbol{\Sigma}^{-1}(\mathbf{y}_k - \boldsymbol{\mu})\right]}{\partial \boldsymbol{\mu}}$$

Now expanding the term in brackets, we get

$$\frac{\partial \ln L}{\partial \boldsymbol{\mu}} = -\frac{1}{2}\sum_{k=1}^{N} \frac{\partial\, \mathrm{tr}\left[\mathbf{y}_k'\boldsymbol{\Sigma}^{-1}\mathbf{y}_k - \boldsymbol{\mu}'\boldsymbol{\Sigma}^{-1}\mathbf{y}_k - \mathbf{y}_k'\boldsymbol{\Sigma}^{-1}\boldsymbol{\mu} + \boldsymbol{\mu}'\boldsymbol{\Sigma}^{-1}\boldsymbol{\mu}\right]}{\partial \boldsymbol{\mu}}$$

$$= -\frac{1}{2}\sum_{k=1}^{N}\left[\frac{\partial\, \mathrm{tr}(\mathbf{y}_k'\boldsymbol{\Sigma}^{-1}\mathbf{y}_k)}{\partial \boldsymbol{\mu}} - \frac{\partial\, \mathrm{tr}(\mathbf{y}_k'\boldsymbol{\Sigma}^{-1}\boldsymbol{\mu}_k)}{\partial \boldsymbol{\mu}} - \frac{\partial\, \mathrm{tr}(\mathbf{y}_k'\boldsymbol{\Sigma}^{-1}\boldsymbol{\mu})}{\partial \boldsymbol{\mu}} + \frac{\partial\, \mathrm{tr}(\boldsymbol{\mu}'_k\boldsymbol{\Sigma}^{-1}\boldsymbol{\mu})}{\partial \boldsymbol{\mu}}\right]$$

The derivative of the first term on the left-hand side inside the brackets is zero, since the expression does not contain $\boldsymbol{\mu}$. The second and third terms are equal because one is the transpose of the other and a scalar. Applying Equations 4.16 and 4.17 to the remaining three terms, we obtain

$$\frac{\partial \ln L}{\partial \boldsymbol{\mu}} = -\frac{1}{2}\sum_{k=1}^{N}\left[-2\boldsymbol{\Sigma}^{-1}\mathbf{y}_k + 2\boldsymbol{\Sigma}^{-1}\boldsymbol{\mu}\right]$$

or, after carrying –1/2 and the summation inside the bracket

$$\frac{\partial \ln L}{\partial \boldsymbol{\mu}} = \sum_{k=1}^{N}\boldsymbol{\Sigma}^{-1}\mathbf{y}_k - N\boldsymbol{\Sigma}^{-1}\boldsymbol{\mu} \tag{5.18}$$

Now we are ready to set the partial derivative of the natural logarithm of the joint likelihood function equal to zero. Doing so, we obtain

$$\boldsymbol{\Sigma}^{-1}\sum_{k=1}^{N}\mathbf{y}_k - N\boldsymbol{\Sigma}^{-1}\boldsymbol{\mu} = \mathbf{0}$$

Or moving the right-hand side term on the left to the right-hand side of the equation, we obtain

$$\boldsymbol{\Sigma}^{-1}\sum_{k=1}^{N}\mathbf{y}_k = N\boldsymbol{\Sigma}^{-1}\boldsymbol{\mu}$$

Now, multiply both sides of the equation by $\mathbf{\Sigma}$, and we then have

$$\sum_{k=1}^{N} \mathbf{y}_k = N\boldsymbol{\mu}$$

or, solving for μ

$$\boldsymbol{\mu} = \frac{1}{N}\sum_{k=1}^{N} \mathbf{y}_k \tag{5.19}$$

which is the maximum-likelihood estimator of the sample mean vector. We will also denote this by $\bar{\mathbf{y}} = (1/N)\sum_{k=1}^{N} \mathbf{y}_k$.

Now, for the purposes of obtaining the derivative of the natural logarithm of the joint likelihood function with respect to $\mathbf{\Sigma}$, we need to write Equation 5.17 as

$$\ln L = -\frac{1}{2}Nn\ln 2\pi - \frac{1}{2}N\ln|\mathbf{\Sigma}| - \frac{1}{2}\sum_{k=1}^{N} \mathrm{tr}(\mathbf{y}_k - \boldsymbol{\mu})'\mathbf{\Sigma}^{-1}(\mathbf{y}_k - \boldsymbol{\mu})$$

or taking advantage of invariance of traces under cyclic permutations

$$\ln L = -\frac{1}{2}Nn\ln 2\pi - \frac{1}{2}N\ln|\mathbf{\Sigma}| - \frac{1}{2}\sum_{k=1}^{N} \mathrm{tr}\left[\mathbf{\Sigma}^{-1}(\mathbf{y}_k - \boldsymbol{\mu})(\mathbf{y}_k - \boldsymbol{\mu})'\right]$$

or

$$\ln L = -\frac{1}{2}Nn\ln 2\pi - \frac{1}{2}N\ln|\mathbf{\Sigma}| - \frac{1}{2}\mathrm{tr}\,\mathbf{\Sigma}^{-1}\sum_{k=1}^{N} (\mathbf{y}_k - \boldsymbol{\mu})(\mathbf{y}_k - \boldsymbol{\mu})'$$

or by letting $\mathbf{W} = \sum_{k=1}^{N}(\mathbf{y}_k - \boldsymbol{\mu})(\mathbf{y}_k - \boldsymbol{\mu})'$

$$\ln L = -\frac{1}{2}Nn\ln 2\pi - \frac{1}{2}N\ln|\mathbf{\Sigma}| - \frac{1}{2}\mathrm{tr}\mathbf{\Sigma}^{-1}\mathbf{W}$$

$$\frac{\partial \ln L}{\partial \mathbf{\Sigma}} = -\frac{1}{2}N\frac{\partial \ln|\mathbf{\Sigma}|}{\partial \mathbf{\Sigma}} - \frac{1}{2}\frac{\partial \mathrm{tr}\mathbf{\Sigma}^{-1}\mathbf{W}}{\partial \mathbf{\Sigma}} \tag{5.20}$$

We will need the expressions for the derivative of a determinant and the derivative of an inverse matrix. Schönemann (1965, p. 19 and 44) gave these as

$$\frac{\partial|\mathbf{Y}|}{\partial \mathbf{Y}} = |\mathbf{Y}|\mathbf{Y}^{-1} \tag{5.21}$$

and

$$\frac{\partial \operatorname{tr} \mathbf{Y}^{-1}\mathbf{A}}{\partial \mathbf{X}} = \frac{-\partial \operatorname{tr}(\mathbf{Y}^{-1}\mathbf{A}\mathbf{Y}^{-1})_c \mathbf{Y}}{\partial \mathbf{X}} \tag{5.22}$$

The subscript "c" indicates that the expression in parentheses is to be treated as a constant for the partial derivative.

Hence,

$$\frac{\partial \ln L}{\partial \boldsymbol{\Sigma}} = -\frac{1}{2}N\frac{1}{|\boldsymbol{\Sigma}|}\frac{\partial |\boldsymbol{\Sigma}|}{\partial \boldsymbol{\Sigma}} - \frac{1}{2}\frac{\partial \operatorname{tr}(\boldsymbol{\Sigma}^{-1}\mathbf{W})}{\partial \boldsymbol{\Sigma}}$$

or, using Equations 5.21 and 5.22,

$$\frac{\partial \ln L}{\partial \boldsymbol{\Sigma}} = -\frac{1}{2}N\frac{1}{|\boldsymbol{\Sigma}|}|\boldsymbol{\Sigma}|\boldsymbol{\Sigma}^{-1} - \frac{1}{2}\frac{-\partial \operatorname{tr}(\boldsymbol{\Sigma}^{-1}\mathbf{W}\boldsymbol{\Sigma}^{-1})_c \boldsymbol{\Sigma}}{\partial \boldsymbol{\Sigma}},$$

which may be simplified further as

$$\frac{\partial \ln L}{\partial \boldsymbol{\Sigma}} = -\frac{1}{2}N\boldsymbol{\Sigma}^{-1} + -\frac{1}{2}(\boldsymbol{\Sigma}^{-1}\mathbf{W}\boldsymbol{\Sigma}^{-1}) \tag{5.23}$$

Setting the partial derivative equal to zero, we obtain

$$-\frac{1}{2}N\boldsymbol{\Sigma}^{-1} + \frac{1}{2}(\boldsymbol{\Sigma}^{-1}\mathbf{W}\boldsymbol{\Sigma}^{-1}) = \mathbf{0}$$

Multiply both sides by 2, and move the left-hand side expression on the left to the right-hand side of the equation.

$$\boldsymbol{\Sigma}^{-1}\mathbf{W}\boldsymbol{\Sigma}^{-1} = N\boldsymbol{\Sigma}^{-1} \tag{5.24}$$

Now, pre- and postmultiply both sides of Equation 5.24 by $\boldsymbol{\Sigma}$

$$\mathbf{W} = N\boldsymbol{\Sigma}$$

Hence

$$\boldsymbol{\Sigma} = \frac{1}{N}\mathbf{W}$$

Or because $\mathbf{W}$ is based on the unknown population mean vector $\boldsymbol{\mu}$, we will replace $\boldsymbol{\mu}$ by the sample estimator $\bar{\mathbf{y}}$, hence the maximum-likelihood estimate of $\boldsymbol{\Sigma}$ is

$$\hat{\boldsymbol{\Sigma}} = (1/N)\sum_{k=1}^{N}(\mathbf{y}_k - \bar{\mathbf{y}})(\mathbf{y}_k - \bar{\mathbf{y}})' = (1/N)\mathbf{W} \tag{5.25}$$

This is not an unbiased estimate, although it is consistent and sufficient. The unbiased estimate is given by

$$\mathbf{S} = \frac{1}{N-1}\mathbf{W} \tag{5.26}$$

5.4.4.1 Distribution of $\bar{\mathbf{y}}$ and S

Given that $\mathbf{Y}$ is an $n \times 1$ random vector of random variables distributed as N_n $(\boldsymbol{\mu}, \boldsymbol{\Sigma})$, then $\bar{\mathbf{y}} = (1/N)\Sigma_{k=1}^{N}\mathbf{y}_k$ is distributed as $N_n(\boldsymbol{\mu}, \boldsymbol{\Sigma}/N)$. However, Rencher (2002) notes that, when $\mathbf{Y}$ does not have a multivariate normal distribution, then as the sample size N increases and becomes very large, $\bar{\mathbf{y}} = (1/N)\Sigma_{k=1}^{N}\mathbf{y}_k$ is approximately distributed as $N_n(\boldsymbol{\mu}, \boldsymbol{\Sigma}/N)$. This is the multivariate generalization of the central limit theorem. Furthermore, Rencher adds that if $\mathbf{Y}$ is not distributed normally, but its population mean vector is $\boldsymbol{\mu}$, then as $N \to \infty$, the distribution of $\sqrt{N}(\bar{\mathbf{y}} - \boldsymbol{\mu})$ approaches $N_n(\mathbf{0}, \boldsymbol{\Sigma})$. This provides justification in large samples for the use of the sample mean vector along with the multivariate normal distribution in making statistical inferences.

On the other hand, given a sample of N observation vectors, $\mathbf{y}_1,\ldots,\mathbf{y}_N$, of the random vector $\mathbf{Y}$ distributed as $N_n(\boldsymbol{\mu}, \boldsymbol{\Sigma})$, then the sample variance–covariance matrix $\mathbf{S} = 1/(N-1)\mathbf{W} = 1/(N-1)\Sigma_{k=1}^{N}(\mathbf{y}_k - \bar{\mathbf{y}})(\mathbf{y}_k - \bar{\mathbf{y}})'$ has a Wishart distribution with $N - 1$ degrees of freedom and $\boldsymbol{\Sigma}$ for its parameters. This distribution is named for its discoverer, Wishart (1928). By way of shorthand, we say that $\mathbf{W} = (N - 1)\mathbf{S}$ is distributed as $W(N - 1, \boldsymbol{\Sigma})$.

The Wishart distribution has a fairly complex form

$$W(\mathbf{W}; N-1, \boldsymbol{\Sigma}) = \frac{|\mathbf{W}|^{(N-n-2)/2} e^{-\mathrm{tr}\,\mathbf{W}\boldsymbol{\Sigma}^{-1}/2}}{2^{n(N-1)/2}\pi^{n(n-1)/4}|\boldsymbol{\Sigma}|^{(n-1)/2}\prod_{i=1}^{n}\Gamma(1/2(N-i))} \tag{5.27*}$$

where $\Gamma(\cdot)$ is the gamma function. According to Johnson and Wichern (1998), the Wishart density does not exist if the sample size N is less than the number of variates n; also $\mathbf{W}$ must be positively definite.

Transformations of $\mathbf{W}$ also have Wishart distributions. Let $\mathbf{Q}$ be an $n \times n$ positive definite transformation matrix. If $\mathbf{W}$ is distributed $W(N-1, \boldsymbol{\Sigma})$ then $\mathbf{QWQ}'$ is distributed $W(N-1, \mathbf{Q\Sigma Q}')$.

Given that $\mathbf{Y}$ is distributed $N_n(\boldsymbol{\mu}, \boldsymbol{\Sigma})$, $\bar{\mathbf{y}}$ and $\mathbf{S}$ are independently distributed.

6

Fundamental Equations of Factor Analysis

6.1 Analysis of a Variable into Components

In the preceding chapters on composite variables and multiple and partial correlations, we considered how a composite variable could be constructed by a process of adding together several component variables. In those cases, a composite variable was not considered given but was said to be dependent for its existence upon an already established set of component variables. In this chapter on the fundamental equations of factor analysis, we will reverse our point of view about what is an independent variable and what is a dependent variable. For example, in factor analysis we may begin with an already given set of variables, such as test scores on a scholastic aptitude test battery, Likert personality rating scales, the prices of several stocks, the measurements of archeological artifacts, or the measurements of the taxonomic characters of several genera of mammals. But factor analysis will lead us to consider such variables as not necessarily fundamental but rather as derivable from some basic set of "unmeasured" or "latent" component variables. Thus factor analysis, at least in the traditional sense, is concerned with the problem of analyzing a variable into components.

So far we have given no hint of what these components might be. As a matter of fact, from a strictly mathematical point of view, there are an infinite number of potential sets of components that might be asserted for a given set of observed variables. Because of this mathematical ambiguity, the history of factor analysis has been filled with controversy over what are the appropriate components into which to analyze a set of variables. These controversies are by no means over today. Their intensity has subsided, however, in recent years, perhaps because factor analysis is no longer the exclusive concern of the scientists of the psychological laboratory who first developed the factor-analytic method but who did not always make the sharp distinction between the formal mathematics of factor analysis and the content of its application. In a later chapter, we will involve ourselves in the issues of these controversies, but for the present there is sufficient consensus among factor analysts regarding the mathematics of the exploratory common factor model.

Before going further, we need to note that exploratory common factor analysis is an exploratory method. This means it is a hypothesis-developing method to be used principally at the outset of a research program designed to suggest hypothetical variables that may be immediate common causes of a set of observed variables. It functions like a series of templates, each representing a common factor model with a specified number of common factors, to be superimposed on a set of data—in this case correlations or covariances between some observed variables and the degree of fit noted. One seeks the template for a given number of common factors that is the best fit to the data.

But the model may not always be appropriate for just any set of data. This is because the causal structure between hypothetical and observed variables may not conform to the common factor model. To understand this, one must remember that there are several ways for two observed variables to be correlated: Suppose *X* and *Y* are two variables that have a nonzero correlation between them. There are many ways this could come about. Consider the diagrams in Figure 6.1. Boxes represent observed variables. Circles are latent, hypothetical variables. Arrows represent causal connections or that the variable pointed to is a linear function of the variable from which the arrow comes. In **a**, *X* is a direct cause of *Y*. In **b**, *Y* is a direct cause of *X*. In **c**, *X* and *Y* have some third variable that is an immediate common cause of both. In **d**, the common cause is mediated through other unmeasured variables. All of these causal structures will produce nonzero correlations between *X* and *Y*.

An assumption made in common factor analysis is that correlations between variables are due to their having immediate common causes and/or correlations between the common causes of variables. In Figure 6.2, we show a graphical diagram illustrative of a common factor model.

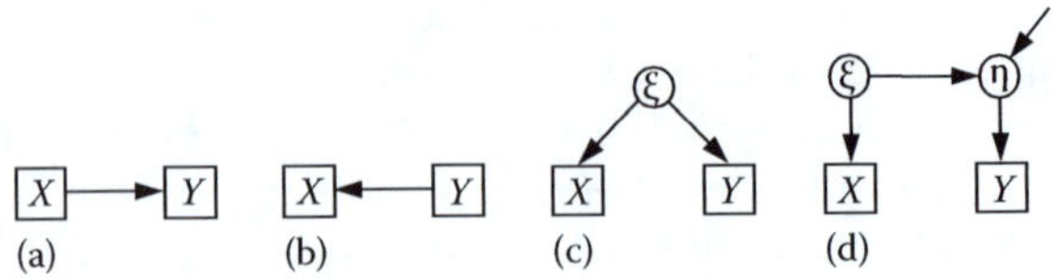

FIGURE 6.1
Correlations between variables may be produced in several ways.

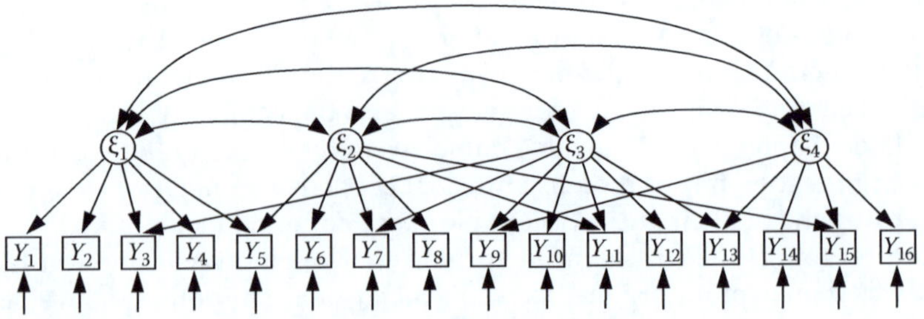

FIGURE 6.2
Causal path diagram of a common factor model for 16 variables.

In Figure 6.2, the boxes denote observed variables (labeled with *Y*'s), while the circles denote latent, common factor variables. Arrows indicate causal paths or functional relations between variables. Two-headed arrows indicate correlations between the latent common factors. The single arrows pointing to an observed variable from below denote unique factor effects, which we do not label here. There are no double-headed arrows between unique factors because they are presumed to be mutually uncorrelated.

Factor analysis does not model causal relations between latent common factors. (That is done in structural equation modeling.) Factor analysis also holds no intrinsic ordering among the variables, e.g., as in time or space. So, observed variables may be arranged in any order and all variables may be thought of as measured at the same time period. Factor analysis requires at least three observed variables be "indicators" of a common factor in order to identify the factor, meaning that the three variables have the common factor as their common cause.

Factor analysis is a formal model about hypothetical component variables that account for the linear relationships that exist between observed variables. To be appropriately applied, the factor-analytic model requires only that the functional relationships between variables be linear and that the variables to be analyzed have some nonzero correlations existing between them. If such correlations do not exist, the observed variables are themselves the only sufficient set of variables to account for their relationships. However, if definite correlations exist among some of the observed variables, then a hypothetical set of component variables can be derived mathematically from the observed variables that may make it possible to explain their interrelationships.

These hypothetical component variables have the following properties: (1) They form a linearly independent set of variables. That is, no hypothetical component variable can be derived from the other hypothetical component variables as a linear combination of them. (2) The hypothetical component variables can be divided into two basic kinds of components: (a) common factors and (b) unique factors. These two kinds of components are distinguished by the patterns of differential weights in the linear equations that derive the observed variables from the hypothetical component variables. A common factor is distinguished by having more than one variable with a nonzero weight associated with the factor. Hence, two or more observed variables depend upon a common factor. A unique factor has a nonzero weight associated with only one observed variable. (3) Common factors are always assumed to be uncorrelated with unique factors. Unique factors are also usually assumed to be mutually uncorrelated. However, as convenience dictates, common factors may or may not be correlated. (4) It is generally assumed that there are fewer common factors than observed variables. The number of unique factors is assumed to equal the number of observed variables.

The above description of the factor-analytic model can be expressed mathematically as follows:

$$Y_1 = \lambda_{11}\xi_1 + \cdots + \lambda_{1r}\xi_r + \psi_1\varepsilon_1$$

$$Y_2 = \lambda_{21}\xi_1 + \cdots + \lambda_{2r}\xi_r + \psi_2\varepsilon_2$$

$$Y_3 = \lambda_{31}\xi_1 + \cdots + \lambda_{3r}\xi_r + \psi_3\varepsilon_3$$

....

$$Y_n = \lambda_{n1}\xi_1 + \cdots + \lambda_{nr}\xi_r + \psi_n\varepsilon_n \tag{6.1}$$

Which are the fundamental equations of factor analysis.

Here $Y_1,\ldots,Y_n$ are observed random variables. $\xi_1,\ldots,\xi_r$ are common factors (ξ is "xi"). $\lambda_{i1},\ldots,\lambda_{ir}$, $i = 1,\ldots,n$, are "factor-pattern loadings," which are weights assigned to a common factor indicating by how much a unit change in the common factor will produce a change in the ith observed variable (λ is "lambda"). $\varepsilon_1,\ldots,\varepsilon_n$ are unique factor variables (ε is "epsilon"). ψ_i, $i=1,\ldots,n$ are unique factor-pattern loadings (ψ is "psi"). We will assume that common factor and unique factor variables have zero means and unit variances, i.e., $E(\xi_j)=0$, $E(\xi_j^2)=1$, $j=1,\ldots,r$ and $E(\varepsilon_i)=0$, $E(\varepsilon_i^2)=1$, $i=1,\ldots,n$. Furthermore, the correlation between any given common factor and any unique factor is zero, i.e., $E(\varepsilon_i\xi_j) = 0$ for all i and j, $i = 1,\ldots,n$ and $j=1,\ldots,r$.

We see in any given equation in Equation 6.1 how any given observed variable is regarded as a composite of one or more common factors and a unique factor.

6.1.1 Components of Variance

Before going into the details of the factor-analysis model, we should give some explanation of the application of the above model. Historically, factor analysis has been concerned with explaining the total "common variance" found between variables in terms of hypothetical common factors. For example, if school grades, aptitude-test scores, and IQ test scores have nonzero correlations between them, this suggests that there exist some hypothetical components common to these variables. In this case, these components could be verbal and symbolic skills as well as rule-inferring ability (Guttman, 1965). These common components would correspond to common factors described above. However, the lack of perfect correlations between these variables would also suggest the presence of unique factors associated with each one of the variables, which give rise to "unique variance." As an example of one contributor to unique variance, consider the presence of random error of measurement in a test score. This error will be uncorrelated with other scores and, hence, will not contribute to the covariance between the scores. Rather, a random error will contribute to "unique variance." Many factor analysts believe that unique variance also contains "true variance," which is uncorrelated with other observed variables "in the set analyzed." This is known as "specific variance." This corresponds to some reliable part of a variable that is found in no other variable in the set. It is possible, however, with the inclusion of additional variables in subsequent studies that some of these additional variables may also have something in common with the original "specific variance" of a variable in the initial set. Then in this case, some of the "specific variance" in a variable may become common factor variance in a subsequent study with these additional variables. To summarize these components consider the following diagram:

$$\text{Total variance} = \overbrace{\text{common variance} + \text{specific variance}}^{\text{True variance}} + \text{error variance}$$

$$\text{specific variance} + \text{error variance} = \text{Unique variance}$$

6.1.2 Variance of a Variable in Terms of Its Factors

Since Equation 6.1 is an equation for a differentially weighted composite variable, we can apply the principles learned in Chapter 3 to the problem of finding an equation for the variance of a variable expressed in terms of its factors. Let Y_i be the jth variable defined in terms of its factors as in Equation 6.1. Then

$$\sigma_j^2 = E(Y_i^2) = E[(\lambda_{i1}\xi_1 + \cdots + \lambda_{ir}\xi_r + \psi_i\varepsilon_i)^2]$$

The expansion of the above expression is tedious but routine. Drawing upon Equation 3.21, we see that this reduces to

$$\sigma_i^2 = \sum_{h=1}^{r}\sum_{k=1}^{r}\lambda_{ih}\lambda_{ik}\rho_{\xi_h\xi_k} + 2\sum_{k=1}^{r}\psi_i\lambda_{ik}\rho_{\varepsilon_i\varepsilon_k} + \psi_i^2$$

The derivation of this expression is partly facilitated by the use of normalized variables. We see that it can be further simplified because the correlation between a common factor and a unique factor is always zero, hence making the middle term above equal to zero. Thus we have the following general formula for the variance of a variable in terms of its factor components:

$$\sigma_i^2 = \sum_{h=1}^{r}\sum_{k=1}^{r}\lambda_{ih}\lambda_{ik}\rho_{\xi_h\xi_k} + \psi_i^2 \quad \text{(formula for variance of a variable in terms of correlated factors)} \qquad (6.2)$$

This formula is a general formula because it permits the common factors to be correlated. However, if we assume that the common factors are uncorrelated, Equation 6.2 reduces to

$$\sigma_i^2 = \sum_{k=1}^{r}\lambda_{ik}^2 + \psi_i^2 \quad \text{(formula for variance of a variable in terms of uncorrelated factors)} \qquad (6.3)$$

Most frequently factor analysts begin a factor analysis by assuming the common factors are uncorrelated. Later, after rotation of the factors to simple structure, the factors may become correlated.

The two terms in the expression on the right-hand side of Equation 6.3 correspond to two important variances in the model of components of variance discussed above. The left-hand side term corresponds to an expression for the "common variance," known generally as the "communality" of the variable. It is designated by most authors as h_i^2 and is defined for the jth variable here as

$$h_i^2 = \sum_{k=1}^{r} \lambda_{ik}^2 \quad \text{(communality formula for uncorrelated factors)} \qquad (6.4)$$

where λ_{ik}^2 is the square of the kth factor loading of the ith variable.

The right-hand side term of Equation 6.3 corresponds to the "unique" variance of the ith variable.

Since we defined the variable Y_i to be expressed in the standard score form, it follows that $\sigma_i^2 = 1$. Consequently, we can form the following equations:

$$h_i^2 = 1 - \psi_i^2 \qquad (6.5)$$

$$\psi_i^2 = 1 - h_i^2 \qquad (6.6)$$

which will be useful later on.

Some authors consider that the unique variance of a variable contains both true specific variance and random error variance unique to the given variable. This can be expressed as

$$\psi_i^2 = s_i^2 + e_i^2 \qquad (6.7)$$

where

s_i^2 is the specific (true) variance of variable i

e_i^2 is random error variance unique to variable i

From these definitions, it follows that the reliability, ρ_{ii}, of variable i is the sum of the two components of true variance

$$\rho_{ii} = h_i^2 + s_i^2 \qquad (6.8)$$

From this we can conclude that the communality of a variable is always less than or equal to the reliability of the variable.

6.1.3 Correlation between Two Variables in Terms of Their Factors

To determine an expression for the correlation between two variables in terms of their factors, we again can draw upon the principles explained in the previous chapter. Define two variables Y_i and Y_m expressed in a normalized form as in Equation 6.1. The correlation, ρ_{im}, between them is equivalent to

$$\rho_{im} = E(Y_i Y_m) \tag{6.9}$$

which, after substitution and expansion, will then simplify to practically the same equation as Equation 3.23, the equation for the correlation between two differentially weighted composite variables

$$\rho_{im} = \sum_{h=1}^{r}\sum_{k=1}^{r} \lambda_{ik}\lambda_{mh}\rho_{\xi_h\xi_k} \quad \text{(correlation between variables in terms of correlated factors)} \tag{6.10}$$

which contains no term corresponding to unique factors because none of the nonzero weights for the unique factors matches with a nonzero weight of another unique factor during the derivation.

As with the case of the variance of a variable, if we may assume that the common factors are uncorrelated, then Equation 6.10 reduces conveniently to the following expression:

$$\rho_{im} = \sum_{h=1}^{r}\sum_{k=1}^{r} \lambda_{ik}\lambda_{mh} \quad \text{(correlation between variables in terms of uncorrelated factors)} \tag{6.11}$$

We now see the advantage of dealing with uncorrelated common factors: The variances as well as the correlations can be accounted for completely by their factor loadings.

6.2 Use of Matrix Notation in Factor Analysis

So far in this chapter, we have developed the method of factor analysis in terms of ordinary linear-algebraic equations using the summational notation. This enabled us to see at an elementary level the relationships between the factors, the factor loadings, and the observed variables. However, the summational notation is very cumbersome to use for further developments of the method of factor analysis. This is because the summational notation does not allow a broader view of what is happening when we must deal with several equations simultaneously. A notation that is far more compact and presents the broader view is matrix notation. Consequently, throughout the remainder of this book, we will rely heavily on matrix notation to express equations.

6.2.1 Fundamental Equation of Factor Analysis

Let $\mathbf{Y}$ be an $n \times 1$ random vector of random variables whose variables are the observed random variables $Y_1, \ldots, Y_n$. Assume that $E(\mathbf{Y}) = \mathbf{0}$ and that

$\mathbf{E(YY')} = \mathbf{R}_{YY}$ is a correlation matrix with unities for the variances in its principal diagonal. Let $\mathbf{X}$ be an $r \times 1$ random vector whose variables are the common factors, $X_1, \ldots, X_r$. Assume that $E(\mathbf{X}) = \mathbf{0}$ and $E(\mathbf{XX'}) = \mathbf{R}_{XX}$ is a correlation matrix, i.e., common factors have means of zero and variances of unity. Next let $\mathbf{E}$ be an $n \times 1$ random vector whose variables are the unique factors $\varepsilon_1, \ldots, \varepsilon_n$, having the property that $E(\mathbf{E}) = \mathbf{0}$ and $E(\mathbf{EE'}) = \mathbf{I}$; that is, the unique factors are normalized to have means of zero and variances of unity and are mutually uncorrelated. Finally, let $\mathbf{\Lambda}$ be an $n \times r$ matrix of common factor-pattern coefficients and $\mathbf{\Psi}$ an $n \times n$ diagonal matrix of unique factor-pattern coefficients whose diagonal elements are $\psi_1, \ldots, \psi_n$. Then

$$\mathbf{Y} = \mathbf{\Lambda X} + \mathbf{\Psi E} \quad \text{(fundamental equation of factor analysis)} \tag{6.12}$$

Equation 6.12 is the "fundamental equation of factor analysis." The equation states that the observed variables in $\mathbf{Y}$ are weighted combinations of the common factors in $\mathbf{X}$ and the unique factors in $\mathbf{E}$.

6.2.2 Fundamental Theorem of Factor Analysis

Since $\mathbf{R}_{YY} = E(\mathbf{YY'})$, we may substitute Equation 6.12 for $\mathbf{Y}$ and write

$$\begin{aligned}
\mathbf{R}_{YY} &= E[(\mathbf{\Lambda X} + \mathbf{\Psi E})(\mathbf{\Lambda X} + \mathbf{\Psi E})'] \\
&= E[(\mathbf{\Lambda X} + \mathbf{\Psi E})(\mathbf{X'\Lambda'} + \mathbf{E'\Psi'})] \\
&= E[\mathbf{\Lambda XX'\Lambda'} + \mathbf{\Lambda XE'\Psi'} + \mathbf{\Psi EX'\Lambda'} + \mathbf{\Psi EE'\Psi'}] \\
&= \mathbf{\Lambda R}_{XX}\mathbf{\Lambda'} + \mathbf{\Lambda R}_{XE}\mathbf{\Psi} + \mathbf{\Psi R}_{EX}\mathbf{\Lambda'} + \mathbf{\Psi}^2
\end{aligned}$$

But $\mathbf{R}_{XE} = \mathbf{R}'_{EX} = \mathbf{0}$ because the common factors are uncorrelated with the unique factors, by assumption. Hence, we may write

$$\mathbf{R}_{YY} = \mathbf{\Lambda R}_{XX}\mathbf{\Lambda'} + \mathbf{\Psi}^2 \quad \text{(fundamental theorem of factor analysis)} \tag{6.13}$$

This is the fundamental theorem of factor analysis in matrix notation. To see what it tells us, suppose we subtract $\mathbf{\Psi}^2$ from both sides of Equation 6.13:

$$\mathbf{R}_{YY} - \mathbf{\Psi}^2 = \mathbf{\Lambda R}_{XX}\mathbf{\Lambda'} \tag{6.14}$$

Because $\mathbf{\Psi}^2$ is a diagonal matrix, subtracting it from $\mathbf{R}_{YY}$ only affects the principal diagonal of $\mathbf{R}_{YY}$, leaving the off-diagonal elements unaffected. Hence, the off-diagonal correlations are due only to the common factors. The principal diagonal of $\mathbf{R}_{YY} - \mathbf{\Psi}^2$ contains the communalities of the variables. These are the variances of the observed variables due to just the common factors. $\mathbf{R}_{YY} - \mathbf{\Psi}^2$ is often called "the reduced correlation matrix," which we may denote simply as $\mathbf{R}_c = \mathbf{R}_{YY} - \mathbf{\Psi}^2$.

On the other hand, if we subtract $\mathbf{\Lambda R}_{XX}\mathbf{\Lambda}'$ from both sides of Equation 6.13, we obtain

$$\mathbf{R}_{YY} - \mathbf{\Lambda R}_{XX}\mathbf{\Lambda}' = \mathbf{\Psi}^2 \tag{6.15}$$

This represents a partial covariance matrix among the dependent variables in **Y**, and when the common factor model holds, it is a diagonal matrix. This means that once all the common-factor parts of the observed variables have been partialled from them, there remain no correlations between the respective residual variables. All correlation is due to common factors.

To see why this is a partial covariance matrix, let us consider regressing the observed variables in **Y** onto the common factors in **X**. The regression weights for predicting variables in **Y** from the common factors in **X**, according to Equation 4.24, is given by $\mathbf{R}_{XX}^{-1}\mathbf{R}_{XY}$. The predicted parts of **Y** obtained from the common factors in **X** is given by

$$\hat{\mathbf{Y}}_X = \mathbf{R}_{YX}\mathbf{R}_{XX}^{-1}\mathbf{X}$$

From this, we may derive the variance–covariance matrix among the predicted parts of **Y** as

$$E(\hat{\mathbf{Y}}_X\hat{\mathbf{Y}}'_X) = E(\mathbf{R}_{YX}\mathbf{R}_{XX}^{-1}\mathbf{XX}'\mathbf{R}_{XX}^{-1}\mathbf{R}_{XY}) = \mathbf{R}_{YX}\mathbf{R}_{XX}^{-1}\mathbf{R}_{XX}\mathbf{R}_{XX}^{-1}\mathbf{R}_{XY}$$
$$= \mathbf{\Lambda R}_{XX}\mathbf{\Lambda}' \tag{6.16}$$

Here we identify $\mathbf{\Lambda} = \mathbf{R}_{YX}\mathbf{R}_{XX}^{-1}$ with the transpose of the regression weight matrix for predicting the observed variables from the common factors. Each of the coefficients in a row of $\mathbf{\Lambda}$ represents how much a unit change in a common factor produces a change in the variable corresponding to that row. And that is what a regression coefficient represents: the degree to which a unit change in the independent variable produces a change in the dependent variable, independent of all other independent variables.

6.2.3 Factor-Pattern and Factor-Structure Matrices

Occasionally, there is a need to distinguish between two kinds of matrices that reveal relationships between the observed variables and the factors. First, we are already familiar with one of these matrices in the matrix $\mathbf{\Lambda} = \mathbf{R}_{YX}\mathbf{R}_{XX}^{-1}$ of the fundamental equation of factor analysis, which is the "factor-pattern" matrix. The coefficients of a factor-pattern matrix are weights to be assigned to the common factors in deriving the observed variables as linear combinations of the common and unique factors. The factor-pattern coefficients are equivalent to regression weights for predicting the observed variables from the common factors. The second kind of matrix is known as a "factor-structure" matrix. The coefficients of this matrix are the covariances (or correlations when all variables are in standard score form) between the observed variables and the factors

$$\mathbf{R}_{YX} = E(\mathbf{YX}') = E[(\mathbf{\Lambda X} + \mathbf{\Psi E})\mathbf{X}'] = E(\mathbf{\Lambda XX}') + E(\mathbf{\Psi EX}')$$
$$= \mathbf{\Lambda R}_{XX} + \mathbf{\Psi R}_{EX} = \mathbf{\Lambda R}_{XX} \tag{6.17}$$

because $\mathbf{R}_{EX} = \mathbf{0}$. We can see how the factor-structure matrix is itself the product of the factor-pattern matrix times the matrix of correlations among the factors. But now, if we postmultiply both sides of Equation 6.17 by $\mathbf{R}_{XX}^{-1}$, we obtain

$$\mathbf{\Lambda} = \mathbf{R}_{YX}\mathbf{R}_{XX}^{-1} \tag{6.18}$$

which provides confirmation of our previous identification of $\mathbf{\Lambda}$ with the transpose of the regression weight matrix for predicting the observed variables from the common factors. That in Equation 6.17 the factor-structure matrix is itself the product of the factor-pattern matrix times the matrix of correlations among the common factors suggests that the factor-structure matrix provides redundant information beyond that given by these other two matrices. For this reason, some factor analysts only rely upon the factor-pattern coefficients and the correlations among the factors for interpretation of the factors.

Finally, the variance–covariance matrix among the common factor parts of the observed variables may be rewritten as

$$\mathbf{R}_c = \mathbf{\Lambda R}_{XX}\mathbf{\Lambda}' = \mathbf{R}_{YX}\mathbf{R}_{XX}^{-1}\mathbf{R}_{XX}\mathbf{\Lambda}' = \mathbf{R}_{YX}\mathbf{\Lambda}' \tag{6.19}$$

the product of the factor-structure matrix times the transpose of the factor-pattern matrix.

One final note on notation: Karl Jöreskog has popularized the use of the capital Greek letter phi $\mathbf{\Phi}$ for the matrix of correlations among the common factors. This does not immediately show that this is a correlation matrix, which was why I used $\mathbf{R}_{XX}$ above. But I will bow to convention later in this book to use $\mathbf{\Phi}$ or $\mathbf{\Phi}_{XX}$ to denote the matrix of correlations among the common factors, and will represent the equation of the fundamental theorem as

$$\mathbf{R}_{YY} = \mathbf{\Lambda\Phi}_{XX}\mathbf{\Lambda}' + \mathbf{\Psi}^2$$

or even as

$$\mathbf{R}_{YY} = \mathbf{\Lambda\Phi\Lambda}' + \mathbf{\Psi}^2$$

7

Methods of Factor Extraction

7.1 Rationale for Finding Factors and Factor Loadings

Most factor analysts are interested in finding the common factors for a reduced correlation matrix $\mathbf{R}_c = (\mathbf{R}_{YY} - \mathbf{\Psi}^2)$. Although there are several different ways to accomplish this, most ways assume, as a matter of convenience, that all factors to be found will initially be uncorrelated with one another. This makes it possible to equate the coefficients in a factor pattern matrix with correlations between the factors and the observed variables. The reason for this is given in Equation 6.17, which gives the factor structure matrix as $\mathbf{R}_{YX} = \mathbf{\Lambda R}_{XX}$. If the factors are uncorrelated, then $\mathbf{R}_{XX} = \mathbf{I}$, and $\mathbf{R}_{YX} = \mathbf{\Lambda}$. The task facing the factor analyst then is to define the factors and to determine the correlations of the variables with them.

One way to do this is to find each factor one at a time. This is not used much any more. Most computer programs today try to find all the common factors at once. But it helps to understand what a factor is if we first look at the one-at-a-time method. When we find a factor, we find the correlations of the variables with it, and then we partial its contribution to the observed scores for the observed-score matrix. Each succeeding factor is derived from the residual observed variables remaining after the previous factors have been partialled from the original observed variables. This procedure guarantees that the factors will be uncorrelated with one another.

Before going into the details of this procedure, let us review some concepts from multiple regression that we will need. Let $\mathbf{Y}$ be a $q \times 1$ vector of q normalized criterion random variables. Let $\mathbf{X}$ be a $p \times 1$ random vector of p normalized predictor random variables. Assume $E(\mathbf{Y}) = \mathbf{0}$ and $E(\mathbf{X}) = \mathbf{0}$. Then the $p \times q$ matrix of covariances between $\mathbf{X}$ and $\mathbf{Y}$ is

$$\mathbf{\Sigma}_{XY} = E(\mathbf{XY}')$$

If we now wish to predict variables in $\mathbf{Y}$ from variables in $\mathbf{X}$, we need to find a multiple-regression weight matrix $\mathbf{W}$ such that $\hat{\mathbf{Y}} = \mathbf{W}'\mathbf{X}$, where $\hat{\mathbf{Y}}$ represents the least squares estimates of the variables in $\mathbf{Y}$ based on the variables in $\mathbf{X}$.

$\mathbf{W}$ is given as

$$\mathbf{W} = \mathbf{\Sigma}_{XX}^{-1}\mathbf{\Sigma}_{XY}$$

where $\mathbf{\Sigma}_{XX}$ is the matrix of covariances among the **X** variables. If we partial the contribution of **X** to **Y** from **Y**, we find a matrix $\mathbf{Y}_1$, such that

$$\mathbf{Y}_1 = \mathbf{Y} - \hat{\mathbf{Y}} = \mathbf{Y} - \mathbf{W}'\mathbf{X} = \mathbf{Y} - \mathbf{\Sigma}_{YX}\mathbf{\Sigma}_{XX}^{-1}\mathbf{X} \tag{7.1}$$

Keep in mind that to partial the effect of one set of variables from another set of variables we need to find the matrix of covariances between the two sets of variables and the inverse of the matrix of covariances among the variables to be partialled out.

7.1.1 General Computing Algorithm for Finding Factors

We will now describe a procedure to derive successive factors from the observed variables. The method described here is due to Guttman (1944) but was derived independently by the author in Mulaik (1972), not knowing of Guttman's paper at the time. The procedure builds on the ideas of composite variables, multiple regression, and partial correlation.

The method forms the framework of several well-known factor-analytic methods such as the diagonal method, the centroid method, and the iterative method of principal axes of Hotelling involving eigenvalues and eigenvectors.

In some respects, one pulls oneself up by one's bootstraps in applying this method. It reminds the author of the joke about the thermodynamic engineer, the physicist, and the statistician who were marooned on a desert island. As food became scarce, one day they observed a wave wash a can of beans ashore. They rushed and took it from the beach ready to eat it. But how were they to open the can to get the beans? The thermodynamic engineer pondered a moment and then he said, "I've got it! Let's build a fire and heat a pile of stones quite hot, cover these with palm leaves, place the can of leaves on top of them and cover it also with palm leaves. The heat of the stones will cause the water in the can to boil and the can will burst, freeing the beans." The physicist did not like that idea because the beans might be spread around from the explosion of the can. "I have a better idea," he said. See that tall palm tree down the beach. It leans over a pile of sharp rocks. I figure it is high enough that if we climb the tree and drop the can onto the rocks, the can will be broken open and we can get the beans." The statistician looked incredulously at the other two. "I don't know why you are going to such elaborate and dangerous steps when the answer is quite simple: first, we assume we have a can opener…."

Well, the can opener here will be the assumption that we know the scores and unique-factor pattern matrix of the unique factors. This will allow us to use multiple regression to illustrate the method of extracting one factor at a time. At a later point, we will show that we do not need the scores on the unique factors at all. All we will need are some reasonable estimates of the unique-factor variances.

Let us first partial the unique factors from the $n \times 1$ random vector **Y** of n observed variables. We will assume that we have the $n \times 1$ random vector **E** of unique-factor variables. Then we need to know the matrix of covariances among the unique factors and the covariance matrix of the observed variables with the unique factors. We already know the matrix of covariances among the unique factors. It is an identity matrix **I** because we assume that the unique factors have zero means and standard deviations of 1.00 and are mutually uncorrelated. The matrix of covariances between the **Y** variables and the unique factors in **E** is given by

$$\begin{aligned}\mathbf{R}_{YE} &= E(\mathbf{YE}') = E[(\mathbf{\Lambda X} + \mathbf{\Psi E})\mathbf{E}'] = \mathbf{\Lambda}E(\mathbf{XE}') + \mathbf{\Psi}E(\mathbf{EE}') \\ &= \mathbf{\Lambda R}_{XE} + \mathbf{\Psi I} = \mathbf{\Psi}\end{aligned}$$

because $\mathbf{R}_{XE} = \mathbf{0}$. The inverse of the identity matrix is still the identity matrix. So, the regression weights are $\mathbf{\Sigma}_{EE}^{-1}\mathbf{\Sigma}_{EY} = \mathbf{I\Psi} = \mathbf{\Psi}$. So, define

$$\mathbf{Y}_c = \mathbf{Y} - \mathbf{\Psi E} \tag{7.2}$$

from which we may obtain

$$\begin{aligned}\mathrm{var}(\mathbf{Y}_c) = \mathbf{R}_c = E(\mathbf{Y}_c\mathbf{Y}_c') &= E[(\mathbf{Y} - \mathbf{\Psi E})(\mathbf{Y} - \mathbf{\Psi E})'] \\ &= E[(\mathbf{Y} - \mathbf{\Psi E})(\mathbf{Y}' - \mathbf{E}'\mathbf{\Psi})] \\ &= E(\mathbf{YY}') - E(\mathbf{YE}'\mathbf{\Psi}) - E(\mathbf{\Psi EY}') + E(\mathbf{\Psi EE}'\mathbf{\Psi}) \\ &= \mathbf{R}_{YY} - \mathbf{R}_{YE}\mathbf{\Psi} - \mathbf{\Psi R}_{EY} + \mathbf{\Psi}\cdot\mathbf{\Psi} \\ &= \mathbf{R}_{YY} - \mathbf{\Psi}\cdot\mathbf{\Psi} - \mathbf{\Psi}\cdot\mathbf{\Psi} + \mathbf{\Psi}\cdot\mathbf{\Psi}\end{aligned}$$

or

$$\mathbf{R}_c = \mathbf{R}_{YY} - \mathbf{\Psi}^2 \tag{7.3}$$

$\mathbf{Y}_c$ contains variables dependent on only the common factors.

We will now use the variables in $\mathbf{Y}_c$ to derive the first common factor, ξ_1.

We must keep in mind that there is no unique-factor solution for which we are searching. We thus may look for a factor with different ideas in mind, such as making a factor account for as much variance as possible, or making a factor coincide with a residual variable, or making a factor coincide with the centroid of the residual variables. In any case, each of these ideas will lead to a different $n \times 1$ weight vector $\mathbf{b}_1$ such that the first common factor ξ_1 will be found as a linear composite of the variables in $\mathbf{Y}_c$.

The reason we can do this is because we assume that $\mathbf{Y}_c$ is a linear function of the r common factors. That means the variables in $\mathbf{Y}_c$ and the r common factors occupy the same vector space of r dimensions. So, there should be at most r linear combinations of any n vectors in this space that are linearly independent. So, we should be able to derive another set of r basis vectors as linear combinations of the n variables in $\mathbf{Y}_c$. These will be our choices for the common factors—although many other choices are possible. By forcing these common factors to be mutually orthogonal, we will guarantee their linear independence.

We thus define our first factor variable to be

$$\xi_1 = \mathbf{b}_1'\mathbf{Y}_c \tag{7.4}$$

But we must also require that

$$E(\xi_1^2) = 1 \tag{7.5}$$

so that ξ_1 will be normalized to have unit variance. To do this, let us first suppose that $\boldsymbol{\beta}_1$ is an $n \times 1$ weight matrix which combines the variables in $\mathbf{Y}_c$ to produce a composite corresponding to the kind of factor we have in mind, with the exception that the standard deviation of the composite does not equal 1.

For example, suppose we find a composite

$$\xi_1^* = \boldsymbol{\beta}_1' \mathbf{Y}_c \tag{7.6}$$

where $\boldsymbol{\beta}_1$ is our concept of the weight vector used to find the appropriate factor composite. If we can find the standard deviation of σ_1 of ξ_1^*, we can then multiply both sides of the preceding equation by $1/\sigma_1$ to obtain

$$\xi_1 = \xi_1^*(1/\sigma_1) = \boldsymbol{\beta}_1'\mathbf{Y}_c(1/\sigma_1)$$

Thus we see that $\mathbf{b}_1 = \boldsymbol{\beta}_1(1/\sigma_1)$. But how can we find σ_1? We need only refer back to Chapter 3 on composite variables to find the answer. The variance of a weighted composite is given by Equation 3.36. Because the variance–covariance matrix of $\mathbf{Y}_c$ is given by $\text{var}(\mathbf{Y}_c) = \mathbf{R}_c = (\mathbf{R}_{YY} - \boldsymbol{\Psi}^2)$

$$\sigma_1^2 = \boldsymbol{\beta}_1' (\mathbf{R}_{YY} - \boldsymbol{\Psi}^2)\boldsymbol{\beta}_1 \tag{7.7}$$

Once we find σ_1^2, we can then define $\mathbf{b}_1$ as

$$\mathbf{b}_1 = \boldsymbol{\beta}_1(1/\sigma_1) \tag{7.8}$$

Having found ξ_1, we now turn our attention to finding the first $n \times 1$ column $\boldsymbol{\lambda}_1$ of the factor pattern matrix $\boldsymbol{\Lambda}$. If we assume that we are working with uncorrelated factors, then the correlations of the variables with ξ_1 will give us $\boldsymbol{\lambda}_1$. In other words,

$$\boldsymbol{\lambda}_1 = E(\mathbf{Y}\xi_1)$$

which, upon substitution from Equation 7.4, becomes

$$\boldsymbol{\lambda}_1 = E(\mathbf{Y}\mathbf{Y}_c'\mathbf{b}_1)$$

which in turn can be modified by substitution form Equation 7.2 to give

$$\boldsymbol{\lambda}_1 = E[\mathbf{Y}(\mathbf{Y}-\boldsymbol{\Psi}\mathbf{E})'\mathbf{b}_1]$$

We can now expand this result by carrying **Y** into the parentheses and taking the expected value to obtain

$$\boldsymbol{\lambda}_1 = E[(\mathbf{Y}\mathbf{Y}' - \mathbf{Y}\mathbf{E}'\boldsymbol{\Psi})'\mathbf{b}_1] = (\mathbf{R}_{YY} - \boldsymbol{\Psi}^2)\mathbf{b}_1 \tag{7.9}$$

The first column, $\boldsymbol{\lambda}_1$, of the factor pattern matrix, $\boldsymbol{\Lambda}$, is equal to an appropriate linear transformation applied to the correlation matrix with the unique variances partialled from it. This suggests that the transformation does not need to be applied directly to any observed scores, if we can get a good estimate for $\boldsymbol{\Psi}^2$. We will eventually learn that this is the case for all of the factors.

Our next problem is to find the second factor ξ_2. But so that ξ_2 will be uncorrelated with ξ_1, let us partial ξ_1 from the vector $\mathbf{Y}_c$. To do this we can rely upon the principle from multiple regression which states that the predicted value of $\mathbf{Y}_c$ from ξ_1 can be found by using a weight matrix which is obtained by multiplying the covariance matrix between $\mathbf{Y}_c$ and ξ_1 by the inverse of the matrix of covariances among the variables in ξ_1. We note that ξ_1 is only a single variable occupying a 1 × 1 variance–covariance matrix, which is thus a scalar. By construction the variance of ξ_1 is 1, hence its inverse is equal to 1 also. So, we can set aside any concern as to what the inverse of the predictor variable variance is and concentrate on the matrix (column vector) of covariances between $\mathbf{Y}_c$ and ξ_1. This covariance matrix is given by

$$\begin{aligned} E(\mathbf{Y}_c\xi_1) &= E[(\mathbf{Y} - \boldsymbol{\Psi}\mathbf{E})\xi_1] \\ &= E(\mathbf{Y}\xi_1 - \boldsymbol{\Psi}\mathbf{E}\xi_1) \end{aligned}$$

But because the correlations between common factors and unique factors are zero

$$E(\mathbf{Y}_c\xi_1) = E(\mathbf{Y}\xi_1) = \boldsymbol{\lambda}_1 \tag{7.10}$$

because of Equation 7.8. Hence, the covariances between $\mathbf{Y}_c$ and ξ_1 are found in $\boldsymbol{\lambda}_1$. Thus, we have the ingredients for the weight matrix for partialling out ξ_1 from $\mathbf{Y}_c$. The weight matrix we seek is simply $\boldsymbol{\lambda}_1$. Therefore, we define the first residual variable vector, $\mathbf{Y}_1$, as

$$\mathbf{Y}_1 = \mathbf{Y}_c - \boldsymbol{\lambda}_1\xi_1 \tag{7.11}$$

The matrix of covariances among the first residuals will be

$$E(\mathbf{Y}_1\mathbf{Y}_1') = E[(\mathbf{Y}_c - \boldsymbol{\lambda}_1\xi_1)(\mathbf{Y}_c' - \xi_1\boldsymbol{\lambda}_1')]$$

$$= E(\mathbf{Y}_c\mathbf{Y}_c') - E(\mathbf{Y}_c\xi_1\boldsymbol{\lambda}_1') - E(\boldsymbol{\lambda}_1\xi_1\mathbf{Y}_c') + E(\boldsymbol{\lambda}_1\xi_1\xi_1\boldsymbol{\lambda}')$$

$$= (\mathbf{R}_{YY} - \boldsymbol{\Psi}^2 - \boldsymbol{\lambda}_1\boldsymbol{\lambda}_1' - \boldsymbol{\lambda}_1\boldsymbol{\lambda}_1' + \boldsymbol{\lambda}_1\boldsymbol{\lambda}_1') = (\mathbf{R}_{YY} - \boldsymbol{\Psi}^2 - \boldsymbol{\lambda}_1\boldsymbol{\lambda}_1') \quad (7.12)$$

We are now ready to find the second factor. As before, this will involve finding some weight matrix that transforms the residual variables into this next factor. That is, if $\mathbf{b}_2$ is the second $n \times 1$ weight vector, then

$$\xi_2 = \mathbf{b}_2'\mathbf{Y}_1$$

The column vector $\mathbf{b}_2$ can be thought of as some other vector $\boldsymbol{\beta}_2$ multiplied by a scalar $1/\sigma_2$, that is, $\mathbf{b}_2 = \boldsymbol{\beta}_2(1/\sigma_2)$, where $\sigma_2{}^2 = \boldsymbol{\beta}_2(\mathbf{R}_{YY} - \boldsymbol{\Psi}^2 - \boldsymbol{\lambda}_1\boldsymbol{\lambda}_1')\boldsymbol{\beta}_2$. We can also show that ξ_2 is uncorrelated with ξ_1:

$$\rho_{\xi 1\xi 2} = E(\xi_1\xi_2) = E(\xi_1\mathbf{Y}_1'\mathbf{b}_2)$$

Upon substitution from Equation 7.10, we get

$$\rho_{\xi_1\xi_2} = E[\xi_1(\mathbf{Y}_c' - \xi_1\boldsymbol{\lambda}_1')\mathbf{b}_2] = [E(\xi_1\mathbf{Y}_c') - E(\xi_1\xi_1)\boldsymbol{\lambda}_1']\mathbf{b}_2$$

and, finally, substituting from Equation 7.10, we obtain

$$\rho_{\xi_1\xi_2} = (\boldsymbol{\lambda}_1' - \boldsymbol{\lambda}_1')\mathbf{b}_2 = 0 \quad (7.13)$$

This shows that the factors are uncorrelated.

Let us now find the second column $\boldsymbol{\lambda}_2$ of the factor pattern matrix $\boldsymbol{\Lambda}$ which for uncorrelated factors gives the correlations between $\mathbf{Y}$ and ξ_2:

$$\boldsymbol{\lambda}_2 = E(\mathbf{Y}\xi_2)$$

$$= E[\mathbf{Y}(\mathbf{Y} - \boldsymbol{\Psi}\mathbf{E} - \boldsymbol{\lambda}_1\xi_1)'\mathbf{b}_2]$$

$$= (\mathbf{R}_{YY} - \boldsymbol{\Psi}^2 - \boldsymbol{\lambda}_1\boldsymbol{\lambda}_1')\mathbf{b}_2 \quad (7.14)$$

In general, as we proceed, extracting one factor at a time, we find that the pth column of the factor-loading matrix $\boldsymbol{\lambda}_p$ for the pth common factor ξ_p extracted from the scores in $\mathbf{Y}_c$ is

$$\boldsymbol{\lambda}_p = (\mathbf{R}_{YY} - \boldsymbol{\Psi}^2 - \lambda_1\lambda'_1 - \cdots - \lambda_{p-1}\lambda'_{p-1})\mathbf{b}_p \quad (7.15)$$

where $\mathbf{b}_p$ is the weight vector required to find the pth factor. The weight vector $\mathbf{b}_p$ is itself defined in terms of another vector $\boldsymbol{\beta}_p$ as

$$\mathbf{b}_p = \boldsymbol{\beta}_p(1/\sigma_p) \quad (7.16)$$

where

$$\sigma_p^2 = \boldsymbol{\beta}_p' (\mathbf{R}_{YY} - \boldsymbol{\Psi}^2 - \boldsymbol{\lambda}_1 \boldsymbol{\lambda}_1 - \cdots - \boldsymbol{\lambda}_{p-1} \boldsymbol{\lambda}_{p-1}) \boldsymbol{\beta}_p \tag{7.17}$$

Early in the history of factor analysis, it was recognized that one did not have to work with the observed variables in **Y** directly but only with the correlation matrix. As each factor was extracted from the residual correlation matrix by partial-correlation methods, operations on the residual correlation could produce the next factor-loading column, which is what is represented by Equation 7.15. This in turn was partialled from the residual correlation matrix, and so on. But the effect of such procedures was equivalent to working with derivations of the factors from the observed variables.

7.2 Diagonal Method of Factoring

The general computing algorithm for finding factors described in the previous section can be adapted to one well-known method for finding factors, known as the square-root or the diagonal method of factoring. The method is attributed to A.L. Cholesky around 1915, based upon publication of his work, after his death in 1924, by Commandant Benoit of the French Navy. Dwyer (1944) is credited with having introduced this algorithm into problems of regression and correlation. The method is not routinely used today, but it does form the part of some multivariate statistics routines, and can be used when variables are known to be strong indicators of a given common factor. Its main value at this point is pedagogical, to show us how to apply the general computing algorithm.

The method can be applied to a correlation matrix, $\mathbf{R}_{YY}$, whose diagonal elements have not been altered by partialling out unique-factor variances. In this case, the result is known as a Cholesky factorization of $\mathbf{R}_{YY}$ into a lower triangular matrix $\mathbf{L}$ where $\mathbf{R}_{YY} = \mathbf{LL}'$. The rapidity with which the matrix $\mathbf{L}$ is computed recommends itself whenever a factoring of a square symmetric matrix of full rank is required.

We shall apply the method to the reduced correlation matrix, $\mathbf{R}_c$. In this method, a factor is defined as a variable collinear with the next variable in order in the matrix. Hence, variables are ordered so that from the first, each successive variable represents a more or less pure measure of an expected common factor. We begin with the weight vector $\boldsymbol{\beta}_1 = [1,0,0,\ldots,0]'$. Applied to the score vector, $\mathbf{Y}_c$, $\boldsymbol{\beta}_1'\mathbf{Y}_c$ "selects out" the first variable in $\mathbf{Y}_c$ as the first factor ξ_1. But since we are bypassing working with raw variables and instead working on the residual correlation matrix, $\mathbf{R}_c$, to find the factor-loading vector, λ_1, for ξ_1, we find that the variance of the composite $\boldsymbol{\beta}_1'\mathbf{Y}_c$ is equal to $\sigma_1^2 = \boldsymbol{\beta}_1'\mathbf{R}_c\boldsymbol{\beta}_1$ and

$\mathbf{b}_1 = \boldsymbol{\beta}_1'(1/\sigma_1)$. In this case, $\mathbf{b}_1' = [1/\sqrt{\rho_{11}}, 0, 0, \ldots, 0]$. Then, the first column vector $\boldsymbol{\lambda}_1$ of the factor pattern matrix $\boldsymbol{\Lambda}$ is given by

$$\boldsymbol{\lambda}_1 = \mathbf{R}_c \mathbf{b}_1$$

Then we partial the first factor ξ_1 from $\mathbf{R}_c$ to find the first residual matrix

$$\mathbf{R}_1 = (\mathbf{R}_c - \boldsymbol{\lambda}_1 \boldsymbol{\lambda}_1')$$

The effect of partialling leaves zeros for each of the elements of the first row and column of $\mathbf{R}_1$. The remaining elements are somewhat reduced. We are then ready to seek the second column, $\boldsymbol{\lambda}_2$, of the factor pattern matrix, $\boldsymbol{\Lambda}$, using the weight vector

$$\boldsymbol{\beta}_2' = [0,1,0,0,\ldots,0]$$

$$\sigma_2^2 = \boldsymbol{\beta}_2' \mathbf{R}_1 \boldsymbol{\beta}_2$$

$$\mathbf{b}_2 = \boldsymbol{\beta}_2(1/\sigma_2)$$

Then $\boldsymbol{\lambda}_2$ is given as

$$\boldsymbol{\lambda}_2 = \mathbf{R}_1 \mathbf{b}_1$$

It should be noted that the first element in the column vector, $\boldsymbol{\lambda}_2$, is zero.

The next residual matrix is $\mathbf{R}_2 = \mathbf{R}_1 - \boldsymbol{\lambda}_2 \boldsymbol{\lambda}_2'$. The first two rows and columns of $\mathbf{R}_2$ now contain only zeros, while the remaining elements are again reduced further. The third weight vector is $\boldsymbol{\beta}_3' = [0,0,1,0,\ldots,0]$. And the method proceeds from here, with a weight matrix, $\boldsymbol{\beta}_p$, being a column vector of 0s with a 1 in the pth row only. The sweep-out of successive columns and rows of the residual matrix is continued until n factors have been extracted (not likely) or a diagonal element becomes negative or 0. In the latter case, it is impossible to compute a square root or its reciprocal. The number of common factors is then found as the number of positive variances, σ_i^2, found for the composites formed with the weight vectors, $\boldsymbol{\beta}_i$.

To summarize the procedure, for the pth iteration

$$\mathbf{b}_p = \left[0,0,\ldots,\frac{1}{\sqrt{_{p-1}\rho_{pp}}},\ldots,0\right] \tag{7.18}$$

$$\boldsymbol{\lambda}_p = \mathbf{R}_{p-1} \mathbf{b}_p \tag{7.19}$$

$$\mathbf{R}_p = \mathbf{R}_{p-1} - \boldsymbol{\lambda}_p \boldsymbol{\lambda}'_p \tag{7.20}$$

where $[1,1,1,1,\ldots,1]'$ is the pth diagonal element of the residual matrix $\mathbf{R}_{p-1}$.

7.3 Centroid Method of Factoring

Until the advent of computers made the principal-axes method of factoring the preferred method, the centroid method was the most frequently used for extracting the factors. This method was developed to approximate the principal-axes method while saving considerable labor in computation, which in those days was done by hand using mechanical calculators. Harman (1960) indicates that Burt was the first to apply this method, but Thurstone (1947) is chiefly credited with its development and popularization.

The centroid method seeks to find a variable that is an unweighted composite of the residual variables after previous factors have been extracted. This composite is the factor sought. It corresponds to an average variable among the residual variables. Geometrically speaking, this variable represents the centroid of the residual score vectors obtained by summing them together by vector summation. Consequently, in the general computing algorithm in the matrix, $\boldsymbol{\beta}_p = [1,1,1,1,\ldots,1]$, some of the 1s may be given negative signs to increase the variance of the resulting composite. In effect, the weight vector is the sum vector, $\mathbf{1} = [1,1,1,1,\ldots,1]'$. To summarize for the pth iteration

$$\boldsymbol{\beta}_p = \mathbf{1} \tag{7.21}$$

$$\sigma_p^2 = \mathbf{1}'\mathbf{R}_{p-1}\mathbf{1} \tag{7.22}$$

$$\mathbf{b}_p = \boldsymbol{\beta}_p(1/\sigma_p) \tag{7.23}$$

$$\boldsymbol{\lambda}_p = \mathbf{R}_{p-1}\mathbf{b}_p \tag{7.24}$$

$$\mathbf{R}_p = \mathbf{R}_{p-1} - \boldsymbol{\lambda}_p\boldsymbol{\lambda}_p' \tag{7.25}$$

If during the computation, σ_p^2 becomes negative, the residual matrix must be premultiplied and postmultiplied by a diagonal matrix containing positive and negative 1s to make the signs predominantly positive. The purpose of this is to make the sum of all the entries in the matrix positive and a relative maximum. Then σ_p^2 is reestimated as before.

7.4 Principal-Axes Methods

Suppose a psychologist has the problem of finding a single score that best represents a subject's performance on a battery of tests. His problem involves finding that particular combination of weights which, when applied to the scores of the tests in the test battery, determines the most representative single score. The solution of this kind of problem is the essence of the principal-axes methods of factoring. A most representative score must have something in it found in almost every score; thus it is an excellent candidate for consideration as a factor to account for what is in the observed scores in a factor-analytic study.

Let us now develop a little more rigorously what we mean by a most representative score that is derived from the observed scores. Let us consider the $n \times 1$ standardized random vector **Y** whose coordinates are n variables. What we wish to find is a linear composite X_1 by weighting the variables in **Y** with a weight vector $\boldsymbol{\beta}$. That is,

$$X_1 = \boldsymbol{\beta}'\mathbf{Y} \tag{7.26}$$

This is the ordinary equation for finding a linear composite. However, let us place some constraints on Equation 7.25 to obtain a particular kind of composite such that

$$\sigma_{X_1}^2 = E(X_1^{\,2}) = \boldsymbol{\beta}'\mathbf{R}_{YY}\boldsymbol{\beta} \tag{7.27}$$

is a maximum under the constraint that

$$\boldsymbol{\beta}'\boldsymbol{\beta} = 1 \tag{7.28}$$

In other words, we want to find a linear composite variable having the greatest possible variance under the restriction that the sum of the squares of the weights used to find the composite equals 1. The constraint on the weights makes the solution possible. Without it, the weights in $\boldsymbol{\beta}$ could be allowed to get larger and larger without bound, allowing the variance to increase without bound. In that case, there would be no definite maximum for the variance.

Intuitively we can see that it will be possible to maximize the variance of a linear composite, under the restriction that the sum of squares of the weights is 1, only if the average contribution of each component variable to the composite is a maximum. In such a case, the resulting composite would be the best single variable to represent that which is found in the components.

Turning this around, we can conclude that the most representative composite will account for the greatest amount of variance among the components in a single dimension. Hence, it has desirable characteristics, making it eligible as a factor.

Let us therefore find a weight vector, $\boldsymbol{\beta}$, that serves as a weight vector as in Equation 7.6. Since this involves finding the maximum of a function, we must resort to derivative calculus. Moreover, since a constraint is placed on the solution, we must use Lagrangian multipliers.

The function, F, of the weights in $\boldsymbol{\beta}$ to maximize is

$$\begin{aligned} F = E(X_1^2) &= \boldsymbol{\beta}'E(\mathbf{YY}')\boldsymbol{\beta} - \gamma(\boldsymbol{\beta}'\boldsymbol{\beta} - 1) \\ &= \boldsymbol{\beta}'\mathbf{R}_{YY}\boldsymbol{\beta} - \gamma(\boldsymbol{\beta}'\boldsymbol{\beta} - 1) \end{aligned} \tag{7.29}$$

where

γ is a Lagrangian multiplier

$\mathbf{R}_{YY}$ is the $n \times n$ correlation matrix of the variables in $\mathbf{Y}$

We now must take the partial derivative of F with respect to the unknown weights in vector $\boldsymbol{\beta}$. Note that $\boldsymbol{\beta}'\mathbf{R}_{YY}\boldsymbol{\beta}=\mathrm{tr}(\boldsymbol{\beta}'\mathbf{R}_{YY}\boldsymbol{\beta}')$. Then by Equation 4.17 applied to $\mathrm{tr}(\boldsymbol{\beta}'\mathbf{R}_{YY}\boldsymbol{\beta})$ and $\boldsymbol{\beta}'\boldsymbol{\beta} = \boldsymbol{\beta}'\mathbf{I}\boldsymbol{\beta}$, the partial derivative of Equation 7.29 is

$$\frac{\partial F}{\partial \boldsymbol{\beta}} = 2\mathbf{R}_{YY}\boldsymbol{\beta} - 2\gamma\boldsymbol{\beta} \tag{7.30}$$

Let us now set the result equal to a null vector, since we wish to solve for the weights when the partial derivatives are zero. This will give us the point where F is a maximum

$$2\mathbf{R}_{YY}\boldsymbol{\beta} - 2\gamma\boldsymbol{\beta} = \mathbf{0} \tag{7.31}$$

or, after multiplying both sides by 1/2 and factoring $\boldsymbol{\beta}$ outside the parentheses

$$(\mathbf{R}_{YY} - \gamma\,\mathbf{I})\boldsymbol{\beta} = \mathbf{0} \tag{7.32}$$

Although this represents a system of equations in n unknowns, we cannot solve this equation for $\boldsymbol{\beta}$ by finding the inverse of $(\mathbf{R}_{YY} - \gamma\mathbf{I})$ and multiplying it times both sides of the equation, as we do with most matrix equations. The solution would be the trivial $\boldsymbol{\beta}=\mathbf{0}$ and it would violate the constraint that $\boldsymbol{\beta}'\boldsymbol{\beta}=1$. So, we must not be able in the first place to obtain the inverse of $(\mathbf{R}_{YY} - \gamma\mathbf{I})$. This will imply that $(\mathbf{R}_{YY} - \gamma\mathbf{I})$ is a singular matrix. To be singular, it must have a determinant

$$|\mathbf{R}_{YY} - \gamma\mathbf{I}| = 0 \tag{7.33}$$

that is

$$\begin{vmatrix} \rho_{11}-\gamma & \rho_{12} & \cdots & \rho_{1n} \\ \rho_{21} & \rho_{22}-\gamma & \cdots & \rho_{2n} \\ \vdots & \vdots & \ddots & \vdots \\ \rho_{n1} & \rho_{n2} & \cdots & \rho_{nn}-\gamma \end{vmatrix} = 0 \tag{7.34}$$

The determinant $|\mathbf{R}_{YY} - \gamma\mathbf{I}|$ can be expanded by the rules for expanding determinants, and the result will be a polynomial equation in γ. The term of highest order in γ in this equation will come from the product of the diagonal elements of the determinant. Thus $(-\gamma)^n$ is the highest-order term of the polynomial, and $|\mathbf{R}_{YY} - \gamma\mathbf{I}|$ is a polynomial of degree n. That is, let $f(\gamma)$ be the polynomial

$$f(\gamma)=(-\gamma)^n+b_{n-1}(-\gamma)^{n-1}+\cdots+b_1(-\gamma)+b_0=0 \tag{7.35}$$

Equation 7.35 is known as the "characteristic equation" of the matrix $\mathbf{R}_{YY}$, and $f(\gamma)$ is known as the "characteristic polynomial" for $\mathbf{R}_{YY}$. The roots $\gamma_1, \gamma_2, \ldots, \gamma_n$ are known as the "characteristic roots or eigenvalues" of $\mathbf{R}_{YY}$. There are n roots of the characteristic equation because, by the fundamental theorem of algebra for any polynomial equation of degree n there are n roots.

The solution for the roots of the characteristic equation by direct methods is a tremendous computing job, even for electronic computers. As a matter of fact, workers in the field of factor analysis early recognized the value of the principal-axes concept but were stymied to apply it because of the formidable problem of solving the characteristic equation. The centroid method usually produced a close approximation and was used for that reason. However, in 1933, Hotelling published a paper in the *Journal of Educational Psychology* that described a method of finding solutions for the characteristic equation by an iterative algorithm. The manner in which he did this involved hard work in the days of the desk, and mechanical calculators, but it was not unfeasible for small problems.

But let us return to Equation 7.32 to multiply both sides of the equation by $\boldsymbol{\beta}'$. We then obtain

$$\boldsymbol{\beta}'(\mathbf{R}_{YY}-\gamma\mathbf{I})\boldsymbol{\beta}=0$$

which simplifies to

$$\boldsymbol{\beta}'\mathbf{R}_{YY}\boldsymbol{\beta}-\gamma\boldsymbol{\beta}'\boldsymbol{\beta}'=0.$$

Move the second term to the right-hand side of the equation.

$$\boldsymbol{\beta}'\mathbf{R}_{YY}\boldsymbol{\beta}=\gamma\boldsymbol{\beta}'\boldsymbol{\beta} \tag{7.36}$$

If we take any $\boldsymbol{\beta}$ satisfying Equation 7.32 and the constraint $\boldsymbol{\beta}'\boldsymbol{\beta}=1$, then this becomes

$$\boldsymbol{\beta}'\mathbf{R}_{YY}\boldsymbol{\beta}=\gamma \tag{7.37}$$

But by Equation 3.36 this is an equation for the variance of a composite random variable, and in this case the variance of the "raw" factor X_1^*. In other words, the variance of a composite variable that we seek equals a characteristic root of the characteristic equation. But which one? Since the roots will vary in magnitude, obviously for our purposes, it would have to be the largest one. Thus, let us designate $\boldsymbol{\beta}_1$ as the vector corresponding to the largest eigenvalue γ_1 of $\mathbf{R}_{YY}$. Hereafter, we will refer to a vector that generates a characteristic root or eigenvalue as a "characteristic vector" or "eigenvector." Sometimes it is called a "proper vector" and the root a "proper root." The word "eigenvector" comes from German where "eigen" means "own," which corresponds to "proper" in the sense that "property" is something someone owns. This comes from Latin, *proprius*, meaning one's own. In this case, it means that it is a property belonging to the matrix in question, here $\mathbf{R}_{YY}$. Since there are n roots and as a consequence n eigenvalues, let us arrange them in descending order of magnitude from the largest to the smallest. Since each different eigenvalue is generated by a different eigenvector, there will be a corresponding eigenvector for each eigenvalue. We will thus also arrange the eigenvectors in the descending order of their corresponding eigenvalues and designate both eigenvalues and eigenvectors by integer subscripts corresponding to their ranks from the largest to smallest of the eigenvalues.

One property of the different eigenvectors is that they are mutually orthogonal when the eigenvalues are distinct in value. Moreover, the different resulting composite variables generated are mutually uncorrelated. If we consider the eigenvectors in the order corresponding to the descending order of magnitude of their associated eigenvalues, the eigenvectors will successively generate uncorrelated composite variables, which account for decreasing amounts of the total variance among the component variables. Hence the eigenvectors may be useful to the factor analyst for generating potential factors.

But so far we have not shown how to find the particular eigenvector that generates the composite with the largest variance. Yet to obtain the eigenvector that generates a composite with the maximum variance we would have to have either a method that proceeds directly to the desired eigenvector or a method which derives all eigenvectors and eigenvalues at once. In the latter case we would then be able to pick from among the eigenvectors the one associated with the largest eigenvalue. As a matter of fact, both kinds of methods are available. The first, which converges to the eigenvector associated with the largest eigenvalue is due to Hotelling (1933). The overall computing procedure of extracting factors used with Hotelling's method is the general computing algorithm presented in Section 7.1. The unique feature of his method is the iterative method of finding eigenvectors and eigenvalues. The second kind of computing method, one which is now exclusively used by most factor analysis programs on computers, computes all of the eigenvectors and eigenvalues simultaneously, using an iterative algorithm. There are two major algorithms used to accomplish this, the Jacobi method and the Q-R method. They are much faster than Hotelling's method. The reason we will describe Hotelling's method is pedagogical, because it is fairly simple, relatively speaking.

7.4.1 Hotelling's Iterative Method

We shall adapt Hotelling's iterative procedure to the general computing algorithm for finding factors. Hotelling's method is distinguished by the manner in which he derives the vector $\boldsymbol{\beta}$ in the general computing algorithm.

Hotelling's method works with any square symmetric matrix having real characteristic roots for its characteristic equation. For our purposes, let us begin with the original correlation matrix, $\mathbf{R}_{YY}$, having unities in the diagonal. We could just as well begin with the reduced correlation matrix or a residual correlation matrix. But we will begin here with $\mathbf{R}_{YY}$ having unities in the diagonal for the simplicity of notation.

Our first task is to find the first eigenvector of $\mathbf{R}_{YY}$, which we will find convenient to designate as $\mathbf{a}_1$. When found, $\mathbf{a}_1$ corresponds to $\boldsymbol{\beta}_1$ in the general computing algorithm. From this we will find γ_1, which is the first and largest eigenvalue of $\mathbf{R}_{YY}$ and which also corresponds to the variance of the composite generated by $\boldsymbol{\beta}_1$. Then we can find the first column $\boldsymbol{\lambda}_1$ of the factor-loading matrix $\boldsymbol{\Lambda}$. After we find $\boldsymbol{\lambda}_1$, we partial the matrix product $\boldsymbol{\lambda}_1\boldsymbol{\lambda}_1'$ from $\mathbf{R}_{YY}$. Then we continue to find the next factor, and so on.

To start the iterations to find the first eigenvector $\mathbf{a}_1$, let us select an arbitrary $n \times 1$ vector $\mathbf{g}_{(0)}$. (The subscripts in parentheses will indicate the iteration number. A good choice is the sum vector [1, 1,…,1]. Let us normalize $\mathbf{g}_{(0)}$ by dividing its elements by its length $|\mathbf{g}_{(0)}|$. We will designate the normalized vector as $\mathbf{a}_{1(0)}$. We will now start an iterative sequence such that during the ith iteration

$$\mathbf{a}_{1(i)} = \frac{\mathbf{g}_{(i)}}{\sqrt{\mathbf{g}_{(i)}'\mathbf{g}_{(i)}}} \tag{7.38}$$

and

$$\mathbf{g}_{(i+1)} = \mathbf{R}_{YY}\mathbf{a}_{1(i)} \tag{7.39}$$

Equation 7.38 normalizes $\mathbf{g}_{(i)}$ by dividing by the square root of the sum of squares of its components (its length). Equation 7.39 finds the next choice for $\mathbf{g}$ by multiplying the matrix $\mathbf{R}_{YY}$ times the normalized column vector $\mathbf{a}_{1(i)}$. Eventually after going through many cycles of iteration

$$\mathbf{a}_{1(i+1)} - \mathbf{a}_{1(i)} \cong \mathbf{0} \tag{7.40}$$

In other words, eventually the process "converges" so that successive vectors of eigenvectors differ by only a very small amount. We stop the iterations when the absolute value of the difference reaches a minimum tolerance level. For example, we may stop if

$$\left|\mathbf{a}_{1(i+1)} - \mathbf{a}_{1(i)}\right| < .00001$$

The last vector found after v iterations, which we will designate as $\mathbf{a}_{1(v)}$, will be taken to be the first eigenvector $\mathbf{a}_1$ of $\mathbf{R}_{YY}$ and

$$\mathbf{R}_{YY}\mathbf{a}_1 = \gamma_1\mathbf{a}_1 \tag{7.41}$$

where γ_1 is the first and largest eigenvalue of $\mathbf{R}_{YY}$. The value for γ_1 can be found by

$$\gamma_1 = \mathbf{a}_1'\mathbf{a}_1\gamma_1 = \mathbf{a}_1'\mathbf{R}_{YY}\mathbf{a}_1 \tag{7.42}$$

since $\mathbf{a}_1'\mathbf{a}_1 = 1$.

That the process always converges to the largest eigenvalue and corresponding eigenvector can be proved mathematically. However, we shall not do so here. A proof of convergence is given by T.W. Anderson (1958, pp. 281–283).

If we equate the first eigenvector, $\mathbf{a}_1$, with $\boldsymbol{\beta}_1$ in the general computing algorithm of factor analysis, we can find the factor loadings on the first principal axis as

$$\boldsymbol{\lambda}_1 = \mathbf{R}_{YY}\mathbf{a}_1 \frac{1}{\sqrt{\gamma_1}} = \mathbf{a}_1\sqrt{\gamma_1} \qquad (7.43)$$

where γ_1 equals the first eigenvalue of $\mathbf{R}_{YY}$ and also the variance of the linear composite which is the principal-axes factor (before being standardized). The simplification on the right-hand side is possible because of Equation 7.41.

In general, for the pth iteration of the general computing algorithm where $\mathbf{a}_p$ is the pth eigenvector found by

$$\boldsymbol{\beta}_p = \mathbf{a}_p$$

$$\sigma_p^2 = \boldsymbol{\beta}_p' \, \mathbf{R}_{p-1} \, \boldsymbol{\beta}_p = \gamma_p$$

$$\mathbf{b}_p = \boldsymbol{\beta}_p \frac{1}{\sigma_p}$$

$$\boldsymbol{\lambda}_p = \mathbf{R}_{p-1}\mathbf{b}_p = \mathbf{a}_p\sqrt{\gamma_p}$$

$$\mathbf{R}_p = \mathbf{R}_{p-1} - \boldsymbol{\lambda}_p\boldsymbol{\lambda}_p' = \mathbf{R}_p - \gamma_p\mathbf{a}_p\mathbf{a}_p'$$

In theory, it is possible to accelerate the convergence of Hotelling's method by raising the matrix $\mathbf{R}$ to a power before beginning the iterations. For example

$$\mathbf{R}^2 = \mathbf{R}\mathbf{R}$$

If in turn we square $\mathbf{R}^2$ by multiplying it times itself, we obtain $\mathbf{R}^4$, and so on. Let us designate the matrix raised to the hth power as $\mathbf{R}^h$. A matrix $\mathbf{R}^h$ can be used instead of $\mathbf{R}$ in Equation 7.39.

However, the largest root γ_1 should not be estimated by taking the iterated vector

$$\mathbf{g}_{(m+1)} = \mathbf{R}^h\mathbf{a}_{(m)}$$

then finding the value

$$\gamma_1^h = \mathbf{g}_{(h+1)}'\mathbf{a}_{(h)}$$

and taking the kth root to find γ_1. Under the squaring procedure to obtain $\mathbf{R}^h$, h will be an even power; hence if the root γ_1 is a negative root, it will be positive when raised to an even power. But any characteristic vector of $\mathbf{R}$ with a corresponding negative root should be avoided, since it generates a composite with an imaginary standard deviation. This is not interpretable. Since most estimates of communalities inserted in the principal diagonal of a correlation matrix before factoring will leave the matrix non-Gramian, and hence give it negative characteristic roots as well as positive roots, this could be a problem and thus should be avoided. The preferred procedure is, therefore, to multiply $\mathbf{a}_{1(h)}$ times the original unpowered $\mathbf{R}$. If the obtained root is negative, extract it from the matrix and continue.

Because Hotelling's matrix depends on raising the matrix to any integral power, it will accumulate round-off error in computers with small word sizes, meaning the computer does not have a large number of significant digits. As a consequence the resulting matrix raised to a power will correspond to some other matrix raised to a power and not that of the original matrix.

7.4.2 Further Properties of Eigenvectors and Eigenvalues

Let $\mathbf{R}$ be a square symmetric matrix having real roots for its characteristic equation. Let $\mathbf{a}$ be an eigenvector of $\mathbf{R}$. Then

$$\mathbf{Ra} = \gamma\mathbf{a} \tag{7.44}$$

where γ is an eigenvalue of $\mathbf{R}$. This is the fundamental equation for relating an eigenvector of $\mathbf{R}$ to the matrix $\mathbf{R}$. It means that premultiplying the eigenvector by the matrix produces the same effect as multiplying the eigenvector by the eigenvalue, which is a scalar. Any unit length vector, $\mathbf{a}$, when postmultiplied times a square symmetric matrix $\mathbf{R}$ that produces a scalar multiple of $\mathbf{a}$ is an eigenvector of $\mathbf{R}$ and the scalar is its corresponding eigenvalue.

We will now prove that two different eigenvectors corresponding to a matrix $\mathbf{R}$ are mutually orthogonal. Let $\mathbf{a}_j$ be one eigenvector and $\mathbf{a}_k$ the other eigenvector of $\mathbf{R}$. Then

$$\mathbf{Ra}_j = \gamma_j\mathbf{a}_j \quad \text{and} \quad \mathbf{Ra}_k = \gamma_k\mathbf{a}_k \tag{7.45}$$

Consider next

$$\mathbf{a}_k'\mathbf{Ra}_j = \gamma_j\mathbf{a}_k'\mathbf{a}_j \quad \text{and} \quad \mathbf{a}_j'\mathbf{Ra}_k = \gamma_k\mathbf{a}_j'\mathbf{a}_k \tag{7.46}$$

But, $\mathbf{a}_k'\mathbf{Ra}_j = \mathbf{a}_j'\mathbf{Ra}_k$ because one is the transpose of the other and both are scalars. So, we can write

$$\mathbf{a}_k'\mathbf{Ra}_j - \mathbf{a}_j'\mathbf{Ra}_k = 0$$

Substituting now from Equation 7.46, we can write

$$\gamma_j \mathbf{a}_k' \mathbf{a}_j - \gamma_k \mathbf{a}_j' \mathbf{a}_k = 0$$

But, $\mathbf{a}_k'\mathbf{a}_j = \mathbf{a}_j'\mathbf{a}_k$ because these are scalars and each the transpose of the other. So now we have

$$(\gamma_j - \gamma_k)\mathbf{a}_j'\mathbf{a}_k = 0 \tag{7.47}$$

But if $\gamma_j \neq \gamma_k$, then to make Equation 7.47 zero, it must be the case that

$$\mathbf{a}_j'\mathbf{a}_k = 0 \tag{7.48}$$

which proves that two eigenvectors corresponding to distinct valued eigenvalues are orthogonal to one another.

Let us next consider the effect of arranging all of the eigenvectors of a matrix $\mathbf{R}$ into an $n \times n$ matrix $\mathbf{A}$, so that $\mathbf{a}_1$ is the first column of $\mathbf{A}$, $\mathbf{a}_2$ the second column of $\mathbf{A}$, and so on. Then, by Equations 7.44 and 7.48

$$\mathbf{RA} = \mathbf{AD} \tag{7.49}$$

where $\mathbf{D}$ is a diagonal matrix having for its diagonal elements the eigenvalues $\gamma_1, \gamma_2, \ldots, \gamma_n$.

If we premultiply both sides of Equation 7.49 by $\mathbf{A}'$, we obtain an interesting result

$$\mathbf{A}'\mathbf{RA} = \mathbf{D} \tag{7.50}$$

The matrix $\mathbf{A}$ is an orthogonal matrix because its columns are unit length and mutually orthogonal with respect to one another. Hence $\mathbf{A}'\mathbf{A} = \mathbf{I}$. But more importantly, we find that premultiplying $\mathbf{R}$ by the transpose of its eigenvector matrix and postmultiplying by the eigenvector matrix itself "diagonalizes" $\mathbf{R}$ to the diagonal matrix of its eigenvalues.

Next, premultiply Equation 7.50 by $\mathbf{A}$ and postmultiply Equation 7.50 by $\mathbf{A}'$:

$$\mathbf{AA}'\mathbf{RAA}' = \mathbf{R} = \mathbf{ADA}' \tag{7.51}$$

We call this result the "spectral composition theorem." The square diagonal matrix is composed of its eigenvector matrix times its diagonal eigenvalue matrix times the transpose of its eigenvector matrix.

From Equation 7.51, we can obtain

$$\mathbf{R}^{-1} = (\mathbf{ADA}')^{-1} = \mathbf{AD}^{-1}\mathbf{A}' \tag{7.52}$$

This follows because $\mathbf{A}' = \mathbf{A}^{-1}$, since $\mathbf{A}$ is an orthonormal matrix, and because, given $\mathbf{F}$, $\mathbf{G}$, and $\mathbf{H}$ are square $n \times n$ matrices

$$(\mathbf{FGH})^{-1} = \mathbf{H}^{-1}\mathbf{G}^{-1}\mathbf{F}^{-1}$$

The implication of this result is that the largest eigenvalue of the inverse matrix, $\mathbf{R}^{-1}$, equals the reciprocal of the smallest eigenvalue γ_n of $\mathbf{R}$. As $\mathbf{R}$ approaches a singular matrix with a zero eigenvalue, the first and largest eigenvalue of $\mathbf{R}^{-1}$ is increasing without bound.

7.4.3 Maximization of Quadratic Forms for Points on the Unit Sphere

The following is suggested by Johnson and Wichern (1998) but with some modification of the argument. Let $\mathbf{R}$ be an $n \times n$ positive definite matrix with distinct eigenvalues $\gamma_1, \gamma_2, \ldots, \gamma_n$ and eigenvectors $\mathbf{a}_1, \mathbf{a}_2, \ldots, \mathbf{a}_n$. Let $\mathbf{x}$ be an arbitrary $n \times 1$ column vector such that $\mathbf{x}'\mathbf{x} = 1$, i.e., having unit length. Then

$$\mathbf{x}'\mathbf{R}\mathbf{x} \le \gamma_1 \quad \text{with equality when } \mathbf{x} = \mathbf{a}_1 \tag{7.53}$$

$$\mathbf{x}'\mathbf{R}\mathbf{x} \ge \gamma_n \quad \text{with equality when } \mathbf{x} = \mathbf{a}_n \tag{7.54}$$

Moreover, if $\mathbf{x}$ is orthogonal to the first h eigenvectors, $\mathbf{a}_1, \mathbf{a}_2, \ldots, \mathbf{a}_h$, but is not orthogonal to the remaining eigenvectors, $\mathbf{a}_{h+1}, \mathbf{a}_2, \ldots, \mathbf{a}_h$, then

$$\mathbf{x}'\mathbf{R}\mathbf{x} \le \gamma_{h+1} \quad \text{with equality when } \mathbf{x} = \mathbf{a}_{h+1} \tag{7.55}$$

Proof: Let $\mathbf{A}$ be the $n \times n$ orthogonal matrix whose columns are the eigenvectors, $\mathbf{a}_1, \mathbf{a}_2, \ldots, \mathbf{a}_n$. Let $\mathbf{D}$ be the diagonal matrix whose diagonal elements are the corresponding eigenvalues $\gamma_1, \gamma_2, \ldots, \gamma_n$ in descending order of magnitude of $\mathbf{R}$. From the spectral composition theorem, $\mathbf{R} = \mathbf{A}\mathbf{D}\mathbf{A}'$. Next, let $\mathbf{y} = \mathbf{A}'\mathbf{x}$. Note that $\mathbf{x} \neq \mathbf{0}$ (because $\mathbf{x}'\mathbf{x} = 1$) and this implies that $\mathbf{y} = \mathbf{A}'\mathbf{x} \neq \mathbf{0}$. Furthermore, $\mathbf{y}'\mathbf{y} = \mathbf{x}'\mathbf{A}\mathbf{A}'\mathbf{x} = \mathbf{x}'\mathbf{x} = 1$.

Now, $\mathbf{x}'\mathbf{R}\mathbf{x} = \mathbf{x}'\mathbf{A}\mathbf{D}\mathbf{A}'\mathbf{x} = \mathbf{y}'\mathbf{D}\mathbf{y}$

We can see what this resolves to if we let $n = 4$:

$$\mathbf{x}'\mathbf{R}\mathbf{x} = \mathbf{y}'\mathbf{D}\mathbf{y}$$

$$\begin{bmatrix} y_1 & y_2 & y_3 & y_4 \end{bmatrix} \begin{bmatrix} \gamma_1 & 0 & 0 & 0 \\ 0 & \gamma_2 & 0 & 0 \\ 0 & 0 & \gamma_3 & 0 \\ 0 & 0 & 0 & \gamma_4 \end{bmatrix} \begin{bmatrix} y_1 \\ y_2 \\ y_3 \\ y_4 \end{bmatrix}$$

$$= \begin{bmatrix} \gamma_1 y_1 & \gamma_2 y_2 & \gamma_3 y_3 & \gamma_4 y_4 \end{bmatrix} \begin{bmatrix} y_1 \\ y_2 \\ y_3 \\ y_4 \end{bmatrix}$$

$$= \sum_{i=1}^{4} \gamma_i y_i^2$$

Next observe that

$$\sum_{i=1}^{n} \gamma_i y_i^2 = \gamma_1 y_1^2 + \gamma_2 y_2^2 + \cdots + \gamma_n y_n^2 \leq \gamma_1 y_1^2 + \gamma_1 y_2^2 + \cdots + \gamma_1 y_n^2$$

or

$$\sum_{i=1}^{n} \gamma_i y_i^2 \leq \gamma_1 \sum_{i=1}^{n} y_i^2 = \gamma_1$$

This follows because $\gamma_1 \geq \gamma_i$, $i = 2,\ldots, n$, that is, the first eigenvalue is always larger than any subsequent eigenvalue, and so for each term $\gamma_1 y_i \geq \gamma_i y_i$, $i = 2,\ldots, n$. Furthermore, we may factor the first eigenvalue γ_1 from the sum on the right-hand side and note that $\mathbf{y}'\mathbf{y} = \sum_{i=1}^{n} y_i^2 = 1$. If the first eigenvalue is strictly greater than all other eigenvalues, and if $\mathbf{x} \neq \mathbf{a}_1$, then $\sum_{i=1}^{n} \gamma_i y_i^2 < \sum_{i=1}^{n} \gamma_1 y_i^2 = \gamma_1$ and strict inequality $\mathbf{x}'\mathbf{R}\mathbf{x} < \gamma_1$ holds.

Finally, if $\mathbf{x} = \mathbf{a}_1$, that is, if $\mathbf{x}$ equals the first eigenvector, then $\mathbf{y} = \mathbf{A}'\mathbf{x}$ is a column vector $\mathbf{y} = [1,0,0,\ldots,0]$ resulting from the fact that the first row of the transposed eigenvector matrix $\mathbf{A}'$ has a scalar product of unity with the first eigenvector $\mathbf{a}_1$, while all other eigenvectors in the remaining rows of $\mathbf{A}'$ are orthogonal with $\mathbf{a}_1$. Thus in this case $\sum_{i=1}^{n} \gamma_i y_i^2 = \gamma_1$. More straightforwardly, $\mathbf{x}'\mathbf{R}\mathbf{x} = \mathbf{a}_1'\mathbf{R}\mathbf{a}_1 = \gamma_1$. This completes the proof of the first part.

The second part of the theorem establishes that $\mathbf{x}'\mathbf{R}\mathbf{x} \geq \gamma_n$ and equality is attained when $\mathbf{x} = \mathbf{a}_n$. This is proven exactly as in the proof of the first part of the theorem, except we establish that

$$\sum_{i=1}^{n} \gamma_i y_i^2 \geq \sum_{i=1}^{n} \gamma_n y_i^2 = \gamma_n$$

since $\gamma_i > \gamma_n$ for $i = 1,\ldots,n-1$. And we show that when $\mathbf{x} = \mathbf{a}_n$, $\mathbf{x}'\mathbf{R}\mathbf{x} = \gamma_n$, the last and smallest eigenvalue, γ_n.

The third part of the theorem states that if a vector $\mathbf{x}$ is orthogonal to the first h eigenvectors of $\mathbf{R}$, then $\mathbf{x}'\mathbf{R}\mathbf{x} \leq \gamma_{h+1}$ and attains its maximum value γ_{h+1} when $\mathbf{x} = \mathbf{a}_{h+1}$. We can show this by noting that if $\mathbf{y} = \mathbf{A}'\mathbf{x}$, and $\mathbf{x}$ is orthogonal to the first h rows of $\mathbf{A}'$, this means that the first h elements of $\mathbf{y}$ must be zeroes. That is to say,

$$\begin{bmatrix} a_{11} & a_{12} & \cdots & a_{1h} & a_{1h+1} & \cdots & a_{1n} \\ a_{21} & a_{22} & \cdots & a_{2h} & a_{2h+1} & \cdots & a_{2n} \\ \vdots & \vdots & \ddots & \vdots & \vdots & \ddots & \vdots \\ a_{h1} & a_{h2} & \cdots & a_{hh} & a_{h,h+1} & \cdots & a_{hn} \\ a_{h+1,1} & a_{h+1,2} & \cdots & a_{h,h+1} & a_{h+1,h+1} & \cdots & a_{h+1,n} \\ \vdots & \vdots & \ddots & \vdots & \vdots & \ddots & \vdots \\ a_{n1} & a_{n2} & \cdots & a_{nh} & a_{n,h+1} & \cdots & a_{nn} \end{bmatrix} \begin{bmatrix} x_1 \\ x_2 \\ \vdots \\ x_h \\ x_{h+1} \\ \vdots \\ x_n \end{bmatrix} = \begin{bmatrix} 0 \\ 0 \\ \vdots \\ 0 \\ y_{h+1} \\ \vdots \\ y_n \end{bmatrix}$$

Then $\mathbf{y}'\mathbf{D}\mathbf{y} = \sum_{i=1}^{n} \gamma_i y_i^2 \leq \sum_{i=1}^{n} \gamma_{h+1} y_i^2 = \gamma_{h+1}$, and the argument proceeds as before because $\gamma_{h+1} \geq \gamma_i$, $i = h + 1, \ldots, n$. Equality and the maximum value is attained when $\mathbf{x} = \mathbf{a}_{h+1}$ by arguments similar to those above.

The importance of this theorem lies in its relevance for establishing upper and lower bounds on the variances of linear combinations of variables. For example, if $X = \mathbf{c}'\mathbf{X}$, where $\mathbf{X}$ is an $n \times 1$ random vector of random variables and $\mathbf{c}$ is a vector of weights whose sum of squares equals 1, i.e., $\mathbf{c}'\mathbf{c} = 1$, then $\sigma^2(X) = \mathbf{c}'\boldsymbol{\Sigma}_{XX}\mathbf{c} \leq \gamma_1$. The variance of a linear combination of random variables using weights whose squares add up to unity is always less than or equal to the largest eigenvalue of the covariance matrix of the component random variables. Principal components, for example, seek a linear combination, $X = \mathbf{a}'\mathbf{Y}$, whose variance $\sigma^2(X) = \mathbf{a}'\boldsymbol{\Sigma}_{YY}\mathbf{a}$ is a maximum under the constraint that the sum of squares of the weight coefficients equals unity, i.e., $\mathbf{a}'\mathbf{a} = 1$. Thus the first eigenvector, $\mathbf{a}_1$, is the solution for the weights of this linear combination, and the variance of this linear combination is the largest eigenvalue, γ_1, of the matrix, $\boldsymbol{\Sigma}_{YY}$, of the component variables.

7.4.4 Diagonalizing the R Matrix into Its Eigenvalues

What we are now about to discuss are algorithms for simultaneously finding all of the common factors at once. All methods today using computers extract all of the common factors at once.

Consider Equation 7.49, which states that $\mathbf{RA} = \mathbf{AD}$, where $\mathbf{A}$ and $\mathbf{D}$ are the eigenvector and eigenvalue matrices of the square, symmetric matrix $\mathbf{R}$, respectively. Recalling that $\mathbf{A}$ is an orthonormal matrix whose inverse is $\mathbf{A}'$, that is, $\mathbf{A}'\mathbf{A} = \mathbf{I}$, we may multiply both sides of the equation by $\mathbf{A}'$ to obtain

$$\mathbf{A}'\mathbf{R}\mathbf{A} = \mathbf{D} \tag{7.56}$$

In effect, the transformation $\mathbf{A}'\mathbf{R}\mathbf{A}$ of $\mathbf{R}$ "diagonalizes" $\mathbf{R}$ into its eigenvalue matrix. Any orthogonal matrix $\mathbf{B}$ that diagonalizes $\mathbf{R}$ is an eigenvector matrix of $\mathbf{R}$. For example, suppose $\mathbf{B}'\mathbf{R}\mathbf{B} = \mathbf{G}$. This implies that $\mathbf{RB} = \mathbf{BG}$, which establishes that $\mathbf{B}$ is an eigenvector matrix of $\mathbf{R}$ and $\mathbf{G}$ an eigenvalue matrix of $\mathbf{R}$. If further, the diagonal elements of $\mathbf{G}$ are in order of descending magnitude, as are also the diagonal elements of $\mathbf{D}$, then $\mathbf{G} = \mathbf{D}$ and $\mathbf{A} = \mathbf{B}$.

We should consider more closely a property of orthogonal matrices. If $\mathbf{Q}$ is an orthogonal square matrix of order n and $\mathbf{P}$ is another square orthogonal matrix of order n, then $\mathbf{QP}$ is an orthogonal matrix, also of order n. The proof follows from premultiplying $\mathbf{QP}$ by its transpose:

$$(\mathbf{QP})'\mathbf{QP} = \mathbf{P}'\mathbf{Q}'\mathbf{QP} = \mathbf{P}'\mathbf{P} = \mathbf{I} \tag{7.57}$$

Since the result is an identity matrix, $\mathbf{QP}$ must be an orthogonal matrix.

Let us now represent the eigenvector matrix, $\mathbf{A}$, as a product series of orthogonal matrices

$$\mathbf{A} = \mathbf{T}_1, \mathbf{T}_2, \mathbf{T}_3, \ldots, \mathbf{T}_s \tag{7.58}$$

Let us now substitute the expression in Equation 7.58 for **A** in Equation 7.56 to obtain

$$\mathbf{T}_s' \ldots \mathbf{T}_3'\mathbf{T}_2'\mathbf{T}_1'\mathbf{R}\mathbf{T}_1\mathbf{T}_2\mathbf{T}_3 \ldots \mathbf{T}_s = \mathbf{D} \tag{7.59}$$

Thus, we see that if we successively apply an appropriate set of orthogonal matrices $\mathbf{T}_1$, $\mathbf{T}_2$, $\mathbf{T}_3, \ldots$, $\mathbf{T}_s$ to the matrix **R** as in Equation 7.59, we should obtain the eigenvalues of **R**.

It follows then that if we find a series of orthogonal matrices that diagonalize **R**, the product of the series must yield the eigenvectors of **R** and the diagonal matrix contains the eigenvalues in its principal diagonal. This forms the basis for the current methods for finding eigenvalues and eigenvectors of square symmetric matrices. These methods find all of the eigenvectors and all of the eigenvalues at the same time. We do not intend to describe these methods in fine detail, since they are now better described in numerical analysis books, or in Mulaik (1972). But, we will describe generally how they work.

7.4.5 Jacobi Method

The idea of the Jacobi method is to find a set of "elementary" orthogonal transformation matrices $\mathbf{T}_1$, $\mathbf{T}_2$, $\mathbf{T}_3, \ldots$, $\mathbf{T}_s$ such that

$$\mathbf{T}_s' \ldots \mathbf{T}_3'\mathbf{T}_2'\mathbf{T}_1'\mathbf{R}\mathbf{T}_1\mathbf{T}_2\mathbf{T}_3 \ldots \mathbf{T}_s = \mathbf{D}$$

The idea is not to find all these transformation matrices at once but simply to find them one at a time. For example, if $\mathbf{T}_1$ is the first elementary orthogonal transformation matrix, then we find a new matrix $\mathbf{R}_1$ such that

$$\mathbf{R}_1 = \mathbf{T}_1' \mathbf{R}\mathbf{T}_1$$

Then we find another elementary orthogonal transformation matrix, $\mathbf{T}_2$, and obtain a new matrix, $\mathbf{R}_2$, by

$$\mathbf{R}_2 = \mathbf{T}_2'\mathbf{R}_1\mathbf{T}_2$$

By successively applying the appropriate orthogonal transformation matrices to the matrix **R**, we eventually arrive at the result

$$\mathbf{D} \approx \mathbf{R}_s = \mathbf{T}_s'\mathbf{R}_{s-1}\mathbf{T}_s$$

which is to say that after the sth transformation, the matrix $\mathbf{R}_s$ is for all practical purposes the diagonal matrix of eigenvalues of **R**.

The problem is to pick the proper orthogonal transformation matrices so that the desired result can be obtained.

Jacobi (1846) found a solution to this problem to be simply to pick the largest off-diagonal element of the matrix **R** and to "annihilate" it to zero by applying a proper orthogonal transformation. Then he looked for the largest remaining off-diagonal element and annihilated that. He applied the procedure again and again until the off-diagonal elements were sufficiently close to zero as to be negligible. The diagonal elements of the matrix then contained very close approximations to the eigenvalues. If the successive transformation matrices were multiplied together, they would produce an accurate approximation to the eigenvector matrix.

Let us therefore consider the largest nonzero, off-diagonal element r_{jk} of the matrix $\mathbf{R}_m$ where m indicates the matrix $\mathbf{R}$ after m iterations of the algorithm. Consider in turn an elementary transformation matrix $\mathbf{T}_{m+1}$ such that

$$\mathbf{T}_{m+1} = \begin{bmatrix} 1 & & & & & & & & & & \\ & \ddots & & & & & & & & & \\ & & 1 & & & & & & & & \\ & & & \cos\theta & & & & \sin\theta & & & \\ & & & & 1 & & & & & & \\ & & & & & \ddots & & & & & \\ & & & & & & 1 & & & & \\ & & & -\sin\theta & & & & \cos\theta & & & \\ & & & & & & & & 1 & & \\ & & & & & & & & & \ddots & \\ & & & & & & & & & & 1 \end{bmatrix} \tag{7.60}$$

This matrix is an identity matrix except for elements t_{jj}, t_{kk}, which equal cos θ and t_{jk} and t_{kj}, which equal $\sin\theta$ and $-\sin\theta$, respectively. The angle θ represents an angle of rotation, which will leave r_{jk} equal to zero. The problem is to find the angle θ for accomplishing this with a given r_{jk}.

If we apply the transformation, $\mathbf{R}_{m+1} = \mathbf{T}'_{m+1}\mathbf{R}_m\mathbf{T}_m$, to the matrix $\mathbf{R}_m$, the new element, r^*_{jk}, in $\mathbf{R}_{m+1}$ will be in terms of the elements of $\mathbf{R}_m$:

$$r^*_{jk} = (r_{jj} - r_{kk})\sin\theta\cos\theta + r_{jk}(\cos^2\theta - \sin^2\theta)$$

Set r^*_{jk} equal to zero and solve this equation in terms of the trigonometric functions, considering the trigonometric identities for sin 2θ and cos 2θ

$$0 = 1/2(r_{jj} - r_{kk})\sin 2\theta + r_{jk}(\cos^2\theta - \sin^2\theta)$$

or

$$\frac{-2r_{jk}}{r_{jj} - r_{kk}} = \frac{\sin 2\theta}{\cos 2\theta} = \tan 2\theta$$

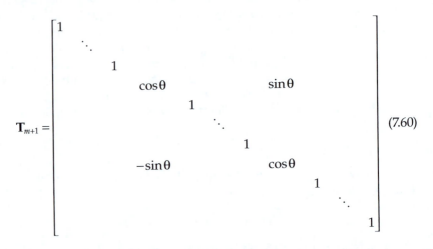

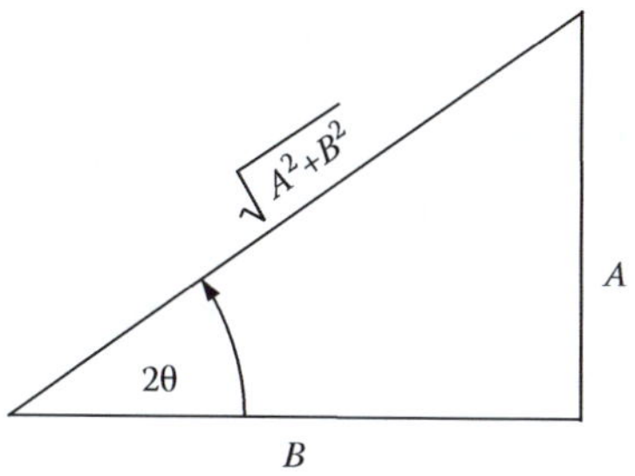

FIGURE 7.1
Right triangle having side opposite of length A, side adjacent B.

Let $A = -2r_{jk}$ and $B = r_{jj} - r_{kk}$; then consider the right triangle in Figure 7.1.

From the fundamental definitions of the trigonometric functions, $\tan 2\theta = A/B$ and $\cos 2\theta = B/\sqrt{A^2 + B^2}$. But by another trigonometric identity, $\sin\theta = \sqrt{(1 - \cos 2\theta)/2}$; hence in terms of $-2r_{jk}$ and $r_{jj} - r_{kk}$, this becomes

$$\sin\theta = \text{sgn}\left(\frac{-2r_{jk}}{r_{jj} - r_{kk}}\right)\sqrt{\frac{1}{2}\left(1 - \frac{r_{jj} - r_{kk}}{\sqrt{4r_{jk}^2 + (r_{jj} - r_{kk})^2}}\right)} \tag{7.61}$$

and

$$\cos\theta = \sqrt{1 - \sin^2\theta} \tag{7.62}$$

Thus we need not find the angle θ itself but can find directly sin θ and cos θ. These values are then inserted in the matrix $\mathbf{T}_{m+1}$ and when the matrix transformation is carried out with $\mathbf{T}_{m+1}$, r_{jk} becomes zero.

Until the late 1950s, Jacobi's original method was not taken very seriously because it was slow when applied to automatic computing devices, because of the tremendous effort to search for the largest off-diagonal element. However, the development of high-speed computers, funded in part by government support to develop nuclear weapons and to decipher ciphers of hostile nations, led mathematicians to review some of the older numerical methods. They found that Jacobi's method was not only feasible but highly accurate. Goldstine et al. (1959) demonstrated that Jacobi's method is extremely stable against round-off errors as long as the square root in Equation 7.62 is taken accurately. Moreover, Pope and Tompkins (1957) discovered that computers could handle the method quite nicely by annihilating not only the largest off-diagonal element but by annihilating in sequence those-off-diagonal elements whose absolute values were above a minimum threshold until no elements remained above the threshold. Then the threshold could be lowered and the procedure continued until the threshold had to be lowered again, and so on. Eventually, the off-diagonal elements would become less in absolute value than a maximum error tolerance level, and the method was stopped.

Subsequently, Corbato (1963) developed a computer procedure for finding the largest off-diagonal element with a minimum of search operations at any point in the execution of the classical Jacobi method. The method is based upon the idea that annihilating an off-diagonal element by an orthogonal transformation matrix $\mathbf{T}_{m+1}$ changes only the elements of two columns and two rows of the matrix **R**. Consequently, if one initially has knowledge for each column of what the magnitude of the largest element is and which row it is in (stored in two n-dimensional arrays of computer memory), one need only search the changed columns and rows for their largest elements, replace these values in the array of values, and then search the array of values for the largest element across all the columns, noting the column and row it is in. The advantage of Corbato's procedure is that it minimizes the number of transformations to be carried out, thus minimizing the effects of error and speeding up the process.

Another method, somewhat different from Jacobi's method, is to first "tridiagonalize" the **R** matrix by a special orthogonal transformation matrix **T** that produces a matrix of the form

$$\mathbf{T'RT} = \begin{bmatrix} x & x & 0 & 0 & 0 & 0 & 0 \\ x & x & x & 0 & 0 & 0 & 0 \\ 0 & x & x & x & 0 & 0 & 0 \\ 0 & 0 & x & x & x & 0 & 0 \\ 0 & 0 & 0 & x & x & x & 0 \\ 0 & 0 & 0 & 0 & x & x & x \\ 0 & 0 & 0 & 0 & 0 & x & x \end{bmatrix}$$

A method for finding **T** by multiplying together a series of $n-2$ simpler orthogonal matrices such that $\mathbf{T} = \mathbf{T}_1\mathbf{T}_2\ldots\mathbf{T}_{n-2}$ was given first by Householder and described by Wilkinson (1960, 1965) and Ralston (1965). At this point, one might begin applying Jacobi's algorithm to the tridiagonalized matrix. But Ortega and Kaiser (1963) discovered that another algorithm, the Q-R algorithm, applied to the tridiagonal matrix would at each iteration leave the resulting matrix tridiagonal. Furthermore, one would only have to attend to the diagonal and immediate off-diagonal elements. The immediate off-diagonal elements would also get smaller while the diagonal elements converged to the eigenvalues. So, this is a very efficient algorithm.

Because the name of the Q-R algorithm refers to the two matrices by which a given real-valued square symmetric tridiagonal matrix is decomposed, we shall retain that notation here to describe briefly this algorithm. Let **M** be a square, symmetric tridiagonal matrix whose eigenvalues and eigenvectors are sought. Let

$$\mathbf{M}_s = \mathbf{Q}_s\mathbf{R}_s \quad s = 0,1,2,\ldots \tag{7.63}$$

where

$\mathbf{Q}_s$ is a certain orthogonal matrix

$\mathbf{R}_s$ is an upper triangular matrix (not to be confused with the original matrix **R** which has no subscript here)

Then

$$\mathbf{M}_{s+1} = \mathbf{R}_s\mathbf{Q}_s$$

Theoretically, the sequence of matrices, $\mathbf{M}_0, \mathbf{M}_1, \mathbf{M}_2, \ldots$, converges to a diagonal matrix of the eigenvalues of **M**, which are also the eigenvalues of the original square symmetric matrix **R**, because **M** is a similarity transformation, $\mathbf{M} = \mathbf{T}'\mathbf{R}\mathbf{T}$ of **R**. Similar matrices have the same eigenvalues.

The affinities of this method to the Jacobi method are seen if we premultiply both sides of Equation 7.63 by $\mathbf{Q}_s'$, which transforms $\mathbf{M}_s$ into the upper triangular matrix

$$\mathbf{R}_s = \mathbf{Q}_s'\mathbf{M}_s \tag{7.64}$$

Then

$$\mathbf{M}_{s+1} = \mathbf{Q}_s'\mathbf{M}_s\mathbf{Q}_s \tag{7.65}$$

and

$$\mathbf{D} = \cdots \mathbf{Q}_3'\mathbf{Q}_2'\mathbf{Q}_1'\mathbf{Q}_0'\mathbf{M}_0\mathbf{Q}_0\mathbf{Q}_1\mathbf{Q}_2\mathbf{Q}_3 \cdots \tag{7.66}$$

where **D** is the diagonal matrix of eigenvalues of $\mathbf{M} = \mathbf{M}_0$ and the product of the **Q** matrices yields the eigenvector matrix **V** of **M**. Then, the eigenvectors of the original square symmetric matrix **R** are given by

$$\mathbf{A} = \mathbf{T}\mathbf{Q}_0\mathbf{Q}_1\mathbf{Q}_2\mathbf{Q}_3 \cdots = \mathbf{T}\mathbf{V} \tag{7.67}$$

The Q-R method differs from the Jacobi method in the manner in which the orthogonal matrices, $\mathbf{Q}_s$, are defined. In the Jacobi method, the orthogonal matrices annihilate a single, off-diagonal element of the matrix at a time, while the orthogonal **Q** matrices produce upper triangular matrices when their transposes are premultiplied times the $\mathbf{M}_s$ matrices.

Ortega and Kaiser (1963) proposed that one obtain $\mathbf{Q}_s$ as the product of $n - 1$ elementary plane transformations applied to the matrix $\mathbf{M}_s$, that is,

$$\mathbf{Q}_s = \mathbf{P}_1\mathbf{P}_2 \cdots \mathbf{P}_{n-1} \tag{7.68}$$

where the hth matrix $\mathbf{P}_h$ has the form

$$P_h = \begin{bmatrix} 1 & & & & & & & \\ & \ddots & & & & & & \\ & & 1 & & & & & \\ & & & c_h & -s_h & & & \\ & & & s_h & c_h & & & \\ & & & & & 1 & & \\ & & & & & & \ddots & \\ & & & & & & & 1 \end{bmatrix} \tag{7.69}$$

We note that these elementary transformation matrices only operate on the principal diagonal and immediate off-diagonal elements of matrix $\mathbf{M}_s$.

Then

$$\mathbf{M}_{s+1} = \mathbf{P}'_{n-1} \cdots \mathbf{P}'_2 \mathbf{P}'_1 \mathbf{M}_s \mathbf{P}_1 \mathbf{P}_2 \cdots \mathbf{P}_{n-1} \tag{7.70}$$

Thus we see that after the initial matrix $\mathbf{R}$ has been tridiagonalized to $\mathbf{M}_0$, the Q-R method can proceed, just as in the method of Jacobi, using successions of elementary, orthogonal plane rotations.

Further details of this method are described in Ortega and Kaiser (1963) and Mulaik (1972).

7.4.6 Powers of Square Symmetric Matrices

Consider that

$$\mathbf{R}^2 = \mathbf{RR}$$

which we may rewrite as

$$\mathbf{R}^2 = \mathbf{ADA'ADA'} = \mathbf{ADDA'} = \mathbf{AD}^2\mathbf{A'}$$

In general

$$\mathbf{R}^n = \mathbf{AD}^n\mathbf{A'} \tag{7.71}$$

This even extends to

$$\mathbf{R}^{1/2} = \mathbf{AD}^{1/2}\mathbf{A'} \tag{7.72}$$

because

$$\mathbf{R} = \mathbf{R}^{1/2}\,\mathbf{R}^{1/2} = \mathbf{AD}^{1/2}\mathbf{A'AD}^{1/2}\mathbf{A'} = \mathbf{AD}^{1/2}\,\mathbf{D}^{1/2}\mathbf{A'} = \mathbf{ADA'}$$

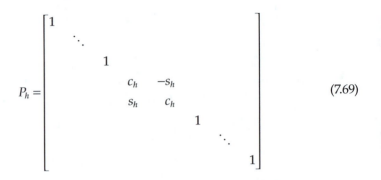

A square symmetric matrix raised to the *n*th power has the same eigenvectors as the original matrix, but with eigenvalues equal to the original eigenvalues raised to the *n*th power.

7.4.7 Factor-Loading Matrix from Eigenvalues and Eigenvectors

Generalizing from the expression for a single column $\boldsymbol{\lambda}_1$ of the factor-loading matrix in Equation 7.43 to the factor-loading matrix $\boldsymbol{\Lambda}$ itself

$$\boldsymbol{\Lambda} = \mathbf{A}_r \mathbf{D}_r^{1/2} \tag{7.73}$$

where

- $\boldsymbol{\Lambda}$ is the $n \times r$ factor-loading matrix of $\mathbf{R}_c = \mathbf{R}_{YY} - \boldsymbol{\Psi}^2$
- $\mathbf{A}_r$ is an $n \times r$ matrix whose r columns are the first r eigenvectors of $\mathbf{R}_c$
- $\mathbf{D}_r^{1/2}$ is the square root of the $r \times r$ diagonal matrix of the first r largest eigenvalues of $\mathbf{R}_c$, where r is determined to be the number of factors

We will consider how this number is determined in Chapter 8.

8

Common-Factor Analysis

8.1 Preliminary Considerations

As we indicated in Chapter 6, factor analysis is a hypothesis-developing method to be used principally at the outset of a research program designed to suggest hypothetical variables that may be immediate, common causes of a set of observed variables. The assumption is that the common causes are "latent" and not directly measured, which is why they are hypothetical. Another assumption is that the common causes are fewer in number than the observed variables studied. As explanatory entities, the common factors need to be fewer because there would be no saving in thought if the number of explanatory entities equaled the number of entities to be explained (Thurstone, 1937). Furthermore, we assume that the common factors (as we will call the latent variables) are not dependent on the particular set of observed variables for their existence, but rather that they exist in their own right, and may be found in studies with other observed variables. Furthermore, "The factorial description of a test must remain invariant when the test is moved from one battery to another" (Thurstone, 1937, p. 75). This implies that a variable's factor pattern coefficients should be invariant regardless of what other variables are embedded with it. The indeterminacy of the common and unique factors of the common-factor model, which mathematically results from there being only n observed variables with which to determine $n + r$ latent variables, allows the latent variables to be more than what may be known of them from the observed variables.

We also indicated in Chapter 6 that the common-factor model is a rather restricted model of causal relations. The model only models causal relations between common factors and observed variables or between unique factors and their counterpart observed variable. No causal relations between the common factors themselves are modeled. The relations between common factors will be correlations, if they exist.

The unique factors are presumed to be uncorrelated, which reflects the idea that if we partial the common factors from the observed variables, they should have no residual covariation between them. This reflects a principle of conditional independence among variables conditioned on their immediate common causes. This also suggests that the number of common factors would be the number of those common factors that when partialled from the original correlation matrix, leave the residual observed variables uncorrelated. So, this gives rise to a criterion for the number of common factors.

8.1.1 Designing a Factor Analytic Study

Factor analysis is about variables. Variables do not exist solely on their own. They represent attributes or states of objects. Any consideration of the variables to be selected for a factor analysis should also take into account the objects, subjects, or persons whose attributes or states are to be measured by the variables. Consideration of the subjects studied is not always thought of by those doing factor analyses. But it is an essential consideration if one wishes the factor analysis to accurately represent something more than an artifactual average mishmash of relations between variables. The subjects chosen should be homogeneous in being influenced by the same common factors, subject to the same factor loadings indicating how much a unit change in a common factor produces a change in the observed variables. The subjects may differ in the scores or values they have on the variables, but the underlying causal structure should be the same for all subjects.

Of course, this is an ideal that is next to impossible to achieve in practice the first time around, but as an ideal it represents a situation to strive for. One approach to deal with this is to strive for homogeneity of subjects. Subjects chosen for study should be as homogeneous as possible on variables other than those chosen as the observed variables for study. That means one should strive to have subjects that are homogeneous in age, homogeneous in education, homogeneous in cultural background and experiences, homogeneous in training, homogeneous in socioeconomic level, homogeneous in language, and so on. Another approach is to ensure that any extraneous variables on which subjects vary are independent of those variables under study—both observed and latent. Any variation in subjects that may give rise to variation in the causal relations between common factors and observed variables should be controlled for. Of course, this seems to put the cart before the horse. But unless one gives some thought to this problem, one will not anticipate that which can compromise and contaminate a factor analytic study and take steps to deal with it. There is a certain aspect of pulling oneself up by one's bootstraps in doing factor analysis.

The restricted nature of the common-factor model suggests that to use optimally its causal structure to identify common causes of observed variables, considerable planning and thought should enter in at the outset of a factor analytic study. One needs to identify the domain of variables to be studied. Identifying the domain of variables to be studied goes with identifying the population of subjects appropriate for the study as already discussed. Once the domain is generally identified, one needs to specify the potential common factors of this domain. This in turn leads to further refinements of the concept of the domain as all those variables that are effects of the specified common factors (Thurstone, 1947). In order to recover these factors, one needs to have 3, 4, or more variables that are effects of these common factors (Thurstone, 1937) to overdetermine the factors. Hence, one needs a clear idea of the kinds of variables that are effected by the factors. This gives rise next to the specification of constitutive rules for selecting or constructing indicators of these common factors. Then the researcher should construct or select indicators that are measures of these latent variables according to these specifications.

L. L. Thurstone (1951), for example, in his studies of mechanical aptitudes, spent months thinking about, designing and constructing problem sets to represent variables that elicit certain aptitudes and abilities he expected mechanical engineers to have—all before any data were collected from subjects. It was essential to Thurstone that tests included in the study be as simple as possible in depending on as few common factors as possible. Otherwise, the factors will be difficult to interpret from observed variables having many factors in common. Only after the tests were built was Thurstone ready to have subjects perform on these tests to obtain scores on the test variables.

The next stage in a factor analytic study concerns collecting measurements of subjects or objects on the observed variables chosen for study. Although rarely explicitly treated in discussions of factor analytic methodology, good measurement requires control of extraneous influences that may affect the responses measured. Measurements obtained at different sites for later analysis should be obtained in as uniform a set of conditions across all sites as possible. Subjects when responding should be free of distractions, should be well-rested, and free of influence from other subjects during the measurements. Instructions to subjects should be clearly understood, uniform, and standard. If responses are later scored by the researcher, a uniform and clearly understood set of instructions for the scoring process should be followed by all scorers.

Once scores are obtained on the variables, they are ready for analysis. Today the analyses are performed automatically by computers, with the researcher only specifying certain options to be followed, e.g., whether to compute correlations or covariances, what kind of factor extraction or estimation method to use, what kind of factor rotation to use, whether to obtain factor scores or not, what magnitude of loadings to display on output, etc. Different programs have different options.

8.2 First Stages in the Factor Analysis

A factor analysis proper begins with a correlation matrix computed and ready for analysis. In some cases the matrix input to the analysis may be a covariance matrix. (There are advantages in working with covariance matrices that are discussed in Chapter 14.)

Two intertwined problems have to be solved at the outset: (1) What are the unique-factor variances (or complementarily, what are the communalities of the observed variables), and (2) how many common factors are there? It will turn out that these are not separate problems. Solving one will turn out to depend on solving the other. Usually reasonable but still provisional solutions for the communality problem will allow a tentative solution for the number-of-factors problem, which in turn may allow for a refinement of the initial solution for the communalities, which in turn may involve a modification of the solution for the number of factors, and so on. To understand these problems further we need to understand certain concepts that were developed early in the development of the methodology of factor analysis.

8.2.1 Concept of Minimum Rank

The indeterminacy of the common-factor-analysis model seems not to have caused great concern for Thurstone who championed this model in the United States in the 1930s through to his death in the mid-1950s. To Thurstone, indeterminacy was not so great as to prevent reasonable if sometimes still approximate solutions for the common and unique factors. Much could be learned from approximations at the outset of a field of study. Moreover, the indeterminacies of factor scores, which we take up in Chapter 13, were not a major concern to him, because he rarely obtained or computed factor scores. What he was interested in was the relations between common factors and observed variables as given by the factor loading coefficients. We shall, see, however in our chapter on factor scores, that some aspects of factor score indeterminacy would present problems not just for getting factor scores but in interpreting the common factors from the factor loadings themselves.

For Thurstone the objective of a common-factor analysis was to obtain fewer common factors than observed variables. If this objective could not be attained, he believed, then there would be very little advantage in performing a factor analysis, because the full set of observable variables would be as good as any set of variables to explain relationships between variables. Setting this requirement for common-factor analysis, Thurstone then proceeded to establish a criterion for common-factor analysis for the number of common factors. His criterion was essentially this: The number of common factors is determined by the rank of the correlation matrix after an appropriate set of communalities is inserted in place of 1s in the principal diagonal of the correlation matrix. By an appropriate set of communalities Thurstone meant that the communalities must leave the reduced correlation matrix Gramian. This means that the reduced correlation matrix must be conceived of as being derived by taking some matrix and multiplying it times its transpose. What he had specially in mind was that the common-factor loading matrix $\mathbf{\Lambda}$, should reproduce the correlation matrix $(\mathbf{R} - \mathbf{\Psi}^2)$, that is

$$(\mathbf{R} - \mathbf{\Psi}^2) = \mathbf{\Lambda}\mathbf{\Lambda}'$$

Although Thurstone stated his criterion of minimum rank in terms of determinants, an easier way of stating this criterion today arises out of the principal-axes methods of factoring. A Gramian matrix has only nonnegative eigenvalues equal in number to the rank of the matrix. The rest of its eigenvalues are zero. Thurstone recognized that, if the communalities of the variables were overestimated, the factors would account for more than just the off-diagonal correlations and not reduce the rank of the matrix sufficiently. It would not be a "common-factor analysis." On the other hand, if the communalities were underestimated, one would also get a matrix with too high rank. This would result from producing a non-Gramian matrix that had negative eigenvalues. (The rank of a square, symmetric matrix equals the number of positive and negative eigenvalues of the matrix.) But factors corresponding to negative eigenvalues would represent variances for the respective common factors that had imaginary values for their variances. This would make no sense. Thurstone believed that there was a point between underestimating and overestimating the communalities where the rank of the reduced correlation matrix would be minimal. This point was to be determined by various trial-and-error techniques.

Thurstone implicitly assumed that the common-factor model holds. This would imply that $\mathbf{\Psi}^2$ is a diagonal, usually positive definite matrix with $n(n-1)/2$ fixed off-diagonal zeros. By "fixed" we mean values that remain unchanged during the analysis or computations.

But suppose, in fact, that $\mathbf{\Psi}^2$ is not diagonal by having nonzero off-diagonal elements. That could come about if there exists pairwise covariation between variables that are due to what factor analysts call "doublet factors." A doublet factor has nonzero loadings on only two variables, the two variables that covary. But in this case for any such doublet factor there are two pattern loadings to determine and only one residual covariance coefficient to determine them with. So, from $\psi_{jk} = \lambda_{js}\lambda_{ks}$, where ψ_{jk} is the off-diagonal residual covariance between variables j and k, while λ_{js} and λ_{ks} are the unknown pattern loadings on their common doublet factor s, the solution for λ_{js} and λ_{ks} in this case is underdetermined, or "underidentified." We will learn more about identification and underidentification when we study confirmatory factor analysis and structural equation modeling in a later course. Potentially there could exist $n(n-1)/2$ doublet factors, one for each pair of covariances in $\mathbf{\Psi}^2$. But now this would mean that the common-factor model postulates r latent factors, each common to three or more variables, at most and additional $n(n-1)/2$ doublet factors and n unique factors.

Furthermore, with doublet factors, the number of parameters to estimate of the resulting expanded common-factor model now exceeds the number of observed correlations by which to determine them. By this we mean that the fundamental theorem, $\mathbf{R} = \mathbf{\Lambda\Lambda}' + \mathbf{\Psi}^2$, implies a system of $n(n+1)/2$ equations, one equation for each distinct observed coefficient in $\mathbf{R}$ that makes the correlation coefficient on the left-hand side, a function of unknown parameters on the right-hand side. We can seek to solve for the parameters on the right-hand side of the equation, by treating this as a system of $n(n + 1)/2$ nonlinear equations with unknown parameters equal to the number of unknowns on the right-hand side of the equation. The number of unknown parameters on the right-hand side in the common-factor model with added doublet factors is $nr - r(r-1)/2 + n(n+1)/2$, which exceeds the number $n(n + 1)/2$ of known parameters on the left-hand side. (The expression $r(r - 1)/2$ represents the number of constraints on the factor pattern matrix needed to achieve identification, which must be subtracted out.)

Recognizing the difficulties with possible doublet factors present, a rational approach to eliminating them is to try to eliminate them from the start by a careful examination of the content of the observed variables to identify any possible way a pair of variables could be correlated other than by their depending on the well-defined common factors. So, one of the pair could be replaced by another variable hypothetically loading on these common factors that does not have any pairwise relations with other variables. Often these pairwise relations are due to formatting similarities or presence of certain words with similar but extraneous meanings in the test items that some respondents react similarly to.

The way to overcome underidentification is to specify fixed values for some of the parameters on the right-hand side so that at most as many unknown parameters need to be estimated as the $n(n + 1)/2$ distinct elements of $\mathbf{R}$. Then they no longer are unknowns. Specifying that $\mathbf{\Psi}^2$ is a diagonal matrix introduces constraints into the model that contribute to making it estimable. We no longer need to estimate $n(n - 1)/2$ off-diagonal elements of $\mathbf{\Psi}^2$, but treat them all as zero. But because we want our model to correspond to reality, we must try to anticipate any ways in which the variables selected would still covary after partialling out our expected common factors and take steps to eliminate those sources of covariation so that $\mathbf{R} - \mathbf{\Lambda\Lambda}' = \mathbf{\Psi}^2$ is indeed a diagonal matrix.

In addition to emphasizing the need for a minimum-rank solution, Thurstone also stressed the importance of overdetermining the factors by an appropriate selection of observed variables pertaining to them. Each factor, Thurstone argued should be determined by more than just two variables; otherwise the factors would have tenuous scientific status. In effect, Thurstone's argument was based upon the scientific principle that the number of experimentally independent observations must exceed the number of independent parameters to be determined. Thurstone then demonstrated how many common factors r would be independently determined by a given number n of variables in a correlation matrix. To do this he considered that the factor-loading matrix $\mathbf{\Lambda}$ would contain a number of parameters (the factor loadings) that would have to be determined (estimated) from a number of observed parameters (the correlation coefficients). Since the communalities were considered unknown and to be determined from the correlation coefficients off the diagonal and the principal diagonal of $\mathbf{R}$, this left only $n(n - 1)/2$ coefficients to determine the solution for $\mathbf{\Lambda}$.

Next, Thurstone turned to the question of how many independent parameters (factor loadings) would have to be determined, if, say, r common factors were to be extracted. To answer this question, Thurstone relied upon the fact that in the diagonal method of factoring with orthogonal factors, each successive factor extracted is collinear with one of the variables remaining after previous factors have been partialled from it. The factor loading matrix

$$\mathbf{\Lambda} = \begin{bmatrix} \mathbf{L} \\ \mathbf{U} \end{bmatrix}$$

obtained with this method has an $r \times r$ lower triangular submatrix $\mathbf{L}$ in its first r rows, meaning $r(r-1)/2$ coefficients above the principal diagonal of these first r rows are necessarily fixed to zero. There are other ways to obtain this lower triangular matrix, but it requires methods covered under factor rotation later in the book. The elements below the first r rows of $\mathbf{\Lambda}$ in submatrix $\mathbf{U}$ are unknown and estimated. Hence the number of coefficients to be estimated in $\mathbf{\Lambda}$ is nr minus the number of fixed zeros, $r(r-1)/2$. A necessary (but not sufficient) condition that a solution exists for the unknown parameters would be that

$$nr - r(r-1)/2 \le n(n-1)/2 \tag{8.1}$$

The expression on the left-hand side is the number of unknown parameters to be estimated and the number on the right-hand side is the number of off-diagonal correlations in $\mathbf{R}$ with which to estimate them.

In words, inequality (Equation 8.1) states that the number of independently determinable common-factor loadings should be always less than or equal to the number of distinct correlation coefficients in the correlation matrix. If we now consider the case where we set the number of distinct determinable parameter loadings to the number of distinct correlation coefficients, which would occur when we have the maximum number of determinable factors, we obtain the following equation:

$$nr - \frac{r(r-1)}{2} = \frac{n(n-1)}{2} \tag{8.2}$$

Let us now bring the right-hand side of Equation 8.2 over to the left-hand side and multiply by 2 to get rid of the denominators. This gives

$$2nr - r(r-1) - n(n-1) = 0$$

or

$$2nr - r^2 + r - n^2 + n = 0$$

Rearranging and solving the quadratic equation for *r*, we find the following formula:

$$r \le \frac{(2n+1)-\sqrt{8n+1}}{2} \tag{8.3}$$

This is a well-known inequality which was derived not only by Thurstone (1947, p. 293), but also earlier by Lederman (1937), (who is honored by this being known as the "Lederman inequality"), and Harman (1960, pp. 69–73), each using slightly different assumptions. In any case the inequality states that the maximum number of determinable common factors *r* is a whole-integer value that is less than or equal to but never greater than the expression on the right-hand side of the inequality in Equation 8.3 based on the number of observed variables.

The inequality in Equation 8.3 can also be rearranged so that we can solve for *n* given a prespecified number of factors *r* sought. The formula is

$$n \ge \frac{(2r+1)+\sqrt{8r+1}}{2} \tag{8.4}$$

The inequality in Equation 8.4 has been popularly cited as an indicator of the minimum number of variables that is necessary but not sufficient to include in a correlation matrix if one is to determine *r* common factors. In Table 8.1

TABLE 8.1

Number of Variables, *n*, Needed to Minimally Overdetermine, *r*, Common Factors

r	*n*	*r*	*n*
1	3	25	33
2	5	30	39
3	6	35	44
4	8	40	50
5	9	45	55
6	10	50	61
7	12	55	66
8	13	60	72
9	14	65	77
10	15	70	83
11	17	75	88
12	18	80	94
13	19	85	99
14	20	90	104
15	21	95	110
20	27	100	115

we have indicated the value of n (the number of variables required) to determine r (the number of factors to be determined) according to Equation 8.4. Notice that to determine one factor one must have three variables. To determine five factors, one needs at least nine variables. However to determine 75 factors one needs only 88 variables. As the number of factors to be determined increases, the ratio of the number of factors to the number of variables required approaches 1.

Equation 8.3, which indicates the number of independently determinable factors, given n tests, has received an erroneous interpretation by some factor analysts. Some have believed that an upper bound to the minimum rank of the correlation matrix would be given by Equation 8.3. That is to say, some factor analysts have believed that by modifying the elements of the principal diagonal of the correlation matrix so that they are between 0 and 1 in value, one obtains a matrix for which the maximum possible rank would be given by Equation 8.3.

However, Guttman (1958) showed that by modifying the principal diagonal elements to have values between 0 and 1, one can still obtain a matrix with a rank of $n - 1$, which is a value greater than that given by Equation 8.3. For example, one could subtract the smallest eigenvalue of the correlation matrix $\mathbf{R}$ from each of the diagonal elements in the correlation matrix, and the rank of the resulting positive, semidefinite reduced correlation matrix would be $n - 1$. At best Equation 8.3 says nothing about maximum possible rank but is concerned only with the question of the number of factors that can be determined from the intercorrelations of a given number of variables as far as the count of independent parameters is concerned.

In Chapter 13 we see how Thurstone's stress of the overdetermination of factors is a good policy. For if one does not select enough variables, say at least three or four distinct variables for each factor to be determined, there is a danger of obtaining a correlation matrix to which an application of the common-factor model would be inappropriate, since in this case the factors obtained by the model would be quite indeterminate. We shall discuss a criterion of the applicability of the model of common-factor analysis to a correlation matrix in connection with a discussion of Guttman's image analysis in Chapter 9.

8.2.2 Systematic Lower-Bound Estimates of Communalities

A great number of methods were reported in the literature up to the 1960s for directly estimating the communalities. Thurstone (1947), for example, recommended as a simple, first approximation using the absolute magnitude of the largest correlation between a variable and other variables as the communality for that variable. He also described many methods of estimating communalities up to his time. Nevertheless none of these methods have succeeded in achieving minimum rank every time. This has resulted in the need to apply trial-and-error corrections to initial communality estimates to achieve some solution for minimum rank. As a matter of fact, trial-and-error efforts will always find some solution for minimum rank, although such a solution may not be unique and may involve a tremendous amount of computing. Several iterative schemes have also been tried, some converging to solutions, but there has been no proof that such solutions produce minimum rank.

Such lack of success with these methods for solving the communality problem led Guttman (1954a,b, 1956) to consider lower-bound estimates of the communalities. Lower-bound estimates of the communalities also determine lower-bound estimates of minimum rank. Thus lower-bound estimates can be useful in providing the minimum number of factors that one should keep in a factor analysis. We will now consider Guttman's lower bounds.

8.2.3 Congruence Transformations

As background for Guttman's proofs of the lower bounds, we need to consider a number of concepts from matrix algebra. In particular, we need to consider what are known as "congruence transformations": Let **R** and **B** be two $n \times n$ square matrices such that there exists another $n \times n$ square, nonsingular matrix **A** which transforms **R** into **B** in the following manner:

$$\mathbf{B} = \mathbf{A'RA} \tag{8.5}$$

When a square matrix **B** is derived from a square matrix **R** by means of a third square matrix **A** by premultiplying **R** by the transpose of **A** and postmultiplying **R** by **A**, the matrix **B** is said to be congruent to **R**.

We shall now give an important theorem concerning congruence transformations of square symmetric matrices: *Every symmetric matrix* **R** *of rank r is congruent to a diagonal matrix* **D** *whose first r diagonal elements (if arranged in descending order of magnitude) are nonzero and all other elements are zero.*

In other words, we can take any square symmetric matrix **R** and find some other nonsingular matrix that transforms **R** into a diagonal matrix by a congruence transformation. If the matrix **R** has a rank of *r*, there will be *r* nonzero elements in the principal diagonal of the diagonal matrix, whereas the rest of the diagonal elements will be zero. A consequence of this theorem is that, if we can find a congruence transformation that transforms a square symmetric matrix **R** to a diagonal matrix **D**, we can determine the rank of **R**. We need only count the number of positive and negative diagonal elements of the diagonal matrix **D** to learn the rank of **R**.

8.2.4 Sylvester's Law of Inertia

However, if we can transform the matrix **R** to a diagonal matrix **D** by a congruence transformation, **D** is not the only diagonal matrix congruent to **R**. There are many nonsingular transformation matrices which will transform **R** into a diagonal matrix by a congruence transformation. Each of these diagonal matrices may be different. But a question which might be raised concerning all the diagonal matrices congruent to **R** is whether the numbers of positive-, negative-, and zero-value diagonal elements of these diagonal matrices are respectively the same. The answer to this question is yes, and it is given by a theorem known as "Sylvester's law of inertia," named after a well-known mathematician of the nineteenth century who pioneered work with matrices. Sylvester's law of inertia can be stated for our purposes as follows: *Every diagonal matrix* **D** *that is congruent to a square symmetric matrix* **R** *contains the same number of positive, negative, and zero diagonal elements.*

So that we can talk about the numbers of positive, negative, and zero diagonal elements of a diagonal matrix **D** congruent to a given square symmetric matrix **R**, let us call the number of positive diagonal elements in **D** the "positive index" of the matrix **R**, the combined number of positive and negative elements in **D** the "nonzero index" of **R**, and the combined number of positive and zero elements in **D** the "nonnegative index" of **R**. In effect the nonzero index of **R** is the rank r. It naturally follows that for any given matrix the positive index of the matrix is always less than or equal to the nonzero index or the nonnegative index of the matrix.

8.2.5 Eigenvector Transformations

The properties of congruence transformations also apply to the eigenvector transformation matrix that transforms a given matrix into a diagonal matrix having eigenvalues in the principal diagonal (cf. Equation 7.56). The eigenvector matrix, however, being an orthogonal matrix, is a unique transformation matrix, and so the resulting diagonal matrix of eigenvalues is always unique. But the diagonal matrix of eigenvalues for a given square symmetric matrix belongs to the set of all diagonal matrices congruent to the given square matrix. This is because all diagonal matrices of eigenvalues for square symmetric matrices are obtained by congruence transformations.

8.2.6 Guttman's Lower Bounds for Minimum Rank

We are ready to define Guttman's lower bounds for the minimum rank of a reduced correlation matrix. Let r_1 be the number of eigenvalues greater than 1.00 of the correlation matrix **R** with 1s in its principal diagonal. This corresponds to the number of positive eigenvalues of the correlation matrix with 0s in its principal diagonal. For proof, define $\mathbf{R}_1$ as the correlation matrix with 0s in its principal diagonal. That is given as

$$\mathbf{R}_1 = \mathbf{R} - \mathbf{I} \tag{8.6}$$

It follows that the eigenvector matrix **A** of **R** is also an eigenvector matrix of $\mathbf{R}_1$. As proof of this, use Equation 7.56 to diagonalize $\mathbf{R}_1$, that is,

$$\mathbf{A}'\mathbf{R}_1\mathbf{A} = \mathbf{A}'(\mathbf{R} - \mathbf{I})\mathbf{A} = \mathbf{A}'\mathbf{R}\mathbf{A} - \mathbf{A}'\mathbf{I}\mathbf{A} = \mathbf{D} - \mathbf{I} \tag{8.7}$$

where **D** is the matrix of eigenvalues of **R**. Since **D** – **I** is also a diagonal matrix, it must also be the eigenvalue matrix of $\mathbf{R}_1$. But each of the eigenvalues of $\mathbf{R}_1$ is an eigenvalue of $\mathbf{R}-1$. Hence all eigenvalues of **R** less than 1 correspond to negative eigenvalues of $\mathbf{R}_1$, and all eigenvalues greater than 1 of **R** correspond to positive eigenvalues of $\mathbf{R}_1$.

Define r_2 as the number of positive eigenvalues of a matrix $\mathbf{R}_2$, where

$$\mathbf{R}_2 = \mathbf{R} - \mathbf{D}_2 \tag{8.8}$$

$\mathbf{D}_2$ is an $n \times n$ diagonal matrix where the jth diagonal element is defined as $d_{jj} = 1 - r_{j\max}^2$, with $r_{j\max}$ the largest off-diagonal correlation in absolute magnitude in the jth column of **R**. In other words, $\mathbf{R}_2$ corresponds to the correlation matrix with the unities in the principal diagonal replaced by the squares of the largest respective correlation coefficients in the corresponding columns.

Define r_3 as the number of positive eigenvalues of a matrix $\mathbf{R}_3$, where

$$\mathbf{R}_3 = \mathbf{R} - \mathbf{S}^2 \tag{8.9}$$

The matrix $\mathbf{S}^2$ is defined as $\mathbf{S}^2 = [\text{diag } \mathbf{R}^{-1}]^{-1}$ (cf. Equation 4.52). $\mathbf{S}^2$ contains the errors of estimate for predicting each respective variable from all the $n - 1$ other variables in the correlation matrix. The jth diagonal element of $\mathbf{R} - \mathbf{S}^2$ is then equal to $R_j^2 = 1 - s_{jj}^2$, the squared multiple correlation for predicting the jth variable from the $n - 1$ other variables in **R**.

8.2.7 Preliminary Theorems for Guttman's Bounds

Guttman's proofs that r_1, r_2, and r_3 are lower bounds to the minimum rank of the reduced correlation matrix are classic and very elegant (to a mathematician's eye for beauty), and we will go through them here.

Guttman first asks us to consider for the sake of argument the following lemma: *Let* **S** *be any real symmetric matrix,* $\mathbf{G}_0$ *be a nonsingular real Gramian matrix, and* **H** *their sum:*

$$\mathbf{H} = \mathbf{S} + \mathbf{G}_0 \tag{8.10}$$

Let h be the positive index of **H**, *and let s be the nonnegative index of* **S**. *Then*

$$h \geq s \tag{8.11}$$

The positive index of **H** is greater than or equal to the nonnegative index of **S**. In other words, the positive index of a matrix which is the sum of a square symmetric matrix and a nonsingular Gramian matrix is always greater than or equal to the nonnegative index of the square symmetric matrix.

For proof of this lemma, Guttman observes that $\mathbf{G}_0$ is a Gramian matrix. Thus, by the definition of a Gramian matrix, there must exist some other nonsingular matrix, call it **F**, such that

$$\mathbf{G}_0 = \mathbf{F}\mathbf{F}' \tag{8.12}$$

He then asks us to consider two other arbitrarily defined matrices. Let

$$\mathbf{H}_0 = \mathbf{F}^{-1}\mathbf{H}(\mathbf{F}')^{-1} \quad \text{and} \quad \mathbf{S}_0 = \mathbf{F}^{-1}\mathbf{S}(\mathbf{F}')^{-1} \tag{8.13}$$

Thus, if we premultiply $\mathbf{H}$ in Equation 8.10 by $\mathbf{F}^{-1}$ and postmultiply $\mathbf{H}$ by $(\mathbf{F}')^{-1}$, we obtain

$$\begin{aligned} \mathbf{H}_0 &= \mathbf{F}^{-1}\mathbf{H}(\mathbf{F}')^{-1} = \mathbf{F}^{-1}\mathbf{S}(\mathbf{F}')^{-1} + \mathbf{F}^{-1}\mathbf{G}_0(\mathbf{F}')^{-1} \\ &= \mathbf{F}^{-1}\mathbf{S}(\mathbf{F}')^{-1} + \mathbf{F}^{-1}\mathbf{F}\mathbf{F}'(\mathbf{F}')^{-1} = \mathbf{S}_0 + \mathbf{I} \end{aligned} \tag{8.14}$$

Now, let us consider the eigenvalues of $\mathbf{H}_0$. These can be found by finding an eigenvector matrix $\mathbf{A}_H$ of $\mathbf{H}_0$ that diagonalizes $\mathbf{H}_0$, that is

$$\mathbf{A}'_H\mathbf{H}_0\mathbf{A}_H = \mathbf{A}'_H(\mathbf{S}_0 + \mathbf{I})\mathbf{A}_H = \mathbf{A}'_H\mathbf{S}_0\mathbf{A}_H + \mathbf{A}'_H\mathbf{A}_H \tag{8.15}$$

But $\mathbf{A}_H$ is an orthogonal matrix, and so its inverse is its transpose. Thus Equation 8.15 reduces to

$$\mathbf{D}_H = \mathbf{D}_S + \mathbf{I} \tag{8.16}$$

where

$\mathbf{D}_H$ is the matrix of eigenvalues of $\mathbf{H}_0$
$\mathbf{D}_S$ is the matrix of eigenvalues of $\mathbf{S}_0$
$\mathbf{I}$ the identity matrix

We know that $\mathbf{D}_S$ contains the eigenvalues of $\mathbf{S}_0$, because $\mathbf{D}_H$ cannot be a diagonal matrix if $\mathbf{D}_S$ is not itself a diagonal matrix and $\mathbf{D}_S$ is obtained by a congruence transformation with an orthogonal matrix $\mathbf{A}_H$.

We are now in a position to discuss the relationship of the eigenvalues of $\mathbf{H}_0$ to the eigenvalues of $\mathbf{S}_0$. Since $\mathbf{D}_H$ differs from $\mathbf{D}_S$ by the identity matrix $\mathbf{I}$, we know that all the positive eigenvalues of $\mathbf{S}_0$ correspond to positive eigenvalues of $\mathbf{H}_0$. Moreover, for every zero eigenvalue of $\mathbf{S}_0$ there corresponds an eigenvalue of 1 of $\mathbf{H}_0$. And for every negative eigenvalue of $\mathbf{S}_0$ greater than -1 there must be a positive eigenvalue for $\mathbf{H}_0$. Thus the combined number of zero and positive eigenvalues of $\mathbf{S}_0$ is always less than or equal to the number of positive eigenvalues of $\mathbf{H}_0$. If we designate the positive index of $\mathbf{H}_0$ as h_+ and the nonnegative index of $\mathbf{S}_0$ as s_{0+}, we have

$$h_+ \geq s_{0+} \tag{8.17}$$

We are now ready to use Sylvester's law of inertia. Consider a transformation matrix

$$\mathbf{P} = (\mathbf{F}')^{-1}\mathbf{A}_H \tag{8.18}$$

It follows from Equations 8.13 and 8.15 that

$$\mathbf{P}'\mathbf{H}\mathbf{P} = \mathbf{D}_H \quad \text{and} \quad \mathbf{P}'\mathbf{S}\mathbf{P} = \mathbf{D}_S \tag{8.19}$$

Thus the diagonal matrices $\mathbf{D}_H$ and $\mathbf{D}_S$ are congruent to the matrices **H** and **S**. By Sylvester's law of inertia it follows that the positive index h of **H** is the same as the positive index h_+ of $\mathbf{H}_0$, and the nonnegative index s of **S** is the nonnegative index s_{0+} of $\mathbf{S}_0$. Thus,

$$h = h_+ \quad \text{and} \quad s = s_{0+} \tag{8.20}$$

and by Equation 8.17 the inequality (Equation 8.11) stated in the lemma is true.

With this lemma proved, Guttman used it to prove the following important theorem: *Let* **R** *be any Gramian matrix, and let* **E** *be any diagonal matrix such that* **G** *is also Gramian, where* **G** is defined by

$$\mathbf{G} = \mathbf{R} - \mathbf{E} \tag{8.21}$$

Let **D** *be any diagonal matrix such that*

$$d_j > e_j \quad \textit{for all } j \tag{8.22}$$

where d_j and e_j are the jth diagonal elements, respectively of **D** *and* **E**. *Let*

$$\mathbf{S} = \mathbf{R} - \mathbf{D} \tag{8.23}$$

Let s be the nonnegative index of **S** *and r the rank of* **G**. *Then,*

$$r \geq s \tag{8.24}$$

The proof of this theorem is as follows: From Equations 8.21 and 8.23

$$\mathbf{G} = \mathbf{S} + (\mathbf{D} - \mathbf{E}) \tag{8.25}$$

The matrix (**D** – **E**) is a real diagonal matrix containing, by Equation 8.22, positive diagonal elements. Moreover, being a diagonal matrix, it is also a Gramian matrix because there will always exist for a diagonal matrix **D** with positive values on the diagonal another diagonal matrix $\mathbf{D}^{1/2}$ such that $\mathbf{D} = \mathbf{D}^{1/2}\,\mathbf{D}^{1/2}$. Thus we can refer to Equation 8.10 in the lemma previously proved to identify **G** with **H**, **S** with **S**, and (**D** – **E**) with $\mathbf{G}_0$. Consequently, the positive index of **G** is always greater than or equal to the nonnegative index of **S**. But the positive index of a Gramian matrix is also equivalent to its rank, because a Gramian matrix has no negative eigenvalues. Therefore, the rank of **G** is always greater than or equal to the nonnegative index of **S**.

Guttman's theorem provides us with the means to prove that the indices r_1, r_2, and r_3 defined in connection with Equations 8.6, 8.8, and 8.9 are lower bounds for estimates of the minimum rank of the reduced correlation matrix. The proof that these are lower bounds will rest upon showing that the communality estimates used with these indices for the minimal rank are themselves lower-bound estimates of the true communalities.

8.2.8 Proof of the First Lower Bound

Let **G** be the Gramian reduced correlation matrix such that

$$\mathbf{G} = \mathbf{R} - \mathbf{\Psi}^2$$

where

R is the original correlation matrix with 1s in the principal diagonal

$\mathbf{\Psi}^2$ is the diagonal matrix of (unknown) unique variances that creates the matrix **G** of minimum rank

Let r be the rank of **G**.

Consider the first lower bound r_1. This is the positive index of $\mathbf{R}_1$, where

$$\mathbf{R}_1 = \mathbf{R} - \mathbf{I} \tag{8.26}$$

$\mathbf{R}_1$ can be identified with **S** in Equation 8.23, $\mathbf{\Psi}^2$ corresponds with **E**, and **G** corresponds with **G** in Equation 8.21. All that remains is to show that the elements of the identity matrix are all greater than any one of the unique variances. This will be so if we restrict ourselves to the case where all communalities are greater than zero. Consequently, by Guttman's theorem, the nonnegative index of $\mathbf{R}_1$ must be less than the positive index of **G**, which is to say that the nonnegative index of $\mathbf{R}_1$ is less than the rank of **G**. It follows that r_1 is always less than or equal to the nonnegative index of $\mathbf{R}_1$. Hence r_1 is always less than or equal to r.

The lower bound r_1 can be seen as the weakest lower bound possible, if we assume that the estimates of the communalities cannot take on negative values. This is because the bound r_1 is based upon estimating every communality as zero. Of course, this is the worst possible estimate one can make of the true communalities. Kaiser (1961) advocated on heuristic grounds the use of this lower bound in factoring correlation matrices with 1s in the principal diagonal since the factors usually retained are usually interpretable whereas those rejected are normally difficult to interpret. With correlation matrices on the order of 75–100 variables the number of factors retained by this lower bound is from one-fourth to one-third of the number of variables. Also factoring indices with 1s in the principal diagonal is not common-factor analysis, but principal components.

8.2.9 Proof of the Third Lower Bound

We will postpone proof that the second bound r_2 is a lower bound to minimum rank of the reduced correlation matrix in order to consider proof of the third bound r_3 as a lower bound. If r_3 is a lower bound, to minimum rank, it can easily be shown that r_2 is a lower bound.

The bound r_3 is the number of positive eigenvalues of the matrix of correlations having in its principal diagonal the squares of the multiple correlations for predicting each variable from the $n - 1$ other variables. The proof that r_3

is a lower bound to minimum rank rests upon showing that the square of the multiple correlation for predicting a variable from the $n - 1$ other variables in the correlation matrix is always less than or equal to the communality of the variable.

Roff (1936) first suggested that the square of the multiple-correlation coefficient for predicting a variable from the $n - 1$ other variables in a correlation matrix is always less than or equal to the communality of the variable. Dwyer (1939) however is credited with a more rigorous proof of this theorem. Guttman (1940) also presented a rigorous proof. Kaiser (1960) claimed to have tested this theorem empirically on a high-speed computer and found it to hold in all cases tried. We will present a proof here based in part upon the proof of Dwyer (1939).

Let us construct a supermatrix $\mathbf{M}$ based upon the matrices $\mathbf{R}_{YY}$, $\mathbf{R}_{YX}$, and $\mathbf{I}_r$ such that

$$\mathbf{M} = \begin{bmatrix} \mathbf{R}_{YY} & \mathbf{R}_{YX} \\ \mathbf{R}_{XY} & \mathbf{I}_r \end{bmatrix} = \begin{bmatrix} \mathbf{R}_{YY} & \mathbf{\Lambda} \\ \mathbf{\Lambda}' & \mathbf{I}_r \end{bmatrix} \tag{8.27}$$

where

$\mathbf{R}_{YY}$ is the $n \times n$ correlation matrix with unities in the principal diagonal

$\mathbf{R}_{YX}$ is the $n \times r$ matrix of correlations between the n observed variables and the r common factors

$\mathbf{I}_r$ is the $r \times r$ correlation matrix among the common factors

In this case, the common factors are mutually orthogonal, meaning $\mathbf{R}_{YX}$ is the same as $\mathbf{\Lambda}$ in this special case. In other words, $\mathbf{M}$ is a correlation matrix for n observed variables and their r common factors.

Let $\mathbf{K}$ be a nonsingular transformation matrix of the same order as $\mathbf{M}$ such that

$$\mathbf{K} = \begin{bmatrix} \mathbf{I}_n & \mathbf{0} \\ -\mathbf{\Lambda} & \mathbf{I}_r \end{bmatrix} \tag{8.28}$$

The determinant of $\mathbf{K}$ is equal to 1 because only the permutation of elements in the diagonal of $\mathbf{K}$ will produce a nonzero product in the expansion of the determinant. Since the diagonal elements of $\mathbf{K}$ are all 1s, their product is 1, and so 1 is the determinant of $\mathbf{K}$.

Now consider the matrix product $\mathbf{MK}$:

$$\mathbf{MK} = \begin{bmatrix} \mathbf{R}_{YY} - \mathbf{\Lambda\Lambda}' & \mathbf{\Lambda} \\ \mathbf{0} & \mathbf{I}_r \end{bmatrix} = \begin{bmatrix} \mathbf{\Psi}^2 & \mathbf{\Lambda} \\ \mathbf{0} & \mathbf{I}_r \end{bmatrix} \tag{8.29}$$

where $\mathbf{\Psi}^2$ is the diagonal matrix of unique variances, obtained by subtracting from $\mathbf{R}_{YY}$ everything due to the common factors.

We can now show that the determinant of **M** is equal to the determinant of **MK**. This is because of the rule that states that if **A** and **B** are two square matrices of order *m* then the determinant of their product is equal to the product of their determinants; that is, if **C** = **AB**, then $|\mathbf{C}| = |\mathbf{A}||\mathbf{B}|$. In the present case, since the determinant of **K** is 1, the determinant of **MK** must be the determinant of the original matrix **M**.

Moreover, we can show that the determinant of **MK** is equal to the determinant of $\boldsymbol{\Psi}^2$. The determinant of **MK** is equal to the product of the diagonal elements of **MK** since this is an upper-triangular matrix. Since these diagonal elements of **MK** are the diagonal elements of $\boldsymbol{\Psi}^2$ and the unities of $\mathbf{I}_r$, the determinant of **MK** is equal to the determinant of $\boldsymbol{\Psi}^2$. Hence

$$|\mathbf{MK}| = |\mathbf{M}| = |\boldsymbol{\Psi}^2| = \psi_1^2\psi_2^2\cdots\psi_{j-1}^2\psi_j^2\psi_{j+1}^2\cdots\psi_n^2 \tag{8.30}$$

where

$|\mathbf{MK}|$, $|\mathbf{M}|$, and $|\boldsymbol{\Psi}^2|$ are the determinants, respectively, of **MK**, **M**, and $\boldsymbol{\Psi}^2$

ψ_j^2 is the unique variance of the *j*th variable found in the principal diagonal of $\boldsymbol{\Psi}^2$

Let M_{jj} be a minor of **M** and $\boldsymbol{\Psi}^2$ obtained by deleting the variable *j* from the *m* variables. The minor M_{jj} in other words, is the determinant of the submatrix of **M** found by excluding the row and column due to variable *j*. Consequently,

$$M_{jj} = \psi_1^2\psi_2^2\cdots\psi_{j-1}^2\psi_{j+1}^2\cdots\psi_n^2 \tag{8.31}$$

But M_{jj} is also a cofactor of **M** found by deleting the *j*th row and column of **M** that intersect in the element r_{jj} of the submatrix of $\mathbf{R}_{YY}$. Therefore, using the determinantal formula (cf. Equation 4.52) for multiple correlation, we have

$$R_j^2 = 1 - \frac{|\mathbf{M}|}{M_{jj}} \tag{8.32}$$

where R_j^2 is the square of the multiple correlation for predicting variable *j* from the $n - 1$ other variables and their *r* common factors. However, by dividing Equation 8.30 by Equation 8.31, we have

$$\frac{|\mathbf{M}|}{M_{jj}} = \psi_j^2 \tag{8.33}$$

Consequently

$$R_j^2 = 1 - \psi_j^2 = h_j^2 \tag{8.34}$$

where h_j^2 is the communality of the *j*th variable in $\mathbf{R}_{YY}$.

In conclusion, the square of the multiple-correlation coefficient for predicting a variable from the $n - 1$ other variables **and** the *r* common factors is equal to the communality of the variable.

To complete our proof we now must show that the square of the multiple-correlation coefficient for predicting a variable j from the $n - 1$ other variables in $\mathbf{R}_{YY}$ is always less than or equal to the square of the multiple correlation for predicting variable j from the $n - 1$ other variables *and* the r common factors, which equals the communality of variable j. Our proof readily follows when we consider that $\mathbf{R}_{YY}$ is a submatrix of $\mathbf{M}$, and thus the prediction of variable j in $\mathbf{R}_{YY}$ from just the $n - 1$ other variables in $\mathbf{R}_{YY}$ should produce a squared multiple correlation that is less than or equal to the squared multiple correlation for predicting variable j from the variables in $\mathbf{R}_{YY}$ and the r common factors, by Yule's equation [cf. 4.62].

We thus have proved the theorem: *The square of the multiple correlation for predicting a variable j from the n – 1 other variables in a correlation matrix is a lower bound to the communality of variable j.*

We now can prove that r_3 is a lower bound to the minimum rank of the reduced correlation matrix. Let $\mathbf{S}^2 = [\text{diag } \mathbf{R}_{YY}^{-1}]^{-1}$ be the diagonal matrix of errors of estimate for the multiple-correlation coefficients for predicting the variables of $\mathbf{R}$, respectively, from the $n - 1$ other variables in $\mathbf{R}$. That is, a single element is

$$s_j^2 = \frac{|\mathbf{R}_{YY}|}{R_{jj}} = 1 - R_j^2 \tag{8.35}$$

From Equation 8.33 and the theorem we have just proved and the definition of s_j^2 in Equation 8.35, we can directly show that

$$s_j^2 \geq \psi_j^2 \tag{8.36}$$

Consequently, if we identify the matrix $\mathbf{G}$ in Equation 8.21 with the reduced correlation matrix $\mathbf{R}_{YY} - \mathbf{\Psi}^2$, $\mathbf{E}$ in Equation 8.21 with $\mathbf{\Psi}^2$, $\mathbf{D}$ in Equation 8.23 with $\mathbf{S}^2 = [\text{diag } \mathbf{R}_{YY}^{-1}]^{-1}$, and $\mathbf{S}$ of Equation 8.23 with $\mathbf{R}_3$ of Equation 8.9, we can show immediately from Guttman's theorem that the positive index r_3 of $\mathbf{R}_3$ is always less than or equal to the minimum rank of the reduced correlation matrix $\mathbf{R}_{YY} - \mathbf{\Psi}^2$.

In conclusion, if we insert the squared multiple-correlation coefficients for predicting each variable from the $n - 1$ other variables into the principal diagonal of the correlation matrix, then the number of principal-axes factors with positive eigenvalues extractable from this correlation matrix is always less than or equal to the minimum rank of the correlation matrix.

8.2.10 Proof of the Second Lower Bound

The proof that r_2, the number of positive eigenvalues of a correlation matrix with the squares of the largest absolute correlation coefficients in each column in the corresponding principal diagonal position, is a lower bound to the minimum rank of the matrix, follows directly from the proof that r_3 is a lower bound for the estimate of the minimum rank of the matrix.

Let $r_{j\max}^2$ be the square of the correlation coefficient having the largest absolute value in column *j* of the correlation matrix. Then,

$$1 - r_{j\max}^2 \geq 1 - R_j^2 \tag{8.37}$$

because a squared multiple-correlation coefficient for predicting variable *j* is always greater than or equal to the squared correlation of one of the predictors with variable *j*. This follows from Yule's equation (Equation 4.62). As in the case of bounds r_1 and r_3, it follows directly that r_2 is also a lower bound to the minimum rank of the reduced correlation matrix.

In carrying out the proofs of these three lower bounds for the estimates of the minimum rank of the reduced correlation matrix, we have indirectly shown that an order exists among these bounds. That is to say

$$r_1 \leq r_2 \leq r_3 \leq r \tag{8.38}$$

We have also shown that a corresponding order exists among the communality estimates, that is,

$$0 \leq r_{j\max}^2 \leq R_j^2 \leq h_j^2 \tag{8.39}$$

We now can conclude that in practice the best systematic lower-bound estimate of minimum rank is given by r_3. It is obtained by entering into each diagonal position of the correlation matrix the corresponding squared multiple correlation for predicting the corresponding variable from all the other variables in the correlation matrix and obtaining the positive index of the resulting matrix. If one takes the diagonal elements of the matrix inverse of the correlation matrix and finds the reciprocals of these elements, he will have the errors of estimate for each of the variables in the matrix. These errors of estimate are close upper bounds to the unique variances and may thus be subtracted from the 1s in the principal diagonal of the correlation matrix to obtain a close lower-bound estimate of the communalities. One can then proceed to factor the resulting reduced correlation matrix. The number of common factors retained will equal the number of positive eigenvalues computed from the matrix.

8.2.11 Heuristic Rules of Thumb for the Number of Factors

Despite Thurstone's theory of minimum rank for a reduced correlation matrix and Guttman's proofs of the lower bounds to minimum rank, a number of heuristic practices not really based on the common-factor model have sprung up for determining the number of common factors. Most of these concentrate on the eigenvalues of the original correlation matrix $\mathbf{R}_{YY}$ and not the reduced correlation matrix $\mathbf{R}_{YY} - \mathbf{\Psi}^2$. We will also point out when we examine Jöreskog's solution for maximum-likelihood factor analysis later in this chapter that his method works with even a different set of eigenvalues, those of the matrix $\mathbf{\Psi}^{-1}\mathbf{R}_{YY}\mathbf{\Psi}^{-1}$.

8.2.11.1 Kaiser's Eigenvalues-Greater-Than-One Rule

Kaiser (1960, 1961) described a heuristic rule for the number of factors to retain in a principal components analysis that has been extended by some to common-factor analysis. This rule is to retain as many factors as there are eigenvalues greater than 1.00 of the correlation matrix $\mathbf{R}_{YY}$. This is Guttman's (1954a,b, 1956), r_1 for the number of factors to retain, which is why this is sometimes called the "Kaiser–Guttman rule." But Guttman would never have recommended its use in common-factor analysis because there it is the weakest lower bound. He would favor the strongest lower bound based on the squared multiple correlations for predicting a variable from the $n - 1$ other variables, r_3. Kaiser's justification for using r_1 was pragmatic and not mathematical: his personal experience was that this usually produced solutions that were interpretable and meaningful. He also argued (Kaiser and Caffrey, 1965, p. 6) that a component factor with a nonstandardized variance greater than 1.00 has a positive generalizability in the sense of Cronbach's alpha coefficient. And somehow a variable that would have less variance than an observed variable (with variance 1 in a correlation matrix) should not be taken seriously. However, he was fully aware that this was the weakest lowest bound in common-factor analysis (personal communication). And it was with some chagrin that he found some persons critical of his rule for taking too few factors (personal communication). Others claim that it takes too many (but they do not relate this to the theory of common-factor analysis). It is essential also to understand that the number of eigenvalues greater than 1.00 of the original correlation matrix is also the number of eigenvalues greater than the average eigenvalue. These eigenvalues sum to n and their average $n/n = 1$. In any case, despite these qualifications, many commercial factor analysis programs have had the effect of popularizing this rule by offering it as a default, if not one of the few criteria for the number of factors to retain. Many researchers seem to use it without a real understanding of the theory behind it or that it is inappropriate for a common-factor analysis.

8.2.11.2 Cattell's Scree Criterion

Another popular "rule-of-thumb" criterion for the number of factors to retain is Cattell's (1966a,b) "scree test." This is an "eyeball" test. What one does is plot the magnitude of the eigenvalues of a correlation matrix, or the reduced correlation matrix against the ordinal number of the eigenvalue. Then one connects the points with straight lines. Now, one needs to know what a "scree" is. In geology, a scree is an accumulation of loose rock or debri at the base of a rocky slope or cliff. The scree begins somewhat abruptly at the base of the steep slope of a cliff or rocky outcrop, where the pile of loose rock begins to fall away in a gradual slope from the hillside. That is an apt metaphor for the eigenvalue plot. According to the "scree plot" the number of factors to retain is where the plot line of the first so-many eigenvalues drops precipitously and for subsequent eigenvalues falls off gradually, almost in a straight line descent with gentle slope. Gorsuch (1983), however, notes that sometimes there are two or more "scree points," making the determination of the number of factors ambiguous. A frequent problem also is that there may be no sharp "scree point," but rather a gradual bending of the curve through several eigenvalue points, so the number of factors to retain is ambiguous (Figure 8.1).

The author himself has used a modified version, which he calls "the ruler method." He takes a ruler and places it along the plot line at the right-hand side where the eigenvalues appear to fall almost on a straight line. He then projects the line to the left-hand side and looks for where the plot line rises above the projected line. Beyond and above that is where the common factors are to be found. There is no strong rationale for this. The author is aware that if one takes a random sample of n variables from a population distributed $N_n(\mathbf{0}, \mathbf{I})$, obtains the correlations of the variables, then the plot of the eigenvalues of the correlation matrix will fall on a straight line that rises gradually to the left-hand side. If the last so many eigenvalues in the full correlation matrix correspond to what would be mostly unique-factor variances of roughly equal magnitude, all less than unity in value, then they should be rising from near zero in a linear fashion from the right-hand side. Principal components that contain both common-factor variance and unique-factor variance should rise noticeably above the line. Obviously this is more indirect than examining the eigenvalues of the reduced correlation matrix directly because common- and unique-factor variances are confounded in a components analysis. Components containing common factors that account

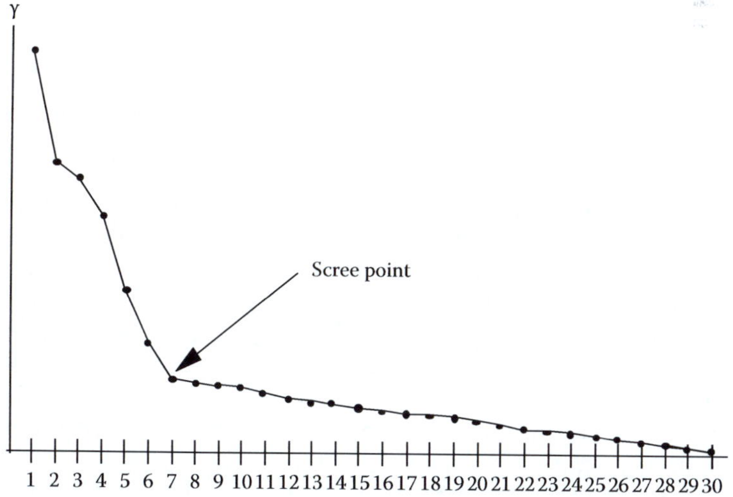

FIGURE 8.1
A scree plot of values of eigenvalues (vertical axis) against the ordinal number of the eigenvalue. The scree point is where the curve abruptly begins a gradual linear decline.

for small portions of the variance and small unique-factor variances may be confused with components that are pure unique variance.

8.2.11.3 Parallel Analysis

John Horn (1965) proposed a now popular method for determining the number of factors known as parallel analysis (PA). Parallel analysis sought to overcome a limitation of Kaiser's eigenvalues-greater-than-one rule, that it was not sensitive to variation in sample eigenvalues with sample size. Hayton, et al. (2004) note that Horn (1965) and Turner (1998) found that the first eigenvalues tended to be larger than their population counterparts, while the last tended to be smaller, all due to sampling error and bias of least-squares estimation. The Kaiser rule would only be appropriate at the level of the population or an infinite sample size. Parallel analysis is supposed to adjust for sampling error of the eigenvalues. Hayton et al. (2004) also indicate that the rationale of parallel analysis is that the eigenvalues of those components from real data satisfying a common-factor model should have larger eigenvalues than those of "parallel components derived from random data having the same sample size and number of variables" (p. 194). More specifically, the eigenvalues from a correlation matrix of real data for n variables and N observations, are compared, eigenvalue by eigenvalue in descending order, to the corresponding eigenvalues of a parallel analysis computed from a sample correlation matrix computed from a sample of N observations on n variables from a population distributed normally as $N_n(\mathbf{0}, \mathbf{I})$. In fact, the parallel analysis does not use actual eigenvalues for a single sample correlation matrix from data distributed $N_n(\mathbf{0}, \mathbf{I})$, but the averages of the respective eigenvalues from 50, 100, or 200 sample correlation matrices from data distributed $N_n(\mathbf{0}, \mathbf{I})$ for random samples of size N. The averages of the eigenvalues from large samples of correlation matrices from data distributed $N_n(\mathbf{0}, \mathbf{I})$ for samples of size N are presumably less affected by sampling error than would be the eigenvalues from a single sample correlation matrix of "random data."

Now, the way the PA criterion works is this: One proceeds, starting with the first eigenvalue of the real data and of the parallel analysis, and compares these two eigenvalues. If the real data eigenvalue is greater than the corresponding PA eigenvalue, one accepts the factor corresponding to that eigenvalue as a factor to be retained. Then one goes to the second eigenvalues for the real data and the parallel analysis, respectively, and compares these. Again if the respective eigenvalue of the real data is larger than (and not equal to) the corresponding PA eigenvalue, one retains the corresponding factor. One proceeds in this order comparing each of the real data eigenvalues to the corresponding PA eigenvalues. Finally at some point the real-data eigenvalues will equal or drop below the PA eigenvalues, and from that point on the factors of the real data analysis are to be rejected as "due to sampling error" (Hayton et al., 2004, p. 194). Actually, in the common-factor model they would not be due to sampling error alone, but to unique-factor variance as well.

Some have suggested that instead of using the mean PA eigenvalue in these comparisons, one use the 95 percentile value from the 100 or 200 samples used to generate the PA eigenvalues.

So, how accurate is the PA method in recovering the number of common factors? A number of Monte Carlo studies (Humphreys and Montanelli, 1975; Zwick and Velicer, 1986; Fabrigar et al., 1999; Glorfeld, 1995; Velicer et al., 2000) have been performed and found the PA method to be quite accurate. However, my impression of these studies is that they have tended to use data sets with common-factor models that have large loadings, uncorrelated factors, and relatively few common factors relative to the number of variables, and small number of variables. So, the PA method may work well in these instances. However, I decided to perform an experiment of my own. I used the EQS program of Peter Bentler to generate sample data for 300 observations from a common-factor model for 15 variables having seven common factors. Some of the loadings were low, and others high, and there were correlations in the .50s between several of the factors.

The model used to generate the "real-data" correlations was the common-factor model whose parameter matrices are shown below. The matrix $\mathbf{\Lambda}$ is the factor pattern matrix, $\mathbf{\Psi}$ is a column (for convenience of display) of unique-factor pattern weights (square roots of unique variances). $\mathbf{R}_{XX}$ contains the correlations among the common factors.

$$
\underset{\mathbf{\Lambda}}{\begin{bmatrix}
.538 & 0 & 0 & 0 & .230 & 0 & 0 \\
.476 & 0 & 0 & 0 & 0 & .238 & 0 \\
.367 & 0 & 0 & .440 & 0 & 0 & 0 \\
.610 & 0 & 0 & 0 & 0 & 0 & .152 \\
0 & .405 & 0 & 0 & 0 & 0 & .324 \\
0 & .454 & 0 & 0 & .302 & 0 & 0 \\
0 & .426 & 0 & 0 & 0 & 0 & 0 \\
0 & .584 & 0 & 0 & 0 & 0 & 0 \\
0 & 0 & .464 & 0 & 0 & 0 & .387 \\
0 & 0 & .639 & 0 & 0 & 0 & 0 \\
0 & 0 & .445 & 0 & .559 & 0 & 0 \\
0 & 0 & .604 & 0 & 0 & .145 & 0 \\
0 & 0 & 0 & .396 & 0 & 0 & 0 \\
0 & 0 & 0 & .227 & .709 & 0 & 0 \\
0 & 0 & 0 & .391 & 0 & .186 & 0
\end{bmatrix}}
\quad
\underset{\mathbf{\Psi}}{\begin{bmatrix}
.803 \\ .792 \\ .714 \\ .727 \\ .832 \\ .819 \\ .905 \\ .811 \\ .746 \\ .769 \\ .685 \\ .731 \\ .918 \\ .650 \\ .885
\end{bmatrix}}
$$

$$
\begin{bmatrix}
1 & & & & & & \\
.30 & 1 & & & & & \\
.40 & .50 & 1 & & & & \\
.50 & .28 & .40 & 1 & & & \\
.05 & .12 & .04 & .07 & 1 & & \\
.40 & .30 & .45 & .20 & .32 & 1 & \\
.41 & .15 & .22 & .10 & .13 & .07 & 1
\end{bmatrix}
$$
$$\mathbf{R}_{XX}$$

The correlation matrix was then computed as $\mathbf{R}_{YY} = \mathbf{\Lambda R}_{XX}\mathbf{\Lambda}' + \mathbf{\Psi}^2$. I then computed the eigenvalues from the resulting correlation matrix among the 15 variables. Next I obtained the corresponding PA averaged eigenvalues for 15 variables and 300 observations from tables prepared by Lautenschlager (1989). These are plotted in Figure 8.2. What the plot in Figure 8.2 shows is that the PA criterion does not always retain the correct number of factors. In this case, only the first three real-data eigenvalues exceeded the corresponding PA eigenvalues. All other real-data eigenvalues were less than the PA eigenvalues. The PA method should have retained seven common factors. All of the factors in the generating model satisfied simple structure (in Thurstone's sense).

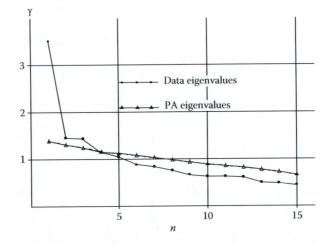

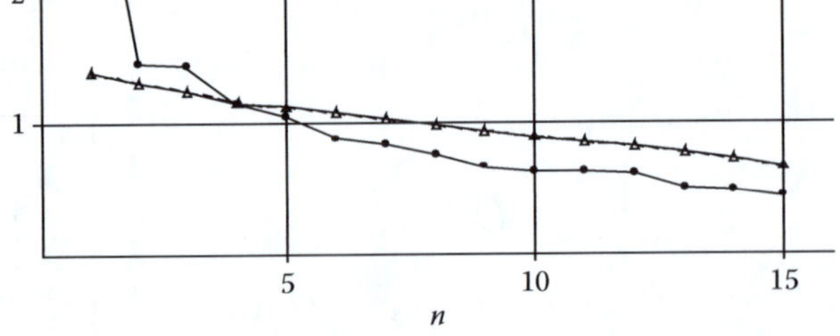

FIGURE 8.2
Plot of eigenvalues from a correlation matrix with $n = 15$ variables and 7 common factors compared to PA eigenvalues (due to Lautenschlager, 1989). Round dots connected by unbroken lines are eigenvalues from the factor analytic model. Triangular dots connected by unbroken lines are PA eigenvalues. This case shows the PA method accepting only three common factors.

Turner (1998) noted that in some cases PA will underestimate the number of common factors, so care needs to be taken in using the PA method. Turner also believed that the size of the first real-data eigenvalue had a strong effect on decisions regarding later eigenvalues. Because the sum of the eigenvalues must add up to n [since $\mathrm{tr}(\mathbf{D}) = \mathrm{tr}(\mathbf{ADA'}) = \mathrm{tr}(\mathbf{R}_{YY}) = n$] a large first eigenvalue γ_1 means later eigenvalues must sum to $n - \gamma_1$, and the larger γ_1 is, the smaller is the sum $n - \gamma_1$, implying that later eigenvalues will be even smaller still. On the other hand, PA eigenvalues seem to decline in almost linear fashion and may decline much more slowly than do real-data eigenvalues. Yet real-data eigenvalues smaller than PA eigenvalues may still represent common-factor variance. I might point out that in formulating the common-factor model to generate the "real data" in my example, I deliberately included moderate correlations among several of the common factors. Not only does the first eigenvalue depend on the magnitude of the loadings on the factors, but also on the correlations among them. So, those Monte Carlo studies of the PA method that only study orthogonal factors do not realistically portray situations frequently encountered with real data.

But the major failing, in the author's opinion, of the PA method, Kaiser's eigenvalues-greater-than-one criterion, and the scree test commonly used and reported in commercial factor analysis programs is that they are based on eigenvalues of the wrong matrix. It ignores the rationale behind Thurstone's minimum rank concept, where rank here refers not to the rank of the original correlation matrix $\mathbf{R}_{YY}$, but the rank of the reduced correlation matrix $\mathbf{R}_{YY} - \mathbf{\Psi}^2$. That means we should be examining the eigenvalues of $\mathbf{R}_{YY} - \mathbf{\Psi}^2$ or a congruent transformation of $\mathbf{R}_{YY} - \mathbf{\Psi}^2$, to determine where the eigenvalues of this matrix change from positive to 0 or even negative.

This may suggest more common factors than we may be able to handle substantively in theory, but it may also represent sources of variance we should be focusing on and eliminating, if possible, if we wish to formulate clean models and measurement. We should note that perhaps a major reason why many exploratory factor analytic studies fail to replicate in confirmatory factor analysis is because they originally were based on using one of these methods for determining the number of common factors based on the eigenvalues of the original correlation matrix. Those methods can underestimate the number of common factors, and the chi-square test of fit used in confirmatory factor analysis will detect variance due to previously omitted factors.

As we will see in the next section, the number of positive eigenvalues of the matrix $\mathbf{R}_{YY} - \mathbf{\Psi}^2$ is also equal to the number of eigenvalues greater than 1.00 of the matrix $\mathbf{\Psi}^{-1}\mathbf{R}_{YY}\mathbf{\Psi}^{-1}$. For proof consider that $\mathbf{\Psi}^{-1}(\mathbf{R}_{YY} - \mathbf{\Psi}^2)\mathbf{\Psi}^{-1} = \mathbf{\Psi}^{-1}\mathbf{R}_{YY}\mathbf{\Psi}^{-1} - \mathbf{I}$. Let $\mathbf{A}$ be the $n \times n$ eigenvector matrix and $\mathbf{D}$ the $n \times n$ diagonal matrix of eigenvalues of $\mathbf{\Psi}^{-1}\mathbf{R}_{YY}\mathbf{\Psi}^{-1}$, then

$$\mathbf{A}'(\mathbf{\Psi}^{-1}\mathbf{R}_{YY}\mathbf{\Psi}^{-1} - \mathbf{I})\mathbf{A} = \mathbf{D} - \mathbf{I} = [\gamma_i - 1]$$

The matrix at the right-hand side is the matrix $\mathbf{D} - \mathbf{I}$ but represents this by showing a typical diagonal element of $(\mathbf{D} - \mathbf{I})$ in brackets, which is an eigenvalue $\gamma_i - 1$. So, any eigenvalue γ_i of $\mathbf{\Psi}^{-1}\mathbf{R}_{YY}\mathbf{\Psi}^{-1}$ that is greater than 1.00 corresponds to a positive eigenvalue $\gamma_i - 1$ of $\mathbf{\Psi}^{-1}\mathbf{R}_{YY}\mathbf{\Psi}^{-1} - \mathbf{I}$, otherwise it corresponds to a zero or negative eigenvalue of $\mathbf{\Psi}^{-1}\mathbf{R}_{YY}\mathbf{\Psi}^{-1} - \mathbf{I}$. Because the matrix $\mathbf{\Psi}^{-1}(\mathbf{R}_{YY} - \mathbf{\Psi}^2)\mathbf{\Psi}^{-1}$ is a congruent transformation of $(\mathbf{R}_{YY} - \mathbf{\Psi}^2)$, $\mathbf{\Psi}^{-1}(\mathbf{R}_{YY} - \mathbf{\Psi}^2)\mathbf{\Psi}^{-1}$ will have the same number of positive, zero, and negative eigenvalues as those of the matrix $(\mathbf{R}_{YY} - \mathbf{\Psi}^2)$, by reason of Sylvester's law of inertia. Thus we can use the number of eigenvalues greater than 1.00 of $\mathbf{\Psi}^{-1}\mathbf{R}_{YY}\mathbf{\Psi}^{-1}$ as an index for the number of common factors.

Using the smallest upper-bound estimate $\mathbf{S}^2 = [\text{diag } \hat{\mathbf{R}}_{YY}^{-1}]^{-1}$ of $\mathbf{\Psi}^2$, I obtained the matrix $\hat{\mathbf{S}}^{-1}\hat{\mathbf{R}}_{YY}\hat{\mathbf{S}}^{-1}$ and then its eigenvalues for the correlation matrix $\hat{\mathbf{R}}_{YY}$ whose raw eigenvalues were shown in Figure 8.2. Seven of the eigenvalues of $\hat{\mathbf{S}}^{-1}\hat{\mathbf{R}}_{YY}\hat{\mathbf{S}}^{-1}$ were greater than 1.00 in value, which is precisely the number of common factors used to generate the sample correlation matrix $\hat{\mathbf{R}}_{YY}$ to begin with.

8.3 Fitting the Common-Factor Model to a Correlation Matrix

Up until the late 1960s, the common-factor model was frequently treated as a tautology to be imposed on any correlation matrix. By that is meant that the model was adjusted until it fit the data perfectly. This means that one could not reject the common-factor model as appropriate for the data. Browne (1967) was among the first to point out that the common-factor model could not be fitted to a correlation matrix if the number of common factors, r, satisfies Lederman's inequality (cf. Equation 8.3). In other words, when the postulated number of common factors, r, is less than $[2n + 1 - (8n + 1)^{1/2}]/2$, it may be impossible with some empirical correlation matrices to find a set of communalities that will leave the reduced correlation matrix Gramian and of rank r. Thus, when r equals $[2n + 1 - (8n + 1)^{1/2}]/2$, this represents a "lower bound" to the common-factor model treated as a tautology. Postulating a smaller number of common factors may possibly lead to a failure of the common-factor model to fit the data. For example, if n equals 100, then one will need to postulate at least 86 common factors to make the common-factor model a tautology that fits necessarily by definition. But if the common-factor model is to be treated as a scientific hypothesis with the potential of being disconfirmed by lack of fit to the data, then one will hypothesize a smaller number of common factors. In this respect, a testable common-factor-analysis model may not be a model with only a few overdetermined common factors, but it has to hypothesize a smaller number of factors than given by the Lederman inequality.

A testable common-factor-analysis model specifies fewer common factors than $[2n + 1 - (8n + 1)^{1/2}]/2$ and also hypothesizes that the common-factor and unique-factor variables are uncorrelated and that the unique factors are mutually uncorrelated. While specifying the number of common factors r introduces testable constraints into the model, also specifying the lack of correlation between common and unique factors and between unique factors themselves introduces other testable constraints. Any test of the fit of the common-factor model for r common factors to a real-data correlation matrix is a test of not only that the number of factors is r but that the other constraints of the model hold. Failure of any of these constraints to be satisfied with the data, will lead to rejection of the joint specifications of the model as a whole, although such rejection does not unambiguously point to any particular constraint as at fault.

But compared to a confirmatory factor analysis, which places additional constraints on the pattern loadings, the correlations among the factors and on values of the unique variances, the exploratory common-factor-analysis model is quite "unrestricted." Beyond a few other constraints on the factor pattern matrix designed to identify the model (which are not by themselves testable) the exploratory common-factor-analysis model is quite unrestricted—which is what makes it good as an exploratory, hypothesis-building method.

8.3.1 Least-Squares Estimation of the Exploratory Common-Factor Model

In what follows, by $\mathbf{R}$ we will mean the correlation matrix among the observed variables, $\mathbf{R}_{YY}$. We will also assume that the common factors are uncorrelated. Hence, the common-factor model for the observed correlation matrix will be $\mathbf{R} = \mathbf{\Lambda\Lambda}' + \mathbf{\Psi}^2$. The least-squares method will seek to find values for $\mathbf{\Lambda}$ and $\mathbf{\Psi}^2$ that minimizes the function

$$f_{LS} = \text{tr}\left[\left(\mathbf{R} - \mathbf{\Lambda\Lambda}' - \mathbf{\Psi}^2\right)\left(\mathbf{R} - \mathbf{\Lambda\Lambda}' - \mathbf{\Psi}^2\right)\right]$$

$$= \text{tr}\left(\mathbf{RR} + \mathbf{\Lambda\Lambda}'\mathbf{\Lambda\Lambda}' + \mathbf{\Psi}^2\mathbf{\Psi}^2 - 2\mathbf{R\Lambda\Lambda}' - 2\mathbf{R\Psi}^2 + 2\mathbf{\Lambda\Lambda}'\mathbf{\Psi}^2\right) \quad (8.40)$$

Note that the matrix $(\mathbf{R} - \mathbf{\Lambda\Lambda}' - \mathbf{\Psi}^2)$ represents the discrepancy between the observed matrix, $\mathbf{R}$, and its reproduction, $\mathbf{\Lambda\Lambda}' + \mathbf{\Psi}^2$, given arbitrary values for the elements of $\mathbf{\Lambda}$ and $\mathbf{\Psi}^2$. Let us denote the discrepancy by the $n \times n$ matrix

$$\mathbf{\Delta} = (\mathbf{R} - \mathbf{\Lambda\Lambda}' - \mathbf{\Psi}^2) \quad (8.41)$$

Each one of the elements of $\mathbf{\Delta}$ represents the discrepancy between a corresponding element of $\mathbf{R}$ and the corresponding element of $\mathbf{\Lambda\Lambda}' + \mathbf{\Psi}^2$. So, if we want to obtain the sum of the squares of all these discrepancies, we need a way to sum the squares in each column and then add up the column sums. This is accomplished by the matrix equation

$$f_{LS} = \text{tr}(\boldsymbol{\Delta}'\boldsymbol{\Delta}) = \text{tr}(\boldsymbol{\Delta}\boldsymbol{\Delta}) \tag{8.42}$$

Because of the symmetry of $\mathbf{R}$, $\boldsymbol{\Lambda}\boldsymbol{\Lambda}' + \boldsymbol{\Psi}^2$, and $\boldsymbol{\Delta}$ we may write the sum of the squares of the elements of $\boldsymbol{\Delta}$ either way as in Equation 8.42. And this is equivalent to Equation 8.40. Equation 8.40 when fully expanded puts us in a position to be able to use matrix derivatives to find an expression for the minimum of Equation 8.40. For a table of the derivatives of traces look back at Equations 4.15 through 4.22.

Let us now break up Equation 8.40 into its distinct expressions after applying the trace operation to each of the square matrices represented by the expressions.

$$f_{LS} = \text{tr}(\mathbf{RR}) + \text{tr}(\boldsymbol{\Lambda}\boldsymbol{\Lambda}'\boldsymbol{\Lambda}\boldsymbol{\Lambda}') + \text{tr}(\boldsymbol{\Psi}^2\boldsymbol{\Psi}^2) - 2\,\text{tr}(\mathbf{R}\boldsymbol{\Lambda}\boldsymbol{\Lambda}') - 2\,\text{tr}(\mathbf{R}\boldsymbol{\Psi}^2) + 2\,\text{tr}\left(\boldsymbol{\Lambda}\boldsymbol{\Lambda}'\boldsymbol{\Psi}^2\right)$$

Taking the partial derivative of each expression with respect to $\boldsymbol{\Lambda}$, we obtain

$$\frac{\partial\,\text{tr}(\mathbf{RR})}{\partial\boldsymbol{\Lambda}} = \mathbf{0}$$

from Equation 4.15.

$$\frac{\partial\,\text{tr}(\boldsymbol{\Lambda}\boldsymbol{\Lambda}'\boldsymbol{\Lambda}\boldsymbol{\Lambda}')}{\partial\boldsymbol{\Lambda}} = 4\boldsymbol{\Lambda}\boldsymbol{\Lambda}'\boldsymbol{\Lambda}$$

from recursive application of the product rule (Equation 4.21), invariance of traces under cyclic permutations (Equation 4.19), invariance of traces under transposition (Equation 4.20), and from Equation 4.16.

$$\frac{\partial\,\text{tr}(\boldsymbol{\Psi}^2\boldsymbol{\Psi}^2)}{\partial\boldsymbol{\Lambda}} = \mathbf{0}$$

because $\boldsymbol{\Lambda}$ does not appear in the expression.

$$\frac{\partial[-2\,\text{tr}(\mathbf{R}\boldsymbol{\Lambda}\boldsymbol{\Lambda}')]}{\partial\boldsymbol{\Lambda}} = \frac{-2\partial\,\text{tr}(\boldsymbol{\Lambda}'\mathbf{R}\boldsymbol{\Lambda})}{\partial\boldsymbol{\Lambda}} = -4\mathbf{R}\boldsymbol{\Lambda}$$

from Equations 4.19 and 4.17 and symmetry of $\mathbf{R}$.

$$\frac{\partial[-2\,\text{tr}(\mathbf{R}\boldsymbol{\Psi}^2)]}{\partial\boldsymbol{\Lambda}} = \mathbf{0}$$

$$\frac{\partial[2\,\text{tr}(\boldsymbol{\Lambda}\boldsymbol{\Lambda}'\boldsymbol{\Psi}^2)]}{\partial\boldsymbol{\Lambda}} = \frac{\partial[2\,\text{tr}(\boldsymbol{\Lambda}'\boldsymbol{\Psi}^2\boldsymbol{\Lambda})]}{\partial\boldsymbol{\Lambda}} = 4\boldsymbol{\Psi}^2\boldsymbol{\Lambda}$$

from Equations 4.19 and 4.17 and symmetry of $\boldsymbol{\Psi}^2$.

Putting these results all together, we obtain

$$\frac{\partial f_{\mathrm{LS}}}{\partial \boldsymbol{\Lambda}} = 4\boldsymbol{\Lambda}\boldsymbol{\Lambda}' - 4\mathbf{R}\boldsymbol{\Lambda} + 4\boldsymbol{\Psi}^2\boldsymbol{\Lambda} \tag{8.43}$$

For the partial derivatives with respect to a diagonal matrix, we need

$$\frac{\partial\ \mathrm{tr}(\mathbf{M})}{\partial(\mathrm{diag}\ \mathbf{X})} = \mathrm{diag}\left(\frac{\partial\ \mathrm{tr}(\mathbf{M})}{\partial \mathbf{X}}\right) \tag{8.44}$$

Hence

$$\frac{\partial f_{\mathrm{LS}}}{\partial \boldsymbol{\Psi}^2} = \mathrm{diag}(2\boldsymbol{\Psi}^2 - 2\mathbf{R} + 2\boldsymbol{\Lambda}\boldsymbol{\Lambda}') \tag{8.45}$$

We now turn our attention to minimizing f_{LS} with respect to $\boldsymbol{\Lambda}$ and $\boldsymbol{\Psi}^2$. First let us set Equation 8.43 equal to zero, divide both sides by 4, and rearrange terms to obtain:

$$(\mathbf{R} - \boldsymbol{\Psi}^2)\boldsymbol{\Lambda} = \boldsymbol{\Lambda}\boldsymbol{\Lambda}'\boldsymbol{\Lambda} \tag{8.46}$$

To provide a unique solution for $\boldsymbol{\Lambda}$, we will consider out of many possible solutions, that $\boldsymbol{\Lambda}'\boldsymbol{\Lambda}$ is an $r{\times}r$ diagonal matrix, which makes the solution unique under linear transformations. We may then treat $\boldsymbol{\Lambda}$ as the matrix $\boldsymbol{\Lambda} = \mathbf{A}_r\mathbf{D}_r^{1/2}$, where $\mathbf{A}_r$ consists of the first r normalized eigenvector columns of the eigenvector matrix $\mathbf{A}$ and $\mathbf{D}_r$ is a diagonal matrix with the corresponding first r eigenvalues of the matrix $\mathbf{R} - \boldsymbol{\Psi}^2$. Then $\boldsymbol{\Lambda}'\boldsymbol{\Lambda} = \mathbf{D}_r^{1/2}\mathbf{A}_r'\mathbf{A}_r\mathbf{D}_r^{1/2} = \mathbf{D}_r$. In this case $\boldsymbol{\Lambda}$ is known as the "principal-axes factor pattern matrix." Thus for a given $\boldsymbol{\Psi}^2$ an $n{\times}r$ matrix $\boldsymbol{\Lambda}$ that conditionally minimizes the least-squares criterion is the pattern matrix of the first r principal-axes factors of $\mathbf{R} - \boldsymbol{\Psi}^2$.

Turning now to solving for $\boldsymbol{\Psi}^2$, let us set Equation 8.45 equal to zero, divide by 2, and rearrange terms to obtain the equation

$$\boldsymbol{\Psi}^2 = \mathrm{diag}(\mathbf{R} - \boldsymbol{\Lambda}\boldsymbol{\Lambda}') \tag{8.47}$$

Again, this is only a conditional solution for $\boldsymbol{\Psi}^2$, conditional in this case on $\boldsymbol{\Lambda}$.

Let us back up a minute. We are trying to minimize the least-squares function $f_{\mathrm{LS}}(\boldsymbol{\Lambda}, \boldsymbol{\Psi}^2)$. We seek the values of both $\boldsymbol{\Lambda}$ and $\boldsymbol{\Psi}^2$ that minimize the least-squares function over all possible values for these matrices. And we part-way minimize the function when we find $\boldsymbol{\Lambda}_{\boldsymbol{\Psi}^2}$, which denotes the lambda matrix that minimizes the least-squares function conditional on a specific choice of $\boldsymbol{\Psi}^2$. Define now the function

$$g(\boldsymbol{\Psi}^2) = \min_{\boldsymbol{\Lambda}} f_{\mathrm{LS}}(\boldsymbol{\Lambda}, \boldsymbol{\Psi}^2) = f_{\mathrm{LS}}(\boldsymbol{\Lambda}_{\boldsymbol{\Psi}^2}, \boldsymbol{\Psi}^2) \tag{8.48}$$

which represents the value of our least-squares function when minimized over values of $\mathbf{\Lambda}$ for a given value of $\mathbf{\Psi}^2$. The function $g(\mathbf{\Psi}^2) = \min_{\mathbf{\Lambda}} f_{\text{LS}}(\mathbf{\Lambda}, \mathbf{\Psi}^2) = f_{\text{LS}}(\mathbf{\Lambda}_{\mathbf{\Psi}^2}, \mathbf{\Psi}^2)$ is minimized when $\mathbf{\Lambda} = \mathbf{A}_r\mathbf{D}_r^{1/2}$ and $\mathbf{A}_r$ and $\mathbf{D}_r$ are the first r eigenvectors and eigenvalues of $\mathbf{R} - \mathbf{\Psi}^2$. But now we wish to minimize this latter function over all possible diagonal matrices $\mathbf{\Psi}^2$. That is, we seek

$$\min_{\mathbf{\Psi}^2} g(\mathbf{\Psi}^2) = \min_{\mathbf{\Lambda},\mathbf{\Psi}^2} f_{\text{LS}}(\mathbf{\Lambda}, \mathbf{\Psi}^2) \tag{8.49}$$

and we can be even further toward this objective if we now find

$$\mathbf{\Psi}^2 = \text{diag}(\mathbf{R} - \mathbf{\Lambda}_{\mathbf{\Psi}^2}\mathbf{\Lambda}'_{\mathbf{\Psi}^2}) \tag{8.50}$$

where the subscript of $\mathbf{\Lambda}_{\mathbf{\Psi}^2}$ refers to the previous value of $\mathbf{\Psi}^2$ on which the current $\mathbf{\Lambda}$ is conditioned.

This suggests now an iterative algorithm that first finds an initial estimate for the number of factors to retain and then finds the absolute minimum for the least-squares function for this number of common factors.

Let

$$\mathbf{\Psi}^2_{(1)} = [\text{diag } \mathbf{R}^{-1}]^{-1} \tag{8.51}$$

which is a heuristic estimate of the diagonal matrix of errors of estimate for predicting each of the n variables in $\mathbf{R}$, respectively, from all the $n - 1$ other variables in $\mathbf{R}$. It is based on what is needed to get Guttman's greatest lower bound r_3.

1. Find the eigenvector and eigenvalue matrices, $\mathbf{A}_{(t)}$ and $\mathbf{D}_{(t)}$, of

$$(\mathbf{R} - \mathbf{\Psi}^2_{(t)}) = \mathbf{A}_{(t)}\mathbf{D}_{(t)}\mathbf{A}'_{(t)} \tag{8.52}$$

   In the first iteration, determine r to be the number of positive eigenvalues of $(\mathbf{R} - \mathbf{\Psi}^2_{(1)})$, and hold r fixed throughout the following iterations.

2. Find

$$\mathbf{\Lambda}_{r(t)} = \mathbf{A}_{r(t)}\mathbf{D}^{1/2}_{r(t)} \tag{8.53}$$

   where

   $\mathbf{A}_{r(t)}$ is an $n \times r$ matrix consisting of the first r eigenvectors in $\mathbf{A}_{(t)}$

   $\mathbf{D}_{r(t)}$ is an $r \times r$ diagonal matrix with corresponding r largest eigenvalues in $\mathbf{D}_{(t)}$

3. Find

$$\mathbf{\Psi}^2_{(t+1)} = \text{diag}(\mathbf{R} - \mathbf{\Lambda}_{r(t)}\mathbf{\Lambda}'_{r(t)}) \tag{8.54}$$

4. Check for convergence. If $\max_i |(\psi^2_{i(t+1)}) - (\psi^2_{i(t)})| < .001$. exit the loop, otherwise go to the beginning of the loop at (1) and proceed again through the loop with the updated value for $\mathbf{\Psi}^2$.

When you have exited the loop, $\hat{\mathbf{\Lambda}}_{r(t)}$ and $\hat{\mathbf{\Psi}}^2_{(t)}$ are the least-squares solutions for the factor pattern matrix and unique-factor variance matrix. The symbol "^" indicates that these are the least-squares estimators in this case.

This is known as the principal-axes factor (PAF) solution in some commercial programs.

8.3.2 Assessing Fit

There is no statistic possible with known distribution by which we may test the goodness of fit of the least-squares solution with probabilistic inference. Evaluation of the least-squares solution usually takes the form of descriptive statistics. The value of the conditional least-squares criterion

$$g(\hat{\mathbf{\Psi}}^2) = F_{LS,r}(\mathbf{\Lambda}_{\hat{\psi}^2}, \hat{\mathbf{\Psi}}^2) = \mathrm{tr}(\mathbf{R} - \mathbf{\Psi}^2 - \mathbf{\Lambda}_{\hat{\psi}^2}\mathbf{\Lambda}'_{\hat{\psi}^2})(\mathbf{R} - \mathbf{\Psi}^2 - \mathbf{\Lambda}_{\hat{\psi}^2}\mathbf{\Lambda}_{\hat{\psi}^2})$$

$$= \mathrm{tr}(\mathbf{R} - \hat{\mathbf{\Psi}}^2 - \mathbf{A}_r\mathbf{D}_r\mathbf{A}'_r)(\mathbf{R} - \hat{\mathbf{\Psi}}^2 - \mathbf{A}_r\mathbf{D}_r\mathbf{A}'_r)$$

$$= \mathrm{tr}(\mathbf{R} - \hat{\mathbf{\Psi}}^2 - \mathbf{A}_r\mathbf{D}_r\mathbf{A}'_r)^2$$

can serve as a criterion of fit. But

$$\mathbf{R} - \hat{\mathbf{\Psi}}^2 = \mathbf{ADA}' = \mathbf{A}_r\mathbf{D}_r\mathbf{A}'_r + \mathbf{A}_{n-r}\mathbf{D}_{n-r}\mathbf{A}'_{n-r}$$

with $\mathbf{A}_{n-r}$ and $\mathbf{D}_{n-r}$ the $n \times r$ and $(n-r) \times (n-r)$ matrices, respectively of remaining eigenvectors and eigenvalues of $\mathbf{R} - \hat{\mathbf{\Psi}}^2$. Hence

$$g(\hat{\mathbf{\Psi}}^2) = \mathrm{tr}(\mathbf{ADA}' - \mathbf{A}_r\mathbf{D}_r\mathbf{A}'_r)^2 = \mathrm{tr}(\mathbf{A}_{\text{n-r}}\mathbf{D}_{n-r}\mathbf{A}'_{n-r})^2 = \mathrm{tr}\,\mathbf{D}^2_{n-r}$$

$$= \sum_{i=r+1}^{n} \gamma_i^2 \tag{8.55}$$

that is, the sum of the remaining $n - r$ squared eigenvalues. Ideally this value would be zero, but small, near zero values still indicate good fit. Or we could eyeball the residual matrix, which is given by $\mathbf{R} - \hat{\mathbf{\Lambda}}\hat{\mathbf{\Lambda}}' - \hat{\mathbf{\Psi}}^2 = \mathbf{A}_{n-r}\mathbf{D}_{n-r}\mathbf{A}'_{n-r}$ and look for the largest absolute value for a residual variance or covariance as an index of fit. If that value is near zero, then we have a good fit.

8.3.3 Example of Least-Squares Common-Factor Analysis

In Table 8.2 we show a correlation matrix among 15 adjective-trait-rating scales obtained from 280 raters of random person descriptions. (See Chapter 15 for a more complete description of this study.)

TABLE 8.2

Correlation Matrix among 15 Trait-Adjective Scales

1.00—Friendly														
.777	1.00—Sympathetic													
.809	.869	1.00—Kind												
.745	.833	.835	1.00—Affectionate											
.176	.123	.123	.112	1.00—Intelligent										
.234	.159	.205	.183	.791	1.00—Capable									
.243	.155	.187	.186	.815	.865	1.00—Competent								
.234	.190	.238	.215	.818	.841	.815	1.00—Smart							
.433	.319	.321	.435	.174	.209	.239	.258	1.00—Talkative						
.473	.480	.410	.527	.220	.274	.269	.261	.744	1.00—Outgoing					
.433	.438	.406	.526	.188	.227	.242	.228	.711	.853	1.00—Gregarious				
.447	.396	.350	.500	.192	.221	.227	.224	.758	.846	.801	1.00—Extrovert			
.649	.693	.697	.694	.283	.344	.370	.365	.443	.552	.514	.473	1.00—Helpful		
.662	.692	.676	.679	.311	.345	.375	.351	.431	.557	.514	.493	.740	1.00—Cooperative	
.558	.543	.510	.632	.213	.289	.287	.287	.745	.886	.820	.830	.631	.626	1.00—Sociable

Source: Carlson, M. and Mulaik, S.A., *Multivar. Behav. Res.*, 28, 111, 1993.

TABLE 8.3

Unrotated Factor-Loading Matrix for Seven Common Factors from the Least-Squares Principal-Axes Factoring and Eigenvalues of (**R** – **S**²)

	Unrotated Factors							
	1	**2**	**3**	**4**	**5**	**6**	**7**	**Eigenvalues**
1.	.7637	.2116	.3657	–.3011	.1122	.0702	–.1469	7.463
2.	.7530	.2929	.4510	.0366	–.0635	–.1447	–.0236	2.730
3.	.7463	.2517	.5100	–.0629	–.0911	–.0328	.0726	1.665
4.	.7886	.2874	.3087	–.0032	–.0940	–.0565	.1018	.124
5.	.4335	–.8043	.0206	–.0059	.1742	–.2380	–.0172	.073
6.	.5016	–.7958	.0389	–.0277	–.2174	.1285	–.0890	.026
7.	.5019	–.7586	.0272	.0037	.0136	.0681	–.0211	.014
8.	.5075	–.7445	.0525	–.0382	–.0280	–.0178	.1431	–.010
9.	.6840	.1260	–.4547	–.1851	.0897	.1003	.1564	–.016
10.	.8121	.1586	–.4560	.0692	–.0550	–.0636	–.1096	–.049
11.	.7582	.1722	–.4246	.0490	–.0579	–.0667	–.0057	–.051
12.	.7468	.1697	–.4826	–.0591	.0072	–.0421	–.0132	–.075
13.	.7949	.0593	.2206	.2259	.1130	.1518	.0482	–.078
14.	.7895	.0506	.2056	.1776	.1341	.0764	–.0536	–.089
15.	.8586	.1753	–.3328	.0507	–.0335	.0256	–.0166	–.103

I subjected this correlation matrix to a principal-axes factor analysis algorithm. The eigenvalues for (**R** – **S**²) obtained are shown in Table 8.3.

There were seven eigenvalues that were greater than 0, and this was chosen to be the number of common factors to extract. This may not be the number we wish to rotate or interpret, but one should acknowledge this, because there seem to be doublet factors in these data for which the common-factor model is not appropriate (see Chapter 9).

8.3.4 Maximum-Likelihood Estimation of the Exploratory Common-Factor Model

As a rule most of the early users and proponents of factor analysis were not statisticians concerned with questions of statistical inference. Perhaps this is because most early factor analysts were psychologists with primarily psychological interests who backed off from the exceedingly complex mathematical problems of statistical inference in factor analysis that initially even the mathematical statisticians found difficult to solve. Nevertheless there were mathematical statisticians who kept trying to solve these problems. In 1940, Lawley (1940) made a major breakthrough with the development of equations for the maximum-likelihood estimation of factor loadings, and he followed up this work with other papers (1941, 1942, 1943, 1949, 1958) that sketched a framework for statistical testing in factor analysis. However, the catch was that to use Lawley's methods it was necessary to have maximum-likelihood estimates of the factor pattern loadings and unique variances, and Lawley's computational recommendations for obtaining these estimates were not practical for problems having the number of variables to which psychologists were accustomed. As a consequence, psychologists continued to obtain their centroid solutions and to use such heuristics as "Treat a factor loading with absolute value less than .30 as nonsignificant" when dealing with statistical problems.

By the 1950s, however, when the first-generation electronic computers were being installed in universities and laboratories across the United States, thus making possible the complex calculations required in multivariate statistics, other statisticians joined Lawley in investigating factor-analytic problems. In particular, Howe (1955) showed that the maximum-likelihood estimators of the factor loadings derived by Lawley could be derived from a model making no distributional assumptions about the variates, and he also provided a Gauss–Seidel computing algorithm that was far superior to Lawley's for obtaining these estimates. Unfortunately, at this time psychologists did not generally take advantage of Howe's algorithm and adapt it for computers. Had they done so, they might have brought maximum-likelihood factor analysis to the fore sooner. Instead they used an algorithm developed by Rao (1955) and programmed on an electronic computer at the University of Illinois, which, though better than Lawley's algorithm, in retrospect (Jennrich and Robinson, 1969) turns out to have been not much better, requiring extensive computing and many iterations with doubtful convergence. Again the psychologists felt the statisticians had not provided the solutions needed, and so, while awaiting further developments, the psychologists abandoned the centroid method of computing and took up variations of components analysis.

In 1967, Harman and Jones (1966) came forth with their Gauss–Seidel minres method which closely approximated the maximum-likelihood method by least-squares estimation providing estimates of the factor loadings. The Harman and Jones algorithm for computing these estimates was far superior to Rao's method developed previously and similar to Howe's. In 1966, Jöreskog (1966) published a paper on testing a simple-structure hypothesis in factor analysis in which he used an algorithm similar to the Fletcher–Powell algorithm to obtain his estimates. According to Jöreskog (1967), Lawley saw this article and wrote to him, suggesting that Jöreskog use this algorithm to minimize a function of the maximum-likelihood criterion for the common-factor-analysis model. Jöreskog did so, improving it by the Fletcher–Powell (1963) innovation, and discovered that the algorithm not only converged to the desired solution but did so far faster than any other algorithm tried previously. Jöreskog and Lawley (1968) collabrated further to complete the theory and method. At this point, one could say that the major computational problem of maximum-likelihood factor analysis was solved and that a computationally feasible method was now available for psychologists to use.

From the point of view of statisticians, maximum-likelihood estimators are usually superior to other estimators in estimating population parameters. When estimating a population parameter, if a sufficient statistic exists to estimate the parameter, the maximum-likelihood estimator is usually based on it. Moreover, the maximum-likelihood estimator is a consistent estimator as well as frequently a minimum-variance estimator. And, finally, in many cases in connection with typical distributions the maximum-likelihood estimators are (approximately) normally distributed with large samples.

The idea of a maximum-likelihood estimator is this: We assume that we know the "general form" of the population distribution from which a sample is drawn. For example, we might assume the population distribution is a multivariate normal distribution. But what we do not know are the population parameters which give this distribution a particular form among all possible multivariate normal distributions. For example, we do not know the population means and the variances and covariances for the variables. But if we did know the values of these parameters for the population, we could determine the density of a sample-observation vector from this population having certain specified values for each of the variables. In the absence of such knowledge, however, we can take arbitrary values and treat them as if they were the population parameters and then ask ourselves what is the likelihood (we will reserve the term "probability" or its counterpart, "density," for the case when the true parameters are known) of observing certain values for the variables on a single observation drawn from such a population? If we have more than one observation, then we can ask what is the joint likelihood of obtaining such a sample of observation vectors? The answer to this question is the product of the likelihoods of the individual observations drawn singly from the population. Finally we can ask: What values for the population parameters make the sample observations have the greatest joint likelihood? When we answer this question, we will take such values to be the maximum-likelihood estimators of the population parameters.

We shall now seek to find maximum-likelihood estimators for the factor pattern loadings and the unique variances of the model $\mathbf{\Sigma}=\mathbf{\Lambda\Lambda}'+\mathbf{\Psi}^2$, where $\mathbf{\Sigma}=\mathbf{\Sigma}_{YY}=E(\mathbf{YY}'-\boldsymbol{\mu\mu}')$. We will assume that we have a sample of N $n\times 1$ observation vectors $\mathbf{y}_1, \mathbf{y}_2, \ldots, \mathbf{y}_N$ from a multivariate normal distribution $N_n(\boldsymbol{\mu}, \mathbf{\Sigma})$. The joint likelihood for these N observations is given by the product

$$L=\prod_{k=1}^{N}\frac{|\mathbf{\Sigma}|^{-1/2}}{(2\pi)^{n/2}}\exp\left[-\frac{1}{2}(\mathbf{y}_k-\boldsymbol{\mu})'\mathbf{\Sigma}^{-1}(\mathbf{y}_k-\boldsymbol{\mu})\right] \tag{8.56}$$

It is more convenient, usually, in developing maximum-likelihood estimates to take the natural logarithm of the likelihood function, because derivatives are easier to obtain in that case:

$$\ln L = -(1/2)Nn\ln(2\pi) - (1/2)N\ln|\mathbf{\Sigma}| - (1/2)\sum_{k=1}^{N}(\mathbf{y}_k - \boldsymbol{\mu})'\mathbf{\Sigma}^{-1}(\mathbf{y}_k - \boldsymbol{\mu}) \quad (8.57)$$

We now need to estimate simultaneously both $\boldsymbol{\mu}$ and $\mathbf{\Sigma}$. We have already developed the maximum-likelihood estimates for these in Equations 5.19 and 5.25:

$$\hat{\boldsymbol{\mu}} = \frac{1}{N}\sum_{k=1}^{N}\mathbf{y}_k \quad (8.58)$$

which we may substitute into Equation 8.57 in place of $\mathbf{\Sigma}$. (Again the "∧" indicates an estimator, in this case, the maximum-likelihood estimator.) Now the sum of the outer products of the sample observation vectors corrected by the sample mean is

$$(N-1)\mathbf{S} = \sum_{k=1}^{N}\mathbf{y}_k\mathbf{y}_k' - N\hat{\boldsymbol{\mu}}\hat{\boldsymbol{\mu}}' \quad (8.59)$$

where $\mathbf{S}$ is in this case the unbiased estimator of $\mathbf{\Sigma}$. We may now rewrite Equation 8.57 as

$$\ln L = -K - (1/2)(N-1)\ln|\mathbf{\Sigma}| - (1/2)(N-1)\mathrm{tr}(\mathbf{\Sigma}^{-1}\mathbf{S}) \quad (8.60)$$

where K is a constant expression in Equation 8.57.

We are now ready to work out the value for $\mathbf{\Sigma} = \mathbf{\Lambda\Lambda}' + \mathbf{\Psi}^2$ that maximizes Equation 8.59. However, Jöreskog (1967) shows that the $\mathbf{\Sigma}$ matrix that maximizes the natural logarithm of the joint likelihood function also minimizes the fit function

$$F_r(\mathbf{\Lambda}, \mathbf{\Psi}^2) = \ln|\mathbf{\Sigma}| + \mathrm{tr}(\mathbf{\Sigma}^{-1}\mathbf{S}) - \ln(\mathbf{S}) - n \quad (8.61)$$

and so, he uses this criterion instead, in developing the equations of his algorithm for finding the maximum-likelihood estimates of $\hat{\mathbf{\Lambda}}$, $\hat{\mathbf{\Psi}}^2$ and in turn $\hat{\mathbf{\Sigma}}$. F_r is considered to be a function of $\mathbf{\Lambda}$ and $\mathbf{\Psi}^2$ because $\mathbf{\Sigma} = \mathbf{\Lambda\Lambda}' + \mathbf{\Psi}^2$. The subscript r indicates that the number of common factors r is regarded as prespecified and known and $\mathbf{\Lambda}$ is an $n \times r$ matrix.

*8.3.4.1 Maximum-Likelihood Estimation Obtained Using Calculus

Now, according to Schönemann (1965, pp. 19 and 44)

$$\frac{\partial|\mathbf{Y}|}{\partial\mathbf{Y}} = |\mathbf{Y}|\mathbf{Y}^{-1} \quad (8.62)$$

and

$$\frac{\partial\,\mathrm{tr}\,\mathbf{Y}^{-1}\mathbf{A}}{\partial\mathbf{X}} = \frac{-\mathrm{tr}(\mathbf{Y}^{-1}\mathbf{A}\mathbf{Y}^{-1})_c\mathbf{Y}}{\partial\mathbf{X}} \quad (8.63)$$

Thus

$$\frac{\partial F}{\partial \mathbf{W}} = \frac{\partial \operatorname{tr}(\boldsymbol{\Sigma}^{-1} - \boldsymbol{\Sigma}^{-1}\mathbf{S}\boldsymbol{\Sigma}^{-1})_c \boldsymbol{\Sigma}}{\partial \mathbf{W}} \tag{8.64}$$

where $\mathbf{W}$ is some arbitrary matrix. Noting that according to our model $\boldsymbol{\Sigma} = \boldsymbol{\Lambda}\boldsymbol{\Lambda}' + \boldsymbol{\Psi}^2$, we may substitute this expression for $\boldsymbol{\Sigma}$ on the right-hand side in Equation 8.64 and then take the partial derivative of the function F with respect to the matrix $\boldsymbol{\Lambda}$

$$\frac{\partial F_r}{\partial \boldsymbol{\Lambda}} = \frac{\partial \operatorname{tr}(\boldsymbol{\Sigma}^{-1} - \boldsymbol{\Sigma}^{-1}\mathbf{S}\boldsymbol{\Sigma}^{-1})_c \boldsymbol{\Lambda}\boldsymbol{\Lambda}'}{\partial \boldsymbol{\Lambda}}$$

dropping the expression due to $\boldsymbol{\Psi}^2$ since it is treated as a constant with respect to variation in $\boldsymbol{\Lambda}$, with a zero for derivative.

Then

$$\begin{aligned}\frac{\partial F_r}{\partial \boldsymbol{\Lambda}} &= 2(\boldsymbol{\Sigma}^{-1} - \boldsymbol{\Sigma}^{-1}\mathbf{S}\boldsymbol{\Sigma}^{-1})\boldsymbol{\Lambda} \\ &= 2\boldsymbol{\Sigma}^{-1}(\boldsymbol{\Sigma} - \mathbf{S})\boldsymbol{\Sigma}^{-1}\boldsymbol{\Lambda}\end{aligned} \tag{8.65}$$

On the other hand, the partial derivative of F with respect to $\boldsymbol{\Psi}^2$ is

$$\begin{aligned}\frac{\partial F_r}{\partial \boldsymbol{\Psi}^2} &= \frac{\partial \operatorname{tr}(\boldsymbol{\Sigma}^{-1} - \boldsymbol{\Sigma}^{-1}\mathbf{S}\boldsymbol{\Sigma}^{-1})\boldsymbol{\Psi}^2}{\partial \boldsymbol{\Psi}^2} \\ &= \operatorname{diag}[\boldsymbol{\Sigma}^{-1}(\boldsymbol{\Sigma} - \mathbf{S})\boldsymbol{\Sigma}^{-1}]\end{aligned} \tag{8.66}$$

Jöreskog (1967) emphasizes, as we did in deriving the algorithm for least-squares estimation, the importance of finding the value of $\boldsymbol{\Lambda}$ that conditionally minimizes the function F_r when diagonal $\boldsymbol{\Psi}^2$ is fixed at some arbitrary value. To do this, we first set Equation 8.64 equal to zero and premultiply by $\boldsymbol{\Sigma}$ to obtain

$$(\boldsymbol{\Sigma} - \mathbf{S})\boldsymbol{\Sigma}^{-1}\boldsymbol{\Lambda} = \mathbf{0} \tag{8.67}$$

The problem at this point is that the matrix $\boldsymbol{\Sigma}^{-1}$ obscures the fact that $\boldsymbol{\Sigma}$ is a function of both $\boldsymbol{\Lambda}$ and $\boldsymbol{\Psi}^2$. Thus Jöreskog (1967) takes advantage of the fact that, according to Lawley and Maxwell (1963, p. 13, Equation 2.9)

$$\boldsymbol{\Sigma}^{-1} = \boldsymbol{\Psi}^{-2} - \boldsymbol{\Psi}^{-2}\boldsymbol{\Lambda}(\mathbf{I} + \boldsymbol{\Lambda}'\boldsymbol{\Psi}^{-2}\boldsymbol{\Lambda})\boldsymbol{\Lambda}'\boldsymbol{\Psi}^{-2} \tag{8.68}$$

which can be substituted for $\boldsymbol{\Sigma}^{-1}$ in Equation 8.66 to obtain

$$(\boldsymbol{\Sigma} - \mathbf{S})\boldsymbol{\Sigma}^{-1}\boldsymbol{\Lambda} = (\boldsymbol{\Sigma} - \mathbf{S})[\boldsymbol{\Psi}^{-2} - \boldsymbol{\Psi}^{-2}\boldsymbol{\Lambda}(\mathbf{I} + \boldsymbol{\Lambda}\boldsymbol{\Psi}^{-2}\boldsymbol{\Lambda})\boldsymbol{\Lambda}'\boldsymbol{\Psi}^{-2}]\boldsymbol{\Lambda} = \mathbf{0}$$

Then carrying the $\mathbf{\Lambda}$ into the brackets, we get

$$= (\mathbf{\Sigma} - \mathbf{S})[\mathbf{\Psi}^{-2}\mathbf{\Lambda} - \mathbf{\Psi}^{-2}\mathbf{\Lambda}(\mathbf{I} + \mathbf{\Lambda}'\mathbf{\Psi}^{-2}\mathbf{\Lambda})\mathbf{\Lambda}'\mathbf{\Psi}^{-2}\mathbf{\Lambda}] = \mathbf{0}$$

Now, factor out the left-hand side multiplier $\mathbf{\Psi}^{-2}\mathbf{\Lambda}$ from the terms in the brackets

$$= (\mathbf{\Sigma} - \mathbf{S})\mathbf{\Psi}^{-2}\mathbf{\Lambda}[\mathbf{I} - (\mathbf{I} + \mathbf{\Lambda}'\mathbf{\Psi}^{-2}\mathbf{\Lambda})^{-1}\mathbf{\Lambda}'\mathbf{\Psi}^{-2}\mathbf{\Lambda}] = \mathbf{0}$$

Consider the identity $\mathbf{I} = (\mathbf{I} + \mathbf{\Lambda}'\mathbf{\Psi}^{-2}\,\mathbf{\Lambda})^{-1}\,(\mathbf{I} + \mathbf{\Lambda}'\mathbf{\Psi}^{-2}\,\mathbf{\Lambda})$. Substitute this identity for the left-most $\mathbf{I}$ in the previous equation to obtain

$$= (\mathbf{\Sigma} - \mathbf{S})\mathbf{\Psi}^{-2}\mathbf{\Lambda}[(\mathbf{I} + \mathbf{\Lambda}'\mathbf{\Psi}^{-2}\mathbf{\Lambda})^{-1}(\mathbf{I} + \mathbf{\Lambda}'\mathbf{\Psi}^{-2}\mathbf{\Lambda}) - (\mathbf{I} + \mathbf{\Lambda}'\mathbf{\Psi}^{-2}\mathbf{\Lambda})^{-1}\mathbf{\Lambda}'\mathbf{\Psi}^{-2}\mathbf{\Lambda}] = \mathbf{0}$$

Now, multiply $(\mathbf{I} + \mathbf{\Lambda}'\mathbf{\Psi}^{-2}\,\mathbf{\Lambda})^{-1}$ times $(\mathbf{I} + \mathbf{\Lambda}'\mathbf{\Psi}^{-2}\,\mathbf{\Lambda})$ in the expression:

$$= (\mathbf{\Sigma} - \mathbf{S})\mathbf{\Psi}^{-2}\mathbf{\Lambda}[(\mathbf{I} + \mathbf{\Lambda}'\mathbf{\Psi}^{-2}\mathbf{\Lambda})^{-1} + (\mathbf{I} + \mathbf{\Lambda}'\mathbf{\Psi}^{-2}\mathbf{\Lambda})\mathbf{\Lambda}'\mathbf{\Psi}^{-2}\mathbf{\Lambda} - (\mathbf{I} + \mathbf{\Lambda}'\mathbf{\Psi}^{-2}\mathbf{\Lambda})^{-1}\mathbf{\Lambda}'\mathbf{\Psi}^{-2}\mathbf{\Lambda}] = \mathbf{0}$$

Notice that the two right-most terms in the brackets are equal and opposite in sign. Hence

$$= (\mathbf{\Sigma} - \mathbf{S})\mathbf{\Psi}^{-2}\mathbf{\Lambda}(\mathbf{I} + \mathbf{\Lambda}'\mathbf{\Psi}^{-2}\mathbf{\Lambda})^{-1} = \mathbf{0} \tag{8.69}$$

Now, if we postmultiply both sides by $(\mathbf{I} + \mathbf{\Lambda}'\mathbf{\Psi}^{-2}\,\mathbf{\Lambda})$, we obtain

$$(\mathbf{\Sigma} - \mathbf{S})\mathbf{\Psi}^{-2}\mathbf{\Lambda} = \mathbf{0} \tag{8.70}$$

Next, substitute $\mathbf{\Lambda\Lambda}' + \mathbf{\Psi}^2$ for $\mathbf{\Sigma}$ in Equation 8.70 to obtain

$$(\mathbf{\Lambda\Lambda}' + \mathbf{\Psi}^2 - \mathbf{S})\mathbf{\Psi}^{-2}\mathbf{\Lambda} = \mathbf{0} \tag{8.71}$$

which, after removal of parentheses, becomes

$$\mathbf{\Lambda\Lambda}'\mathbf{\Psi}^{-2}\mathbf{\Lambda} + \mathbf{\Lambda} - \mathbf{S}\mathbf{\Psi}^{-2}\mathbf{\Lambda} = \mathbf{0}$$

Moving now $\mathbf{S}\mathbf{\Psi}^{-2}\,\mathbf{\Lambda}$ to the other side of the equation, factoring out $\mathbf{\Lambda}$, and rearranging, we get

$$\mathbf{S}\mathbf{\Psi}^{-2}\mathbf{\Lambda} = \mathbf{\Lambda}(\mathbf{I} + \mathbf{\Lambda}'\mathbf{\Psi}^{-2}\mathbf{\Lambda}) \tag{8.72}$$

If we now premultiply both sides of Equation 8.71 by $\mathbf{\Psi}^{-1}$ and do some similar rearranging, we get

$$\mathbf{\Psi}^{-1}\mathbf{S}\mathbf{\Psi}^{-1}(\mathbf{\Psi}^{-1}\mathbf{\Lambda}) = (\mathbf{\Psi}^{-1}\mathbf{\Lambda})(\mathbf{I} + \mathbf{\Lambda}'\mathbf{\Psi}^{-2}\mathbf{\Lambda}) \tag{8.73}$$

In Equation 8.73, Jöreskog (1967) says that it would be convenient to regard $\mathbf{\Lambda}'\mathbf{\Psi}^{-2}\,\mathbf{\Lambda}$ to be diagonal. We can do this, Jöreskog tells us, because F_r is unaffected by postmultiplication of $\mathbf{\Lambda}$ by an orthogonal matrix. This follows from the fundamental theorem of common-factor analysis with orthogonal factors

$$\mathbf{\Sigma} = \mathbf{\Lambda}\mathbf{\Lambda} + \mathbf{\Psi}^2 = \mathbf{\Lambda}\mathbf{Q}\mathbf{Q}'\mathbf{I}\mathbf{Q}\mathbf{Q}'\mathbf{\Lambda}' + \mathbf{\Psi}^2 \tag{8.74a}$$

where $\mathbf{Q}$ is an arbitrary orthogonal matrix. Any linear transformation matrix postmultiplying a factor pattern matrix must have the inverse of that transformation applied to the common factors and in turn the correlation matrix among the factors, which here is $\mathbf{I}$. With $\mathbf{Q}\mathbf{Q}' = \mathbf{I}$, and $\mathbf{Q}\mathbf{Q}'\mathbf{I}\mathbf{Q}\mathbf{Q}' = \mathbf{I}$ and so the right-most expression in Equation 8.74a equals $\mathbf{\Lambda}\mathbf{\Lambda}' + \mathbf{\Psi}^2$, and in turn $\mathbf{\Sigma}$. With $\mathbf{\Sigma}$ invariant under the transformation, F_r is invariant.

In effect, given an arbitrary $\mathbf{\Lambda}^*$, and a corresponding nondiagonal $\mathbf{\Lambda}'^*\,\mathbf{\Psi}^{-2}\,\mathbf{\Lambda}^*$, then let $\mathbf{Q}$ be an orthogonal transformation matrix and define $\mathbf{\Lambda} = \mathbf{\Lambda}^*\,\mathbf{Q}$. If $\mathbf{Q}$ is chosen to be the orthogonal eigenvector matrix of $\mathbf{\Lambda}'^*\,\mathbf{\Psi}^{-2}\mathbf{\Lambda}^*$, then $\mathbf{\Lambda}\mathbf{\Psi}^{-2}\,\mathbf{\Lambda} = \mathbf{Q}'\mathbf{\Lambda}^*\,\mathbf{\Psi}^{-2}\,\mathbf{\Lambda}^*\,\mathbf{Q} = \mathbf{V}$, where $\mathbf{V}$ contains the eigenvalues of $\mathbf{\Lambda}'^*\,\mathbf{\Psi}^{-2}\,\mathbf{\Lambda}^*$. So, we can choose for convenience the solution for $\mathbf{\Lambda}$ that makes $\mathbf{\Lambda}'\mathbf{\Psi}^{-2}\,\mathbf{\Lambda}$ a unique diagonal matrix, $\mathbf{V}$. Equation 8.73 then becomes

$$\mathbf{\Psi}^{-1}\mathbf{S}\mathbf{\Psi}^{-1}(\mathbf{\Psi}^{-1}\mathbf{\Lambda}) = (\mathbf{\Psi}^{-1}\mathbf{\Lambda})(\mathbf{I} + \mathbf{V}) \tag{8.74b}$$

Now, if $\mathbf{V}$ is diagonal, so is $(\mathbf{I} + \mathbf{V})$. And because post multiplying $\mathbf{\Psi}^{-1}\mathbf{S}\mathbf{\Psi}^{-1}$ by $(\mathbf{\Psi}^{-1}\mathbf{\Lambda})$ is the same as postmultiplying $(\mathbf{\Psi}^{-1}\,\mathbf{\Lambda})$ by a diagonal matrix, this equation is of the form of an "eigen equation," $\mathbf{MA} = \mathbf{AD}$, where $\mathbf{M}$ is square, symmetric, $\mathbf{A}$ an eigenvector matrix, and $\mathbf{D}$ a diagonal matrix of corresponding eigenvalues. So, $(\mathbf{\Psi}^{-1}\,\mathbf{\Lambda})$ with r columns, has column vectors that are proportional to the eigenvectors of $\mathbf{\Psi}^{-1}\mathbf{S}\mathbf{\Psi}^{-1}$ while $(\mathbf{I} + \mathbf{V})$ is proportional to the first r eigenvalues of $\mathbf{\Psi}^{-1}\mathbf{S}\mathbf{\Psi}^{-1}$.

Now consider the matrix $\mathbf{\Psi}^{-1}\mathbf{S}\mathbf{\Psi}^{-1} - \mathbf{I}$. Evidently this matrix has the same eigenvectors as those of $\mathbf{\Psi}^{-1}\mathbf{S}\mathbf{\Psi}^{-1}$, but its eigenvalues are the eigenvalues of $\mathbf{\Psi}^{-1}\mathbf{S}\mathbf{\Psi}^{-1}$ with 1 subtracted from each. Consider now that we may express $\mathbf{\Psi}^{-1}\mathbf{S}\mathbf{\Psi}^{-1} - \mathbf{I}$ as the matrix $\mathbf{\Psi}^{-1}\,(\mathbf{S} - \mathbf{\Psi}^2)\mathbf{\Psi}^{-1}$. Thus if we postmultiply $\mathbf{\Psi}^{-1}(\mathbf{S} - \mathbf{\Psi}^2)$ $\mathbf{\Psi}^{-1}$ by $(\mathbf{\Psi}^{-1}\,\mathbf{\Psi})$, we get

$$\begin{aligned}\mathbf{\Psi}^{-1}(\mathbf{S} - \mathbf{\Psi}^2)\mathbf{\Psi}^{-1}(\mathbf{\Psi}^{-1}\mathbf{\Lambda}) &= (\mathbf{\Psi}^{-1}\mathbf{S}\mathbf{\Psi}^{-1} - \mathbf{I})\mathbf{\Psi}^{-1}\mathbf{\Lambda}\\ &= (\mathbf{\Psi}^{-1}\mathbf{S}\mathbf{\Psi}^{-1}\mathbf{\Psi}^{-1}\mathbf{\Lambda} - \mathbf{\Psi}^{-1}\mathbf{\Lambda})\\ &= (\mathbf{\Psi}^{-1}\mathbf{\Lambda})(\mathbf{I} + \mathbf{\Lambda}'\mathbf{\Psi}^{-2}\mathbf{\Lambda}) + (\mathbf{\Psi}^{-1}\mathbf{\Lambda})\end{aligned}$$

from Equation 8.74b

$$= (\mathbf{\Psi}^{-1}\mathbf{\Lambda})[(\mathbf{I} + \mathbf{\Lambda}'\mathbf{\Psi}^{-2}\mathbf{\Lambda}) - \mathbf{I}]$$

or

$$\boldsymbol{\Psi}^{-1}(\mathbf{S}-\boldsymbol{\Psi}^2)\boldsymbol{\Psi}^{-1}(\boldsymbol{\Psi}^{-1}\boldsymbol{\Lambda})=\boldsymbol{\Psi}^{-1}\boldsymbol{\Lambda}(\boldsymbol{\Lambda}'\boldsymbol{\Psi}^{-2}\boldsymbol{\Lambda}) \tag{8.75}$$

8.3.5 Maximum-Likelihood Estimates

Let the diagonal matrix $\mathbf{V} = \boldsymbol{\Lambda}'\boldsymbol{\Psi}^{-2}\boldsymbol{\Lambda}$ be scaled to have in its diagonal, eigenvalues of the matrix $\boldsymbol{\Psi}^{-1}(\mathbf{S} - \boldsymbol{\Psi}^2)\boldsymbol{\Psi}^{-1} = \boldsymbol{\Psi}^{-1}\mathbf{S}\boldsymbol{\Psi}^{-1} - \mathbf{I}$. Then this means that the matrix $(\boldsymbol{\Psi}^{-1}\boldsymbol{\Lambda})$ must be proportional to r normalized eigenvectors of $\boldsymbol{\Psi}^{-1}(\mathbf{S} - \boldsymbol{\Psi}^2)\boldsymbol{\Psi}^{-1}$ by the square root of the corresponding eigenvalues of this same matrix. But the matrix $\boldsymbol{\Psi}^{-1}(\mathbf{S} - \boldsymbol{\Psi}^2)\boldsymbol{\Psi}^{-1}$ has the same eigenvectors as the matrix $\boldsymbol{\Psi}^{-1}\mathbf{S}\boldsymbol{\Psi}^{-1}$, whereas the eigenvalues of $\boldsymbol{\Psi}^{-1}(\mathbf{S} - \boldsymbol{\Psi}^2)\boldsymbol{\Psi}^{-1}$ are each less by 1 than the eigenvalues of $\boldsymbol{\Psi}^{-1}\mathbf{S}\boldsymbol{\Psi}^{-1}$. Thus let $\mathbf{A}$ be the $n{\times}n$ matrix of normalized eigenvectors of $\boldsymbol{\Psi}^{-1}\mathbf{S}\boldsymbol{\Psi}^{-1}$. Let $[\gamma_i]$ be the $n{\times}n$ matrix of eigenvalues of $\boldsymbol{\Psi}^{-1}\mathbf{S}\boldsymbol{\Psi}^{-1}$. Then in this case $\boldsymbol{\Psi}^{-1}\boldsymbol{\Lambda} = \mathbf{A}_r[\gamma_i]_r^{1/2}$ and

$$\boldsymbol{\Lambda} = \boldsymbol{\Psi}\mathbf{A}_r[\gamma_i - 1]_r^{1/2} \tag{8.76}$$

after premultiplying both sides by $\boldsymbol{\Psi}$, where the subscript r denotes the matrix obtained from the respectively named matrix by taking the first r rows and columns or diagonal elements, as the case may be. Equation 8.75 is analogous to Equation 8.53 in the derivation of the least-squares estimates for the parameter matrices of the common-factor model.

Jöreskog (1967) indicates, that conditional on a given $\boldsymbol{\Psi}^2$ matrix, the criterion F is minimized when the matrix $\boldsymbol{\Lambda}$ is chosen to be $\boldsymbol{\Lambda} = \boldsymbol{\Psi}\mathbf{A}_r[\gamma_i - 1]_r^{1/2}$, where $[\gamma_i - 1]_r$ contains the r largest eigenvalues of $\boldsymbol{\Psi}^{-1}(\mathbf{S} - \boldsymbol{\Psi}^2)\boldsymbol{\Psi}^{-1}$ and $\mathbf{A}_r$ contains the corresponding eigenvectors. Note that $\mathbf{A}_r$ also corresponds to the r largest eigenvalues $[\gamma_i]_r$ of $\boldsymbol{\Psi}^{-1}\mathbf{S}\boldsymbol{\Psi}^{-1}$. If it should turn out that some of the r largest eigenvalues of $\boldsymbol{\Psi}^{-1}\mathbf{S}\boldsymbol{\Psi}^{-1}$ are less than 1, this means that the corresponding eigenvalues of the matrix $\boldsymbol{\Psi}^{-1}(\mathbf{S} - \boldsymbol{\Psi}^2)\boldsymbol{\Psi}^{-1}$ are negative, which would lead to a factor matrix $\boldsymbol{\Lambda}$ with complex numbers, the interpretation of which is dubious since it would involve taking the square roots of negative numbers.

By the function $g(\boldsymbol{\Psi}^2)$ let us mean the function

$$g(\boldsymbol{\Psi}^2) = \min_{\boldsymbol{\Lambda}} F(\boldsymbol{\Lambda}, \boldsymbol{\Psi}^2) = F(\boldsymbol{\Lambda}_{\boldsymbol{\Psi}^2}, \boldsymbol{\Psi}^2)$$

where $\boldsymbol{\Lambda}_{\boldsymbol{\Psi}^2} = \boldsymbol{\Psi}\mathbf{A}_r[\gamma_i - 1]_r^{1/2}$. The function, $g(\boldsymbol{\Psi}^2)$, is analogous to the function g in the discussion of least-squares fitting of the exploratory common-factor model. Now, Jöreskog (1967, p. 448) shows that for a given

$$g(\boldsymbol{\Psi}^2) = -\sum_{j=r+1}^{n} \ln \gamma_j + \sum_{j=r+1}^{n} \gamma_j - (n - r) \tag{8.77}$$

where the γ_j, $j = r + 1,\ldots,n$, are the $n - r$ residual eigenvalues of the matrix $\mathbf{\Psi}^{-1}\mathbf{S}\mathbf{\Psi}^{-1}$.

We shall now seek to minimize $g(\mathbf{\Psi}^2)$ by finding the appropriate $\mathbf{\Psi}^2$ that accomplishes this. When $g(\mathbf{\Psi}^2)$ is minimized, then $F(\mathbf{\Lambda}, \mathbf{\Psi}^2)$ is minimized. The partial derivative of $g(\mathbf{\Psi}^2)$ with respect to $\mathbf{\Psi}^2$ evaluated for a particular $\mathbf{\Psi}^2$ is equivalent to the partial derivative of F with respect to $\mathbf{\Psi}^2$ evaluated for the particular $\mathbf{\Psi}^2$ matrix with $\mathbf{\Lambda} = \mathbf{\Lambda}_{\mathbf{\Psi}^2} = \mathbf{\Psi}\mathbf{A}_r[\gamma_j - 1]_r^{1/2}$, where $\mathbf{\Psi}$ is the square root of the matrix $\mathbf{\Psi}^2$ in question, $\mathbf{A}_r$ is an $n \times r$ matrix of the first r eigenvectors corresponding to the r largest eigenvalues, $\gamma_1,\ldots, \gamma_r$ of the matrix $\mathbf{\Psi}^{-1}\mathbf{S}\mathbf{\Psi}^{-1}$, and $[\gamma_j - 1]_r$ an $r \times r$ diagonal matrix of eigenvalues formed from these eigenvalues by subtracting 1 from each. Thus let us substitute Equation 8.68 for $\mathbf{\Psi}^{-1}$ in Equation 8.66, with the $\mathbf{\Lambda}$ matrix of $\mathbf{\Sigma} = (\mathbf{\Lambda}\mathbf{\Lambda}' + \mathbf{\Psi}^2)$ set equal to $\mathbf{\Lambda}_{\mathbf{\Psi}^2}$, and then use Equation 8.67 to obtain, after simplification (Jöreskog, 1967, p. 450),

$$\frac{\partial g}{\partial \mathbf{\Psi}^2} = \text{diag}[\mathbf{\Psi}^{-2}(\Sigma - \mathbf{S})\mathbf{\Psi}^{-2}]$$

$$= \text{diag}[\mathbf{\Psi}^{-2}(\mathbf{\Lambda}_{\mathbf{\Psi}^2}\mathbf{\Lambda}'_{\mathbf{\Psi}^2} + \mathbf{\Psi}^2 - \mathbf{S})\mathbf{\Psi}^{-2}] \tag{8.78}$$

At this point Jöreskog (1967) used the then new algorithm of Fletcher and Powell (1963) to solve iteratively for the diagonal elements of $\mathbf{\Psi}^2$. Before going further with this, we need to consider this algorithm, since it became eventually also the basis for solution of the confirmatory factor analysis and structural equation programs produced by Jöreskog, as well as problems of nonlinear optimization generally.

*8.3.6 Fletcher–Powell Algorithm

Until algorithms like the Fletcher–Powell algorithm that became available in the 1960s, another algorithm seemed superior for the solution of simultaneous nonlinear equations, the method of Newton–Raphson. The algorithm is named for Isaac Newton (1643–1727), the great seventeenth century English mathematician, physicist, and alchemist who first proposed it after inventing the calculus, and Joseph Raphson (1648–1715), an English mathematician, who is credited with independently developing this algorithm. We will consider multivariable versions of the algorithm and then show how the Fletcher–Powell algorithm is a modification of it.

Multivariable optimization problems, such as least-squares and maximum-likelihood estimation, generate systems of simultaneous nonlinear equations when the derivatives of the various functions with respect to the function parameters are set to zero. The aim is to solve these equations for the values of the parameters when the derivatives are set equal to zero. For example, consider the partial derivatives of the least-squares fit function (Equations 8.43 and 8.45), when set to zero, yield the equations

$$4(\mathbf{\Lambda\Lambda}')\mathbf{\Lambda} - 4\mathbf{R\Lambda} + 4\mathbf{\Psi}^2\mathbf{\Lambda} = \mathbf{0} \tag{8.79}$$

and

$$\text{diag}(2\mathbf{\Psi}^2 - 2\mathbf{R} + 2\mathbf{\Lambda\Lambda}') = \mathbf{0} \tag{8.80}$$

Or again consider the partial derivatives of the maximum-likelihood fit function for the common-factor analysis problem of Equations 8.65 and 8.66 when set to zero:

$$2\mathbf{\Sigma}^{-1}(\mathbf{\Sigma} - \mathbf{S})\mathbf{\Sigma}^{-1}\mathbf{\Lambda} = \mathbf{0} \tag{8.81}$$

and

$$\text{diag}[\mathbf{\Sigma}^{-1}(\mathbf{\Sigma} - \mathbf{S})\mathbf{\Sigma}^{-1}] = \mathbf{0} \tag{8.82}$$

Usually these systems of equations cannot be solved algebraically. Their solution then is obtained using iterative numerical algorithms. It is for problems like these that the Newton–Raphson method has been developed.

In general, the Newton–Raphson method works as follows: Consider a set of nonlinear, simultaneous equations, $f_1,\ldots, f_n$, in n unknowns, $x_1,\ldots, x_n$, represented by a vector $\mathbf{f}(\mathbf{x})$ of functions where each coordinate $f_i(\mathbf{x}), i = 1,\ldots,n$ of $\mathbf{f}(\mathbf{x})$ takes on the value of the corresponding function for the simultaneous values of the coordinates of the vector $\mathbf{x}$, that is $f_i(\mathbf{x}) = f_i(x_1,\ldots, x_n)$. Our problem is to find that set of values for the vector $\mathbf{x}$ such that $\mathbf{f}(\mathbf{x}) = \mathbf{0}$. To do this, the Newton–Raphson method finds an $n \times n$ matrix $\mathbf{J} = [\partial f_j/\partial x_k]$ by arranging the partial derivatives of each function with respect to each unknown in the $n \times n$ matrix $\mathbf{J}$. The matrix $\mathbf{J}$ is itself a function of the vector $\mathbf{x}$, with the elements of $\mathbf{J}$ taking on the values of the respective partial derivatives of the function evaluated at the point whose coordinates are the current values of the vector $\mathbf{x}$. The Newton–Raphson method then proceeds as follows: Beginning with some initial (best guess) approximation $\mathbf{x}_{(0)}$ to the final solution vector $\mathbf{x}$, improved approximations to the solution vector are obtained iteratively by the formula

$$\mathbf{x}_{(i+1)} = \mathbf{x}_{(i)} - \mathbf{J}(\mathbf{x}_{(i)})^{-1}\mathbf{f}(\mathbf{x}_{(i)}) \tag{8.83}$$

where $\mathbf{x}_{(i)}$ and $\mathbf{x}_{(i+1)}$ are the ith and $(i + 1)$th approximation of the solution vector $\mathbf{x}$. If $\mathbf{x}_{(0)}$ is chosen appropriately, the method converges to a very good approximation of the final solution after only a relatively small number of iterations. As one can see, the method involves both finding the second derivatives of the function to optimize as well as the inverse of $\mathbf{J}$, which places an upper limit in computers on the number of parameters that can be varied in seeking the optimum. This has been the major drawback of using the Newton–Raphson method in multivariable problems. Nevertheless Jennrich and Robinson (1969) demonstrated the feasibility of a Newton–Raphson algorithm for finding the maximum-likelihood estimates of the exploratory common-factor-analysis model.

In nonlinear optimization problems, **f**(**x**) corresponds to the gradient vector $\nabla(\mathbf{x})$ of first derivatives of F (function to be optimized) with respect to each of the parameters, while **J** then becomes the matrix of second derivatives (the derivatives of the first derivatives, known as the Hessian matrix) of this function.

The Fletcher–Powell (1963) algorithm has an iterative equation similar in form to that of the Newton–Raphson algorithm in Equation 8.83:

$$\mathbf{x}_{(j+1)} = \mathbf{x}_{(j)} - \alpha_{(j)}\mathbf{H}_{(j)}\nabla\left(\mathbf{x}_{(j)}\right) \tag{8.84}$$

The matrix $\mathbf{H}_{(j)}$ is $n \times n$, where $\mathbf{H}_{(0)} = \mathbf{I}$ is a good starting value for the iterations, making the first iteration equivalent to a method of steepest descent iteration. $\nabla(\mathbf{x}_{(j)})$ is the $n \times 1$ gradient vector of first derivatives of the function with respect to the n parameters evaluated at the current values of $\mathbf{x}_{(j)}$. The starting values, $\mathbf{x}_{(0)}$, depend on the problem. $\mathbf{d}_{(j)} = -\mathbf{H}_{(j)}\nabla(\mathbf{x}_{(j)})$ is an $n \times 1$ vector that when added to $\mathbf{x}_{(j)}$ will move it in the most "downhill" direction toward a minimum. $\alpha_{(j)}$ is a scalar that scales the quantities in $\mathbf{d}_{(j)} = -\mathbf{H}_{(j)}\nabla(\mathbf{x}_{(j)})$ up or down so that the move of $\mathbf{x}_{(j)}$ to a new value will move it to the lowest value along the line $\mathbf{x}_{(j)} + \beta\mathbf{d}_{(j)}$. Starting at $\mathbf{x}_{(j)}$ with $\beta = 0$ and taking increasing positive values of β, the fit function F initially decreases, reaches a minimum and then increases. Much of the computation at each iteration concerns the relatively simple problem of finding the value $\alpha_{(j)}$ for β along the line that is either the minimum point along the line or a very close approximation to it. I describe in detail this "line search" method used by Jöreskog (1967) to find $\alpha_{(j)}$ in Chapter 15, so we will skip over the details here.

Once we have found $\alpha_{(j)}$, we may then proceed to find $\mathbf{x}_{(j+1)}$ by Equation 8.84. At the new point $\mathbf{x}_{(j+1)}$ we must again evaluate the function F and its gradient vector $\nabla(\mathbf{x}_{(j+1)})$. Next, we must find $\mathbf{H}_{(j+1)}$ for the next iteration. To do this, let

$$\mathbf{w}_{(j)} = \mathbf{x}_{(j+1)} - \mathbf{x}_{(j)} = -\alpha_{(j)}\mathbf{H}_{(j)}\nabla(\mathbf{x}_{(j)}) \tag{8.85}$$

$$\mathbf{q}_{(j)} = \nabla(\mathbf{x}_{(j+1)}) - \nabla(\mathbf{x}_{(j)}) \tag{8.86}$$

and

$$\mathbf{t}_{(j)} = \mathbf{H}_{(j)}\mathbf{q}_{(j)} \tag{8.87}$$

Then, the matrix $\mathbf{H}_{(j+1)}$ is obtained as

$$\mathbf{H}_{(j+1)} = \mathbf{H}_{(j)} + \frac{\mathbf{w}_{(j)}\mathbf{w}'_{(j)}}{\mathbf{w}'_{(j)}\mathbf{q}_{(j)}} - \frac{\mathbf{t}_{(j)}\mathbf{t}'_{(j)}}{\mathbf{q}'_{(j)}\mathbf{t}_{(j)}} \tag{8.88}$$

As the number of iterations increases without bound, $\mathbf{x}_{(j)}$ converges to the minimizing parameters, $\mathbf{H}_{(j+1)}$ converges to the inverse of the Jacobian matrix $\mathbf{J}(\mathbf{x}_{(j)})^{-1}$ of the Newton–Raphson method, and $\nabla(\mathbf{x}_{(j)})$ converges to a null vector or (when the model fails to hold) to an unchanging vector. Of course, we stop the iterations before letting the iterations run forever, when the largest element in the vector difference $\mathbf{x}_{(j+1)} - \mathbf{x}_{(j)}$ is so small in absolute magnitude that it is less than some preset very small, near-zero quantity, such as .00001.

There are two major advantages to the algorithm of Fletcher and Powell (1963): (1) There is no need to work out analytically the matrix of second derivatives of the fit function with respect to the parameters. (2) And one does not need to invert a matrix. The algorithm is furthermore quite efficient and rapid in convergence. Other similar algorithms have been developed, and we describe one of these in a later chapter. As a class of algorithms they are known as quasi-Newton algorithms because of their similarity to the Newton–Raphson method.

*8.3.7 Applying the Fletcher–Powell Algorithm to Maximum-Likelihood Exploratory Factor Analysis

Let us return now to Equation 8.78, since this concerns the elements of a diagonal matrix, we may consider them individually. For the initial iteration, Jöreskog (1967) recommended setting

$$\mathbf{\Psi}^2_{(0)} = \left(1 - \frac{r}{2n}\right)[\text{diag } \mathbf{S}^{-1}]^{-1} = (1 - r/2n)\left[(1/s^{ii})\right],$$

where s^{ii} is the ith diagonal element of the inverse $\mathbf{S}^{-1}$ of the sample variance–covariance matrix $\mathbf{S}$

$$[(1/s^{ii})] = [\text{diag } \mathbf{S}^{-1}]^{-1}$$

Let $\tilde{\mathbf{A}}$ be the eigenvector matrix and $\tilde{\mathbf{D}}$ be the eigenvalue matrix of $\tilde{\mathbf{\Psi}}^{-1}\mathbf{S}\tilde{\mathbf{\Psi}}^{-1}$, where $\tilde{\mathbf{\Psi}}^2 = \mathbf{\Psi}^2_{(j)} = [\psi^2_{ii}]_{(j)}$ are diagonal matrices of the values of unique variances for the jth iteration. Then Equation 8.78 may be rewritten as

$$\frac{\partial g_r}{\partial \psi^2_{ii}} = \left(\frac{1}{\psi^2_{ii}}\right)\left[\sum_{m=1}^{r} (\tilde{\gamma}_m - 1)\tilde{a}^2_{im} + 1 - s_{ii}/\psi^2_{ii}\right], \quad i = 1, 2, \ldots, n \tag{8.89}$$

where $\tilde{\gamma}_1, \tilde{\gamma}_2, \ldots, \tilde{\gamma}_r$ are eigenvalues and $\tilde{a}_{im}$ are elements, respectively, of the eigenvectors $\mathbf{a}_1, \mathbf{a}_2, \ldots, \mathbf{a}_r$, of the matrix $\tilde{\mathbf{\Psi}}^{-1}\mathbf{S}\tilde{\mathbf{\Psi}}^{-1}$. The tildes "~" indicate that these elements are conditional on a given $\tilde{\mathbf{\Psi}}^2$, in this case $\mathbf{\Psi}_2^{-1}\mathbf{S}_{22\cdot 1}\mathbf{\Psi}_2^{-1}$. The element s_{ii} is the ith diagonal element of the sample variance-covariance matrix $\mathbf{S}$. ψ_{ii} is the current value of the ith diagonal element of $\psi^2_{(j)}$. We can now arrange the values of these partial derivatives at any iteration j in an $n \times 1$ column vector

$$\nabla(\psi^2_{(j)}) = \left[\frac{\partial g_r}{\partial \psi^2_{ii}}\right]_{(j)} \tag{8.90}$$

Then, the Fletcher–Powell equation for this case is given by

$$\mathbf{\Psi}_{(j+1)} = \mathbf{\Psi}_{(j)} - \alpha_{(j)}\mathbf{H}_{(j)}\nabla(\mathbf{\Psi}_{(j)}) \tag{8.91}$$

and the update to $\mathbf{H}_{(j+1)}$ is given by Equations 8.85 through 8.88. Next, $\mathbf{\Psi}^2_{(j+1)}$ is the diagonal matrix formed from the elements of $\mathbf{\Psi}_{(j+1)}$, and the new gradient vector $\nabla(\mathbf{\Psi}_{(j+1)})$ is computed using Equation 8.89. The Fletcher–Powell algorithm is then cycled until convergence as indicated by extremely small changes from $\mathbf{\Psi}_{(j)}$ to $\mathbf{\Psi}_{(j+1)}$.

Up to this point we have implicitly assumed that the elements of $\psi^2_{(j)}$, being variances, are always all positive. But with real data this is not realistic. It is possible in some cases for the partial derivatives in Equation 8.90 to vanish when diagonal elements of $\psi^2_{(j)}$ are negative or zero. Negative values imply imaginary values for standard deviations. With zero values we cannot obtain $\mathbf{\Psi}^{-1}$. So, we must keep the values for $\psi^2_{(j)}$ within the positive region by assuring that $\psi^2_{ii} > \varepsilon$, for all i, with ε some arbitrary very small positive value, for example, $\varepsilon = .005$. At any point when $\psi^2_{ii} \le 0$, we can set $\psi^2_{ii} = \varepsilon$, fix the corresponding partial derivative of g with respect to ψ^2_{ii} to zero for the next iteration, and continue the iterations. The value of ψ^2_{ii} may move away from the boundary. But Jöreskog (1967) indicates that if the final solution is in the interior of the positive region the solution is a "proper solution." But if it is on the border, the solution is "improper."

If an improper solution occurs, Jöreskog (1967) recommends partitioning the variables into two sets and reanalyzing. In the first set, place the m variables, $m < r$, with near-zero unique variances on the boundary. In the second set, place the $n - m$ remaining variables. Now, partition the sample covariance matrix, $\mathbf{S}$, according to these two sets of variables so that

$$\mathbf{S} = \begin{bmatrix} \mathbf{S}_{11} & \mathbf{S}_{12} \\ \mathbf{S}_{21} & \mathbf{S}_{22} \end{bmatrix}$$

The matrix $\mathbf{S}_{11}$ is $m \times m$, $\mathbf{S}_{12}$ is $m \times (n - m)$, and the matrix $\mathbf{S}_{22}$ is $(n - m) \times (n - m)$. Next, extract the first m Cholesky factors from the matrix $\mathbf{S}$. This in effect treats the variables in the first set as having zero unique variances and takes out all variance due to these first m variables from the matrix $\mathbf{S}$. At this point the matrix $\mathbf{S}$ becomes

$$\mathbf{S}_{\cdot 1} = \begin{bmatrix} \mathbf{0} & \mathbf{0} \\ \mathbf{0} & \mathbf{S}_{22\cdot 1} \end{bmatrix}$$

The associated $n \times m$ of factor pattern loadings on the Cholesky factors is given by

$$\hat{\mathbf{\Lambda}}_1 = \begin{bmatrix} \hat{\mathbf{\Lambda}}_{11} \\ \hat{\mathbf{\Lambda}}_{21} \end{bmatrix}$$

These are indeed maximum-likelihood estimates of the common-factor loadings of the n variables on the m factors which are completely determined by the first set of variables. Note in this special case that $\hat{\mathbf{\Lambda}}_{11}$ is a lower triangular matrix.

Now, using the Fletcher–Powell algorithm, analyze $\mathbf{S}_{22\cdot 1}$ to find estimates of the unique variances and the common-factor pattern/structure loadings for the residual variates in the second set of variables. The result will include a second set of common-factor pattern/structure loadings on the $n \times (r-m)$ matrix

$$\hat{\mathbf{\Lambda}}_2 = \begin{bmatrix} \mathbf{0} \\ \hat{\mathbf{\Lambda}}_{22} \end{bmatrix}$$

Thus, the two sets of factor pattern/structure loadings may be combined to give an estimate of the common-factor loadings, $\hat{\mathbf{\Lambda}} = [\hat{\mathbf{\Lambda}}_1, \hat{\mathbf{\Lambda}}_2]$, and the estimated population covariance matrix

$$\hat{\mathbf{\Sigma}} = \begin{bmatrix} \hat{\mathbf{\Sigma}}_{11} & \hat{\mathbf{\Sigma}}_{12} \\ \hat{\mathbf{\Sigma}}_{21} & \hat{\mathbf{\Sigma}}_{22} \end{bmatrix} = \begin{bmatrix} \hat{\mathbf{\Lambda}}_{11} & \hat{\mathbf{\Lambda}}_{12} \\ \hat{\mathbf{\Lambda}}_{21} & \hat{\mathbf{\Lambda}}_{22} \end{bmatrix} \begin{bmatrix} \mathbf{\Lambda}_{11} & \hat{\mathbf{\Lambda}}_{12} \\ \hat{\mathbf{\Lambda}}_{21} & \hat{\mathbf{\Lambda}}_{22} \end{bmatrix}' + \begin{bmatrix} \mathbf{0} & \mathbf{0} \\ \mathbf{0} & \hat{\mathbf{\Psi}}_2^2 \end{bmatrix}$$

In applying the Fletcher–Powell (1963) algorithm to the second set of residual variates, the function, F, to minimize is

$$F_r^* = \ln|\mathbf{\Sigma}_{22\cdot 1}| + \mathrm{tr}(\mathbf{\Sigma}_{22\cdot 1}^{-1}\mathbf{S}_{22\cdot 1}) - \ln|\mathbf{S}_{22\cdot 1}| - (n-m) \tag{8.92}$$

which means that we treat the same covariance matrix $\mathbf{S}_{22\cdot 1}$ as a distinct matrix to analyze.

Jöreskog (1967) reported that improper solutions with variables having near-zero unique variances are quite common and are found with all sizes of covariance matrices.

8.3.8 Testing the Goodness of Fit of the Maximum-Likelihood Estimates

When we have obtained the estimates $\hat{\mathbf{\Lambda}}$ and $\hat{\mathbf{\Psi}}^2$, which maximize the likelihood criterion by applying the Fletcher–Powell algorithm to the function F, we may wish to test the goodness of fit of these parameters to the data. To do this, we first let the population covariance matrix under the model be estimated by $\hat{\mathbf{\Sigma}} = \hat{\mathbf{\Lambda}}\hat{\mathbf{\Lambda}} + \hat{\mathbf{\Psi}}^2$. Then we calculate the likelihood ratio statistic

$$U_r = (N-1)\left[\ln|\mathbf{\Sigma}| - \ln|\mathbf{S}| + \text{tr}(\mathbf{\Sigma}^{-1}\mathbf{S}) - n\right] \tag{8.93}$$

which Jöreskog (1967, p. 457) points out is $N - 1$ times the minimum value of the function F minimized by the Fletcher–Powell algorithm. Jöreskog attributes Equation 7.82 to Lawley (1940) and Rippe (1953). The exact distribution of U_r is not known, but when N is large this statistic is approximately distributed as the chi-square distribution with degrees of freedom d_r:

$$d_r = \frac{(n-r)^2 - (n+r)}{2} \tag{8.94}$$

Crudely speaking, Equation 8.94 is a test of whether the residual eigenvalues of the population matrix, $\mathbf{\Psi}^{-1}\mathbf{\Sigma}\mathbf{\Psi}^{-1}$, are all equal. The reason for this is the fact that $\mathbf{\Psi}^{-1}\mathbf{\Sigma}\mathbf{\Psi}^{-1} = \mathbf{\Psi}^{-1}(\mathbf{\Sigma} - \mathbf{\Psi}^2)\mathbf{\Psi}^{-1} = \mathbf{\Psi}^{-1}\mathbf{\Lambda}\mathbf{\Lambda}'\mathbf{\Psi}^{-1} + \mathbf{I}$ when the common-factor model applies. Since the matrix, $\mathbf{\Psi}^{-1}(\mathbf{\Sigma} - \mathbf{\Psi}^2)\mathbf{\Psi}^{-1} = \mathbf{\Psi}^{-1}\mathbf{\Lambda}\hat{\mathbf{\Lambda}}\mathbf{\Psi}^{-1}$, has in this case at most r positive eigenvalues, with the remaining eigenvalues all equal to 0, the matrix, $\mathbf{\Psi}^{-1}\mathbf{\Sigma}\mathbf{\Psi}^{-1}$, must have at most r eigenvalues greater than 1, and $n - r$ residual eigenvalues all equal to 1. If the hypothesis of equal residual eigenvalues is rejected, then one must postulate at least $r + 1$ common factors to account for the population matrix $\mathbf{\Sigma}$ or consider possibly that the common-factor model with r common factors does not apply.

Jöreskog (1967) points out that in exploratory factor analyses one has difficulty specifying a hypothesis for the number r of common factors. Thus one might want a statistical procedure for testing for the number of "significant" factors. Jöreskog (1962) developed such a procedure (discussed in detail in Chapter 9 in connection with his image-factor-analysis model). To implement this procedure, one hypothesizes that the number of common factors is less than or equal to the smallest number of factors which are expected to account for the population correlation matrix. Then one computes the maximum-likelihood estimates $\hat{\mathbf{\Lambda}}$ and $\hat{\mathbf{\Psi}}^2$ for this number of factors and applies the test in Equation 8.93. If the hypothesis is rejected, the hypothesized number of factors is increased by 1 and the procedure repeated. This is done over and over until the hypothesis for the number of factors is accepted. This number is then taken to be the number of the common factors for the correlation matrix. (Sometimes, when N is large, the test becomes quite sensitive to slight departures from equality for residual eigenvalues of the population covariance matrix, and one may end up with many more factors than expected. While many researchers think this is a flaw of the chi-square test, actually it may be an indication that the common-factor model is not exactly appropriate for the data. Among other things, there may be other minor common factors or doublet factors present that the model is not appropriate for, and theory has not developed sufficiently to anticipate and/or control for these effects.) This may not be so serious that we cannot extract something from the analysis that will lead to a hypothesis for this kind of data, but it will mean that to get clean data, we will have to understand the source of these "uninterpretable" factors and take steps in future studies to control for them.

It should be pointed out, however, that this procedure does not represent a series of independent tests. Jöreskog (1962, 1967) indicates that if r^* is the "true" number of common factors, and if the final decision for the series of tests is that there are r common factors, then the probability that $r > r^*$ is equal to or less than the significance level of the test.

In the special case, when an improper solution has been found initially and a reanalysis carried out on the residual portions of those variates shown to have nonzero unique variances, we must reformulate the statistical test used for the goodness of fit of the model. In this case, we assume that the unique variances of a certain subset of the variables are zero, and our statistic is conditional on this assumption. The statistic to use is $N - 1$ times the minimum value of F_r^* (defined in Equation 8.92), which computationally is given by modifying Equation 8.77 to be

$$g(\mathbf{\Psi}^2) = -\sum_{j=r+1}^{n} \ln \gamma_j + \sum_{j=r+1}^{n} \gamma_j - (n - r) \tag{8.95}$$

where the elements γ_j $j = (r + 1),\ldots,n$ are the eigenvalues of the residual $(n - m) \times (n - m)$ matrix $\mathbf{\Psi}_2^{-1}\mathbf{S}_{22\cdot 1}\mathbf{\Psi}_2^{-1}$ with $\mathbf{\Psi}_2^2$ containing the unique variances for the second set of variables. (We assume for purposes of labeling that the variables and eigenvalues of the second set are labeled $m + 1$ through to n.)

8.3.9 Optimality of Maximum-Likelihood Estimators

A criticism occasionally given of the maximum-likelihood method of factor analysis is that it is inapplicable to situations where one is ignorant of the distribution of the observed variables. This occurs frequently in factor-analytic research. Many distributions studied are definitely not multivariate normal because they apply to populations that are highly selected on the variables studied. Although such criticism may apply to the statistical tests that rely on the variables having a multivariate normal distribution, they need not apply to the usefulness of the maximum-likelihood estimators of the factor loadings and the unique variances. Howe (1953) showed that the maximum-likelihood estimates of the factor loadings and unique factors can also be derived from a model that makes no assumptions about the distributions of the observed variables. In this model all one seeks to maximize is the determinant

$$\left|\hat{\mathbf{R}}_{YY\cdot X}\right| = \left|\hat{\mathbf{\Psi}}^{-1}(\mathbf{S}_{YY} - \mathbf{\Lambda\Lambda}')\hat{\mathbf{\Psi}}^{-1}\right|$$

which is the determinant of the partial correlations among the residual variables after r common factors have been partialled from the original covariance matrix. As the matrix $\hat{\mathbf{R}}_{YY \cdot X}$ approaches an identity matrix, the determinant of this matrix approaches a maximum value of 1. Estimates of $\mathbf{\Psi}^2$ and $\mathbf{\Lambda}$ are identical to the corresponding estimates obtained with the maximum-likelihood estimator. We will not develop the proof of Howe's assertion here, although it is found in Mulaik (1972).

As a note of historical interest, Robert H. Brown (Brown, 1961) writes that in 1953 L. L. Thurstone first recommended, on intuitive grounds, the maximization of the determinant of the matrix of partial correlations as a method for estimating the factor loadings and unique variances of a correlation matrix. Brown indicates that the method was applied to several studies in the Psychometric Laboratory in Chapel Hill, North Carolina, and in Frankfurt, Germany (Bargmann, 1955). Howe (1955) subsequently studied this method in its relationship to maximum-likelihood estimation in connection with his doctoral dissertation in mathematical statistics at the University of North Carolina in Chapel Hill, North Carolina.

8.3.10 Example of Maximum-Likelihood Factor Analysis

The correlation matrix in Table 8.4 was subjected to maximum-likelihood factor analysis. A preliminary analysis of the matrix, $\mathbf{S}^{-1}\mathbf{R}\mathbf{S}^{-1}$, yielded seven

TABLE 8.4

Unrotated Maximum-Likelihood Factor Loadings for Correlation Matrix in Table 8.2 and Eigenvalues of the Matrix, $\mathbf{S}^{-1}\mathbf{R}\mathbf{S}^{-1}$

	Unrotated Factors							
	1	2	3	4	5	6	7	Eigenvalues
1.	.77545	.63024	−.01969	−.00875	.00171	−.00021	−.00025	37.484
2.	.59954	.50659	.14968	.44966	−.19258	−.11613	.01212	14.467
3.	.61986	.52954	.06756	.44900	−.08490	−.00464	.07379	9.531
4.	.57322	.49103	.26072	.41122	−.11454	.07840	.13292	1.436
5.	.75709	−.65246	−.00606	−.00245	−.00851	−.00004	.00009	1.289
6.	.66695	−.44626	.07984	.14137	.50264	−.05956	.01990	1.116
7.	.68721	−.45759	.06864	.09202	.37314	.05468	−.04874	1.065
8.	.68338	−.46593	.07124	.15334	.30383	.05762	.06907	.949
9.	.40851	.20153	.63869	−.20328	.04374	.22141	.11836	.921
10.	.46807	.19887	.81710	−.08900	−.01432	−.11146	−.04166	.767
11.	.42017	.19308	.75487	−.05129	−.03639	.02856	.05412	.735
12.	.43102	.19997	.74264	−.16561	−.02717	.09421	.10356	.720
13.	.61984	.28135	.30026	.35307	−.00197	.17591	−.25221	.591
14.	.64553	.26911	.28446	.29129	−.03826	.11396	−.22635	.563
15.	.51933	.26921	.73117	.00014	.01271	.07379	.00094	.546

eigenvalues greater than 1.0, which corresponds to the number of positive eigenvalues of the matrix $\mathbf{R} - \mathbf{S}^2$. For the maximum-likelihood analysis, the number of factors was then fixed at 7. Subsequently, at the end of the analysis we computed the chi-square statistic to determine if the common-factor model fit the data for seven factors. The chi-square was not significant at the .05 level. The chi-square test for an analysis with six factors was significant with $p = .03$.

9

Other Models of Factor Analysis

9.1 Introduction

In Chapter 8, we discussed the model of common-factor analysis, pointing out that its chief defect lies in the indeterminacy of the common and unique portions of variables. In this chapter, we consider several models of factor analysis which keep some of the essential ideas of the common-factor-analysis model while introducing new ideas to provide more determinacy. Specifically we will consider (1) the model of component analysis, which makes no distinction between common and unique variance before analyzing variables into components; (2) image analysis, which defines the common part of a variable as that part predictable by multiple correlation from all the other variables in the analysis; (3) canonical-factor analysis, which seeks to find the particular linear combinations of the original variables that would be maximally correlated with a set of common factors for the variables; (4) image-factor analysis; (5) models that control for doublets, and (6) alpha factor analysis, which is a variant of the common-factor-analysis model which rescales the variables before analysis into the metric of the common parts of the variables. We will also consider the question of psychometric inference in factor analysis, because this will have some bearing on our evaluation of these different models.

9.2 Component Analysis

Component analysis is most frequently thought of as a factor analysis of a correlation matrix with 1s instead of communalities in its principal diagonal. Rather than trying to make a more determinate distinction between common and unique variance in variables, the component-analysis model avoids this distinction altogether and instead concentrates upon analyzing the variables into a linearly independent set of component variables from which the original variables can be derived. In general, if there are n variables in a correlation matrix to be analyzed by component analysis, n components will be needed to account for all the variance among the n original variables. That there are not fewer components than variables—in theory at least—is because ordinarily each of the n original variables contains error of measurement which makes the n original variables a linearly independent set. In other words, the error in each variable cannot be derived from the other variables; as a result there must be n components to account for the total variance of all the variables.

Many factor analysts resort to component analysis in practice as an approximation to common-factor analysis. In doing so they believe they avoid the computationally cumbersome problem of estimating communalities and in return obtain results that often do not differ greatly from those they would have gotten by using a common-factor-analysis model. Moreover, being a completely determinate model, component analysis can directly compute factor scores without the need to estimate them, as is the case in common-factor analysis. This feature seems to be a particular advantage in some studies where scores on the factors are needed.

However, there are theoretical drawbacks to the use of component analysis. In component analysis, the "factors" are linear combinations of the observed variables. The factors occupy the same vector space as do the observed variables. In common-factor analysis, the common factors are conceived as existing independently of the observed variables. They may have influences on other variables not among the present set of observed variables. So, the common factors are not simple linear combinations of the observed variables in a given study. Thurstone (1937) argued that "The factorial description of a test must remain invariant when the test is moved from one battery to another. This simply means that if a test calls for a certain amount of one ability and a certain amount of another ability, then that test should be so described irrespective of the other tests in the battery. If we ask the subjects to take some more tests at a later time, and if we add these tests to the battery, then the principal components and the centroid axes will be altered, and the factorial description of each test will also be altered…" (Thurstone, 1937, p. 75). So, one should expect the factor loadings of variables to remain invariant whenever the tests are embedded with new sets of variables from a domain containing the same common factors influencing the tests as in the original study. You cannot generally have this with a component analysis because the components, as linear combinations of the observed variables, change with the addition or change of the variables.

On the other hand, component analysis is especially useful in its own right when the objective of the analysis is not to account for just the correlations among the variables but rather to summarize the major part of the information contained in them in a smaller number of (usually orthogonal) variables. For example, in multiple regression one might subject the independent (predictor) variables to a component analysis and use the subset of components accounting for the greatest proportion of variance among the variables as a new set of predictor variables. With fewer variables in the predictor set one loses fewer degrees of freedom than when the original, full set of variables is used. Similarly one can carry out a component analysis of a set of dependent variables in a multivariate analysis of variance and use the components accounting for the major part of the variance among the independent variables as a new set of dependent variables and thereby minimize losses in degrees of freedom in the analysis.

9.2.1 Principal-Components Analysis

A component analysis can be conducted with any of the methods of factor extraction: diagonal method, centroid method, or principal-axes method. The only requirement is that the matrix to be operated on by these methods be initially the unreduced correlation matrix with 1s in its principal diagonal. However, in practice, with computers doing the extra work involved, most factor analysts use the principal-axes method for extracting components. This is because after any arbitrary number of principal components have been extracted, the remaining variance to account for is a minimum for this number of components extracted. As a consequence the procedure of principal-axes extraction of components has acquired a special name, "principal-components analysis." The advantages of this method over the diagonal method or centroid method lie in the mathematical convenience of working with the eigenvalues and eigenvectors to determine the number of factors to retain and to compute factor scores.

We will now consider a more explicit treatment of the mathematics of principal-components analysis. Let **Y** be an $n \times 1$ random vector whose coordinates are n standardized variables with means of zero and standard deviations of 1. Let **X** be an $n \times 1$ random vector whose coordinates are the n principal components for the n variables standardized to have unit variances. Let $\mathbf{\Lambda}$ be an $n \times n$ square matrix of principal-axes factor loadings of the n variables on the n principal components. Then the fundamental equation of principal-components analysis is

$$\mathbf{Y} = \mathbf{\Lambda X} \tag{9.1}$$

By definition, the matrix $\mathbf{\Lambda}$ is given as

$$\mathbf{\Lambda} = \mathbf{AD}^{1/2} \tag{9.2}$$

where

A is the $n \times n$ eigenvector matrix

D the $n \times n$ diagonal eigenvalue matrix for the matrix **R** which contains the correlations between the n variables

That is

$$\mathbf{R} = E(\mathbf{YY}') \tag{9.3}$$

and

$$\mathbf{A'RA} = \mathbf{D} \tag{9.4}$$

The coordinates of **X**, representing the principal components, are mutually orthogonal so that

$$E(\mathbf{XX'}) = \mathbf{I} \tag{9.5}$$

Consequently

$$\mathbf{R} = \mathbf{\Lambda\Lambda'} = \mathbf{AD}^{1/2}\mathbf{D}^{1/2}\mathbf{A'} = \mathbf{ADA'} \tag{9.6}$$

9.2.2 Selecting Fewer Components than Variables

Although in theory n principal components must be used to account for all the variance of n variables, in practice most factor analysts retain fewer than n principal components. This practice is a carryover from the model of common-factor analysis which postulates fewer factors than variables.

Several rationales are used to justify this procedure. One rationale is that one is interested only in the most important principal components because they account for the major features of the phenomenon being studied. As a rule of thumb, then, one retains only enough principal components to account for, say, 95% of the total variance. This can be accomplished by cumulatively summing the descending series of eigenvalues until the sum divided by n equals .95 or greater. The basis for this procedure is the fact that the eigenvalue of a principal component divided by n, the number of variables, gives the proportion of the total variance accounted for by that principal component. The sum of the eigenvalues of a group of principal components divided by n thus gives the proportion of the total variance accounted for by that group.

Bartlett (1950, 1951) proposed a "significance test" for the number of principal components to retain. The idea behind such a test requires some rather stringent assumptions, which are more compatible with a common-factor-analysis model than with the principal-components model. The basic assumption is that the population correlation matrix can be analyzed into two matrices

$$\mathbf{R} = \mathbf{\Lambda\Lambda'} + \theta\mathbf{I} \tag{9.7}$$

where

R is the population correlation matrix

$\mathbf{\Lambda\Lambda'}$ is a covariance matrix based on r "significant" components

$\theta\mathbf{I}$ is a diagonal matrix of error variances all of which are equal to θ

In other words, after r components have been extracted from the matrix the remaining components are assumed due only to error variance, and the variances of these $n-r$ remaining error components are equal. Thus with a sample correlation matrix a test for the presence of error components involves a test for the equality of the variances of the principal components (before they are rescaled to have variances of unity). Since the eigenvalues of the sample correlation matrix represent these variances, the test reduces to a test of the equality of the sample eigenvalues.

In practice, the sample principal components are arranged in the order of the descending magnitude of their eigenvalues. Then a series of tests is begun: First one tests the hypothesis that all the population eigenvalues are equal except for sampling error. If this hypothesis is rejected, the first principal component is accepted as significant. Then one tests the hypothesis that the $n-1$ remaining eigenvalues are equal except for random fluctuations, etc. The point at which one cannot reject the hypothesis that the remaining eigenvalues are equal except for fluctuations due to chance fixes the number of significant principal components.

The particular test that is used for each of these hypotheses is based upon the ratio of the product of the remaining eigenvalues to the $(n-r)$ th power of the arithmetic mean of the remaining eigenvalues, where r is the number of significant principal components hypothesized at that point in the series of tests. In mathematical terms this ratio will be denoted by Q and defined as follows, where $\tilde{\gamma}_i$, is the ith eigenvalue of the sample correlation matrix.

$$Q = \left(\frac{\gamma_{r+1} + \cdots + \gamma_n}{\left(\frac{\gamma_{r+1} + \cdots + \gamma_n}{n-r} \right)^{n-r}} \right) \tag{9.8}$$

The natural logarithm of Q is then obtained and multiplied by the following coefficient K

$$K = -(N-1) + \frac{2n+5}{6} + \frac{2r}{3} \tag{9.9}$$

where

N is the number of subjects in the sample
n is the number of tests
r is the number of significant principal components accepted to that point in the series of tests

The multiplication of the natural logarithm of Q by the coefficient K results in a coefficient which is a very close approximation of chi-square, that is,

$$\chi^2 = K \ln Q \tag{9.10}$$

The number of degrees of freedom used with this version of chi-square is $(n-r-1)(n-r+2)/2$, where n is the number of variables and r the hypothesized number of principal components. Hence for a given hypothesis for the number of significant principal components one can compare the value of chi-square computed from the data with the tabular value of chi-square corresponding to the appropriate degrees of freedom and significance level under the null hypothesis. If the value of chi-square thus computed from the data is greater than the tabular value of chi-square, then the hypothesis that all the significant components have been found and that the remaining eigenvalues are equal except for sampling error is rejected.

9.2.3 Determining the Reliability of Principal Components

When accurate estimates of the reliabilities of the individual variables are available in a principal-component analysis, we can also obtain estimates of the reliabilities of the principal components. To do this we draw upon the distinction in classical reliability theory between the "true-score" and "error-score" parts of an observed variable. This is to say, consider that a set of n observed variables may be partitioned into a set of n true-score variables and a set of n error score variables as expressed by the following equation

$$\mathbf{Y} = \mathbf{Y}_t + \mathbf{Y}_e \tag{9.11}$$

where

- $\mathbf{Y}$ is an $n \times 1$ random vector of n standardized observed variables
- $\mathbf{Y}_t$ is the $n \times 1$ random vector of true-score parts of the observed variables
- $\mathbf{Y}_e$ is the $n \times 1$ random vector of error-score parts of the observed variables

We will assume that $E(\mathbf{YY}') = \mathbf{R}$, a correlation matrix; $E(\mathbf{Y}_e\mathbf{Y}_e') = \mathbf{E}^2$, a diagonal matrix of error variances (also signifying that errors of different variables are uncorrelated); and $E(\mathbf{Y}_t\mathbf{Y}_e') = \mathbf{0}$, a null matrix (signifying that error and true components of variables are uncorrelated).

Let $\mathbf{a}_j$ be the jth eigenvector of the correlation matrix $\mathbf{R}$. From discussion of the principal-axes method of factor extraction in Chapter 6 we learned that an eigenvector of the correlation matrix provides a set of weights for deriving a composite variable from the original variables that is collinear with the jth principal-component variable. The variance of the composite variable is equal to the jth eigenvalue γ_j of $\mathbf{R}$. Mathematically, this is

$$\mathbf{X}_j = \mathbf{a}_j'\mathbf{Y} \tag{9.12}$$

and

$$\gamma_j = E(\mathbf{a}_j'\mathbf{YY}'\mathbf{a}_j) = \mathbf{a}_j'\mathbf{R}\mathbf{a}_j \tag{9.13}$$

If we substitute the expression for **Y** in Equation 9.10 into Equation 9.12, we obtain

$$\gamma_j = \left[\mathbf{a}_j'(\mathbf{Y}_t + \mathbf{Y}_e)(\mathbf{Y}_t + \mathbf{Y}_e)'\mathbf{a}_j\right] \qquad (9.14)$$

which upon expanding becomes

$$\gamma_j = E[\mathbf{a}_j'(\mathbf{Y}_t\mathbf{Y}_t' + \mathbf{Y}_t\mathbf{Y}_e' + \mathbf{Y}_e\mathbf{Y}_t' + \mathbf{Y}_e\mathbf{Y}_e')\mathbf{a}_j] \qquad (9.15)$$

In the population, classical reliability theory postulates that the covariance between true-score and error-score parts is zero, that is $E(\mathbf{Y}_t\mathbf{Y}_e') = \mathbf{0}$. Hence the middle two matrix products within the parentheses of Equation 9.14 are zero and can be dropped. The remaining two matrix products in Equation 9.14 lead to the covariances among true-score and error-score parts of the observed variables, respectively. When the true-score and error-score parts of the variables are standardized in the metric of the observed score standard deviations, the covariance among true-score parts of the variables reduces to the correlation matrix with reliability coefficients in the principal diagonal, whereas the covariance among their error-score parts reduces to a diagonal matrix with errors of measurement in the principal diagonal. That is, Equation 9.15 reduces to

$$\gamma_j = \mathbf{a}_j'(\mathbf{R} - \mathbf{E}^2)\mathbf{a}_j + \mathbf{a}_j'\mathbf{E}^2\mathbf{a}_j \qquad (9.16)$$

where $\mathbf{E}^2$ is a diagonal matrix containing errors of measurement of the n variables.

There are several ways to express reliability. One way is to find an expression for the true variance and divide it by an expression for total variance. Another way, which will be more convenient for us to use here, is to subtract from 1 the expression for error variance divided by the expression for total variance.

In Equation 9.16, the expression $\mathbf{a}_j'\mathbf{E}^2\mathbf{a}_j$ represents the error entering the jth principal component. Hence the reliability r_j of the jth principal component is given as

$$r_j = 1 - \frac{\mathbf{a}_j'\mathbf{E}^2\mathbf{a}_j}{\gamma_j} \qquad (9.17)$$

which reduces to the algebraic expression

$$r_j = 1 - \frac{\sum_{i=1}^{n} a_{ij}^2\sigma_{ie}^2}{\gamma_j} \qquad (9.18)$$

where

n is the number of variables

a_{ij} refers to the ith element of the jth eigenvector

σ_{ie}^2 is the ith variable's error of measurement

γ_j is the jth eigenvalue

9.2.4 Principal Components of True Components

The presence of error of measurement in scores on tests used to measure the correlation between traits has the effect of producing correlations that are "attenuated" from what they would be were the tests perfectly reliable measures. Since the aim of most factor analyses is to determine factors that account for relationships among traits and not simply among fallible measures of traits, one desires a means for correcting correlation coefficients for this "attenuation" due to unreliability of measurement. Classic reliability theory provides the means for correcting this correlation coefficient. If r_{jk} is the correlation between tests j and k, then the correlation coefficient r_{jk} corrected for attenuation is given as

$$r'_{jk} = \frac{r_{jk}}{\sqrt{r_{jj}}\sqrt{r_{kk}}} \tag{9.19}$$

where r_{jj} and r_{kk} are the reliability coefficients of tests k and j, respectively. What a correlation coefficient corrected for attenuation amounts to is the correlation between two variables based upon their true scores. We have already shown in the discussion of the reliability of principal components how the matrix of covariances among the true scores of variables is the correlation matrix with reliability coefficients in the principal diagonal, when the true scores are standardized in the metric of the observed scores. That is

$$(\mathbf{R} - \mathbf{E}^2) = E(\mathbf{Y}_t\mathbf{Y}_t') \tag{9.20}$$

where

$\mathbf{R}$ is the observed correlation matrix

$\mathbf{E}^2$ is a diagonal matrix of errors of measurement

$\mathbf{Y}_t$ is an $n \times 1$ vector of true-score parts of n variables standardized in the metric of observed variables

To obtain a correlation matrix from a covariance matrix one must premultiply and postmultiply the covariance matrix by a diagonal matrix containing the reciprocals of the square roots of the diagonal elements of the covariance matrix.

Hence define $\mathbf{K}$ to be a diagonal matrix such that

$$\mathbf{K}^2 = [\text{diag}(\mathbf{R} - \mathbf{E}^2)] \tag{9.21}$$

Since the diagonal elements of the matrix $(\mathbf{R} - \mathbf{E}^2)$ contain the reliability coefficients of the variables in $\mathbf{R}$, the diagonal elements of $\mathbf{K}$ also contain these reliability coefficients. The matrix $\mathbf{R}_{tt}$ of correlations among true scores (the matrix equation for correction for attenuation) is

$$\mathbf{R}_{tt} = \mathbf{K}^{-1}(\mathbf{R} - \mathbf{E}^2)\mathbf{K}^{-1} \tag{9.22}$$

A detailed examination of the elements of $\mathbf{R}_{tt}$ will reveal that they are correlation coefficients of $\mathbf{R}$ corrected for attenuation as by Equation 9.19. The diagonal elements of $\mathbf{R}_{tt}$ are all 1s.

Having obtained the matrix $\mathbf{R}_{tt}$ one can then go on to analyze it with one of the various models of factor analysis. A common-factor analyst would seek to obtain the communalities of the variables from the matrix $\mathbf{R}_{tt}$ and then proceed with an analysis of a reduced correlation matrix based on $\mathbf{R}_{tt}$. However, it is also possible to conduct a principal-components analysis on the matrix $\mathbf{R}_{tt}$. Hotelling, who is generally credited with formulating the major features of the principal-components model and with giving it its name, discussed such an analysis in his classic paper (Hotelling, 1933) in the *Journal of Educational Psychology*. In fact, Hotelling preferred such an analysis to a principal-components analysis of observed scores. He also provided a cumbersome test for the significance of principal components of true scores, although such a test has rarely been used by others. We yet await the development of a convenient test for the significance of principal components for true scores.

In the absence of convenient tests for significance of principal components of true scores, one is forced to fall back on other procedures for selecting components to retain. Since, in practice, sampling error leaves the sample matrix $\mathbf{R}_{tt}$ generally non-Gramian, one might save all principal components with positive eigenvalues. This would generally be more than most factor analysts desire, and so other rules of thumb must be investigated to determine the number of components to retain.

Although the model of principal-components analysis applied to observed scores is completely determinate, the model applied to true scores is not determinate. The chief reason for this indeterminateness is because there is no way to identify exactly the partitioning of an observed variable into true-score and error-score parts. Moreover, the reliability coefficients that one would use in practice are only estimates of the reliability coefficients for the variables in the population of subjects. Also, the manner in which these coefficients are obtained are independent of the manner in which the correlation matrix is obtained; consequently, estimated reliability coefficients inserted in the diagonal of the correlation matrix leave the matrix non-Gramian. Even if one had perfect estimates of the reliabilities in the population, their application to a sample correlation matrix would leave the matrix non-Gramian in most cases because sampling error would make the sample correlation matrix deviate from the population correlation matrix, making the true reliabilities inconsistent with the sample correlations. Only when one had a correlation matrix based upon the population of subjects could one use perfectly accurate estimates of the reliability coefficients to obtain a Gramian correlation matrix among true-scores parts of the variables. But even achieving this would not overcome the essential indeterminateness because of the inability to identify true and error parts of observed scores.

9.2.5 Weighted Principal Components

Although we cannot determine exactly the error in variables so that we might partial this error from the variables before we conduct component analyses of them, we may nevertheless conduct component analyses in which the earlier extracted components are relatively free of error whereas the later extracted components are more exclusively error. In such component analyses, which are carried out by the method of weighted principal components, salvaging the first so-many components and discarding the remainder usually result in the retention of the most important, relatively error-free components. We shall now discuss how to obtain such error-free components.

Consider again that a set of n observed variables may be partitioned into a set of n true-score variables and a set of n error-score variables as in Equation 9.11; that is

$$\mathbf{Y} = \mathbf{Y}_t + \mathbf{Y}_e \tag{9.11}$$

Now, an arbitrary linear combination $Y^* = \mathbf{b}'\mathbf{Y} = \mathbf{b}'(\mathbf{Y}_t + \mathbf{Y}_e) = \mathbf{b}'\mathbf{Y}_t + \mathbf{b}'\mathbf{Y}_e$ of the n observed variables may also be seen as consisting of a true-score part, $\mathbf{b}'\mathbf{Y}_t$, and an error-score part, $\mathbf{b}'\mathbf{Y}_e$. The total variance of $\mathbf{Y}^*$ also consists of two nonoverlapping parts, that is

$$\sigma_{Y^*}^2 = \mathbf{b}'\mathbf{R}\mathbf{b} = \mathbf{b}'(\mathbf{R} - \mathbf{E}^2)\mathbf{b} + \mathbf{b}'\mathbf{E}^2\mathbf{b} \tag{9.23}$$

where

$\mathbf{b}'(\mathbf{R} - \mathbf{E}^2)\mathbf{b}$ is the true variance of Y^*

$\mathbf{b}'\mathbf{E}^2\mathbf{b}$ is the error variance of Y^*

The proportion of the total variance that is error variance is given by the ratio

$$\frac{\mathbf{b}'\mathbf{E}^2\mathbf{b}}{\mathbf{b}'\mathbf{R}\mathbf{b}} \tag{9.24}$$

Minimizing this ratio may be seen to be essentially equivalent to maximizing the value of $\mathbf{b}'\mathbf{R}\mathbf{b}$ under the restriction that $\mathbf{b}'\mathbf{E}^2\mathbf{b} = 1$. Let

$$F = \mathbf{b}'\mathbf{R}\mathbf{b} - \gamma(\mathbf{b}'\mathbf{E}^2\mathbf{b} - 1) \tag{9.25}$$

where γ is a Lagrangian multiplier multiplied times the equation of restraint. A necessary condition that $\mathbf{b'Rb}$ be a maximum under the restraint on $\mathbf{b}$ that $\mathbf{b'E^2b} = 1$ is that the partial derivative of F with respect to $\mathbf{b}$ equals zero, that is,

$$\frac{\partial F}{\partial \mathbf{b}} = 2\mathbf{Rb} - 2\gamma\mathbf{E}^2\mathbf{b} = \mathbf{0}$$

$$= (\mathbf{R} - \gamma\mathbf{E}^2)\mathbf{b} = \mathbf{0} \tag{9.26}$$

Equation 9.26 has similarities to Equation 7.11 involved in the search for eigenvectors and eigenvalues of a square symmetric matrix. They are especially similar in that to have nontrivial solutions for the weight vector involved, the premultiplying matrix of each equation must be singular, which in the present case means that

$$|\mathbf{R} - \gamma\mathbf{E}^2| = 0 \tag{9.27}$$

The problem is to find the values of γ that make the determinant equal to zero. To do this, we can use the trick of converting Equation 9.24 into another equation which we know has the same solutions for γ and which is solvable by methods at hand.

Consider the determinant of $\mathbf{E}^{-1}(\mathbf{R} - \gamma\mathbf{E}^2)\mathbf{E}^{-1} = \mathbf{E}^{-1}\mathbf{R}\mathbf{E}^{-1} - \gamma\mathbf{I}$ set equal to zero:

$$|\mathbf{E}^{-1}\mathbf{R}\mathbf{E}^{-1} - \gamma\mathbf{I}| = 0 \tag{9.28}$$

This determinant is analogous to the determinant leading to the characteristic equation for a square symmetric matrix shown in Equation 7.33. Any solution for γ_i that makes the determinant in Equation 9.28 equal to zero will also make the determinant in Equation 9.27 equal to zero. For proof, consider that the matrix $\mathbf{E}^2$ of error variances contains only positive diagonal elements. Then it follows that $|\mathbf{E}^{-1}| \neq 0$. We can next set up the equation

$$|\mathbf{E}^{-1}||\mathbf{R} - \gamma\mathbf{E}^2||\mathbf{E}^{-1}| = 0 \tag{9.29}$$

We see that if $|\mathbf{E}^{-1}| \neq 0$, then the only way for this equation to equal zero is for an appropriate choice of γ to be found to make the determinant $|\mathbf{R} - \gamma\mathbf{E}^2|$ equal to zero. In other words, for the determinant in Equation 9.28 to be zero, the determinant in Equation 9.27 must be zero. But the determinantal expression in Equation 9.28 is equivalent to

$$|\mathbf{E}^{-1}(\mathbf{R} - \gamma\mathbf{E}^2)\mathbf{E}^{-1}| = 0 \tag{9.30}$$

Carrying the matrices $\mathbf{E}^{-1}$ inside the parentheses reduces Equation 9.30 to Equation 9.28, and we have completed the proof.

Not only are the eigenvalues of $\mathbf{E}^{-1}\mathbf{R}\mathbf{E}^{-1}$ the values of γ that we seek, but an eigenvector, **a**, of this matrix is simply related to a weight vector, **b**, satisfying Equation 9.24 by the equations

$$\mathbf{a} = \mathbf{E}\mathbf{b} \quad \text{or} \quad \mathbf{b} = \mathbf{E}^{-1}\mathbf{a} \tag{9.31}$$

For proof, consider that the restraint on **b** is satisfied by

$$\mathbf{b}'\mathbf{E}^2\mathbf{b} = \mathbf{a}'\mathbf{E}^{-1}\mathbf{E}^2\mathbf{E}^{-1}\mathbf{a} = \mathbf{a}'\mathbf{a} = 1$$

by reason of the fact that the vector **a** is a normalized eigenvector of the matrix $\mathbf{E}^{-1}\mathbf{R}\mathbf{E}^{-1}$. On the other hand,

$$\mathbf{b}'\mathbf{R}\mathbf{b} = \mathbf{a}'\mathbf{E}^{-1}\mathbf{R}\mathbf{E}^{-1}\mathbf{a} = \gamma$$

is a maximum if the eigenvector **a** corresponds to the largest eigenvalue of the matrix $\mathbf{E}^{-1}\mathbf{R}\mathbf{E}^{-1}$.

Let $\mathbf{X}^*$ be an $n \times 1$ random vector of unnormalized component variables defined by

$$\mathbf{X}^* = \mathbf{B}'\mathbf{Y} = \mathbf{A}'\mathbf{E}^{-1}\mathbf{Y}$$

where

$\mathbf{B} = \mathbf{E}^{-1}\mathbf{A}$ is an $n \times n$ matrix

A is an $n \times n$ orthogonal matrix having as columns the eigenvectors of $\mathbf{E}^{-1}\mathbf{R}\mathbf{E}^{-1}$

$\mathbf{E}^2$ is the diagonal matrix of error variances of the variables in **Y**

Furthermore

$$E(\mathbf{X}^*\mathbf{X}^{*\prime}) = \mathbf{A}'\mathbf{E}^{-1}\mathbf{R}\mathbf{E}^{-1}\mathbf{A} = [\gamma]$$

where $[\gamma]$ denotes the diagonal matrix of eigenvalues of the matrix $\mathbf{E}^{-1}\mathbf{R}\mathbf{E}^{-1}$. The unnormalized component variables in $\mathbf{X}^*$ are thus orthogonal to one another, and when ordered according to the descending values of the eigenvalues of $\mathbf{E}^{-1}\mathbf{R}\mathbf{E}^{-1}$, they represent unnormalized component variables having successively greater proportions of error variance in them.

The unnormalized component variables in $\mathbf{X}^*$ represent principal components of the unnormalized variables of $\mathbf{E}^{-1}\mathbf{Y}$. Whereas the usual principal-components analysis involves the normalized variables in **Y**, the weighted-principal-components analysis of the variables in $\mathbf{E}^{-1}\mathbf{Y}$ involves a weighting of the respective variables in **Y** before carrying out the analysis. The effect of premultiplying **Y** by $\mathbf{E}^{-1}$ is to give much greater positive weight to variables having little error variance in them in the extraction of the components.

TABLE 9.1

Values of e_{ii}^{-1} Corresponding to Values of e_{ii}^2

e_{ii}^2	e_{ii}	e_{ii}^{-1}
.0001	.01	100
.01	.10	10
.10	.316	3.16
.20	.446	2.24
.30	.547	1.83
.40	.632	1.68
.50	.707	1.41
.60	.775	1.29
.70	.837	1.19
.80	.895	1.12
.90	.95	1.05
.99	.995	1.001

To see why this is so, refer to Table 9.1 which gives values of e_{ii}^{-1} corresponding to select values of e_{ii}^2 (representing the proportion of variance due to error in variable i). If the proportion of variance due to error in a variable is .90, that variable receives a weight of 1.05 in the component-extraction process. On the other hand, if the proportion of variance due to error in a variable is only .01, then that variable receives a weight of 10 in the extraction process.

Since the aim of the principal-axes method of extracting factors is to find linear combinations of the variables having maximum variance, the principal-axes method tends to give greater weight in determining the components extracted to variables which are strongly intercorrelated and have high variances. When the variables are all normalized to have the same variance, the principal-axes method initially is influenced only by the intercorrelations among variables. However, if the variables do not all have equal variances, then the principal-axes factor-extraction process is influenced by those variables that not only are strongly intercorrelated but also have high variances. When we assign differential weights to normalized variables, we in effect increase the variances of some variables relative to the variances of other variables; the result is that variables with greater variances have greater influence on the extraction of earlier principal components than upon later components.

The unnormalized-principal-component variables in $\mathbf{X}^*$ may be normalized by premultiplying them by a diagonal matrix containing the reciprocals of the standard deviations of the variables in $\mathbf{X}^*$; that is, let

$$\mathbf{X} = [\gamma]^{-1/2}\mathbf{X}^* = [\gamma]^{-1/2}\mathbf{A}'\mathbf{E}^{-1}\mathbf{Y} \tag{9.32}$$

where

$\mathbf{X}$ is an $n \times 1$ random vector of normalized principal components of the weighted variables in $\mathbf{E}^{-1}\mathbf{Y}$, with $\mathbf{A}$ the $n \times n$ matrix of eigenvectors

$[\gamma]$ the $n \times n$ diagonal matrix of eigenvalues of the matrix $\mathbf{E}^{-1}\mathbf{R}\mathbf{E}^{-1} - \mathbf{I}$

Ultimately, we are interested in how the unweighted variables in $\mathbf{Y}$ relate to the component variables in $\mathbf{X}$. This may be learned from the structure matrix

$$E(\mathbf{YX}') = E(\mathbf{YY}'\mathbf{E}^{-1}\mathbf{A}[\gamma]^{-1/2}) = \mathbf{RE}^{-1}\mathbf{A}[\gamma]^{-1/2} \tag{9.33}$$

The expression in Equation 9.33 may be simplified to a more useful expression if we recall the effect of postmultiplying a square symmetric matrix by its eigenvector matrix. Note that

$$\mathbf{E}^{-1}\mathbf{R}\mathbf{E}^{-1}\mathbf{A} = \mathbf{A}[\gamma] \tag{9.34}$$

Evidently if we premultiply both sides of Equation 9.34 by **E** we obtain

$$\mathbf{R}\mathbf{E}^{-1}\mathbf{A} = \mathbf{E}\mathbf{A}[\gamma] \tag{9.35}$$

Substituting Equation 9.35 into Equation 9.33 and simplifying, we obtain

$$E(\mathbf{Y}\mathbf{X}') = \mathbf{E}\mathbf{A}[\gamma]^{1/2} \tag{9.36}$$

The factor-structure matrix in Equation 9.36 in many ways represents a more desirable factoring of the matrix **R** than that given by an unweighted principal-components analysis of **R**, for the weighted principal components corresponding to the larger eigenvalues of $\mathbf{E}^{-1}\mathbf{R}\mathbf{E}^{-1}$ contain proportionally less error variance than do the unweighted principal components of **R** corresponding to the larger eigenvalues of **R**. However, if there is any disadvantage to factoring **R** by the method of weighted principal components, it lies in the lack of a convenient point at which to discard the residual components. In unweighted-principal-components analysis with normalized variables we can always rely upon the rule: Save only those components with corresponding eigenvalues of the matrix **R** that are greater than 1. But in weighted principal components saving all components with corresponding eigenvalues of the matrix $\mathbf{E}^{-1}\mathbf{R}\mathbf{E}^{-1}$ greater than 1 will result in saving many more components than the number of unweighted components with eigenvalues of **R** greater than 1. In subsequent factor rotations, retaining too many components will have the effect of spreading the variables too thin across the factors to allow us to overdetermine some of the factors.

However, there is promise of a solution to the problem of how to retain components extracted by the method of weighted-principal-components analysis in the relationships of this method to other methods such as image analysis and canonical-factor analysis. In fact, an examination of these other methods will be valuable not only from the point of view of suggesting a solution to this problem but also from the point of view of suggesting a solution to this problem but also from the point of view of suggesting approximate estimates of the matrix, $\mathbf{E}^2$ which in practice is almost never known exactly for any given set of variables. To these other models we now turn.

9.3 Image Analysis

In the view of some researchers, the primary defect of the model of common-factor analysis is that it fails to provide explicit definitions for the common and unique parts of variables. It is true that common-factor analysis provides postulates of how these parts would be related if one were able to identify them, but the model does not indicate for any given set of data how one could identify these parts, except perhaps by trial-and-error techniques. Moreover, trial-and-error techniques do not guarantee a unique fitting of the data to the model. And at best, in practice, one normally makes only an approximate fit. It would seem for these researchers, therefore, that if the common-factor-analysis model is to be improved, one must provide explicit definitions for the common and unique parts of variables.

Louis Guttman, who was long a critic of the indeterminacy of the common-factor-analysis model, provided a determinate alternative, known as "image analysis," which still retains many of the features of common-factor analysis.

In developing image analysis Guttman relates the method to two infinitely large aggregates: (1) the aggregate of subjects, which is familiar to most statisticians as the population, and (2) the aggregate of tests (thought of as all the tests that logically could be constructed to measure some domain of characteristics of the subjects). Image analysis is concerned with both these aggregates. The population of subjects is used to define ultimate values for the correlation coefficients between tests, which are only estimated when one has a finite sample of subjects. The universe of tests is used to define ultimate values for the common and unique parts of test scores, which are only estimated when one works with a finite collection of tests.

The distinguishing feature of image analysis lies in Guttman's explicit definition of the common part of a test score: "The common part of a variable is that part which is predictable by linear multiple correlation from all the other variables in the universe of variables." This common part of the variable is designated by Guttman as the image of the variable based on all the other variables in the universe of variables. The unique part of a variable is the part remaining which cannot be predicted by all the other variables in the universe of variables by multiple correlation. This unique part is designated as the anti-image of the variable.

Note that the images and anti-images are defined with respect to a universe of variables. Obviously, in practice, we can never deal with the whole universe of variables corresponding to a particular domain of characteristics to measure. Consequently, we can never determine with complete accuracy, by multiple-correlation techniques, the images and anti-images of the variables in a finite set of variables to be analyzed. But the value of the image and anti-image concepts lies in their being ideal constructs, which, as we will see eventually, allow us to formulate a better model of factor analysis as well as to unify concepts of factor analysis with those in the area of measurement in the behavioral sciences.

9.3.1 Partial-Image Analysis

Although in practice we can never determine the total images of a set of variables by multiple correlation, we can nevertheless formulate completely determinate approximations to these images, still using multiple correlation. Let $\mathbf{Y}$ be an $n \times 1$ random vector whose coordinates are n random variables. Now define a $\mathbf{Y}_p$ such that

$$\mathbf{Y}_p = \mathbf{W}'\mathbf{Y} \tag{9.37}$$

where $\mathbf{W}$ is the $n \times n$ multiple-correlation weight matrix for predicting each variable in $\mathbf{Y}$ for a given subject from the $n-1$ other variables in $\mathbf{Y}$. In other words, a coordinate of $\mathbf{Y}_p$ will represent the predicted component of the comparable coordinate variable of $\mathbf{Y}$ based on the $n-1$ other variables of $\mathbf{Y}$ as predictors of it. We will designate the vector $\mathbf{Y}_p$ as the vector of "partial-image variables."

Now let us define an $n \times 1$ random vector $\mathbf{Y}_u$ such that

$$\mathbf{Y}_u = \mathbf{Y} - \mathbf{Y}_p = (\mathbf{Y} - \mathbf{W}'\mathbf{Y}) = (\mathbf{I} - \mathbf{W}')\mathbf{Y} \tag{9.38}$$

$\mathbf{Y}_u$ corresponds to the vector of unpredictable parts of the variables in $\mathbf{Y}$. We will designate the vector $\mathbf{Y}_u$ as the vector of "partial-anti-image variables."

From the above definitions it follows by moving $\mathbf{Y}_p$ to the left-hand side of Equation 9.38 and rearranging that

$$\mathbf{Y} = \mathbf{Y}_p + \mathbf{Y}_u \tag{9.39}$$

The observed variables are the sum of the image and anti-image variables. This equation is the fundamental equation of Guttman's image analysis. When this equation is based upon a finite number of tests it represents a partial-image and anti-image analysis of the scores in $\mathbf{Y}$. In effect, what might be image parts of test scores in the universe of tests will be partly bound up in the partial-anti-image score vector $\mathbf{Y}_u$ when the analysis is carried out with a finite number of tests.

The matrix $\mathbf{W}$ in Equations 9.37 and 9.38 is given as

$$\mathbf{W} = (\mathbf{I} - \mathbf{R}^{-1}\mathbf{S}^2) \tag{9.40}$$

where $\mathbf{R}$ is the $n \times n$ correlation matrix for the n tests in $\mathbf{Y}$ and

$$\mathbf{S}^2 = [\operatorname{diag}\mathbf{R}^{-1}]^{-1} \tag{9.41}$$

($\mathbf{S}^2$ here should not be confused with the sample variance–covariance matrix). For a derivation of the last two equations refer to Section 4.4.

In image analysis we are interested in two matrices: (1) the matrix **G** of covariances among the n different images and (2) the matrix **Q** of covariances among the n different anti-images.

The matrix **G** of image covariances is given as

$$\begin{aligned}\mathbf{G} &= E(\mathbf{Y}_p\mathbf{Y}_p') = E(\mathbf{W}'\mathbf{Y}\mathbf{Y}'\mathbf{W}) = \mathbf{W}'\mathbf{R}\mathbf{W} \\ &= (\mathbf{I}-\mathbf{S}^2\mathbf{R}^{-1})\mathbf{R}(\mathbf{I}-\mathbf{R}^{-1}\mathbf{S}^2) \\ &= \mathbf{R}+\mathbf{S}^2\mathbf{R}^{-1}\mathbf{S}^2-2\mathbf{S}^2\end{aligned} \tag{9.42}$$

The matrix **Q** of anti-image covariances is given as

$$\begin{aligned}\mathbf{Q} &= E(\mathbf{Y}_u\mathbf{Y}_u') = (\mathbf{I}-\mathbf{W}')\mathbf{Y}\mathbf{Y}'(\mathbf{I}-\mathbf{W}) \\ &= (\mathbf{I}-\mathbf{W}')\mathbf{R}(\mathbf{I}-\mathbf{W}) \\ &= [\mathbf{I}-(\mathbf{I}-\mathbf{S}^2\mathbf{R}^{-1})\mathbf{R}(\mathbf{I}-\mathbf{R}^{-1}\mathbf{S}^2)] \\ &= \mathbf{S}^2\mathbf{R}^{-1}\mathbf{R}\mathbf{R}^{-1}\mathbf{S}^2 \\ &= \mathbf{S}^2\mathbf{R}^{-1}\mathbf{S}^2\end{aligned} \tag{9.43}$$

It is interesting to note from Equation 9.43 that the diagonal elements of **Q** are $\mathbf{S}^2[\text{diag }\mathbf{R}^{-1}]\mathbf{S}^2$. But by definition $\mathbf{S}^2 = [\text{diag }\mathbf{R}^{-1}]^{-1}$, which leads to the conclusion that

$$[\text{diag }\mathbf{Q}] = \mathbf{S}^2 \tag{9.44}$$

This gives us an expression for the anti-image variances.

Another interesting observation is that the expression for **Q** also appears in the expression for **G** (Equation 9.42) so that we may write

$$\mathbf{G} = \mathbf{R}+\mathbf{Q}-2\mathbf{S}^2 = \mathbf{R}-\mathbf{S}(-\mathbf{S}\mathbf{R}^{-1}\mathbf{S}+2\mathbf{I})\mathbf{S} \tag{9.45}$$

From this, we can conclude that the matrix **G** approximates a reduced correlation matrix, $(\mathbf{R}-\mathbf{S}^2)$, in common-factor analysis that has the squares of the multiple-correlation coefficients for predicting each variable from the $n-1$ other variables in the matrix inserted in the principal diagonal. But, unlike this reduced correlation matrix, the off-diagonal elements of **R** are readjusted (Kaiser, 1963, p. 161 reports that they are usually slightly reduced) to make the matrix, **G**, Gramian. In a common-factor sense, this would be a model that corrects for doublets.

We in turn can solve Equation 9.44 for **R** to yield

$$\mathbf{R} = \mathbf{G} - \mathbf{Q} + 2\mathbf{S}^2 \tag{9.46}$$

Equation 9.45 is known as the fundamental theorem of image analysis. It expresses the relationship that the correlation matrix of observed test scores is equal to the covariance matrix of image scores minus the covariance matrix of anti-image scores plus two times the diagonal matrix of anti-image variances.

We are now in a position to determine the relationships between total scores, the image scores, and the anti-image scores in the partial-image analysis model.

First, what is the matrix of covariances between image variables and observed variables? That is given by

$$\begin{aligned} E(\mathbf{Y}_p\mathbf{Y}') &= E(\mathbf{W}'\mathbf{Y}\mathbf{Y}') = \mathbf{W}'\mathbf{R} \\ &= (\mathbf{I} - \mathbf{S}^2\mathbf{R}^{-1})\mathbf{R} \\ &= (\mathbf{R} - \mathbf{S}^2) \end{aligned} \tag{9.47}$$

The matrix of covariances between partial-image scores and total scores is a correlation matrix with squares of the multiple-correlation coefficients for predicting each variable from the $n-1$ other variables in the matrix inserted in the principal diagonal. This matrix, however, is not a Gramian matrix, although it is symmetric.

What is the matrix of covariances between anti-image scores and total scores? This is given as

$$\begin{aligned} E(\mathbf{Y}_u\mathbf{Y}') &= E[(\mathbf{I} - \mathbf{W}')\mathbf{Y}\mathbf{Y}' = (\mathbf{I} - \mathbf{W}')\mathbf{R} \\ &= \mathbf{R} - \mathbf{W}'\mathbf{R} = \mathbf{R} - (\mathbf{I} - \mathbf{S}^2\mathbf{R}^{-1})\mathbf{R} \\ &= \mathbf{R} - (\mathbf{R} - \mathbf{S}^2) \\ &= \mathbf{S}^2 \end{aligned} \tag{9.48}$$

The matrix of covariances between partial-anti-image scores and total scores is a diagonal matrix containing the partial-anti-image variances. Moreover, the correlation of the anti-image scores of one variable with the total scores of another is zero.

Finally, what is the matrix of covariances between the partial-anti-image variables and the partial-image variables? This is given as

$$E(\mathbf{Y}_u\mathbf{Y}_p') = E[(\mathbf{I}-\mathbf{W}')\mathbf{Y}\mathbf{Y}'\mathbf{W}] = (\mathbf{I}-\mathbf{W}')\mathbf{R}\mathbf{W}$$
$$= \mathbf{R}\mathbf{W} - \mathbf{W}'\mathbf{R}\mathbf{W} = \mathbf{R}\mathbf{W} - \mathbf{G}$$
$$= \mathbf{R}(\mathbf{I}-\mathbf{R}^{-1}\mathbf{S}^2) - (\mathbf{R}+\mathbf{Q}-2\mathbf{S}^2)$$
$$= -\mathbf{Q}+\mathbf{S}^2 \tag{9.49}$$

We know from Equation 9.44 that [diag $\mathbf{Q}$] = $\mathbf{S}^2$; hence the matrix of covariances of partial-anti-image variables with partial-image variables is the negative of the matrix of partial-anti-image covariances with zeros in its principal diagonal. This immediately suggests that for any given test its partial-anti-image part is uncorrelated with its partial-image part, since the covariance would appear in the principal diagonal of the matrix of covariances between partial-anti-image and partial-image variables.

But of greater interest to us are the off-diagonal elements of the matrix of covariances between partial-anti-image variables and image variables in Equation 9.49. Are these elements zero? The answer to this question depends upon a determination of the off-diagonal elements of the matrix $\mathbf{Q}$, the matrix of covariances among the partial anti-images. If they are not zero, then we have a characteristic of the partial-image-analysis model that differs from the common-factor-analysis model: namely, the so-called unique parts of different variables can be correlated with each other as well as with the common parts of other variables.

Greater insight into the off-diagonal elements of the matrix $\mathbf{Q}$ can be obtained if we concern ourselves with the "correlation matrix" for the partial anti-images. As usual, we can obtain the correlation matrix for a set of variables by premultiplying and postmultiplying the covariance matrix by the inverse of the square root of the diagonal matrix having as diagonal elements the diagonal elements of the covariance matrix. Applied to the matrix $\mathbf{Q}$ this principle yields

$$\mathbf{R}_{Y_uY_u} = [\text{diag}\,\mathbf{Q}]^{-1/2}\mathbf{Q}[\text{diag}\,\mathbf{Q}]^{-1/2}$$
$$= [\mathbf{S}^2]^{-1/2}\mathbf{Q}[\mathbf{S}^2]^{-1/2} = \mathbf{S}^{-1}\mathbf{S}^2\mathbf{R}^{-1}\mathbf{S}^2\mathbf{S}^{-1}$$
$$= \mathbf{S}\mathbf{R}^{-1}\mathbf{S} \tag{9.50}$$

We can see that the only reasonable way in which the off-diagonal elements of either $\mathbf{Q}$ or $\mathbf{R}_{Y_uY_u}$ can be zero is if $\mathbf{R}^{-1}$ is diagonal, and this would be possible only if the matrix $\mathbf{R}$ were diagonal, which usually it is not. But further insights await us if we examine a typical off-diagonal element of $\mathbf{R}_{Y_uY_u}$ in terms of the determinantal formulas for multiple and partial correlation. We can represent these in determinantal form of the matrices $\mathbf{R}^{-1}$ and $\mathbf{S}^2$ using Equations 4.53 and 4.54 and the relationship of the inverse of a symmetric matrix to the determinant of the matrix and its cofactors via the adjoint matrix (see Chapter 2).

An off-diagonal element $r_{jk\mathrm{u}}$ of the matrix $\mathbf{R}_{Y_\mathrm{u}Y_\mathrm{u}}$ is given from Equation 9.50 as

$$r_{jk\mathrm{u}} = s_j r^{jk} s_k \tag{9.51}$$

where

s_j and s_k are the standard deviations of the jth and kth anti-image

r^{jk} is the j,kth element of the matrix $\mathbf{R}^{-1}$

Using Equation 4.54 as a guide, we can represent these elements in terms of determinantal expressions, that is,

$$r^{jk} = \frac{R_{jk}}{|\mathbf{R}|}$$

$$s_j = \frac{|\mathbf{R}|^{1/2}}{R_{jj}}$$

and

$$s_k = \frac{|\mathbf{R}|^{1/2}}{R_{kk}}$$

where

$|\mathbf{R}|$ is the determinant of the correlation matrix $\mathbf{R}$

R_{jk}, R_{jj}, and R_{kk} are the cofactors of the matrix $\mathbf{R}$

Substituting these expressions into Equation 9.51, we obtain

$$r_{jk\mathrm{u}} = \frac{R_{jk}}{\sqrt{R_{jj}}\sqrt{R_{kk}}} \tag{9.52}$$

But this is the "negative" value of the partial correlation between tests j and k with the $n-2$ other variables held constant, given in determinantal form as in Equation 4.46.

The fact that the correlations between the partial anti-images are negative values of partial correlations between the original tests leads us to conclude that the matrix $\mathbf{Q}$ of covariances between the partial anti-images is the negative of the matrix of partial "covariances" between the original variables with the $n-2$ other variables held constant.

It further follows that the covariance between the partial-image part of one variable and the partial-anti-image part of another variables must represent the partial covariance between these two variables with the $n-2$ other variables held constant. This is an immediate consequence of applying the previous conclusions to Equation 9.51 which defines the covariance matrix of the partial-image scores with the partial-anti-image scores. This final conclusion represents one of the lesser known paradoxes of multiple correlation: After we have partialled out the "common part"—the partial image—from a first variable, the remaining unique part—the partial anti-image—in that variable acts with respect to a second variable's common part as if the second variable were not in the prediction equation for predicting the common part of the first variable. Partialling the common parts from n variables still makes possible pairwise relationships to remain between the variables.

We now turn to the question of using partial-image analysis as a basis for a factor analysis. In practice, the matrix that we want to analyze into factors is the image covariance matrix **G**. This may be done, preferably, with the method of principal axes. However, Kaiser (1963, p. 161) pointed out that invariably the matrix **G** is positive definite, having only positive, nonzero eigenvalues, many of which are very small. This creates a problem for objectively determining the number of image factors to retain and rotate. This problem remained such a formidable one that Kaiser claims he tentatively abandoned image analysis until a paper by Harris (1962) appeared in *Psychometrika* and provided a solution. We will take up this solution later in this chapter when we deal with canonical-factor analysis and the problem of doublets.

9.3.2 Image Analysis and Common-Factor Analysis

Guttman's model of image analysis might have remained a mathematical curiosity, considering the popularity of the common-factor-analysis model, had Guttman not demonstrated profound relationships between total-image analysis and common-factor analysis. Fundamental to his demonstrating these relationships is his insistence that the common-factor-analysis model, to be ultimately useful, must be made determinate. If factor analysts are prone to overlook the seriousness of this point, Guttman takes great pains to make awareness of indeterminacy unavoidable. In one of his classic papers (Guttman, 1955) he shows how it is not sufficient to find just a matrix of factor loadings from a correlation matrix. One has to be able to link up the original variables with the factors he believes he has discovered. This involves being able to pin down factor scores for the subjects with considerable accuracy.

Whether Guttman's insistence on avoiding indeterminacy is reasonable is a different question. My view is that Guttman simply reflected the Zeitgeist of his time when logical positivism and empiricism dominated American and British philosophy of science and psychology as well. This made the notion of unobservables and latent variables suspicious. If a concept could not be completely determined by experience, then its epistemological status was dubious. But this came to be regarded as violating common sense. Our concepts of the world are not totally devoid of grounding in experience while being generally underdetermined from experience (Garrison, 1986). Frequently our cognitive apparatus provides some of the structure in our concepts (Lakoff and Johnson, 1999). More than one concept may be drawn from a given pattern in experience. Philosophers like Ludwig Wittgenstein (1953) liked to demonstrate this with an ambiguous line figure derived from Jastrow (1900) that could be interpreted either as a rabbit or a duck (cf. Mulaik, 1993). Furthermore, more than one curve may be drawn through a given set of observed points, anyone of these curves being an interpretation of how to generalize to future points supposedly being generated by the same process (Mulaik, 1993). Moreover, objects are more than what we know about them from our experiences of them up to any given point in time. New experience may change our concepts of them. Thus common factors as latent and not directly observed variables may be tied to given indicator variables of them, while at the same time being not completely determined by these indicator variables. We can conceptualize about the world in terms of unobservables, such as a planet we cannot see, but which we infer from the influence of its gravity on the orbit of another planet as long as there are further observable consequences of these concepts that can be tested at some point.

The basic procedure for linking variables to factor scores in common-factor analysis is to estimate the factor scores for each of the factors by multiple correlation, using the factor-structure matrix (containing the correlations between variables and factors) and the original correlation matrix to find the regression weights required. The success with which one is able to estimate these is gauged by the multiple-correlation coefficients between the n variables and these factors. For example, Thomson (1956, p. 335) cites the squares of the multiple-correlation coefficients for estimating nine factors of Thurstone's "primary mental abilities." The average value for these coefficients is .595, representing multiple R's in the high .70s. Guttman (1955) points out, however, that as the multiple R for predicting a factor goes down, the more one is able to find two sets of factor scores for the same variable, both of which reproduce the factor loadings nicely if correlated with the original variables, but which are very little, if not even negatively, correlated with one another. For example, Guttman shows that if the multiple R for predicting a factor from the original variables is .707 then two sets of factor scores could be derived for that factor that would be correlated zero with each other. Heerman (1963) makes this point clearer by geometrical proofs. The upshot of this demonstration is that it would frequently be possible, when interpreting the factors, for one to hypothesize two different processes which would be uncorrelated with each other (if not negatively correlated) to explain the same factor loadings for a factor. This is just a special case of the fitting of more than one curve through a given set of points.

But for our purposes here we are interested in Guttman's formula (Guttman, 1956, p. 276) for the square of the multiple-correlation coefficient for predicting the unique factors from the n variables in the correlation matrix. Let ψ_j^2 be the unique variance for the jth variable. Let s_j^2 be the error of estimate for predicting the jth variable from the $n-1$ other variables by multiple correlation. In other words, s_j^2 is the jth diagonal element of the matrix $\mathbf{S}^2$, where $\mathbf{S}^2$ is defined in Equation 9.40. Then the square of the multiple-correlation coefficient $R_{j(\psi)}^2$ for predicting the jth unique factor is

$$R_{j(\psi)}^2 = \frac{\psi_j^2}{s_j^2} \tag{9.53}$$

Normally, the unique variance ψ_j^2 is less than the error of estimate s_j^2. But by this equation Guttman (1956, p. 276) then shows that, if the common-factor-analysis model is to be determinate for a particular set of data, $R_{j(\psi)}^2$ must equal 1 for every j, thus providing perfect prediction of factors. This will be possible only if

$$\psi_j^2 = s_j^2 \tag{9.54}$$

In other words, for the common-factor-analysis model to be determinate, the unique variance of each variable must equal the error of estimate in predicting the variable from the $n-1$ other variables.

Guttman then considers the application of the common-factor-analysis model to a universe of tests, to make inferences about which he suggests factor-analytic studies are performed. He shows that if the number of common factors in a universe of tests is finite then the common-factor-analysis model would be determinate in that universe, and the unique variances of all except perhaps a zero proportion of the tests in the universe of tests would equal the errors of estimate for predicting each test from all the other tests.

Several powerful consequences result from being able to assume that in the universe of tests the model of common-factor analysis is determinate. First, one can accept the common-factor model as uniquely interpretable in that universe. In other words, one would not have to deal with the possibility that rival interpretations would equally well account for the factor-pattern loadings and correlations among factors. Second, one can derive several powerful theorems that link not only image analysis but common-factor analysis and other factor-analytic models as well to the classic theory of reliability. We shall, for the present, consider first the impact of this assumption on image analysis and common-factor analysis.

If the model of common-factor analysis is determinate in a universe of tests, we can assume that

$$\lim_{n\to\infty} s_j^2 = \psi_j^2 \tag{9.55}$$

In a determinate universe of tests, as the number of tests increases without bound, the error of estimate for predicting a test from the $n-1$ other tests approaches as a limit the unique variance of the test.

If the limit in Equation 9.53 holds, then the following limit also holds

$$\lim_{n\to\infty} R_j^2 = h_j^2 \tag{9.56}$$

In a determinate universe of tests, as the number of tests increases without bound, the square of the multiple-correlation coefficient for predicting a test from the $n-1$ other tests approaches as a limit the communality of the test.

The immediate consequence of these two theorems is that in the universe of tests image analysis and common-factor analysis are the same. The basis for this conclusion is that if the square of the multiple-correlation coefficient and the corresponding error of estimate equal the communality and uniqueness, respectively, then the images and anti-images must behave as the common parts and unique parts in common-factor analysis.

More specifically, if in a universe of tests the model of common-factor analysis is determinate, then

$$\lim_{n\to\infty} \mathbf{S}^2 = \mathbf{\Psi}^2 \tag{9.57}$$

That is, as the number of tests increases without bound, the diagonal matrix of errors of estimate approaches as a limit the diagonal matrix of unique variances. This is a matrix version of Equation 9.53. But we can also conclude that

$$\lim_{n\to\infty} \mathbf{G} = \mathbf{R} - \mathbf{\Psi}^2 \tag{9.58}$$

In other words, as the number of tests increases without bound, the image covariance matrix approaches as a limit the reduced correlation matrix with communality coefficients in the principal diagonal. And finally

$$\lim_{n\to\infty} \mathbf{Q} = \mathbf{S}^2 = \mathbf{\Psi}^2 \tag{9.59}$$

Or, as the number of tests increases without bound, the anti-image covariance matrix approaches as a limit the diagonal matrix of errors of estimate and in turn the diagonal matrix of unique variances.

The fact that in the universe of tests the anti-image covariance matrix approaches as a limit the diagonal matrix of unique variances led Guttman (1953, p. 295) to suggest a test, based upon partial-image analysis, of whether for a finite sample of tests from the universe of tests the model of common-factor analysis can be fitted to the tests with a high degree of determinateness. This test involves examining the partial-anti-image covariance matrix $\mathbf{Q}$ to see whether the off-diagonal elements are close to zero. If they are, then $\mathbf{Q}$

approximates a diagonal matrix of unique variances, and one can then assume that the model of common-factor analysis would be appropriate for the sample of tests. If **Q** is not approximately diagonal, then the sample of tests is not a good sample to which to try to apply the model of common-factor analysis.

In lieu of the matrix **Q**, one can examine the partial correlation matrix $\mathbf{SR}^{-1}\mathbf{S}$ of partial correlations between pairs of variables with the other $n-2$ variables partialled out to see if there are doublets between the variables. Rarely is this done. But one might be led to discover the sources of these doublets in the items. Their presence in large magnitudes can undermine the common-factor model. In this regard, Kaiser (1974), developed a measure of whether a sample of variables from the universe of variables was appropriate for common-factor analysis. Basically, the idea is to compare the size of the original correlations in **R** to the size of the partial correlations between pairs of variables with the $n-2$ other variables partialled from them in $\mathbf{SR}^{-1}\mathbf{S}$ (see Equation 9.50). He provide two indices. One, KMO, is an overall measure of "sampling adequacy" for the complete set of variables in the analysis. The other provides an index for each variable MSA. These indices are as follows:

$$\mathrm{KMO} = \frac{\sum_{i \neq j}^{n}\sum^{n} r_{ij}^2}{\sum_{i \neq j}^{n}\sum^{n} r_{ij}^2 + \sum_{i \neq j}^{n}\sum^{n} a_{ij}^2}$$

where r_{ij}^2 is the square of the i,jth observed correlation between variables i and j, while a_{ij}^2 is the anti-image correlation coefficient or partial correlation coefficient between variables i and j with the $n-2$ other variables held constant. One can see that if the original correlations are large and their corresponding partial correlations are near zero, that the KMO index would approach 1.00. Kaiser regarded KMO measures in the .90s as "marvelous," in the .80s as "meritorious," in the .70s as "middling," in the .60s as "mediocre" and in the .50s, or below as "miserable." Obviously he recognized these categories were arbitrary, but the idea was that the closer to 1.00, the better the set of variables was for factor analysis.

$$\mathrm{MSA}(i) = \frac{\sum_{j \neq i}^{n} r_{ij}^2}{\sum_{j \neq i}^{n} r_{ij}^2 + \sum_{j \neq i}^{n} a_{ij}^2}$$

is the "measure of sampling adequacy" for variable i. The coefficients are defined the same as for KMO. Again Kaiser's categories for the magnitude of MSA(i) apply. Variables that have low MSA(i)'s should be dropped from the analysis.

In some respects the KMO and MSA index are rather lenient because they are comparing squared correlations rather than the absolute magnitudes of them. Squaring shrinks the correlations toward smaller values and those correlations that are low (but not zero) approach zero much more rapidly than larger correlations. So, this could have the effect of more rapidly diminishing the partial correlations between pairs of variables with $n-2$ other variables partialled out than the larger correlations between observed variables. Hence, the second term involving the sum of the squared a's will tend to be smaller relative to the sum of the squared r's. This means these indices will fall away from 1.00 slowly. A possible alternative is to obtain the square roots of the squared correlations, which effectively compares them in terms of their absolute magnitudes. This is shown in the following equations:

$$\text{KMO2} = \frac{\sum_{i \neq j}^{n}\sum^{n} \sqrt{r_{ij}^2}}{\sum_{i \neq j}^{n}\sum^{n} \sqrt{r_{ij}^2} + \sum_{i \neq j}^{n}\sum^{n} \sqrt{a_{ij}^2}} \qquad \text{MSA2}(i) = \frac{\sum_{j \neq i}^{n} \sqrt{r_{ij}^2}}{\sum_{j \neq i}^{n} \sqrt{r_{ij}^2} + \sum_{j \neq i}^{n} \sqrt{a_{ij}^2}}$$

By the Kaiser, Meyer, and Olkin criteria (Kaiser, 1970), the correlations in Table 9.1 are adequate for common-factor analysis. The more stringent criteria of KMO2 and MSA2 suggest there may be problems in applying the common-factor model to the correlations, especially if we take the full number of common factors for rotation.

An approximate standard error for a partial correlation coefficient is given by the standard error of a Fisher's Z-transformation of the partial correlation coefficient, given in this case by $\sigma_Z = 1/\sqrt{N-3-(n-2)}$ (Hayes, 1988, p. 619). For the present problem $N = 280$ and $n = 15$. Hence σ_Z for the partial correlations in Table 9.2 is .0615, and two standard errors is then .1231. The Fisher Z-transformation of correlations less than .40 is negligible, and so it is sufficient to know that a partial correlation of .123 or larger would be significant at the .046 level for a two-tail test of the hypothesis that the population partial correlation equals zero. We have highlighted the significant correlations by showing them in boldface. Of course, the partial correlations in the table are not independent of one another. It is interesting that the significant partial correlations are among clusters of variables that otherwise would be regarded as having a factor in common within each cluster. For example, Friendly, Sympathetic, Kind, and Affectionate seem to have overlaps in meaning. And their original correlations were all in the high .70s and .80s. And pairs of these variables like Friendly and Kind or Sympathetic and Kind have significant partial correlations between them. I know of nothing now in the literature that explains this, but it may be that it reflects the fact that in this case only four variables are indicators of a common factor, and so the squared multiple correlations may be considerably less than the corresponding communalities. meaning the common-factor model is somewhat underdetermined in this case.

TABLE 9.2

Negative Partial Correlations between Variables of Table 8.2 with $n-2$ Other Variables Partialled Out

.937—Friendly														
−.188	.906—Sympathetic													
−.370	**−.403**	.891—Kind												
−.023	**−.318**	**−.316**	.940—Affectionate											
−.045	−.102	.118	.077	.877—Intelligent										
−.035	.074	−.069	.025	−.103	.862—Capable									
−.079	.072	.085	−.033	**−.318**	**−.492**	.873—Competent	KMO = .920		KMO2 = .809					
.131	.013	**−.132**	−.026	**−.409**	**−.385**	**−.139**	.873—Smart							
−.192	**.176**	.015	−.024	.078	**.134**	−.047	**−.173**	.935—Talkative						
.041	**−.210**	.096	**.141**	−.019	−.090	.037	.048	**−.127**	.902—Outgoing					
.116	.046	−.088	−.073	−.029	.043	−.053	.043	−.108	**−.368**	.946—Gregarious				
−.090	.055	.105	−.111	−.077	.009	.045	.026	**−.239**	**−.267**	**−.183**	.902—Extrovert			
.001	−.092	**−.133**	−.062	.031	.025	−.107	−.047	−.024	−.027	−.003	.067	.946—Helpful		
−.086	**−.126**	−.059	−.037	−.079	.021	−.096	.041	.035	−.027	.003	−.019	**−.304**	.940—Cooperative	
−.034	.048	.005	**−.161**	.079	−.061	.010	−.024	**−.142**	**−.426**	**−.129**	**−.217**	**−.132**	−.097	.964—Sociable
							MSA2							
.827	.776	.769	.832	.744	.768	.766	.765	.804	.792	.843	.817	.876	.878	.834

Note: MSA values of individual variables are shown on the principal diagonal. MSA2 values are shown below the partial correlations.

9.3.3 Partial-Image Analysis as Approximation of Common-Factor Analysis

Guttman (1953, p. 295) also recommended partial-image analysis as an approximation of common-factor analysis. This approximation he suggested could be quite good because multiple correlations tend to approach an asymptotic value quite rapidly as the number of predictors increases. In his opinion, with the number of variables greater than 10 or 15, partial images and anti-images will be so close to their corresponding total images and anti-images that the differences will be negligible. In effect, then, the number of variables can be regarded as virtually "infinite." However, he was quick to point out that, as the number of variables increases, sampling errors will become very large, and so he tempers these observations with the caution that they apply to sample data only if the number of observations is large. The major advantage of using partial-image analysis is in the complete determinateness of the model, permitting one to compute directly factor scores which are uncorrelated with each other when based on orthogonal-factor solutions. We will discuss this problem more in a later chapter concerning the computation of factor scores.

On the other hand, when science hypothesizes latent constructs that are not fully observed but indicated indirectly by finite numbers of observed indicator variables, such constructs will necessarily be indeterminate. Philosophers of science have come to accommodate themselves to empirical underdetermination for their theories from data (Garrison, 1986). Objectivity, furthermore, presumes that a scientific construct will be invariant across not just those observations used in formulating the construct, but also across future observations pertinent to the construct (Mulaik, 2004). The possibility of rejecting an "invariant" construct in future observations is the basis for hypothesis testing (Mulaik, 2004). In factor analysis this will mean that factor-pattern loadings found for a given indicator variable with respect to a given common factor will be invariant as the indicator is embedded with other variables supposedly measuring the same factor (Thurstone, 1937).

So, the concept of a common-factor model being determinate in a universe of tests is a regulative ideal for thought. This ideal forms the basis for the assumption that underlies hypothesis testing, of the invariance of a construct in its relationships to observations. But it is not ultimately necessary for science and objectivity that local indeterminacy for common factors relative to a current set of variables be regarded as a bad thing. After all, exploratory factor analysis is principally a hypothesis-generating method, and objectivity and invariance is established later on by confirmatory factor analyses. Objectivity as a demonstration of an invariant across data used in formulating a hypothesis generalized to additional data not used in that formulation, is a provisional determination (Mulaik, 2004). Science always leaves open the possibility that future conceptualization and experience will overturn a current "objective" concept in favor of a new "objective" synthesis of not only previous experiences but the new ones as well. That is how science progresses to produce "objective" knowledge that encompasses and spans a greater variety of experience than previously. So, in this context, factor indeterminacy is an essential property of human thought relative to experience, which allows concepts to be generalized beyond a given set of observations and observers.

9.4 Canonical-Factor Analysis

The development of the theory of canonical correlation (Hotelling, 1936) has given rise to an interesting application of it to the theory of common-factor analysis. Canonical correlation is a kind of simultaneous component analysis of two sets of variables, having as its aim the discovery of orthogonal components in each set which are so oriented as to be respectively maximally correlated between the sets. In other words, canonical correlation seeks to find those linear combinations of variables in each set having the greatest possible correspondence between the two sets of variables. In the present case, canonical correlation will be applied to the problem of finding a set of linear components of the observed variables that are maximally correlated with a corresponding set of linear components of the common-factor variables.

Let $\mathbf{Y}$ be an $n \times 1$ random vector with coordinates the n observed variables. Assume that $E(\mathbf{Y})=\mathbf{0}$ and $E(\mathbf{YY}')=\mathbf{R}$, a correlation matrix. Let $\mathbf{X}$ be an $r \times 1$ random vector with coordinates the r common-factor random variables. Assume $E(\mathbf{X})=\mathbf{0}$ and $E(\mathbf{XX}')=\mathbf{I}$. Arrange the vectors $\mathbf{Y}$ and $\mathbf{X}$ as the partitions of an $(n+r) \times 1$ partitioned vector

$$\begin{bmatrix} \mathbf{Y} \\ \mathbf{X} \end{bmatrix}$$

A correlation matrix may be obtained from this partitioned vector by finding

$$E\left\{\begin{bmatrix} \mathbf{Y} \\ \mathbf{X} \end{bmatrix}\begin{bmatrix} \mathbf{Y} & \mathbf{X} \end{bmatrix}\right\} = \begin{bmatrix} \mathbf{R} & \boldsymbol{\Lambda} \\ \boldsymbol{\Lambda}' & \mathbf{I} \end{bmatrix}$$

where $\boldsymbol{\Lambda}$ is an $n \times r$ common-factor-structure matrix of correlations between the observed variables and the orthogonal common factors.

Now, let us consider the problem of finding a linear combination X^* of the variables in **Y** which will be maximally correlated with one linear composite of the common-factor variables. Define

$$X^* = \mathbf{b}'\mathbf{Y} \tag{9.60}$$

where **b** is an $n \times 1$ vector of weights for deriving X^* from **Y**.

According to the basic formulas of canonical correlation, the weight vector **b** may be found by solving the following equation dealing with a partitioned correlation matrix showing the correlations between two sets of variables:

$$(\mathbf{R}_{pq}\mathbf{R}_{qq}^{-1}\mathbf{R}_{qp} - \nu\mathbf{R}_{pp})\mathbf{b} = \mathbf{0} \tag{9.61}$$

In our problem, $\mathbf{R}_{pq} = \mathbf{\Lambda}$, $\mathbf{R}_{qq} = \mathbf{I}$, $\mathbf{R}_{pp} = \mathbf{R}$. The Greek letter ν denotes a Lagrangian multiplier that is also equal to the square of the canonical correlation corresponding to the weight vector **b**. Substituting **Λ**, **I**, and **R** into Equation 9.61, we have

$$(\mathbf{\Lambda\Lambda}' - \nu\mathbf{R})\mathbf{b} = \mathbf{0} \tag{9.62}$$

Since we are concerned with the model of common-factor analysis, the expression $\mathbf{\Lambda\Lambda}'$ represents the reproduced correlation matrix with communalities in the diagonal. That is,

$$\mathbf{\Lambda\Lambda}' = (\mathbf{R} - \mathbf{\Psi}^2) \tag{9.63}$$

where $\mathbf{\Psi}^2$ is the diagonal matrix of unique variances. Substituting Equation 9.63 into Equation 9.62, we have

$$(\mathbf{R} - \mathbf{\Psi}^2 - \nu\mathbf{R})\mathbf{b} = \mathbf{0} \tag{9.64}$$

But the solution to this equation will be unchanged if we premultiply it by $\mathbf{R}^{-1}$ to obtain

$$(\mathbf{I} - \mathbf{R}^{-1}\mathbf{\Psi}^2 - \nu\mathbf{I})\mathbf{b} = \mathbf{0} \tag{9.65}$$

We can rearrange Equation 9.65 to become

$$[(\mathbf{I} - \nu\mathbf{I}) - \mathbf{R}^{-1}\mathbf{\Psi}^2]\mathbf{b} = \mathbf{0} \tag{9.66}$$

Now, let us use a trick to simplify Equation 9.66. Define

$$\gamma\mathbf{I} = (\mathbf{I} - \nu\mathbf{I})^{-1} \tag{9.67}$$

If we premultiply Equation 9.66 by $\gamma\mathbf{I}$, we obtain

$$(\mathbf{I} - \gamma\mathbf{R}^{-1}\mathbf{\Psi}^2)\mathbf{b} = \mathbf{0} \tag{9.68}$$

Finally, we can premultiply Equation 9.68 by **R** to obtain

$$(\mathbf{R} - \gamma\mathbf{\Psi}^2)\mathbf{b} = \mathbf{0} \tag{9.69}$$

You may notice that Equation 9.69 is practically identical to Equation 9.24 in the problem of weighted principal components. Actually the present canonical-factor-analysis problem and the weighted-principal-components problem can be seen as essentially the same, since the unique factors of the variables in the common-factor-analysis model behave with respect to the common factors and observed variable in the same manner as do the error parts of the observed variables with respect to the true parts and observed variables in the error and true-score model. The only difference between the common-factor-analysis model and the error-score and true-score models is that these models deal with slightly different but overlapping portions of the variables. In any case we may solve Equation 9.69 by substituting $\mathbf{\Psi}^2$ for $\mathbf{E}^2$ in Equation 9.24 and subsequent equations to arrive at the following solution for the weight vector **b**:

$$\mathbf{b} = \mathbf{\Psi}^{-1}\mathbf{a} \quad \text{or} \quad \mathbf{a} = \mathbf{\Psi}\mathbf{b} \tag{9.70}$$

where **a** is the eigenvector corresponding to the largest eigenvalue of the matrix $\mathbf{\Psi}^{-1}\mathbf{R}\mathbf{\Psi}^{-1}$.

Moreover, drawing upon developments in connection with weighted principal components, we can formulate a complete set of normalized "canonical components" which would have maximal correlations with their respective counterparts in a corresponding set of canonical common factors. These canonical components are given by the equation:

$$\tilde{\mathbf{X}} = [\gamma]^{-1/2}\mathbf{A}'\mathbf{\Psi}^{-1}\mathbf{Y} \tag{9.71}$$

which corresponds to Equation 9.31. Furthermore, we can find the factor-structure matrix $\mathbf{\Lambda}$ of correlations between the observed variables in **Y** and the normalized canonical components in **Y** by the equation:

$$\mathbf{\Lambda} = \mathbf{\Psi}\mathbf{A}[\gamma]^{1/2} \tag{9.72}$$

which corresponds to Equation 9.35.

But not all canonical components in Equation 9.72 correspond to canonical common factors. If there are only $r < n$ common factors, there can be only r canonical common factors, and consequently only r canonical components. How do we determine r? Note that, according to Equation 9.67 an eigenvalue γ_j of the matrix $\mathbf{\Psi}^{-1}\mathbf{R}\mathbf{\Psi}^{-1}$ is related to the Lagrangian multiplier and canonical correlation coefficient ν_j by

$$\gamma_j = (1 - \nu_j)^{-1} \tag{9.73}$$

Solving for v_j, we obtain

$$v_j = \frac{(\gamma_j - 1)}{\gamma_j} \tag{9.74}$$

Evidently, if γ_j is less than 1, then v_j is negative. But v_j corresponds to the square of a correlation coefficient. Hence we should discard canonical correlations with corresponding eigenvalues of the matrix $\mathbf{\Psi}^{-1}\mathbf{R}\mathbf{\Psi}^{-1}$ that are less than 1 because they lead to negative squared correlations with the common factors. These would be imaginary numbers for which we have no empirical interpretation. Therefore let $\mathbf{\Lambda}_r$ be the factor-structure matrix of correlations of the n variables with the first r canonical components corresponding to eigenvalues of the matrix $\mathbf{\Psi}^{-1}\mathbf{R}\mathbf{\Psi}^{-1}$ greater than 1. Then,

$$\Lambda_r = \mathbf{\Psi}\mathbf{A}_r[\gamma]_r^{1/2} \tag{9.75}$$

with

$\mathbf{A}_r$ an $n \times r$ matrix consisting of the first r columns of $\mathbf{A}$
$[\gamma]_r$ an $r \times r$ diagonal matrix of the r largest eigenvalues of the matrix $\mathbf{\Psi}^{-1}\mathbf{R}\mathbf{\Psi}^{-1}$

But we are also interested in finding the factor-structure matrix $\mathbf{\Lambda}$ that represents the correlations of the variables in $\mathbf{Y}$ with the r canonical common factors most related to their counterparts among the first r canonical components in Equation 9.71. Consider that $\mathbf{Y}_c = \mathbf{Y} - \mathbf{\Psi}\mathbf{E}$ represents the common parts of the variables in $\mathbf{Y}$ after their unique-factor parts have been extracted from them. Let $\mathbf{B} = \mathbf{\Psi}^{-1}\mathbf{A}$ be the matrix whose column vectors correspond to weight vectors in $\mathbf{b}$ satisfying Equation 9.69. Then define

$$\mathbf{X}_c^* = \mathbf{B}'\mathbf{Y} = \mathbf{A}'\mathbf{\Psi}\mathbf{Y}_c$$

The coordinates of the random vector $\mathbf{X}^*$ represent the unnormalized common parts of the canonical components.

The matrix of covariances between variables in $\mathbf{Y}$ and the common components in $\mathbf{X}_c^*$ is given by

$$E(\mathbf{Z}\mathbf{X}_c^{*\prime}) = E(\mathbf{Y}\mathbf{Y}_c'\mathbf{\Psi}^{-1}\mathbf{A}) = (\mathbf{R} - \mathbf{\Psi}^2)\mathbf{\Psi}^{-1}\mathbf{A} \tag{9.76}$$

The value of this expression is even less apparent than the one leading to Equation 9.72, yet it has a similar solution. Its key is based on a paper by Harris (1962).

Harris (1962, p. 254) cites the following relationships between the eigenvectors and eigenvalues of $\mathbf{\Psi}^{-1}\mathbf{R}\mathbf{\Psi}^{-1}$ and those of related matrices, where

$\mathbf{A}$ is the $n \times n$ eigenvector matrix and $[\gamma_i]$ the diagonal matrix of eigenvalues of $\mathbf{\Psi}^{-1}\mathbf{R}\mathbf{\Psi}^{-1}$:

$$\mathbf{\Psi}^{-1}\mathbf{R}\mathbf{\Psi}^{-1} = \mathbf{A}[\gamma_i]\mathbf{A}' \tag{9.77}$$

$$\mathbf{R} = \mathbf{\Psi}\mathbf{A}[\gamma_i]\mathbf{A}'\mathbf{\Psi} \tag{9.78}$$

$$(\mathbf{\Psi}^{-1}\mathbf{R}\mathbf{\Psi}^{-1})^{-1} = \mathbf{A}\left[\frac{1}{\gamma_i}\right]\mathbf{A}' \tag{9.79}$$

$$\mathbf{R}^{-1} = \mathbf{\Psi}^{-1}\mathbf{A}\left[\frac{1}{\gamma_i}\right]\mathbf{A}'\mathbf{\Psi}^{-1} \tag{9.80}$$

$$\mathbf{\Psi}^{2} = \mathbf{\Psi}^{-1}\mathbf{A}[1]\mathbf{A}'\mathbf{\Psi}^{-1} \tag{9.81}$$

$$\mathbf{I} = \mathbf{A}[1]\mathbf{A}' \tag{9.82}$$

With respect to Equation 9.76 we can show that

$$\begin{aligned}\mathbf{\Psi}^{-1}(\mathbf{R} - \mathbf{\Psi}^{2})\mathbf{\Psi}^{-1} &= \mathbf{\Psi}^{-1}\mathbf{R}\mathbf{\Psi}^{-1} - \mathbf{I} = \mathbf{A}[\gamma_i]\mathbf{A}' - \mathbf{A}[1]\mathbf{A}' \\ &= \mathbf{A}[\gamma_i - 1]\mathbf{A}'\end{aligned} \tag{9.83}$$

Hence with $\mathbf{a}$ as a column of $\mathbf{A}$, we can find the solution for an $n \times r$ common-factor-structure matrix to be

$$\mathbf{\Lambda}_{\mathrm{c}} = \mathbf{\Psi}\mathbf{A}_r[\gamma_i - 1]_r^{1/2} \tag{9.84}$$

But this is the same solution as in maximum-likelihood exploratory factor analysis by Jöreskog (1967). Hence one can use the maximum-likelihood algorithm to find the solution.

9.4.1 Relation to Image Analysis

Harris (1962) also demonstrates a close relationship between canonical-factor analysis and Guttman's image analysis. If we substitute for $\mathbf{\Psi}^2$ the matrix $\mathbf{S}^2 = [\text{diag } \mathbf{R}^{-1}]^{-1}$ in Equations 9.77 through 9.82, then the covariance matrix of the images of Guttman's analysis is $\mathbf{G} = \mathbf{R} + \mathbf{S}^2\mathbf{R}^{-1}\mathbf{S}^2 - 2\mathbf{S}^2$ by definition. Now consider a matrix $\mathbf{G}^*$

$$\mathbf{G}^* = \mathbf{S}^{-1}\mathbf{G}\mathbf{S}^{-1}$$

The premultiplication and postmultiplication of the complete expression for $\mathbf{G}$ and $\mathbf{S}^{-1}$ yields

$$\mathbf{G}^* = \mathbf{S}^{-1}\mathbf{R}\mathbf{S}^{-1} - \mathbf{S}\mathbf{R}^{-1}\mathbf{S} - 2\mathbf{I} \tag{9.85}$$

If we look closely at the terms in Equation 9.85, we will notice that $\mathbf{SR}^{-1}\mathbf{S}$ is equivalent to $[\mathbf{S}^{-1}\mathbf{RS}^{-1}]^{-1}$. Thus we can substitute Equations 9.77, 9.79, and 9.82 into Equation 9.85 to obtain

$$\mathbf{G}^* = \mathbf{A}[\gamma_i]\mathbf{A}' + \mathbf{A}\left[\frac{1}{\gamma_i}\right]\mathbf{A}' - 2\mathbf{A}[1]\mathbf{A}' \tag{9.86}$$

Equation 9.86 can be rewritten as

$$\begin{aligned}\mathbf{G}^* &= \mathbf{A}\left[\gamma_i + \frac{1}{\gamma_i} - 2\right]\mathbf{A}' \\ &= \mathbf{A}\left[\frac{\gamma_i^2 - 2\gamma_i + 1}{\gamma_i}\right]\mathbf{A}' = \mathbf{A}\left[\frac{(\gamma_i - 1)^2}{\gamma_i}\right]\mathbf{A}'\end{aligned} \tag{9.87}$$

Now we can readily see that the eigenvalues of $\mathbf{G}^*$ are related to the eigenvalues of $\mathbf{S}^{-1}\mathbf{RS}^{-1}$ and $\mathbf{S}^{-1}(\mathbf{R}-\mathbf{S}^2)\mathbf{S}^{-1}$. We know that in canonical-factor analysis, the number of factors would be determined by the number of eigenvalues of $\mathbf{S}^{-1}\mathbf{RS}^{-1}$ greater than 1. This corresponds to the number of positive eigenvalues of $\mathbf{S}^{-1}(\mathbf{R}-\mathbf{S}^2)\mathbf{S}^{-1}$. We will similarly find that this corresponds to the number of meaningful factors of $\mathbf{G}^*$ that we might retain. Notice that if the roots of $\mathbf{S}^{-1}\mathbf{RS}^{-1}$ are less than 1 and approach zero in size, the roots of $\mathbf{G}^*$ will be positive but increase in size without bound, that is, approach infinitely large eigenvalues in size. This seems to correspond to a spurious situation that we ought to avoid. Thus again we are led to accepting eigenvalues of $\mathbf{S}^{-1}\mathbf{RS}^{-1}$ greater than 1 as the determiner of the number of factors to retain. But the eigenvalues of $\mathbf{S}^{-1}(\mathbf{R}-\mathbf{S}^2)\mathbf{S}^{-1}$ and $\mathbf{G}^*$ differ, in that the eigenvalues of the retained factors of $\mathbf{G}^*$ are smaller and approach zero as the roots of $\mathbf{S}^{-1}\mathbf{RS}^{-1}$ approach 1 in size. Thus image analysis will tend to scale down those factors that approach the point of being rejected.

Kaiser (1963) proposed a factor-structure matrix for $\mathbf{G}$ from those of $\mathbf{G}^*$. It was defined as

$$\boldsymbol{\Lambda}_g = \mathbf{SA}_r\left[\frac{(\gamma_i - 1)^2}{\gamma_\iota}\right]_r^{1/2} \tag{9.88}$$

But what Kaiser (1963) overlooked was that, unlike in common-factor analysis where premultiplying by $\boldsymbol{\Psi}$ in $\boldsymbol{\Lambda} = \boldsymbol{\Psi}\mathbf{A}_r[\gamma_\iota - 1]_r^{1/2}$ restored the metric for the observed variables in common-factor analysis to their original standard score form implying that $\boldsymbol{\Lambda}$ is a correlation matrix between observed variables and the common factors (with unit variances), premultiplying by $\mathbf{S}$ in Equation 9.88 only restores the metric for the variables to that of the image variables, implying that $\boldsymbol{\Lambda}_g$ is a "covariance matrix" between the image scores and the orthogonal image components. In other words, $\boldsymbol{\Lambda}_g = E(\mathbf{Y}_p\tilde{\mathbf{X}}')$ and not

$E(\mathbf{Y}\tilde{\mathbf{X}})$, where $\tilde{\mathbf{X}}$ contains the mutually orthogonal image-factor components, $\mathbf{Y}_p$ is the matrix of image variables, and $\mathbf{Y}$ the original observed variables. But most factor analyses seek correlations between the observed variables and the component variables.

It turns out that

$$\tilde{\mathbf{X}} = [\gamma_i]_r^{-1/2}\mathbf{A}_r'\mathbf{S}^{-1}\mathbf{Y} \tag{9.89}$$

where

$\mathbf{A}_r$ contains the first r eigenvectors

$[\gamma_i]_r$ the first, r largest eigenvalues of the matrix $\mathbf{S}^{-1}\mathbf{R}\mathbf{S}^{-1}$

For proof, we use multiple regression to determine the image component factors from the image variables.

$$\Lambda_g' = E(\mathbf{X}\mathbf{Y}_p')$$

$$E(\mathbf{Y}_p\mathbf{Y}_p') = \mathbf{G} = \mathbf{S}\mathbf{A}\left[\frac{(\gamma_i - 1)^2}{\gamma_\iota}\right]\mathbf{A}'\mathbf{S}$$

and

$$\mathbf{G}^{-1} = \mathbf{S}^{-1}\mathbf{A}\left[\frac{\gamma_i}{(\gamma_i - 1)^2}\right]\mathbf{A}'\mathbf{S}^{-1}$$

Hence

$$\tilde{\mathbf{X}} = \Lambda_g\mathbf{G}^{-1}\mathbf{Y}_p \tag{9.90}$$

The squared multiple R's are 1s because the image component factors are error-free linear combinations of the image variables. Ultimately, the image variables are in turn linear combinations of the observed variables. Hence, the image component variables can be expressed as linear combinations of the observed variables:

$$\mathbf{Y}_p = \mathbf{W}'\mathbf{Y} = (\mathbf{I} - \mathbf{S}^2\mathbf{R}^{-1})\mathbf{Y}$$

So

$$\begin{aligned}\tilde{\mathbf{X}} &= \Lambda_g\mathbf{G}^{-1}\mathbf{W}'\mathbf{Y}\\ &= \left(\left[\frac{(\gamma_i - 1)^2}{\gamma_i}\right]_r \mathbf{A}_r'\mathbf{S}\right)\left(\mathbf{S}^{-1}\mathbf{A}\left[\frac{\gamma_i}{(\gamma_i - 1)^2}\right]\mathbf{A}\mathbf{S}^{-1}\right)(\mathbf{I} - \mathbf{S}^2\mathbf{R}^{-1})\mathbf{S}\mathbf{S}^{-1}\mathbf{Y}\\ &= [\gamma_i]^{-1/2}\mathbf{A}_r'\mathbf{S}^{-1}\mathbf{Y}\end{aligned} \tag{9.91}$$

Then

$$\mathbf{\Lambda} = E(\mathbf{Y}\tilde{\mathbf{X}}') = E(\mathbf{YY'S}^{-1}\mathbf{A}_r[\gamma_\iota]^{-1/2}) = \mathbf{RS}^{-1}\mathbf{A}_r[\gamma_\iota]^{-1/2}$$
$$= \mathbf{SA}[\gamma_i]\mathbf{A'SS}^{-1}\mathbf{A}_r[\gamma_\iota]_r^{-1/2}$$
$$= \mathbf{SA}_r[\gamma_i]_r^{1/2} \tag{9.92}$$

This is exactly the same solution for $\mathbf{\Lambda}$ obtained in weighted principal components using $\mathbf{S}^{-1}$ instead of $\mathbf{E}^{-1}$ for the weight matrix (Table 9.3). In other words, weighted principal components and image analysis are the same when solutions for $\mathbf{\Lambda}$ are expressed in the metric of the observed variables and the image component factors having unit variances. This means that instead of using algorithms to compute image analysis in the manner of Kaiser (1963), we need only use the simpler algorithm and theory of weighted principal components applied to the matrix $\mathbf{S}^{-1}\mathbf{RS}^{-1}$. Unfortunately, most commercial programs that provide image analysis options to their factor analysis programs, currently use the improper Kaiser formula and do not offer weighted-principal-components analysis of the matrix $\mathbf{S}^{-1}\mathbf{RS}^{-1}$.

There is, however, a rationale for using the $\mathbf{G}$ matrix of image analysis without the use of image theory. We will discuss that when we discuss the problem of doublets in factor analysis later in the chapter.

TABLE 9.3

Unrotated Factor Loadings for Seven Factors from the $\mathbf{S}^{-1}\mathbf{RS}^{-1}$ Matrix According to the Weighted Principal Components Solution for Image Analysis $\mathbf{\Lambda} = \mathbf{SA}_r[\gamma_i]_r^{1/2}$

	1	2	3	4	5	6	7
1	.755	.195	.389	−.213	.234	−.230	.178
2	.757	.292	.485	.038	−.173	−.129	−.023
3	.745	.246	.552	−.105	−.034	.057	.033
4	.799	.285	.356	−.081	−.082	.188	−.088
5	.415	−.808	.038	.005	−.078	−.253	−.208
6	.487	−.806	.058	−.020	−.046	.097	.216
7	.491	−.804	.048	.041	.081	−.002	.125
8	.497	−.781	.085	−.109	−.053	.106	−.198
9	.698	.106	−.440	−.285	.308	.060	−.151
10	.836	.142	−.450	.076	−.115	−.079	.075
11	.786	.160	−.428	.021	−.156	.029	−.018
12	.771	.155	−.484	−.128	.034	−.094	−.063
13	.792	.043	.241	.336	.222	.142	−.108
14	.790	.036	.227	.353	.226	−.110	−.085
15	.882	.159	−.329	.053	.025	.102	.056

9.4.2 Kaiser's Rule for the Number of Harris Factors

In proposing a "Second-generation Little Jiffy" program for factor analyzing Harris' (1962) weighted covariance matrix $\mathbf{S}^{-1}\mathbf{R}\mathbf{S}^{-1}$, Henry F. Kaiser (1970) suggested using the rule, "Retain only factors with eigenvalues of $\mathbf{S}^{-1}\mathbf{R}\mathbf{S}^{-1}$ greater than the average eigenvalue $\bar{\gamma}$ of $\mathbf{S}^{-1}\mathbf{R}\mathbf{S}^{-1}$." $\bar{\gamma} = (1/n)\text{tr}[\gamma_i]$, where $\text{tr}[\gamma_i] = \text{tr}[\mathbf{S}^{-1}\mathbf{R}\mathbf{S}^{-1}]$ is the sum of the n eigenvalues as well as the sum of the diagonal elements of $\mathbf{S}^{-1}\mathbf{R}\mathbf{S}^{-1}$. He found this worked well with numerous sets of real data. This rule is analogical to his earlier "Retain only factors with eigenvalues greater than 1 of $\mathbf{R}$," because 1 is the average of the eigenvalues in that case. He thought it would always produce a number of factors less than given by the basic rule "Retain only factors with eigenvalues of $\mathbf{S}^{-1}\mathbf{R}\mathbf{S}^{-1}$ greater than the average eigenvalue." But he could not prove it. He asked colleagues around the world to see if they could prove it. No one could. But I (Mulaik) produced a counterexample with an artificial simplex-like correlation matrix. Nevertheless he reports that with "real data," the rule works well. What it seems to do is prune out doublet and quasi-singlet factors that rotation programs do not handle well because so few variables determine these factors. But the rule still lacks a more solid rationale and is only a heuristic supported by experience.

9.4.3 Quickie, Single-Pass Approximation for Common-Factor Analysis

Another spin-off from analyzing Harris' $\mathbf{S}^{-1}\mathbf{R}\mathbf{S}^{-1}$ matrix is a single-pass approximation to the iterated maximum-likelihood solution. The solution for the unrotated factor-pattern matrix is given by

$$\mathbf{\Lambda} = \mathbf{S}\mathbf{A}_r[\gamma_i - 1]_r^{1/2} \tag{9.93}$$

The maximum-likelihood solution goes on from there to replace $\mathbf{S}$ with improving approximations to $\mathbf{\Psi}$ until the program converges on a final solution.

9.5 Problem of Doublet Factors

Butler and Hook (1966) and Butler (1968) claim that not all sources of common variance in a test battery would have been considered by Thurstone to be desirable variance to analyze by common-factor analysis. As Butler (1968, p. 357) puts it:

> The principle to be used for identifying undesirable components of the test variance is that of indeterminacy. Indeterminate test variance is that which cannot determine a test battery common factor. One type of

> test variance unable to determine a common factor is singlet variance orthogonal to $n-1$ tests. Another type is doublet variance orthogonal to $n-2$ tests and common to just two tests. Taken together these components comprise those that cannot determine test battery common factors. Being indeterminate, these components should be eliminated or lessened in effect in the process of analysis.

In other words, we should not be interested in singlet, or unique, variance or that common variance which is due to correlations between just two variables. That Thurstone would likely have supported Butler's contention regarding the undesirability of doublet variance is suggested by references Thurstone made to the need to isolate and then ignore doublet factors which turn up in analyses (cf. Thurstone, 1947, pp. 441–442).

There are several routes along which to proceed to deal with this. We will first consider Butler's (1968) solution and then one based on equations from image analysis.

9.5.1 Butler's Descriptive-Factor-Analysis Solution

Butler's (1968) solution to this problem is to find a provisional estimate of the combined unique and doublet variance in each of the variables, and from that form a weight matrix from which a solution similar to that of the quickie, one-pass solution for common-factor analysis described above may be found. But his solution will find fewer factors when doublets are present than the common-factor solution. Butler (1968) describes his method as "descriptive-factor analysis" in contrast to a statistical solution such as maximum likelihood.

He reasons as follows. Suppose, for argument's sake, we do a full principal-components analysis as in Equations 9.1 and 9.2, with as many factors as variables, and obtain a solution for the principal-axes factors of the $n \times n$ correlation matrix $\mathbf{R}$. The unrotated solution is

$$\mathbf{\Lambda} = \mathbf{A}\mathbf{D}^{1/2}$$

where

$\mathbf{\Lambda}$ is the $n \times n$ factor loading matrix
$\mathbf{A}$ is the $n \times n$ eigenvector matrix
$\mathbf{D}$ is the diagonal matrix of the n eigenvalues of $\mathbf{R}$

Suppose we seek now an orthogonal rotation $\mathbf{T}'\mathbf{X}$ of the orthogonal factors that aligns each factor, respectively, with one of the observed variables. Here we seek to do so in such a way that

$$\begin{aligned} g(\mathbf{T}) &= \text{tr}\, E[(\mathbf{Y} - \mathbf{T}'\mathbf{X})(\mathbf{Y} - \mathbf{T}'\mathbf{X})'] \\ &= \text{tr}(\mathbf{R}_{YY} - \mathbf{R}_{YX}\mathbf{T} - \mathbf{T}'\mathbf{R}_{XY} + \mathbf{I}) \end{aligned}$$

is a minimum. Since $\mathrm{tr}(\mathbf{R}_{YY})$ and $\mathrm{tr}(\mathbf{I})$ are positive constants and $\mathrm{tr}(\mathbf{R}_{YX}\mathbf{T}) = \mathrm{tr}(\mathbf{T}'\mathbf{R}_{XY})$, then minimizing $g(\mathbf{T})$ is equivalent to maximizing $h(\mathbf{T}) = \mathrm{tr}(\mathbf{R}_{YX}\mathbf{T}) = \mathrm{tr}(\mathbf{\Lambda T})$. To make a long story short, the solution to this problem was first developed by Green (1952), and is given in full in Mulaik (1972). Intuitively what we want to do is make $\mathbf{R}_{YX}\mathbf{T} = \mathbf{\Lambda T}$ be as much like $\mathbf{R}_{YY}$ as possible, while at the same time making it as close to an identity matrix as possible. It happens in this case that the solution for $\mathbf{T}$ is $\mathbf{A}'$, which means that the rotated solution for the component loading matrix is

$$\mathbf{\Lambda}^* = \mathbf{AD}^{1/2}\mathbf{A}' = \mathbf{R}^{1/2}$$

This is a symmetric matrix and happens to be the square root of the correlation matrix. It also represents the projection cosines of the observed variables onto the orthogonal factors that maximally fit, respectively, each one of the observed variables. The trace of the diagonal elements of $\mathbf{AD}^{1/2}\mathbf{A}'$ is thus a maximum, meaning that by and large, the largest elements in $\mathbf{AD}^{1/2}\mathbf{A}'$ are to be found mostly in the diagonal, and since the sum of squares of the elements in each row add up to 1, they will all be values less than or equal to 1.

Now let $\mathbf{L}^2 = [\mathrm{diag}(\mathbf{AD}^{1/2}\mathbf{A}']^2$ be a diagonal matrix of the squares of the diagonal elements of $\mathbf{AD}^{1/2}\mathbf{A}'$. These values too will be less than or equal to 1.

Butler (1968) then notes that

$$\mathrm{tr}(\mathbf{L}^2) + \mathrm{tr}(\mathbf{I} - \mathbf{L}^2) = \mathrm{tr}(\mathbf{L}^2 + \mathbf{I} - \mathbf{L}^2) = \mathrm{tr}(\mathbf{I}) = n$$

so that since $\mathrm{tr}(\mathbf{L}^2)$ is a maximum and n is a constant, $\mathrm{tr}(\mathbf{I} - \mathbf{L}^2)$ is a minimum. So, $\mathrm{tr}(\mathbf{L}^2)/n$ measures the extent to which $\mathbf{AD}^{1/2}\mathbf{A}'$ differs from an identity matrix, while $\mathrm{tr}(\mathbf{I} - \mathbf{L}^2)/n$ measures the degree to which $\mathbf{AD}^{1/2}\mathbf{A}'$ is close to being an identity matrix.

Keeping the above in mind, let us turn next to the inverse of $\mathbf{AD}^{1/2}\mathbf{A}'$, which happens to be $(\mathbf{AD}^{1/2}\mathbf{A}')^{-1} = \mathbf{AD}^{-1.2}\mathbf{A}'$. This also happens to equal $\mathbf{R}^{-1/2}$ because $\mathbf{AD}^{-1.2}\mathbf{A}'\mathbf{AD}^{-1.2}\mathbf{A}' = \mathbf{AD}^{-1}\mathbf{A}' = \mathbf{R}^{-1}$. Let us now normalize the rows of $\mathbf{AD}^{-1/2}\mathbf{A}'$ so that they have unit length. To accomplish this we will need to find the current lengths of the rows, which we will find to be the square roots of the diagonal elements of $[(\mathbf{AD}^{-1/2}\mathbf{A}')(\mathbf{AD}^{-1/2}\mathbf{A}')'] = \mathbf{R}^{-1}$. Then let $\mathbf{S}^{-2} = \mathrm{diag}\ \mathbf{R}^{-1}$, so that the diagonal matrix which, premultiplied times $\mathbf{AD}^{-1/2}\mathbf{A}'$ to give it unit length rows, is $\mathbf{S} = [\mathrm{diag}\ \mathbf{R}^{-1}]^{-1/2}$. Let $\mathbf{M} = \mathbf{SAD}^{-1/2}\mathbf{A}'$. Let us define a new set of component factors $\mathbf{X}^{\dagger} = \mathbf{MX}^* = \mathbf{SAD}^{-1/2}\mathbf{A}'\mathbf{X}^*$, where $\mathbf{X}^* = \mathbf{AX}_0$. Then

$$E(\mathbf{X}^{\dagger}\mathbf{X}^{\dagger\prime}) = E(\mathbf{MX}^*\ \mathbf{X}^{*\prime}\ \mathbf{M}') = \mathbf{SAD}^{-1/2}\mathbf{A}'\mathbf{IAD}^{-1/2}\mathbf{AS} = \mathbf{SR}^{-1}\mathbf{S}$$

We should now recognize this to be the matrix of negative partial correlations between each pair of variables with the $n-2$ other observed variables partialled out. On the other hand, the corresponding inverse transformation applied to the matrix $\mathbf{\Lambda}^*$ yields

$$\mathbf{\Lambda}^{\dagger} = \mathbf{\Lambda}^*\ \mathbf{M}^{-1} = \mathbf{AD}^{1/2}\mathbf{A}'\mathbf{AD}^{1/2}\mathbf{A}'\mathbf{S}^{-1} = \mathbf{ADA}'\mathbf{S}^{-1} = \mathbf{RS}^{-1}$$

The new component factor-structure matrix of correlations between the observed variables and the component factors is given by

$$E(\mathbf{YX}') = \Lambda^{\dagger}\Phi_{XX} = \mathbf{RS}^{-1}\mathbf{SR}^{-1}\mathbf{S} = \mathbf{S}$$

Butler (1968) now identifies the correlation matrix $\mathbf{SR}^{-1}\mathbf{S}$ with the matrix of singlet component factors, but notes that they also have pairwise correlations, meaning that this matrix also contains doublets. The matrix **S** on the other hand gives the projections of the n observed variables onto the "singlet" components. Since **S** is diagonal, this indicates that each singlet factor is orthogonal to $n-1$ of the observed variables.

The doublet variance in a set of variables is approximated by the partial correlations between pairs of variables after the variance due to the $n-2$ other variables in the set has been partialled from the pairs. We may observe the negative values of these partial correlations by examining the matrix $\mathbf{SR}^{-1}\mathbf{S}$, which is the matrix of correlations among the partial anti-images. Any nonzero off-diagonal elements of this matrix indicate (approximately) doublet variance present in the correlation matrix. Potentially, there are $n(n-1)/2$ sources of doublet variance (the number of off-diagonal elements of $\mathbf{SR}^{-1}\mathbf{S}$). In other words, there are many more possible doublet factors than there are tests to determine them, and so doublet factors must be indeterminate. Butler (1968, p. 362) emphasizes that all the doublet variance cannot be isolated by factor rotation. Although some of the doublet variance will appear in doublet factors, the rest (if present) will remain hidden in the common-factor space until a different rotation is carried out.

Since the diagonal matrix **S** represents the correlations of the respective observed variables with their corresponding singlet components, the squares of these correlations in $\mathbf{S}^2$ represent the proportions of variance in the respective observed variables due to the respective singlet components. But we must also consider the variance in the observed variables due to doublets.

What Butler (1968) then seeks to do is to get some sense of the proportion of the variance in the correlation matrix among the component factors in $\mathbf{X}^{\dagger}$ that is due to doublets. "True" singlets are mutually uncorrelated. But if singlets also have doublet correlations between pairs of singlets, they will be correlated. An estimate of the doublet variance in each variable could be obtained by rotating the orthogonal factors in $\mathbf{X}^*$ to a new orthogonal set of n factors in Ξ having a factor-structure matrix with respect to **Y** equal to $E(\mathbf{Y}\Xi') = (\mathbf{SR}^{-1}\mathbf{S})^{1/2}$. Again what he does here is analogous to the way he used the trace of the squares of the elements of the principal diagonal of the square root of **R**, $\mathbf{AD}^{1/2}\mathbf{A}'$, in $\mathbf{L}^2 = [\text{diag}(\mathbf{AD}^{1/2}\mathbf{A}')]^2$ to indicate the degree to which the observed variables deviated from the identity matrix **I**. In the present case let $\mathbf{K} = \text{diag}(\mathbf{SR}^{-1}\mathbf{S})^{1/2}$. This contains in the principal diagonal the maximum correlations (cosines) of the observed variables with an orthogonal set Ξ of variables maximally aligned with the correlated singlets in $\mathbf{SR}^{-1}\mathbf{S}$. So the component variables in Ξ are like "true" singlets in $\mathbf{X}^{\dagger}$. And the squares of the correlations in **K** in $\mathbf{K}^2 = [\text{diag}(\mathbf{SR}^{-1}\mathbf{S})^{1/2}]^2$ indicate the proportion of true singlet variance in each observed variable. Consequently

$$\mathbf{U} = \mathbf{I} - \mathbf{K}^2$$

contain minimum values "... due to the presence of doublet factors that result in nonzero correlations among the singlet vectors" (Butler, 1968, p. 363).

Now, let

$$\mathbf{V}^2 = \mathbf{I} - \mathbf{K}^2 + \mathbf{S}^2$$

be a diagonal matrix giving the variance in the normalized anti-image variables attributable respectively to both doublet and singlet components.

The trace of the diagonal matrix,

$$\mathbf{I} - \mathbf{V}^2 = \mathbf{K}^2 - \mathbf{S}^2$$

Butler (1968) construes to be overdetermined common variance. The value of the ith diagonal element in $\mathbf{K}^2 - \mathbf{S}^2$ is the observed common variance in the ith variable.

Butler notes that the ith nonzero element in $\mathbf{K}^2$ should be greater than the ith nonzero element in $\mathbf{S}^2$. If k_{ii}^2 approaches s_{ii}^2 in value, Butler recommends that the ith variable be discarded from the study because too much of its variance is singlet and doublet variance.

If $\mathbf{I} - \mathbf{S}^2$ contains in its principal diagonal the squared multiple correlations for predicting each variable from the $n-1$ other variables, then

$$k_{ii}^2 - s_{ii}^2 \le 1 - s_{ii}^2 = R_i^2$$

If doublet variance is present, the overdetermined common variance in the ith variable, $k_{ii}{}^2 - s_{ii}{}^2$ will be less than the squared multiple correlation R_i^2.

While it would be possible to analyze the reduced matrix $\mathbf{R} - \mathbf{V}^2$ to find its eigenvectors **A** and eigenvalues **D**, and then a principal-axes factor loading matrix $\mathbf{\Lambda} = \mathbf{A}_r\mathbf{D}_r^{1/2}$, for r factors with eigenvalues greater than 0, Butler (1968) recommends instead obtaining a weighted solution: Analyze into its eigenvectors and eigenvalues the matrix

$$\mathbf{V}^{-1}(\mathbf{R} - \mathbf{V}^2)\mathbf{V}^{-1} = \mathbf{V}^{-1}\mathbf{R}\mathbf{V}^{-1} - \mathbf{I}$$

Recall that in connection with an analogous equation for analyzing the common-factor model with a weighted solution, we need to only obtain here the eigenvectors and eigenvalues of $\mathbf{V}^{-1}\mathbf{R}\mathbf{V}^{-1}$, and then subtract 1s from the eigenvalues obtained (Table 9.4). Consequently the solution for the principal-axes factor solution in the weighted case is given by

$$\mathbf{\Lambda}_r = \mathbf{V}\mathbf{A}_r(\mathbf{D}_r - \mathbf{I}_r)^{1/2}$$

TABLE 9.4

Eigenvalues of $\mathbf{V}^{-1}\mathbf{RV}^{-1}$ and Unrotated Factors According to Weighted Descriptive-Factor Analysis (Butler, 1968) Applied to Correlation Matrix in Table 8.1 from Carlson and Mulaik (1993)

	1	2	3	4	Eigenvalues
1	.746	.179	.348	−.072	21.586
2	−.740	.259	.430	−.013	7.953
3	−.729	.217	.480	−.036	5.301
4	−.784	.261	.313	−.031	1.091
5	−.405	−.766	.018	−.006	.901
6	−.470	−.752	.034	−.013	.749
7	−.477	−.750	.027	.004	.635
8	−.483	−.735	.053	−.026	.591
9	−.686	.108	−.423	−.059	.569
10	−.805	.134	−.401	.014	.513
11	−.765	.155	−.400	.010	.440
12	−.751	.151	−.448	−.022	.403
13	−.786	.036	.227	.105	.309
14	−.785	.028	.216	.107	.308
15	−.854	.150	−.295	.014	.262

where r is the number of retained factors. These should correspond to eigenvalues $\mathbf{D}$ of $\mathbf{V}^{-1}\mathbf{RV}^{-1}$ greater than 1.00, because in that case those eigenvalues of $\mathbf{D}-\mathbf{I}$ will be positive. This now gives us a criterion and rationale for taking fewer factors than suggested by the number of eigenvalues greater than 1 of the matrix $\mathbf{S}^{-1}\mathbf{RS}^{-1}$ when estimating the number of factors to retain in common-factor analysis. However, if there are no doublets present, then Butler's descriptive-factor analysis and common-factor analysis will retain the same number of factors.

9.5.2 Model That Includes Doublets Explicitly

Consider the following model equation

$$\mathbf{Y} = \mathbf{\Lambda X} + \mathbf{K\Delta} + \mathbf{\Psi E} \tag{9.94}$$

where

- $\mathbf{Y}$ is an $n \times 1$ vector of observed random variables
- $\mathbf{\Lambda}$ is an $n \times r$ common-factor-pattern matrix
- $\mathbf{X}$ is an $r \times 1$ vector of common factors
- $\mathbf{K}$ is an $n \times k$ doublet pattern matrix
- $\mathbf{\Delta}$ is a $k \times 1$ vector of doublet factors, $k = n(n-1)/2$
- $\mathbf{\Psi}$ is an $n \times n$ diagonal matrix of unique-factor-pattern coefficients with $\mathbf{E}$ an $n \times 1$ vector of unique- or singlet-factor variables

We assume the follow relationships between the random variables of the model:

$$E(\mathbf{X}\boldsymbol{\Delta}') = \mathbf{0}$$

$$E(\boldsymbol{\Delta}\mathbf{E}') = \mathbf{0} \tag{9.95}$$

$$E(\mathbf{X}\mathbf{E}') = \mathbf{0}$$

$$E(\mathbf{X}\mathbf{X}') = \mathbf{I}$$

We will further assume that all variables have zero means and unit variances. Then the matrix of correlations among the observed variables is given by

$$\mathbf{R}_{YY} = \boldsymbol{\Lambda}\boldsymbol{\Phi}_{XX}\boldsymbol{\Lambda}' + \mathbf{K}\mathbf{K}' + \boldsymbol{\Psi}^2 \tag{9.96}$$

The matrix **K**, as a factor-pattern matrix, is unique in that in each column all loadings are zero, except two. Its nonzero elements are not identified. Furthermore, each variable in **Y** is paired with each other variable with nonzero loadings in one of the columns of **K**. The matrix $\mathbf{KK}'$ is the matrix of doublet variances and covariances. $\boldsymbol{\Psi}^2$ is the diagonal matrix of unique-factor variances. A doublet factor is correlated with only two variables in **Y** and orthogonal with all the rest. We further stipulate that in each column of $\boldsymbol{\Lambda}$ there are at least three nonzero loadings. So, the common factors are minimally determined if not overdetermined from the observed variables.

We desire that the variance due to the common factors be separated from that due to the doublet and singlet factors. Thus we will seek the matrix $\mathbf{R}_{YY} - \mathbf{KK}' - \boldsymbol{\Psi}^2 = \boldsymbol{\Lambda}\boldsymbol{\Phi}_{XX}\boldsymbol{\Lambda}'$. Factoring the matrix on the left-hand side of the equality will yield a solution on the right (with suitable constraints). The problem is, how can we get an expression for $\mathbf{KK}' + \boldsymbol{\Psi}^2$? An approximate solution will be given us as a by-product of image analysis. In image analysis we know that the matrix $\mathbf{SR}^{-1}\mathbf{S}$ is the matrix of negative partial correlations between pairs of variables in **Y** with $n-2$ other variables partialled out. We can convert this matrix to a matrix of partial correlations

$$\mathbf{P} = -\mathbf{SR}^{-1}\mathbf{S} + 2\mathbf{I} \tag{9.97}$$

Multiplying $\mathbf{SR}^{-1}\mathbf{S}$ by -1 to get the proper partial correlations also multiplies the diagonal 1s by -1. So, to make this a proper correlation matrix we need to replace the -1s in the diagonal with positive 1s, which is accomplished by adding $2\mathbf{I}$ to $-\mathbf{SR}^{-1}\mathbf{S}$. Now, while **P** contains the (approximate) information about doublet correlations between pairs of variables, we do not know the variances of the combined doublet and singlet variance–covariance matrix.

But if we can find a diagonal matrix **U** to pre- and postmultiply times **P**, we may be able to write

$$\mathbf{KK}' + \mathbf{\Psi}^2 = \mathbf{U}(-\mathbf{SR}^{-1}\mathbf{S} + 2\mathbf{I})\mathbf{U} = \mathbf{UPU} \tag{9.98}$$

Then the matrix we want to analyze by finding its eigenvectors and eigenvalues and obtaining an orthogonal solution for the factors is

$$\mathbf{R}_{YY} - \mathbf{UPU} = \mathbf{\Lambda\Lambda}' \tag{9.99}$$

The matrix **P** is easily obtained in Equation 9.98. The solution for **U** may require an iterative algorithm. However, we may begin with a good approximation for **U** in the matrix **S**. Then

$$\mathbf{R}_{YY} - \mathbf{SPS} = \mathbf{R}_{YY} + \mathbf{S}^2\mathbf{R}^{-1}\mathbf{S}^2 - 2\mathbf{S}^2 = \mathbf{G} \tag{9.100}$$

This is the **G** matrix of image analysis. But in this case this is regarded not as the covariance matrix of covariances among the images but as the covariances among the observed variables with the doublet and singlet variance (approximately) removed. This suggests that factoring **G** will yield eigenvalues that may tell us how many factors to retain.

The unrotated first principal-axes factors of the matrix **G** of Equation 9.100 of image analysis calculated from the correlation matrix in Table 8.1 is shown in Table 9.5, along with the eigenvalues of this matrix. Note that all of the eigenvalues are positive, and most are near zero except for three that are

TABLE 9.5

First Unrotated Principal-Axes Factors of the Image Matrix **G** with Associated Eigenvalues Derived from Correlation Matrix in Table 8.1 Reported by Carlson and Mulaik (1993)

Variable	1	2	3	Eigenvalues	Elements of S^2
1	−.740	.193	.336	7.263	.276
2	−.742	.281	.417	2.542	.173
3	−.731	.236	.477	1.493	.157
4	−.780	.276	.305	.084	.208
5	−.418	−.747	.016	.059	.246
6	−.489	−.742	.031	.056	.188
7	−.490	−.741	.025	.039	.190
8	−.498	−.718	.060	.032	.207
9	−.669	.114	−.415	.021	.337
10	−.801	.150	−.416	.016	.139
11	−.754	.167	−.398	.014	.232
12	−.738	.162	−.450	.003	.214
13	−.776	.052	.201	.001	.329
14	−.773	.045	.188	.001	.337
15	−.848	.165	−.314	.001	.144

TABLE 9.6

Iterated Principal-Axes Solution for **Λ**, Eigenvectors and Elements of $\mathbf{U}^2$ for Model with Doublets in Equation 9.99 Derived for Correlation Matrix in Table 8.1 from Carlson and Mulaik (1993)

Variable	1	2	3	Eigenvectors	u_{ii}^2
1	−.736	.188	.333	7.193	.312
2	−.735	.273	.405	2.461	.221
3	−.720	.225	.453	1.420	.226
4	−.779	.273	.308	.076	.224
5	−.415	−.738	.013	.050	.282
6	−.485	−.730	.026	.043	.231
7	−.486	−.729	.022	.013	.231
8	−.493	−.705	.057	.003	.257
9	−.667	.114	−.403	−.002	.380
10	−.795	.147	−.395	−.012	.191
11	−.754	.167	−.393	−.018	.249
12	−.738	.162	−.444	−.025	.233
13	−.773	.050	.202	−.040	.359
14	−.771	.043	.188	−.042	.369
15	−.847	.165	−.307	−.046	.161

all larger than 1.00. The decision was to retain three common factors with eigenvalues greater than 1.00.

We show the results of an iterated solution for $\mathbf{U}^2$ and **Λ** in Table 9.6. The iterative solution for the matrix **U** is as follows.

Given $\mathbf{R}_{YY}$, compute $\mathbf{R}_{YY}^{-1}$ and $\mathbf{S}^2 = [\text{diag } \mathbf{R}_{YY}^{-1}]^{-1}$. Find $\mathbf{S} = (\mathbf{S}^2)^{1/2}$. Then compute $\mathbf{P} = -\mathbf{SR}^{-1}\mathbf{S} + 2\mathbf{I}$. Now let $\mathbf{U}_{(0)}^2 = \mathbf{S}^2$. Compute **G** in Equation 9.100 and determine the number of factors to retain from its eigenvalues. Let this number be r and do not change it afterward during the iterations. Now, the iterations are as follows:

Step 1. Obtain the eigenvectors **A** and the eigenvalues **D** of

$$\mathbf{C}_{(i)} = \mathbf{R}_{YY} - \mathbf{U}_{(i)}\mathbf{P}\mathbf{U}_{(i)}$$

Compute $\mathbf{\Lambda}_{r(i)} = \mathbf{A}_{r(i)}\mathbf{D}_{r(i)}{}^{1/2}$.

Step 2. Compute $\mathbf{W}_{(i)} = \mathbf{R}_{YY} - \mathbf{\Lambda}_{r(i)}\, \mathbf{\Lambda}'_{r(i)}$

Step 3. Compute $\mathbf{U}_{(i+1)}{}^2 = \text{diag}(\mathbf{W}_{(i)})$. Compute $\mathbf{U}_{(i+1)}$.

By $u(j)_{(i)}{}^2$ denote the value of the jth diagonal element of $\mathbf{U}_{(i)}{}^2$ in the ith iteration. Then if $\left|u(j)_{(t+1)}^2 - u(j)_{(t)}^2\right| > .0001$ for any j, $j = 1$, n go to step 1, else quit. Print the last $\mathbf{\Lambda}_{r(i)}$.

The researcher should examine the matrix **P** of partial correlations to discover the sources of doublet variance. They may be due to method effects common to just two variables, similar use of words or phrasing, underdetermined additional common factors, etc. A great many factor analyses may fail to replicate in confirmatory factor analysis because of the presence of doublet factors which have been "swept under the rug" so to speak by taking only a few factors in preceding exploratory factor analyses. The approximation should improve with more indicator variables per factor.

9.6 Metric Invariance Properties

A unique property of the factors of image analysis, canonical-factor analysis, and maximum-likelihood exploratory factor analysis are invariant under changes of the original scaling of the variables. If we define $\mathbf{S}^2 = [\text{diag } \mathbf{R}^{-1}]^{-1}$, then no matter what scaling method we use with the original variables in $\mathbf{Y}$, we will always arrive at the same factor solution.

Consider an arbitrary diagonal scaling matrix $\mathbf{K}$ (where the diagonal elements of $\mathbf{K}$ contain, for example, values to be divided into the scores in $\mathbf{Y}$). We could find a new matrix of covariances among the variables as

$$\mathbf{R}_+ = E(\mathbf{KYY'K})$$

Now, let us define

$$\mathbf{S}_+^2 = [\text{diag } \mathbf{R}_+^{-1}]^{-1}$$

However

$$\mathbf{R}_+^{-1} = (\mathbf{KRK})^{-1} = \mathbf{K}^{-1}\mathbf{R}\mathbf{K}^{-1}$$

Since $\mathbf{K}$ is a diagonal matrix, the diagonal matrix formed from the diagonal elements of $\mathbf{R}_+^{-1}$ will be $\mathbf{K}^{-1}[\text{diag } \mathbf{R}^{-1}]\mathbf{K}^{-1}$. This is none other than $[\text{diag } \mathbf{R}^{-1}]\mathbf{K}^{-2} = [\text{diag } \mathbf{R}_+^{-1}]$. Thus taking into account our definition for $\mathbf{S}^2$, $\mathbf{S}_+^2 = \mathbf{K}^2\mathbf{S}^2$.

As a consequence

$$\mathbf{S}_+^{-1} = \mathbf{K}^{-1}\mathbf{S}^{-1}$$

Thus if we proceed to find the corresponding matrix for $\mathbf{S}^{-1}\mathbf{R}\mathbf{S}^{-1}$, we get

$$\mathbf{S}_+^{-1}\mathbf{R}_+\mathbf{S}_+^{-1} = \mathbf{S}^{-1}\mathbf{K}^{-1}(\mathbf{KRK})\mathbf{K}^{-1}\mathbf{S}^{-1} = \mathbf{S}^{-1}\mathbf{R}\mathbf{S}^{-1}$$

which is the same matrix as we would have obtained had we not performed this scaling. The factors of $\mathbf{R}_+$ would thus be simply related to the factors of $\mathbf{R}$. That is, if $\mathbf{\Lambda}_r = \mathbf{SA}_r[\gamma]_r^{1/2}$, then $\mathbf{\Lambda}_+ = \mathbf{KSA}_r[\gamma_i]_r^{1/2}$. The same factors are represented, only they differ in that the rows of $\mathbf{\Lambda}_r$ have been scaled by the matrix **K** to obtain Λ_+. Kaiser (1963, p. 162) comments on this by saying, "The importance of this metric invariance cannot be overemphasized: We are freed from the traditional agnostic confession of ignorance implied by standardizing the observable tests... Here this standardization is merely a convenience to which we are in no way tied."

9.7 Image-Factor Analysis

Although similar in many respects to Guttman's image analysis, image-factor analysis is an innovation due to Jöreskog (1962, 1963). Jöreskog develops a model based upon the fundamental equation

$$\mathbf{R} = \Lambda_c \Lambda_c' + \theta \mathbf{S}^2 \tag{9.101}$$

where

- **R** is the population correlation matrix
- Λ_c is a population factor-pattern/structure matrix (because of orthogonal factors)
- θ is a scalar parameter determined by the matrix **R**
- $\mathbf{S}^2 = [\text{diag } \mathbf{R}^{-1}]^{-1}$

The distinguishing feature of this equation is the diagonal matrix $\theta\mathbf{S}^2$, which is analogous to the diagonal matrix $\boldsymbol{\Psi}^2$ of common-factor analysis. The idea behind this matrix is an attempt to construct a more determinate model to approximate the model of common-factor analysis, by providing a close approximation to the diagonal matrix of unique variances $\boldsymbol{\Psi}^2$.

We have already shown in an earlier chapter that s_j^2, the jth diagonal element of the matrix $\mathbf{S}^2$, is always greater than or equal to the jth unique variances ψ_j^2. This suggests that, if we were to rescale the jth element s_j^2 by premultiplying it by an appropriate number θ_j between zero and 1, we could make s_j^2 become ψ_j^2. Of course, we would have to find a different θ_j for each pair of values for s_j^2 and ψ_j^2. But we can see that if we define a diagonal matrix $\boldsymbol{\Theta}$ such that

$$\boldsymbol{\Theta} = [\theta_j] \tag{9.102}$$

where θ_j represents a typical element in the diagonal of $\boldsymbol{\Theta}$, then

$$\boldsymbol{\Psi}^2 = \boldsymbol{\Theta}\mathbf{S}^2 \tag{9.103}$$

The difficulty with this is that the elements ψ_j^2 are basically indeterminate; hence the elements θ_j are indeterminate. But Jöreskog proposed to reduce the indeterminacy in his model by assuming, arbitrarily, that all the θ_j are equal to a single value θ. Consequently, he ends up with the model equation in Equation 9.101.

Jöreskog's model is a very good approximation to common-factor analysis. In fact the approximation to common-factor analysis improves as the number of variables included in the correlation matrix **R** increases. When n becomes indefinitely large the elements s_j^2 approach the elements ψ_j^2 as a limit, and as a consequence θ can then be thought of as having the value 1. Hence, Jöreskog's model and the common-factor analysis model converge to the same model when the number of variables increases indefinitely.

The reduced correlation matrix from which the common-factor-loading matrix is derived is given as

$$\mathbf{\Lambda}_c\mathbf{\Lambda}_c = \mathbf{R} - \theta\mathbf{S}^2 \tag{9.104}$$

Since the matrix $\mathbf{S}^2$ is completely determinate, premultiply and postmultiply Equation 9.91 by $\mathbf{S}^{-1}$ to obtain

$$\mathbf{S}^{-1}\mathbf{\Lambda}_c\mathbf{\Lambda}_c'\mathbf{S}^{-1} = \mathbf{S}^{-1}\mathbf{R}\mathbf{S}^{-1} - \theta\mathbf{I} \tag{9.105}$$

This equation is in a form very much like the one Bartlett used to develop his test of significance for principal components. However, unlike Bartlett's matrix **R** of correlations, the matrix $\mathbf{S}^{-1}\mathbf{R}\mathbf{S}^{-1}$ is a covariance matrix which is invariant with respect to the units used to standardize the original scores. Hence, although there are questions in the literature concerning Bartlett's test in regard to the effect of standardizing scores upon the degrees of freedom to use, no such problem exists with Jöreskog's model because of this property of metric invariance in the case of $\mathbf{S}^{-1}\mathbf{R}\mathbf{S}^{-1}$.

9.7.1 Testing Image Factors for Significance

In practice, one begins an image-factor analysis with the $n \times n$ sample correlation matrix $\tilde{\mathbf{R}}$ based on N observations. Then one finds the sample matrix $\tilde{\mathbf{S}}^2$ and then the matrix $\tilde{\mathbf{S}}^{-1}\tilde{\mathbf{R}}\tilde{\mathbf{S}}^{-1}$. Next one proceeds to factor this last matrix by the method of principal axes to find its eigenvectors **A** and eigenvalues $[\gamma_i]$ arranged in descending order of magnitude of the eigenvalues. Then one begins a series of tests for the number of significant factors, following a procedure very much like the one in Bartlett's test for the number of significant principal components. The essence of these tests for significance is the hypothesis that after r principal axes have been extracted from the matrix the eigenvalues of the remaining principal axes do not differ significantly from each other. The basis for the test is that one compares the arithmetic mean of the $n - r$ remaining eigenvalues with their geometric mean. If the arithmetic mean of the $n - r$ remaining eigenvalues is given as

$$\theta = \frac{\gamma_{r+1} + \cdots + \gamma_n}{n - r} \tag{9.106}$$

then one finds the coefficient

$$\chi^2_{\text{df}} = -(N-1)[(\gamma_{r+1} + \cdots + \gamma_n) - (n-r)\ln\theta] \tag{9.107}$$

which is distributed approximately as chi-square with degrees of freedom $\text{df} = (n - r + 2)(n - r - 1)/2$. If N is moderate, Jöreskog recommended multiplying the expression within brackets in Equation 9.107 by

$$N - \frac{1}{6}\left[2n + 4r + 7 + \frac{2}{(n-r)}\right]$$

instead of $(N-1)$. If the value of χ^2_{df} exceeds the tabular value for chi-square for the respective degrees of freedom, then one rejects the hypothesis that all the significant factors have been found and proceeds to test the hypothesis that there are $r+1$ common factors. Jöreskog points out that, strictly speaking, this sequence of tests violates the rule that the tests must be independent. However, if the level of significance is 5%, one can still assert, with a risk of error of only 5%, that the number of common factors is at least as large as that discovered by this sequence of tests (Jöreskog, 1962, p. 348).

Assume that the matrix $\tilde{\mathbf{A}}_r$ contains for columns the first r eigenvectors corresponding to the r significant factors. Then the sample factor-loading matrix $\tilde{\mathbf{\Lambda}}$ is given by

$$\tilde{\mathbf{\Lambda}}_c = \tilde{\mathbf{S}}\tilde{\mathbf{A}}_r[\tilde{\gamma}_i - \theta]^{1/2} \tag{9.108}$$

where $[\tilde{\gamma}_i - \theta]$ is an $r \times r$ diagonal matrix of the first r largest eigenvalues of $\tilde{\mathbf{S}}^{-1}\tilde{\mathbf{R}}\tilde{\mathbf{S}}^{-1}$ each reduced by the average θ of the $n-r$ remaining eigenvalues obtained by Equation 9.96. The resulting factor loading matrix is a very good approximation to the maximum-likelihood solution for $\mathbf{\Lambda}$ in common-factor analysis, and the approximation should improve as the number of variables increases.

9.8 Psychometric Inference in Factor Analysis

As was pointed out in the chapter on the model of common-factor analysis, the main purpose of this model is to study those factors which account for the overlap or covariance between variables. In psychological applications of this model, one is concerned with finding sources of covariance between performances of persons on various tests, which one might hypothesize are higher integrative processes within persons. For example, in this vein Spearman (1904) went so far as to claim, on the basis of an analysis of performances of students on a few tests, that all measures of intellective or sensory functions could be regarded as measures of a general intellective function, common to all such measures, and a specific function associated uniquely with each measure. What Spearman's claim represented was a psychometric inference, made from a few tests to the whole domain of human-performance tests.

But surprisingly, in the development of the theory and practice of common-factor analysis, there has been very little attention paid to the nature of inferences to be drawn from a particular factor analysis. Generally, the inferences made from a particular analysis are not simply about the particular variables studied but are extended to a universe of other variables that potentially could have been studied but which were not. As far as the present author can tell, Hotelling (1933, p. 504) was the first to publish remarks concerning the problem of psychometric inference in factor analysis when he wrote

> Instead of regarding the analysis of a particular set of tests as our ultimate goal, we may look upon these merely as a sample of a hypothetical larger aggregate of possible tests. Our aim then is to learn something of the situation portrayed by the large aggregate. We are thus brought to a type of sampling theory quite distinct from that which we have heretofore considered. Instead of dealing with the degree of instability of functions of the correlations of the observed tests arising from the smallness of the number of persons, we are now concerned with the degree of instability resulting from the limited number of tests whose correlations enter into our analysis.

From the point of view of a larger aggregate of tests, then, the problem of inference in common-factor analysis is to infer the common factors that would be found in the larger aggregate of tests from a particular sample of tests.

Methods for making such inferences, however, have been developed heretofore chiefly in connection with the problem of reliability. Among the first to publish in this area was Tryon whose classic summary of his viewpoint on domain reliability and validity appeared in 1957 (Tryon, 1957). Shortly thereafter he also published papers (Tryon, 1957, 1959) on a domain-sampling approach to the communality problem in factor analysis, which, however, never seemed to affect greatly the practice of factor analysts, except in terms of focusing attention on the problems of domain sampling. Meanwhile similar developments with regard to reliability and validity were being made from the point of view of analysis of variance, leading to Cronbach's papers on domain generalizability (Cronbach, 1951; Cronbach et al., 1963). All these developments culminated in a single coefficient for the domain reliability, validity, and/or generalizability of a test. This coefficient is perhaps best known as the Kuder-Richardson formula 20 for the internal reliability of a test, although Cronbach renamed it "alpha," the coefficient of generalizability.

The viewpoint of these more recent writers on the problem of domain reliability and validity departs considerably from the viewpoint of classic reliability. The domain-generalizability viewpoint sees the problem of reliability as the problem of inferring the consistency of a score being obtained across a variety of conditions and circumstances. For example, if k judges are asked to rate a group of subjects known to them on the trait "likeableness," one is likely to find that for any given subject the judges vary in the likeableness score that they give him. From the point of view of classic reliability theory one is concerned only with the precision with which any one judge makes his ratings as he sees things. Hence the differences between judges may be due not simply to sloppiness of rating but rather to actual differences in the subject rated. A given subject could conceivably behave differently in the presence of different judges. School children, for example, act differently in the presence of their homeroom teacher and the school principal. Hence variance between judges could be true variance as well as error (inaccuracy) of measurement. But recent writers on domain generalizability point out that in most situations one is not concerned with the accuracy of a single rating. Rather, one would like to know how consistent or generalizable is a rating across all possible raters (who may have different points of view). That would be important to know if one were to make inferences about the consistency with which a given rating represents a subject in many different circumstances. One is thus led to using a subject's average rating across all possible judges as the most representative rating of him, and subsequently one is concerned with the degree to which a single rating is similar to this average rating.

One way of defining the generalizability of a randomly selected single rater's ratings is to let this generalizability be the expected squared correlation between a single rater's ratings and the average ratings across raters. Another index of generalizability is the expected correlation between the ratings of pairs of randomly selected raters from the population of raters. The latter index may be generalized as follows: Given any single-valued, estimable function of a random sample of k random variables from a universe of random variables, the generalizability of that function is the expected value of the correlation between values of the function computed from two independent samples of k variables.

However, the domain-generalizability concept is not confined to dealing with the variability in scores due to the variability in judges. One could hold judges and subjects constant and then study the variability in ratings due to variability in the traits selected to be rated. In this case one might hypothesize that the traits to be rated represent samples of traits from a universe of traits representing repeated attempts to measure a limited number of personality dimensions. Hence in random samples of n traits one would expect to find some personality dimensions appearing over and over again.

Suppose we want to pick a single trait measure, on the basis of a sample of n trait measures, that would have maximum generalizability. This would imply finding a trait measure having a maximum alpha coefficient. So that we need not limit ourselves to considering just the raw trait measures in our search for a trait measure having maximum generalizability, consider that we could find a linear composite trait measure derived from the n original trait measures.

Lord (1958) considers such a problem as we have just outlined here. He begins with the general formula for Cronbach's coefficient alpha, which is

$$\alpha = \frac{n}{n-1}\left(1 - \frac{\sum_{i=1}^{n} V_i}{V_t}\right) \tag{9.109}$$

where
n is the number of components in the composite variable
V_t is the variance of the composite
V_i is the variance of the weighted scores on individual component i

We have already shown how to express in matrix notation the total variance of a composite variable in terms of its component variables, which we will give here as $V_t = \mathbf{w}'\boldsymbol{\Sigma}\mathbf{w}$, where $\mathbf{w}$ is the $n \times 1$ weight vector containing the weights to assign each component variable and $\boldsymbol{\Sigma}$ is the $n \times n$ covariance matrix for the n component variables. If v_i^2 denotes the variance of the ith-component variable, then, after weighting, one can show that $V_i = w_i v_i^2$ where w_i is the weight assigned to the ith component variable. If we designate a matrix $\mathbf{V}^2$ as $[v_i^2]$, we have an $n \times n$ diagonal matrix with the v_i^2 as the elements in the principal diagonal. Then the expression for the sum of the V_i in Equation 9.99 can be expressed in matrix notation as $\mathbf{w}'\mathbf{V}^2\mathbf{w}$. Using these new expressions, we substitute them into Equation 9.99 to obtain

$$\alpha = \frac{n}{n-1}\left(1 - \frac{\mathbf{w}'\mathbf{V}^2\mathbf{w}}{\mathbf{w}'\boldsymbol{\Sigma}\mathbf{w}}\right) \tag{9.110}$$

The problem that now faces us is to find the weight vector, $\mathbf{w}$, which will maximize the alpha coefficient. In order to solve this problem, we must place the restriction on Equation 9.100 that $\mathbf{w}'\boldsymbol{\Sigma}\mathbf{w}$ is a constant. Kaiser and Caffrey (1965, p. 6) show that, to maximize Equation 9.110 under this restriction, it is sufficient to maximize the reciprocal expression

$$\gamma^2 = \frac{\mathbf{w}'\boldsymbol{\Sigma}\mathbf{w}}{\mathbf{w}'\mathbf{V}^2\mathbf{w}}$$

Then they show that

$$\frac{\partial \gamma^2}{\partial \mathbf{w}} = \frac{2(\boldsymbol{\Sigma}\mathbf{w} - \gamma^2\mathbf{V}^2\mathbf{w})}{\mathbf{w}'\mathbf{V}^2\mathbf{w}} \tag{9.111}$$

Since we wish to find the maximum, we set Equation 9.111 equal to zero and obtain the equation

$$(\boldsymbol{\Sigma} - \gamma^2\mathbf{V}^2)\mathbf{w} = \mathbf{0} \tag{9.112}$$

This equation is of the same form as Equation 9.24, derived in connection with weighted principal components. But in connection with Equation 9.24 we found a more convenient equation to solve for, which applied here would be

$$(\mathbf{V}^{-1}\boldsymbol{\Sigma}\mathbf{V}^{-1} - \gamma^2\mathbf{I})\mathbf{m} = \mathbf{0} \tag{9.113}$$

where $\mathbf{m} = \mathbf{V}\mathbf{w}$ as in Equation 9.29. But this is equivalent to

$$(\mathbf{R} - \gamma^2\mathbf{I})\mathbf{m} = 0 \tag{9.114}$$

which is the equation to solve to find the eigenvectors and eigenvalues of the matrix **R**, the correlation matrix. Since in most psychological problems the variance of a variable has no intrinsic meaning, we can assume that the variances of the n variables are arbitrarily set equal to 1. Then Equation 9.113 is equal to Equation 9.114 regardless, and **m** is equivalent to **w**.

From these considerations Lord (1958, p. 294) concludes that a composite variable equal to the unstandardized first principal component of the correlation matrix for n variables would have the maximum generalizability of any composite variable derivable from the original n variables. Moreover, the generalizability coefficient for this principal component would be (Kaiser and Caffrey, 1965, p. 6)

$$\alpha = \frac{n}{n-1}\left(1 - \frac{1}{\gamma^2}\right) \tag{9.115}$$

From Equation 9.115, we see that if γ^2 is less than 1 then $1/\gamma^2$ is greater than 1, and, consequently, the alpha coefficient would be negative, indicating a negative generalizability. Hence this suggests that in a principal-components analysis, we should retain those principal components having a positive generalizability in the sense of Cronbach's alpha coefficient, which would be all those having eigenvalues greater than 1. This conclusion adds further foundation to using Guttman's first lower bound as a basis for retaining principal components.

The meaning of this conclusion is that, to maximize the generalizability of a single variable representing n variables, one should make the single variable equivalent to the most general component among the n variables. The alpha coefficient corresponding to this general component indicates the degree to which one would expect to find this component in random samples of n trait variables taken from the infinite domain of trait variables. Although it has not yet been demonstrated mathematically, the present author has the hunch that the alpha coefficient corresponding to the first principal component may represent the average correlation between all pairs of first principal components in pairs of random samples of n variables taken from the domain of variables. However, the interpretation to be given to the alpha coefficients for subsequent principal components is not as obvious, although they may well be analogous.

9.8.1 Alpha Factor Analysis

Kaiser and Caffrey (1965) capitalized on the work of Lord (1958) in connection with the relationship of Cronbach's alpha coefficient to principal components in order to develop a new method for conducting common-factor analysis. If the matrix $\mathbf{\Sigma}$ in Equation 9.113 is taken to be the covariance matrix of common-factor scores $(\mathbf{R}-\mathbf{\Psi}^2)$, we have a means for estimating the generalizability of common factors as with principal components. Since the matrix $\mathbf{V}^2=[\text{diag } \Sigma]$ in Equation 9.112, this would correspond in the present application to the diagonal matrix $\mathbf{H}^2$ of communalities. Thus we would have the equation

$$[\mathbf{H}^{-1}(\mathbf{R}-\mathbf{\Psi}^2)\mathbf{H}^{-1}-\gamma^2\mathbf{I}]\mathbf{m}=\mathbf{0} \tag{9.116}$$

and would have to solve for γ^2 and $\mathbf{m}$. In this case $\mathbf{m}=\mathbf{Hw}$, or $\mathbf{w}=\mathbf{H}^{-1}\mathbf{m}$, analogous to Equation 9.24.

A principal-axes factor-loading matrix $\mathbf{\Lambda}_r$, for r generalizable common factors is given, analogous to Equation 9.72, as

$$\mathbf{\Lambda}_\alpha=\mathbf{HM}[\gamma] \tag{9.117}$$

where

- $\mathbf{M}$ is a matrix whose columns $\mathbf{m}$, are the first r eigenvectors of the matrix $\mathbf{H}^{-1}(\mathbf{R}-\mathbf{\Psi}^2)\mathbf{H}^{-1}$ having eigenvalues γ_i^2 greater than 1
- $[\gamma]$ is a diagonal matrix with the square roots of the first r eigenvalues in the principal diagonal

The communalities in the matrix $\mathbf{H}^2$ are, of course, unknown. Kaiser and Caffrey (1965, p. 7) suggest an iterative procedure to solve for them. They first invert the correlation matrix $\mathbf{R}$ and, from the result, derive the squares of the multiple-correlation coefficients for predicting each variable from the $n-1$ other variables in the matrix to serve as initial estimates for the communalities. They then use these values for the first $\mathbf{H}^2_{(0)}$ and solve Equation 9.116, obtaining finally an $n\times r$ "factor matrix" $\mathbf{K}_{(0)}=\mathbf{M}_{(0)}[\gamma]_{(0)}$, where r corresponds to the first r largest eigenvalues of $\mathbf{H}^{-1}_{(0)}$ $(\mathbf{R}-\mathbf{\Psi}^2)$ $\mathbf{H}^{-1}_{(0)}$ greater than 1. [According to Sylvester's law of inertia (cf. Chapter 7) r will not change in subsequent iterations.] In subsequent iterations new estimates of the communality matrix are given as

$$\mathbf{H}^2_{(j+1)} = \mathbf{H}^2_{(j)}[\text{diag}\,\mathbf{KK}']$$

These are applied anew to the matrix **R** as in Equation 9.116, until the row sums of squares of **K**, all converge to 1 (or sufficiently close to 1). Then one finds the final factor-loading matrix in the metric of the original variables by Equation 9.107. Kaiser and Caffrey (1965, pp. 12–13) give further details of this computational procedure.

9.8.2 Communality in a Universe of Tests

Although Kaiser and Caffrey (1965) are concerned with the problem of inferring the number of common factors in the universe of tests, they do not consider the problem of the communality of a test with respect to the larger aggregate of tests. But determining the number of common factors in samples of n tests depends upon estimating the universe common variance in the n tests.

To get a better grasp of the problem, consider that the correlations among a particular sample of n tests taken from the larger aggregate of tests are arranged into a correlation matrix. Consider also that all the tests in the large aggregate are arranged into a "super" correlation matrix. Then it follows that the correlation matrix for the particular sample of n tests is a submatrix of the larger correlation matrix. Now, if the submatrix representing the sample of n tests is factor-analyzed separately, communalities inserted in its principal diagonal will be determined by just the tests included in the sample. But if the submatrix is inserted into the supermatrix, the communalities of the n tests will be different. In fact, these communalities will be based upon all the possible tests in the aggregate and will be generally larger in value than those in the submatrix analyzed separately. They will never be less in value, because common variance cannot be taken away from a test by the inclusion of additional tests in the matrix. From this we can conclude that, "when inferring with respect to a universe of tests, the communalities of tests in a sample of tests taken from the universe of tests are lower-bound estimates of the communalities that would be found for the same tests if the whole universe of tests were factor-analyzed."

The above conclusion suggests that in practice most factor analyses underestimate the common variance of tests in the universe. An important consequence of this could be that the number of common factors is also underestimated for a given sample of n tests.

What follows is an argument outlined previously by the author (Mulaik, 1965, 1966), which states that, given a set of assumptions about the nature of a universe of tests, a logical analysis of the upper and lower bounds that might be established a priori for the communality of a test in that universe will converge on a unique value.

First, we must establish the properties of a universe of tests. By a universe of tests we mean an aggregate of all the logically possible tests that might be assembled to measure performance in some given domain of behavior. Here the universe of tests serves as a construct, just as an infinite population of subjects serves as a construct.

Now, we must make clear what we mean by "all the logically possible tests" in the above definition. For one thing, this concerns the eligibility of tests to enter the universe of tests. Ghiselli (1964, pp. 242–245) considers this problem in greater detail than most and concludes that there is no simple unambiguous rule whereby one can determine whether a given test can be eligible or not for inclusion in a universe of tests. But he concludes that perhaps the best definition of a universe of tests is given by whatever sample of tests one chooses to study for a given domain of performance. By this he implies that characteristics of the sample of tests are typical of the universe of tests. Although the present author generally concurs with Ghiselli's analysis of a universe of tests, he is not willing to go so far as to say that a given sample of tests is typical of the universe of tests, since this involves making assumptions about the sampling process for selecting tests for study, which to be validated require just the knowledge about the whole universe of tests that our present ambiguity concerning the eligibility of tests precludes. The present author simply assumes that whatever set of tests that one chooses for study represents a sample, however typical, from the universe of tests measuring the domain of performance in question.

Given a particular sample of tests, what could we infer about other tests' possible eligibility for inclusion in the universe of tests? We can infer the least about the eligibility of tests that are least like any of those in the sample. But it seems reasonable to assume that any additional tests that are practically identical to any test in the sample have the greatest likelihood of being eligible for inclusion in the universe of tests.

Up to now we have not considered tests that are alike in having the same true scores for all individuals but which differ in their error-score parts. But it seems logical that all such tests would be eligible for inclusion in the universe of tests. Moreover, it also seems logical that, for every family of tests that all have the same true score, there is at least one such test that has an error-variance term almost equal to zero, that is, a perfectly reliable test. And it also seems not unreasonable to suppose that there logically exists an unlimited number of tests that could be constructed having the same true scores but with different error-variance terms. All such tests are eligible for inclusion in the universe of tests.

We are now in a position to establish more rigorously assumptions about a universe of tests. These assumptions are as follows:

Assumption 1. The number of true-score dimensions in the universe of tests is fixed and finite.

Assumption 2. For every test in the universe of tests there exists at least one other test in the universe of tests that differs from it with respect to true-score parts by only an infinitesimally small amount.

Assumption 3. For every test in the universe of tests there exists an unlimited number of other tests that have the same true-score parts but which differ from one another in having varying degrees of error variance, from almost zero error variance to appreciable error variance.

These assumptions permit us to deduce the following: For every (almost) perfectly reliable test in the universe of tests there is at least one other (almost) perfectly reliable test correlated (except for an infinitesimally small discrepancy) perfectly with it, and for every fallible test there is a corresponding perfectly reliable test.

We also can deduce a definition for a universe of tests: A universe of tests is a collection of all logically possible tests associated by a psychometrician with a given domain of performance. The number of tests in the universe of tests is unlimited, and for every test in the universe of tests there exists at least one other test that is (almost) perfectly reliable and that has the same true-score part.

Since we are concerned with making inferences about the common dimensions in the universe of tests from a finite battery of tests, we must now consider establishing the upper and lower bounds for the communality of tests in such a universe. To do this we shall use a roundabout procedure, wherein we will first determine the bounds for the multiple-correlation coefficient for predicting a test in the universe of tests from all other tests in the universe of tests.

The upper bound to the multiple-correlation coefficient for predicting a test from the other tests in the universe of tests is the index of reliability of the test (cf. Gulliksen, 1950, p. 23).

A multiple-correlation coefficient is never less than the absolute magnitude of the largest correlation between the criterion and its predictors. This establishes the lower bound to the multiple-correlation coefficient for predicting a test in a universe of tests from all other tests in the universe of tests. Since for every test there exists, by the definition of a universe of tests, at least one other test that measures exactly the same behavior as the test in question but with (almost) perfect reliability, this lower bound can be established as equivalent to the index of reliability.

Since the upper and lower bounds are the same, we can conclude that the multiple-correlation coefficient for predicting a test from all other tests in the universe of tests is the index of reliability of the predicted test.

But since the index of reliability is the square root of the reliability coefficient, we can also conclude that the square of the multiple-correlation coefficient for predicting a test from all other tests in a universe of tests is the reliability coefficient of the predicted test.

Finally, we can rely upon Guttman's theorem (cf. Equation 9.56) that the square of the multiple-correlation coefficient for predicting a test from all other tests in a battery of tests approaches the communality of the predicted test as a limit as the number of tests in the battery increases without bound. By this theorem we establish the equivalence of the communality of a test in a universe of tests with the reliability coefficient of the test.

9.8.3 Consequences for Factor Analysis

Since in a universe of tests the communality of a test is also the reliability of the test, we have the following interesting result: In a universe of tests common-factor analysis, image analysis, canonical-factor analysis, and image-factor analysis all converge to the model of classic reliability theory which partitions variables into true-score and error-score parts.

Moreover, consider the following novel definition for reliability: The reliability of a test is the square of the multiple-correlation coefficient for predicting the test from all the other tests in the universe of tests. LaForge (1965) discusses this concept, crediting Ledyard Tucker with its origin. In a way this definition can be used as an arbitrary postulate about the nature of reliability for a test in a universe of tests. Here we have derived the definition from the classic theory of reliability and certain assumptions about the eligibility of tests to be included in a universe of tests.

But despite these remarkable relationships between the various factor-analytic models and classic reliability theory in a universe of tests, one may question whether the principal aim of factor analysis is to make inferences about the factors in a universe of tests. Many factor-analytic studies seem concerned not with discovering all the latent factors in a domain of tests but rather with isolating just those latent factors which are common to certain distinct tests. In these studies the researcher has some idea of the kinds of factors he wants to study and constructs or selects his tests so as to have more than, say, three tests for each pertinent common factor and no more than one test for any irrelevant factor. Irrelevant factors thus appear only in the unique variance of the set of tests studied. If any psychometric inference is involved in such a study, it is concerned not with all the factors in a domain but with just those factors that are common to certain classes of tests.

As pointed out earlier in this chapter, much of the concern with domains and universes of tests arises out of the hope that such domains and universes will permit a completely determinate common-factor model. Although this may be true, the difficulty is that a common-factor model for the domain may have little relevance to the empirical questions with which most factor-analytic researchers are concerned. On the other hand, the factor-analytic models that relate to these questions are indeterminate. What can the researcher do when faced with a choice between an irrelevant determinate model and a relevant indeterminate model? The answer is simple: Choose the relevant but indeterminate model.

Choosing the relevant but indeterminate model does not mean the problem of indeterminacy is unimportant. Indeterminacy is a serious problem with which one must struggle. But it is a problem with which the researcher must live if he is to use factor analysis for relevant theoretical research that generalizes beyond a given set of variables and data.

10

Factor Rotation

10.1 Introduction

In Chapters 8 and 9, we considered obtaining solutions for the factors and corresponding factor-pattern matrices based on the eigenvectors and eigenvalues of a correlation or covariance matrix. Such solutions for the factor-pattern matrix are characterized by almost all of the variables loading substantially on the first common factor, which is considered a "general factor," since it is a factor found in almost all of the observed variables. On subsequent factors, the variables have decreasing magnitudes for their loadings, and each factor accounts for a decreasing amount of common variance among a smaller and smaller number of variables. The problem is, how shall we interpret these factors?

The earliest factor analysts during the first three decades of the twentieth century were British psychologists who were intellectual descendents of Charles Spearman, who had invented the model of common-factor analysis in 1904. They tended to favor a hierarchical interpretation of the factors. With tests of mental abilities, they interpreted the first factor, as Spearman's general factor g, found among all or almost all of the variables. Succeeding factors were found among decreasing numbers of observed variables. For tests of mental ability, this was the way, they believed, the mind was organized. Later they modified the solution, in what was known as the multiple groups solutions, in which the first factor was still a general factor, but subsequent factors were associated with somewhat nonoverlapping sets of variables, known as "group factors."

But the early factor analysts also became aware of another problem, one that was not evident when there was only one common factor as in Spearman's case. The solution for the common factors is not mathematically unique. There is an indeterminacy in the common-factor model. The common factors are basis vectors of a vector space. The observed variables are different linear combinations of the common-factor vectors. But any set of basis vectors can be derived from any other by a linear transformation. In fact, the number of sets of basis vectors is infinite. So, which basis set of factors shall we use when we seek to provide interpretations for the factors as, say, common causes of the observed variables in the world?

This form of indeterminacy, known as "rotational indeterminacy," is represented in the common-factor model as follows: The common-factor model equation is $\mathbf{Y} = \mathbf{\Lambda}^*\mathbf{X}^* + \mathbf{\Psi}\mathbf{E}$, where $\mathbf{Y}$ denotes the $n \times 1$ random vector of observed variables, $\mathbf{\Lambda}^*$ is the $n \times r$ matrix of unrotated (e.g., principal-axes) factor-pattern coefficients, $\mathbf{X}^*$ is the $r \times 1$ random vector of unrotated common factors, $\mathbf{\Psi}$ is the $n \times n$ diagonal matrix of unique pattern coefficients, and $\mathbf{E}$ is the $n \times 1$ random vector of unique-factor variables. Then it is possible to find infinitely many distinct nonsingular $r \times r$ transformation matrices $\mathbf{M}$ such that

$$\mathbf{Y} = \mathbf{\Lambda}^*\mathbf{X}^* + \mathbf{\Psi}\mathbf{E} = \mathbf{\Lambda}^*\mathbf{M}\mathbf{M}^{-1}\mathbf{X}^* + \mathbf{\Psi}\mathbf{E} = \mathbf{\Lambda}\mathbf{X} + \mathbf{\Psi}\mathbf{E}$$

where

$$\mathbf{\Lambda} = \mathbf{\Lambda}^*\mathbf{M}$$
$$\mathbf{X} = \mathbf{M}^{-1}\mathbf{X}^*$$

In other words, the common-factor model is "rotationally indeterminate." We will call a nonsingular linear transformation of the unrotated factor variables a "rotation" of the factors. The term comes from the idea that the factors have a common origin, and their positions in the vector space may be altered by rotating them around the origin, like the hands of a clock. Also, as the factors are rotated, an inverse of the transformation is applied to the factor-pattern matrix. Now, since there are infinitely many possible rotations that would produce factors that satisfy the common-factor model, which is the "correct" solution? Mathematically alone, there is no unique solution. That being the case, do the factors represent anything in the world other than arbitrary artifacts of superimposing the common-factor model on some data? The problem became more acute when it was recognized that the principal-axes factors and centroid factors would not yield the same solutions, if some of the observed variables are replaced with others from the domain of variables spanned by the same set of common factors.

10.2 Thurstone's Concept of a Simple Structure

Factor analysis came into the forefront in the United States in the 1930s in the laboratory of L.L. Thurstone at the University of Chicago. He took a different approach to the problem of defining the most desirable factor solution. What he sought to emphasize was not a demonstration of hierarchical organization of factors, but rather an analysis of the factors into elemental components. The solution he produced was known as the "simple-structure solution." The essence of simple structure, as Thurstone regarded it, was parsimony in scientific explanation. The common-factor model was most appropriately applied when, for any given observed variable, the model only used the smallest number of factors to account for the common variance of the variable. Thus, if in a factor analysis of n variables r common factors were obtained, then Thurstone deemed the factor solution ideal when each observed variable required fewer than r common factors to account for its common variance. By the same token, when it came to interpreting the common factors by noting the observed variables associated with each respective factor, parsimony of interpretation could be achieved when each factor was associated with only a few of the observed variables. Since 1960 American and British factor analysts have come to accept the simple-structure solution as the solution that resolves most of the problems of rotational indeterminacy.

The basic idea that in a simple-structure solution each observed variable is accounted for by fewer than the total number of common factors is not hard to understand. What is frequently difficult to understand at first is Thurstone's technical implementation of the simple-structure concept, because this involves several new terms and ideas from the geometry of vectors spaces of more than three dimensions. Actually this implementation need not be difficult to master if one realizes that it follows directly from the basic idea that one must account for the common part of each variable by a linear combination of fewer than r common factors. Where will the observed variables be, then, in the common-factor space in that case?

They will be in subspaces of the common-factor space known as "coordinate hyperplanes," with each of the common factors at the intersection of $r-1$ coordinate hyperplanes. Each common factor would be a basis vector in the set of basis vectors for each of these $r-1$ coordinate hyperplanes intersecting with it. This solution made it possible for the common factors of the simple-structure solution to be an objective solution for the factors that was not dependent on the particular set of variables selected for analysis. Different selections of variables from the same domain would yield the same factors.

Thurstone believed that with simple structure there was a way to determine factors that would be invariant across different selections of variables from a variable domain spanned by a set of common factors. In contrast, he believed that defining factors as, say, the principal axes (based on eigenvectors and eigenvalues of the reduced correlation matrix) would not provide an invariant definition for the common factors of the domain, since the principal axes would change from one selection of variables to another from the domain.

Thurstone argued that a domain of variables is defined as all those variables that are functions of a common set of common factors, $X_1,\ldots,X_r$. Now, if one could assume that in a domain of variables, no observed variable is a function of all the common factors of the domain, then in the common-factor space of any selection of observed variables from the domain, the observed variables would be contained completely within different subspaces of $r-1$ dimensions or less of the common-factor space for the observed variables.

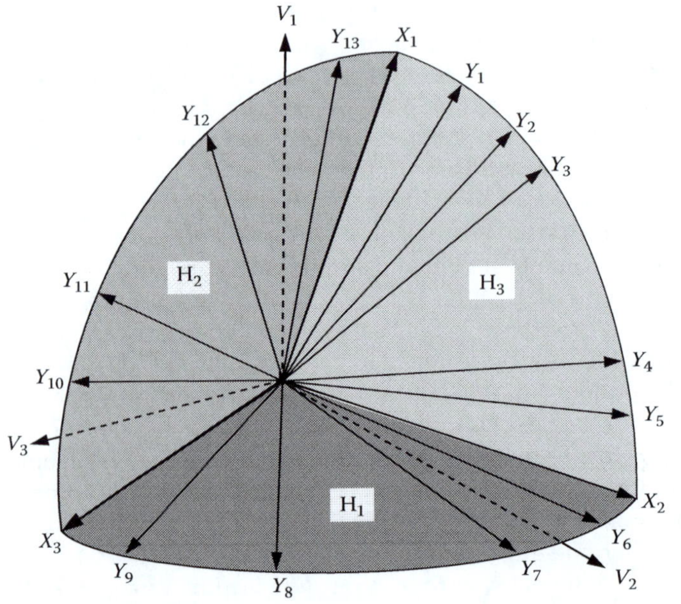

FIGURE 10.1
Three-dimensional example of an ideal simple-structure factor solution. Vectors X_1, X_2, and X_3 are "common factors" at intersections of "coordinate hyperplanes" H_1, H_2, and H_3, which in turn are orthogonal, respectively, to "reference vectors" V_1, V_2, and V_3. Vectors $Y_1,\ldots,Y_{13}$ are common parts of observed variables, each of which is in one of the three hyperplanes, implying they are dependent on no more than two factors.

Furthermore, the various subspaces containing the observed variables would be spanned by different subsets of the common factors. For example, if variable Y_1 is a function of just factors X_1, X_2, and X_3 of the common-factor space spanned by factors $X_1,\ldots,X_r$, then it is contained completely within the subspace spanned by X_1, X_2, and X_3. But any such subspace would be contained in at least one subspace of $r-1$ dimensions, where r is the number of common factors (Figure 10.1).

Thurstone then sought to identify a specific set of subspaces of $r-1$ dimensions each, known as the set of "coordinate hyperplanes." A coordinate hyperplane would be a subspace of the common-factor space that was spanned by all but one of the common factors of the common-factor space. There would be r such coordinate hyperplanes, one for each factor not contained by the respective hyperplane.

In other words, take a given coordinate hyperplane, H_i, say, spanning $r-1$ dimensions. We may identify H_i by the one common factor, X_i, not contained in the set of $r-1$ common factors $X_1,\ldots,X_{i-1}$, $X_{i+1},\ldots,X_r$ serving as basis vectors for the hyperplane H_i. Now, while X_i is missing from H_i, X_i will be found in all of the remaining $r-1$ coordinate hyperplanes $H_1,\ldots,H_{i-1},H_{i+1},\ldots,H_r$.

This is because X_i will be in each of them since it cannot also be excluded from any of these hyperplanes without reducing the dimensionality of the hyperplane below $r-1$. Since the factors spanning these respective $r-1$ coordinate hyperplanes are basis vectors for these hyperplanes (subspaces), then factor X_i is a basis vector of each one of these coordinate hyperplanes and must therefore lie at the intersection of the $r-1$ coordinate hyperplanes $H_1,\ldots,H_{i-1},H_{i+1},\ldots,H_r$.

Thurstone reasoned that if he could identify the r coordinate hyperplanes of the common-factor space, he could identify any one of the factors as that vector at the intersection of some set of $r-1$ of the coordinate hyperplanes. The device he settled upon for identifying the coordinate hyperplanes was a "reference vector," which was a hypothetical vector V_j inserted into the common-factor space onto which one could project the observed-variable vectors to discover those variables whose projections onto the reference vector were zero in length, indicating that they were orthogonal to the reference vector. He would insert the reference vector into the space and move it around until he found a position for it such that subgroups of the observed variables would have zero projections onto it, indicated by their having zero cosines for the angles between them and the reference vector. Then the subspace of $r-1$ dimensions orthogonal to the reference vector would be a candidate coordinate hyperplane. He would use the reference vector method to identify r different subsets of the variables that were orthogonal, respectively, to some one of r reference vectors $V_1,\ldots,V_r$ in the common-factor space, and these reference vectors would in turn identify the various coordinate hyperplanes as those subspaces of $r-1$ dimensions orthogonal to the respective reference vectors. The factors would then be those vectors at the intersections of each of the r distinct subsets of $r-1$ coordinate hyperplanes.

Thurstone reasoned that if the domain were defined by a specific set of r factors, and no variable sampled from the domain ever depended on more than $r-1$ of these factors, then any sample of such variables from the domain would always satisfy a simple structure. The variables from the domain would fall in the same r coordinate hyperplanes. Consequently, any such sample of variables would always define the same hyperplanes, even though in different samples the variables defining these hyperplanes would be different. As a result, the same set of common factors would be defined as those vectors at the respective intersections of each of the r sets of $r-1$ coordinate hyperplanes.

Consequently, variables satisfying the condition of simple structure invariantly determine the same set of common factors of the domain, and the resulting factors thus have an objective status in not depending on any one sample of such variables from the domain. It must be realized, however, that the assumption that a domain of variables satisfies simple structure is a rather strong assumption that might not hold with some domains of variables. Thurstone was simply banking on the possibility that nature is relatively simple in having no phenomenon depend simultaneously upon all the explanatory constructs that might be invoked to understand its composition. Chemists, for example, have long assumed that there are few or no compounds in nature that contain every element in the periodic table. Actually, to identify the coordinate hyperplanes of a simple structure, it is not necessary that all variables depend on fewer than $r-1$ common factors of the domain, only that most of them do.

10.2.1 Implementing the Simple-Structure Concept

Let us see now how the simple-structure concept is implemented: Given the common-factor model equation $\mathbf{Y}=\mathbf{\Lambda}^*\mathbf{X}^* + \mathbf{\Psi}\mathbf{E}$, where $\mathbf{Y}$ denotes the $n\times 1$ random vector of observed variables, $\mathbf{\Lambda}^*$ is the $n\times r$ matrix of unrotated (e.g., principal-axes) factor-pattern coefficients, $\mathbf{X}^*$ is the $r\times 1$ random vector of unrotated common factors, $\mathbf{\Psi}$ is the $n\times n$ diagonal matrix of unique pattern coefficients, and $\mathbf{E}$ is the $n\times 1$ random vector of unique-factor variables, let us now define the common-factor part of each observed variable as $\dot{\mathbf{Y}}_i$ so that we may write $\dot{\mathbf{Y}}=\mathbf{\Lambda}^*\mathbf{X}^*$. Since each one of the r distinct coordinate hyperplanes is associated with one of the reference vectors, the reference vectors themselves also form a basis for the common-factor space. So, we may write $\mathbf{V}=\mathbf{L}'\mathbf{X}^*$ to indicate that the reference vector variables in the constructed random vector $\mathbf{V}$ are linear transformations of the unrotated common-factor variables in $\mathbf{X}^*$, where $\mathbf{L}$ is an $r\times r$ transformation matrix of full rank. Presuming that our variables are in deviation score form (having zero means), we may write the $n\times r$ reference-structure matrix, or the matrix of cosines (correlations) between the reference vectors and the original variables as $\mathbf{V}_\mathrm{R} = E(\mathbf{Y}\mathbf{V}') = E[(\mathbf{\Lambda}^*\mathbf{X}^*+\mathbf{\Psi}\mathbf{E})\mathbf{X}^{*\prime}\mathbf{L}] = \mathbf{\Lambda}^*\mathbf{R}_{X^*X^*}\mathbf{L}+\mathbf{\Psi}\mathbf{R}_{EX^*}\mathbf{L} = \mathbf{\Lambda}^*\mathbf{L}$, with $\mathbf{R}_{X^*X^*} = \mathbf{I}$,that is, the unrotated principal-axes common factors are presumed to be mutually uncorrelated and to have unit variances, while, on the other hand, $\mathbf{R}_{EX^*} = \mathbf{0}$, because unique and common factors are uncorrelated by assumption of the model. Evidently the reference-structure matrix $\mathbf{V}_\mathrm{R}$ is a linear transformation $\mathbf{V}_\mathrm{R}=\mathbf{\Lambda}^*\mathbf{L}$ of the original unrotated factor-pattern matrix (we use "**V**" in the name of this matrix because it has become tradition to do so; it should not be confused with the vector of reference vector variables **V**).

Thurstone sought the transformation matrix **L** that would yield a reference-structure matrix $\mathbf{V}_\mathrm{R}$ satisfying certain criteria:

1. Each row of $\mathbf{V}_\mathrm{R}$ should have at least one zero. Each variable should be in a subspace of at most $r-1$ dimensions.
2. Each column k of $\mathbf{V}_\mathrm{R}$ should have a set of at least r linearly independent observed variables whose correlations with the kth reference axis are zero. This overdetermines the reference vector $\mathbf{y}_k$.
3. For every pair of columns in $\mathbf{V}_\mathrm{R}$ there should be several zero entries in one column corresponding to nonzero entries in the other column. This further ensures the distinctness of the reference axes.

4. When four or more common factors are found, each pair of columns of $\mathbf{V}_R$ should have a large proportion of corresponding zero entries. This implies each reference vector will be related to only a small number of the observed variables and further forces the separation of the variables into distinct clusters.
5. For every pair of columns there should be only a small number of corresponding entries in both columns that do not vanish. Variables should be factorially as simple as possible.

Once the reference vectors $\mathbf{V}$ are found, the common factors may be found as a linear transformation of the reference vectors, i.e., $\mathbf{X} = \mathbf{T}'\mathbf{V}$. It is essential that the resulting variables in $\mathbf{X}$ have unit variances, since they are common factors. Now, $\mathbf{R}_{XX} = E(\mathbf{X}\mathbf{X}') = E(\mathbf{T}'\mathbf{V}\mathbf{V}'\mathbf{T}) = \mathbf{T}'\mathbf{R}_{VV}\mathbf{T}$. Hence $[\text{diag } \mathbf{R}_{XX}] = [\text{diag } \mathbf{T}'E(\mathbf{V}\mathbf{V}')\mathbf{T}] = [\text{diag } \mathbf{T}'\mathbf{R}_{VV}\mathbf{T}] = \mathbf{I}$. Define $\mathbf{D}_V^2 \equiv [\text{diag } \mathbf{R}_{VV}^{-1}]$. Then the $r \times r$ matrix $\mathbf{T}'$ satisfying this constraint is given by

$$\mathbf{T}' = \mathbf{D}_V^{-1}\mathbf{R}_{VV}^{-1}$$

because $[\text{diag} \mathbf{T}'\mathbf{R}_{VV}\mathbf{T}] = [\text{diag } \mathbf{D}_V^{-1}\mathbf{R}_{VV}^{-1}\mathbf{R}_{VV}\mathbf{R}_{VV}^{-1}\mathbf{D}_V^{-1}] = [\text{diag } \mathbf{D}_V^{-1}\mathbf{R}_{VV}^{-1}\mathbf{D}_V^{-1}] = \mathbf{I}$. Now, $\mathbf{R}_{VV} = E(\mathbf{V}\mathbf{V}') = E(\mathbf{L}'\mathbf{X}^*\mathbf{X}^{*\prime}\mathbf{L}) = \mathbf{L}'\mathbf{L}$, because the initial unrotated factors in $\mathbf{X}^*$ are mutually orthogonal with unit variances. Then $\mathbf{D}_V^2 = [\text{diag}(\mathbf{L}'\mathbf{L})^{-1}]$. We may simplify this further using the definition for $\mathbf{T}'$ above, as $\mathbf{R}_{XX} = \mathbf{D}_V^{-1}\mathbf{R}_{VV}\mathbf{D}_V^{-1}$. Furthermore, $\mathbf{R}_{XV} = E(\mathbf{X}\mathbf{V}') = E(\mathbf{T}'\mathbf{V}\mathbf{V}') = \mathbf{T}'\mathbf{R}_{VV} = \mathbf{D}_V^{-1}\mathbf{R}_{VV}^{-1}\mathbf{R}_{VV} = \mathbf{D}_V^{-1}$. This implies that the common factors and the reference axes are "biorthogonal" sets of variables, meaning they are two sets of variables that have the property that any vector in one set is orthogonal to all the vectors in the other set except one. If the variances of the reference vector variables are unity, this implies that the sums of squares of the columns of $\mathbf{L}$ must equal unity, i.e., $[\text{diag } \mathbf{R}_{VV}] = [\text{diag } \mathbf{L}'\mathbf{L}] = \mathbf{I}$, and the diagonal elements of $\mathbf{D}_V^{-1}$ are correlations while the elements of $\mathbf{V}_R$ are also correlations.

Let us now consider how to find the rotated factor-pattern matrix $\mathbf{\Lambda}$. We will provisionally consider that the reference axes could be regarded as a basis for the common-factor space, but we have already established that they represent a linear transformation of the original unrotated factors, i.e., $\mathbf{V} = \mathbf{L}'\mathbf{X}^*$. So let us rewrite the fundamental equation of factor analysis in terms of $\mathbf{V}$ as

$$\mathbf{Y} = \mathbf{\Lambda}^*\mathbf{X}^* + \mathbf{\Psi}\mathbf{E} = (\mathbf{\Lambda}^*\mathbf{L}'^{-1})(\mathbf{L}'\mathbf{X}^*) + \mathbf{\Psi}\mathbf{E} = \mathbf{\Lambda}_R\mathbf{V} + \mathbf{\Psi}\mathbf{E}$$

where we shall call $\mathbf{\Lambda}_R$ the reference pattern matrix. Well, we can turn around and define the common-factor model in terms of the final rotated solution, which we obtain from the reference axes solution by a transformation of the reference axes:

$$\mathbf{Y} = \mathbf{\Lambda}_R\mathbf{V} + \mathbf{\Psi}\mathbf{E} = (\mathbf{\Lambda}_R\mathbf{T}'^{-1})(\mathbf{T}'\mathbf{V}) + \mathbf{\Psi}\mathbf{E} = \mathbf{\Lambda}\mathbf{X} + \mathbf{\Psi}\mathbf{E}$$

So, $\mathbf{\Lambda} = (\mathbf{\Lambda}_R\mathbf{T}'^{-1}) = (\mathbf{\Lambda}^*\mathbf{L}'^{-1}\mathbf{T}'^{-1})$. But because

$$\mathbf{T}' = \mathbf{D}_V^{-1}\mathbf{R}_{VV}^{-1}$$

$$\mathbf{T}'^{-1} = [\mathbf{D}_V^{-1}\mathbf{R}_{VV}^{-1}]^{-1} = \mathbf{R}_{VV}\mathbf{D}_V = \mathbf{L}'\mathbf{L}[\text{diag}(\mathbf{L}'\mathbf{L})^{-1}]^{1/2}$$

So,

$$\begin{aligned}\mathbf{\Lambda} &= (\mathbf{\Lambda}^*\mathbf{L}'^{-1}\mathbf{T}'^{-1})\\ &= \mathbf{\Lambda}^*\mathbf{L}'^{-1}\mathbf{L}'\mathbf{L}[\text{diag}(\mathbf{L}'\mathbf{L})^{-1}]^{1/2}\\ &= \mathbf{\Lambda}^*\mathbf{L}[\text{diag}(\mathbf{L}'\mathbf{L})^{-1}]^{1/2}\end{aligned}$$

Now, let us go back and consider the derivation for the reference-structure matrix as a function of the original, unrotated factors:

$$\mathbf{V}_R = \mathbf{\Lambda}^*\mathbf{L}$$

Compare that with

$$\mathbf{\Lambda} = \mathbf{\Lambda}^*\mathbf{L}[\text{diag}(\mathbf{L}'\mathbf{L})^{-1}]^{1/2}$$

Evidently, then, $\mathbf{\Lambda} = \mathbf{V}_R[\text{diag}(\mathbf{L}'\mathbf{L})^{-1}]^{1/2} = \mathbf{V}_R\mathbf{D}_V$.

Although not immediately obvious, what this equation implies is that every 0 entry in $\mathbf{V}_R$ corresponds to a zero entry in $\mathbf{\Lambda}$ because the effect of the diagonal matrix $\mathbf{D}_V$ postmultiplied times $\mathbf{V}_R$ is to multiply the respective diagonal elements of $\mathbf{D}_V$ times each of the respective column elements of $\mathbf{V}_R$. And since $\mathbf{V}_R$ is determined in such a way that there must be zero elements in each of the columns in order to determine the reference axes as vectors orthogonal to subsets of the variables, then everywhere there is a zero in $\mathbf{V}_R$ there will be a zero in $\mathbf{\Lambda}$. This suggests that any procedure that tries to find $\mathbf{\Lambda}$ directly by a solution determining zero entries in its columns appropriate for a simple structure will also work. Jennrich and Sampson's direct oblimin method of oblique rotation accomplishes just this without an intermediate solution for reference axes. So, the idea of reference axes made perspicuous the criteria for a simple structure, but they are not really needed to actually find it.

10.2.2 Question of Correlated Factors

In our earlier discussions of the extraction of factors, we treated the factors as if they were mutually orthogonal. As a result the reader may have received the impression that in factor analysis the factors must always be mutually orthogonal. However, we have just seen that, if we seriously accept the criterion of simple structure for choosing a set of factors, we implicitly accept the possibility that the matrix $\mathbf{R}_{XX}$ is not diagonal, that is, the factors are correlated. Thus, even though correlated factors are linearly independent, one may raise the question of whether explaining the intercorrelations among the manifest variables in terms of correlated factors does not simply create new problems while solving old ones. In other words, does one not then have the problem of accounting for the intercorrelations among the factors?

Although this is true, let us put this question in a proper perspective by considering it in terms of the objectives of factor analysis. Ultimately, regardless of the kind of factor solution obtained—whether with correlated or uncorrelated factors—one must interpret the obtained factors in terms of a hypothetical domain of latent variables that one believes act upon the variables studied. It is desirable to regard this hypothetical domain of variables as consisting of univocal elemental variables, which is to say relative to the original set of variables, variables that themselves do not imply the existence of other variables. Such variables, of course, would be mutually orthogonal.

Now, relative to the variables selected for analysis, the uncorrelated unique factors are certainly univocal and elemental. They may be interpreted in terms of whatever is unique to each observed variable.

Next, if one assumes the existence of a set of latent, elemental variables pertinent to the observed variables, it is quite possible for the common-factor space of the observed variables to have fewer dimensions than the number of latent elemental variables that are common to these variables. This will occur whenever, in the set of observed variables studied, some of the latent, elemental variables are always confounded with other latent, elemental variables. Consequently, the common factors obtained for the observed variables will represent independent linear combinations of the latent, elemental variables.

For example, suppose eight latent, elemental variables $\{S_1,\ldots,S_5, T_1, T_2, T_3\}$ are involved in a set of observed variables in such a way that the observed variables yield only five common factors. Suppose, furthermore, that a set of common factors $X_1,\ldots,X_5$ obtained for the observed variables are related to the eight latent, elemental variables by the following equations:

$$X_1 = T_1 + S_1$$

$$X_2 = T_1 + S_2$$

$$X_3 = T_1 + T_2 + S_3$$

$$X_4 = T_2 + T_3 + S_4$$

$$X_5 = T_3 + S_5$$

Then, although the eight latent, elemental variables $S_1,\ldots,S_5$, T_1, T_2, T_3 are mutually orthogonal, the five independently obtained common factors, $X_1,\ldots,X_5$, are not. For example, X_1, X_2, and X_3 are intercorrelated, having in common T_1. Furthermore, X_3 and X_4 are intercorrelated, sharing in common T_2, and X_4 and X_5 are in turn intercorrelated because of T_3. On the other hand, X_1 and X_4 are uncorrelated, because they share no latent, elemental variables in common.

Further, observe that in this example each of the common factors $X_1,\ldots,X_5$ contains a latent, elemental variable unique to it. For example, X_1 contains S_1, which is not associated with any other factor; X_2 contains S_2; X_3 contains S_3; and so on. One might suppose that, if one performed a factor analysis of the correlations among the factors $X_1,\ldots,X_5$, the unique factors of this new analysis would be the latent, elemental variables $S_1,\ldots,S_5$, whereas the common factors of this analysis would be the latent, elemental variables T_1, T_2, and T_3. We will concern ourselves further with such analyses of intercorrelations among factors presently.

Before leaving this example, however, let us also see what would happen if the latent, elemental variables s_1 and s_2 were absent from the variables studied. In this case, factors X_1 and X_2 would collapse into the single factor T_1. Still, the resulting factor T_1 would correlate with factor X_3, because T_1 also appears in that factor. On the other hand, the factors $X_3,\ldots,X_5$ would remain the same and would display the same interrelationships.

Additionally, consider what would happen if X_4 did not contain T_3: then X_5 would be uncorrelated with X_4 and all the other factors as well. In effect, X_5 would then consist of two elemental variables, T_3 and S_5, which would be so totally confounded as to be inseparable in the variables studied.

Let us now move from the abstract to an empirical example. Suppose one administers several tests of addition and multiplication of numbers to young elementary-school children who have learned addition and multiplication by a process of memorizing what number is to be associated with a given pair of numbers under a given operation. For example, 5 is to be associated with the pair (2,3) under addition, and 6 is to be associated with (2,3) under multiplication. One would expect the skills of addition and multiplication learned under these circumstances to be independent. Therefore, a factor analysis of intercorrelations among the tests of these skills should yield two well-defined factors, which, after rotation to a simple structure, should align themselves with the addition- and multiplication-test variables, respectively. However, one should not be surprised if an oblique simple-structure solution indicates that these two factors are slightly correlated, either positively or negatively. What interpretation should we give the intercorrelations among the factors?

Consider that in taking these tests a child not only needs to know the relatively specific skills of addition and multiplication but also needs the general skill of discriminating among number symbols. If he or she occasionally confuses 8s with 3s and 5s, or 6s with 9s, this will tend to affect his performance similarly on both addition and multiplication tests. On the other hand, we should also expect interference effects transferring from one kind of test to the other. For example, having learned to respond under addition to the pair (2,3) with 5, the child may also tend to respond with 5 to this pair under multiplication. In this case, there would be a tendency for negative correlation between addition and multiplication tests. But in this set of tests, the number-discriminating ability and the interference-of-transfer effects may be so confounded with the more specific addition and multiplication abilities as to be inestimable as dimensions separate from them. As a consequence the simple-structure factors may not represent pure addition and multiplication abilities but rather these respective abilities confounded with the number-discriminating ability and the interference-of-transfer effects. Since the factors have these common variables confounded with them, they will be positively or negatively intercorrelated, depending upon which of these effects predominates.

It is always possible to obtain an arbitrary solution for the primary factors, which may or may not be correlated. But the interpretation of these factors in terms of latent, elemental variables will be highly arbitrary. On the other hand, in a simple-structure solution the principle of parsimony requires each variable to be accounted for by fewer than the total number of common factors. This ensures that the resulting factors consist of functionally independent common components which are distinguished by their absence from various subsets of the variables. When identifiability conditions are met, this means that in examining the intercorrelations among simple-structure factors one can interpret what is unique to a common factor as a latent elemental variable.

The intercorrelations among factors obtained from simple-structure solutions thus indicate the presence of latent variables which are confounded with other latent variables. Hence the intercorrelations among the factors will be useful in interpretation of the factors. On the other hand, this information is lost in orthogonal approximations to simple structure.

The intercorrelations among the factors can also lend themselves to what is known as a second-order factor analysis, where one finds the common factors among the first-order (first-derived) factors by factor-analyzing their intercorrelation matrix. If the resulting second-order factors are rotated to an oblique simple structure, they, in turn, may be intercorrelated, suggesting in turn other more general latent variables, which are confounded in the second- and first-order factors and which may be extracted by a third factor analysis of the intercorrelations among the second-order factors, and so on. Of course, the interpretation of these higher-order factors is facilitated by the researcher's having an a priori model concerning his observed variables that predicts the possible existence of such higher-order factors.

When one expects factors to be correlated but finds that they are not, this suggests that an expected confounding variable has little effect on the factors. Knowledge of this fact may be important. In the example cited above, we might expect for somewhat older school children that the addition and multiplication factors are not as strongly intercorrelated because, with more practice in discriminating among numbers, they have reached a level where they have almost all mastered this skill and thus do not vary much in it.

For a fuller discussion of methods to be used in connection with second- and higher-order factor analyses the reader should consult articles by Schmid and Leiman (1957) and Humphreys et al. (1970).

10.3 Oblique Graphical Rotation

Since factor analysis was developed before computers were invented in the 1950s, and rotation to the simple structure was developed by Thurstone in the 1930s, researchers were obliged, until the 1960s, to perform rotations of factors, two at a time, by hand, using graphical plots of the variables against pairs of the factors. This was a very arduous, time-consuming task. When I was a graduate student at the University of Utah of Calvin W. Taylor (himself a student of L. L. Thurstone at the University of Chicago), we were told that there was a data set stored in the back of the laboratory that we must guard with our lives. It was Taylor's factor analysis study for his dissertation, and it had taken a large amount of his sweat and blood to conduct it. A large part of that involved several months of rotating factors by hand. Today one performs rotations by computers in a manner of seconds.

Although I devoted a half chapter of the first edition of this book (Mulaik, 1972) to graphical rotation, describing in detail how it was done, I will now only give a partial description of the process here to just provide an overview. Computers have completely taken over the process. And we should concentrate on how those are done. Nevertheless, one advantage of seeing a graphical rotation done is that it gives the student a concrete, geometric/graphical image of reference vectors and hyperplanes, which seeing the computer output of the final loadings does not show you.

For the sake of illustrating the process of oblique graphical rotation to the simple structure, the author has prepared an artificial problem involving a set of 10 variables having an exact simple-structure solution. The matrix of correlations among these 10 variables (with communalities in the diagonal) is given in Table 10.1. The orthogonal principal-axes solution of this correlation matrix yields exactly three common factors, and the principal-axes factor-structure matrix, $\mathbf{\Lambda}^*$, is given in Table 10.2. Since in all orthogonal solutions the reference-axes vectors are collinear with the common factors, we may treat this principal-axes factor-structure matrix as an initial reference-axes structure matrix, to be designated $\mathbf{V}_{R(0)}$, where the subscript in parentheses indicates the particular iteration of the rotation process in which this matrix is obtained. As for the initial reference axes, we shall let $\mathbf{V}_{(0)}$ be a random vector the rows of which correspond to the reference vectors $V_{1(0)},\ldots,V_{r(0)}$, with

TABLE 10.1

Correlation Matrix with Communalities in the Diagonal for Example of Rotation

1	.874									
2	.642	.472								
3	.741	.549	.765							
4	.487	.346	.389	.765						
5	.502	.360	.327	.531	.472					
6	.400	.283	.273	.609	.457	.501				
7	.233	.171	.463	.483	.133	.283	.902			
8	.463	.342	.637	.501	.226	.311	.831	.850		
9	.387	.278	.449	.668	.362	.480	.722	.696	.734	
10	.670	.488	.633	.642	.472	.489	.532	.634	.624	.704

TABLE 10.2

Principal-Axes Solution, $\mathbf{V}_0$, for Example of Rotation

1	.759	.498	−.225
2	.553	.364	−.185
3	.747	.157	−.427
4	.763	.007	.427
5	.533	.326	.286
6	.571	.115	.403
7	.696	−.639	−.103
8	.797	−.394	−.244
9	.774	−.317	.185
10	.836	.076	.007

the right-hand side subscript of these vectors referring to the iteration number and the left-hand side subscript to the number of the vector. The aim of the rotation process is to find a nonsingular linear-transformation matrix $\mathbf{L}$ that will transform the initial reference axes into the final reference axes of the simple-structure solution.

In other words, if $\mathbf{V}$ is the vector of reference axes that determine the simple-structure solution, then $\mathbf{V} = \mathbf{L'V}_{(0)}$. On the other hand, the same transformation matrix may also be used to derive the final reference-structure matrix $\mathbf{V}_R$ from the initial reference-structure matrix $\mathbf{V}_{R(0)}$. As proof, consider that $\mathbf{V}_R = E(\mathbf{YV'}) = \mathbf{R}_{YV}$, by definition, but that this may be rewritten, by substituting $\mathbf{V}'_{(0)}\mathbf{L}$ for $\mathbf{V}$, as $\mathbf{V}_R = E(\mathbf{YV}'_{(0)})\mathbf{L}$. But $E(\mathbf{YV}'_{(0)}) = \mathbf{V}_{R(0)}$ from the definition of $\mathbf{V}_{R(0)}$ as the matrix of correlations between the observed variables and the initial set of reference axes (assuming all observed variables and reference axes to have unit length). Hence, $\mathbf{V}_R = \mathbf{V}_{R(0)}\mathbf{L}$.

We will now consider the oblique-graphical-rotation process as a step-by-step procedure. Initially, we will describe this process in connection with the artificial problem mentioned above. Later, we will summarize this process as an iterative process which is the same for each iteration.

1. The first step in a graphical-rotation procedure is to make graphical plots of the projections of the observed variables onto the planes defined by pairs of reference axes. In these plots, the variables are represented by points having as their coordinates the coefficients for the respective variables in the columns of the reference-structure matrix, $\mathbf{V}_{R(0)}$, in Table 10.2.

   For example, in plotting the 10 variables of the artificial problem mentioned above onto the plane defined by reference axes $V_{1(0)}$ and $V_{2(0)}$, variable 1 takes the coordinates (.759, .498), and variable 7 takes coordinates (.696, −.639). The plot of these two variables along with the rest, may be seen in Figure 10.2. In carrying out this plotting, the reference vectors are always treated as if they are uncorrelated, even though they may not be (which will most likely be true in subsequent plots for new reference vectors). Moreover, the points plotted must

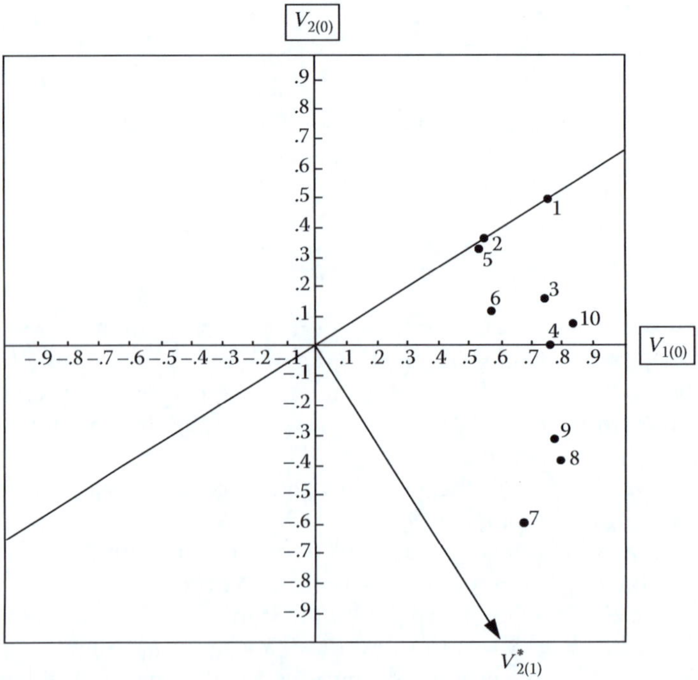

FIGURE 10.2
Reference axes $V_{1(0)}$ and $V_{2(0)}$ are plotted as if orthogonal to one another. Variables are plotted as points whose coordinates are the current reference structure coefficients. References axis $V_{2(0)}$ is rotated independently of $V_{1(0)}$ so that the coordinate hyperplane orthogonal to the resulting reference axis, $V^*_{2(1)}$, lines up with variables 1, 2, and 5, isolating a subset of variables orthogonal to the reference axis.

be regarded not as the variables themselves but as the projections or images of the variables (perhaps "shadows" will convey the idea) of the variable vectors onto the planes defined by pairs of reference vectors. This distinction between the plotted points and the vectors is essential, for it helps to try to visualize the location of the variables in a multidimensional space when trying to locate hyperplanes, which will determine the reference axes of a simple-structure solution (Figure 10.3).

In this third plot, we appear to be looking down upon a triangular formation of points. In effect, if we try to visualize how these points are located in three-dimensional space, they appear to be located within three planes which form an inverted pyramid formation, the apex of which is at the origin. One plane passes through points 1, 2, and 5; another through points 4, 6, 9, and 7; and the third passes through 3, 8, and 7. Thus in Figure 10.4, we get a very good idea of which variables are to determine the final hyperplanes of the simple-structure solution; as a consequence we should attempt to make our subsequent choices of hyperplanes as much like these hyperplanes as possible. Therefore, although our initial choice of the radial streak

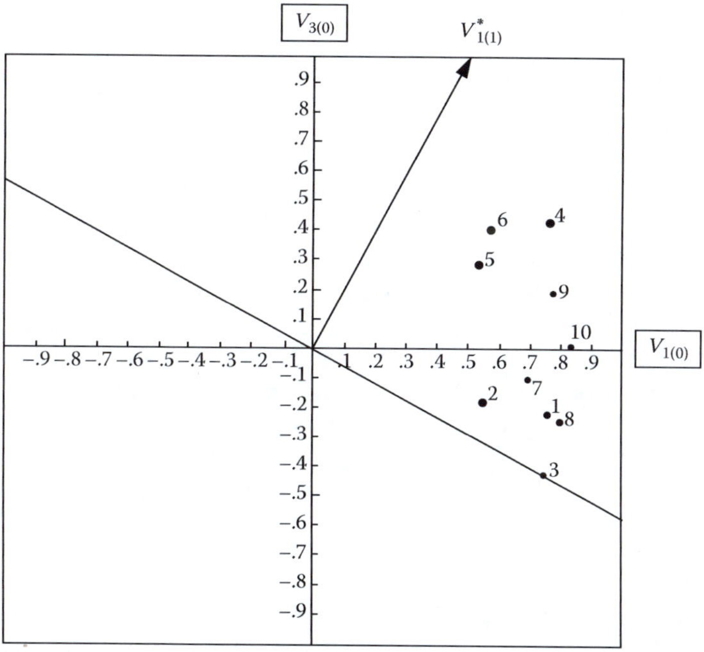

FIGURE 10.3
Resulting reference vector $V_{1(1)}$ is positioned so that its coordinate hyperplane orthogonal to it, contains variable 3.

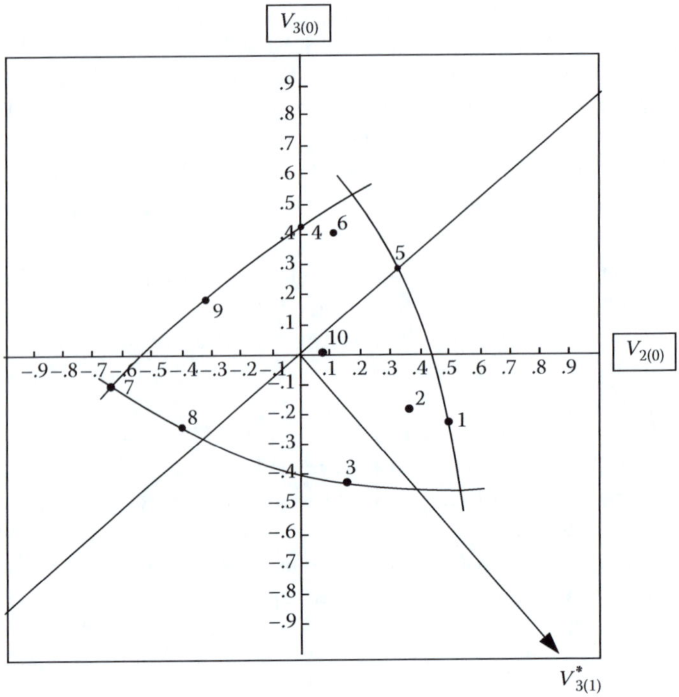

FIGURE 10.4
Resulting reference vector $V^*_{3(1)}$ is positioned so that its coordinate hyperplane orthogonal to it is parallel with variables 4, 9, and 7, in anticipation of positioning a hyperplane through them in subsequent rotations.

through points 1, 2, and 5 in Figure 10.2 appears appropriate as a place for a bounding hyperplane, we must choose our other two hyperplanes so as to approximate those suggested by Figure 10.4.

Thus we see that locating a bounding hyperplane through points 4, 5, and 6 in Figure 10.3 would not be a good procedure, for it would suggest that these variables fall in the same hyperplane of the simple-structure solution, which is not what the plot in Figure 10.4 suggests. Actually, the best location for a bounding hyperplane in Figure 10.3 is through the origin and variable 3. This location for a hyperplane is justified because it approximates the plane suggested by points 3, 8, and 7 in Figure 10.4. As for the third hyperplane, no good bounding hyperplane is to be found; however, to help in later rotations we pick a plane which is parallel to the projections of points 4, 6, 9, and 7 passing through the origin in Figure 10.4.

2. The next step is to construct vectors from the origin at right angles to the planes and to extend them until they pass through lines drawn

parallel to the reference axes at unit distance from these axes. Where these vectors intersect these lines we will note their coordinates in the reference-axes system; these coordinates will be used to define a new set of provisional reference axes, which, although not of unit length, are collinear with the desired new set of reference axes for the next plots. For example, the provisional new reference axis, $V^*_{1(1)}$, has coordinates (.58, .00, 1.00) in Figure 10.3 and may be obtained from the initial reference axes by the equation $V^*_{1(1)} = .58V_{1(0)} + 0V_{2(0)} + 1.00V_{3(0)}$. Similarly, the provisional new reference axis, $V^*_{2(1)}$, may be obtained from its coordinates by the equation $V^*_{2(1)} = .66V_{1(0)} - 1.00V_{2(0)} + 0V^6_{3(0)}$, and the provisional new reference axis, $V^*_{3(1)}$, is obtained by the equation $V^*_{3(1)} = 0V_{1(0)} + .68V_{2(0)} - 1.00V_{3(0)}$. These three vectors may be obtained from the initial reference axes by the matrix equation

$$\begin{bmatrix} V^*_{1(1)} \\ V^*_{2(1)} \\ V^*_{3(1)} \end{bmatrix} = \begin{bmatrix} .58 & .00 & 1.00 \\ .66 & -1.00 & .00 \\ .00 & .68 & -1.00 \end{bmatrix} \begin{bmatrix} V_{1(0)} \\ V_{2(0)} \\ V_{3(0)} \end{bmatrix}$$

$$\mathbf{V}^*_{(1)} = \mathbf{N}'_{01} \cdot \mathbf{V}_{(0)}$$

where

$\mathbf{V}^*_{(1)}$ is the vector of new provisional reference axes
$\mathbf{N}'_{01}$ a transformation matrix of direction numbers
$\mathbf{V}_{(0)}$ the column vector of initial reference-axes variables

3. We are now ready to determine the new unit-length reference axes, $V_{1(1)}$, $V_{2(1)}$, and $V_{3(1)}$, for the next iteration of rotation. These unit-length reference axes will be derived from the non-unit-length, provisional reference axes, $V^*_{1(1)}$, $V^*_{2(1)}$, and $V^*_{3(1)}$, by normalizing the latter to unit length, using the usual procedures for normalization of vectors: Let

$$\mathbf{D}^2_{01} = \left[\operatorname{diag} E(\mathbf{V}^*_1\mathbf{V}^{*\prime}_1)\right] = \left[\operatorname{diag} \mathbf{N}'_{01} E(\mathbf{V}_{(0)}\mathbf{V}'_{(0)})\mathbf{N}_{01}\right]$$
$$= \left[\operatorname{diag} \mathbf{N}'_{01}\mathbf{N}_{01}\right]$$

where it must be recalled that $E(\mathbf{V}_{(0)}\mathbf{V}'_{(0)}) = \mathbf{I}$ because the initial reference axes are orthonormal. Then the new unit-length reference axes are found by $\mathbf{V}_{(1)} = \mathbf{D}^{-1}_{01}\mathbf{V}^*_{(1)} = \mathbf{D}^{-1}_{01}\mathbf{N}'_{01}\mathbf{V}_{(0)}$. Suppose we define a transformation matrix $\mathbf{L}'_{01} = \mathbf{D}^{-1}_{01}\mathbf{N}'_{01}$; then we may write $\mathbf{V}_{(1)} = \mathbf{L}'_{01}\mathbf{V}_{(0)}$, or, in words, the new unit-length reference axes are a linear transformation of the old unit-length reference axes.

4. At this point we must find the new reference-structure matrix $\mathbf{V}_{R(1)} = E(\mathbf{Y}\mathbf{V}'_{(1)})$, which contains the coefficients of correlation between the observed variables and the new unit-length reference axes. Since $\mathbf{V}_{(1)} = \mathbf{L}'_{01}\mathbf{V}_{(0)}$, then $\mathbf{V}_{R(1)} = E(\mathbf{Y}\mathbf{V}'_{(0)})\mathbf{L}_{01} = \mathbf{V}_{R(0)}\mathbf{L}_{01}$, where $\mathbf{V}_{R(0)} = \mathbf{\Lambda}_{(0)}$ is the initial reference-structure matrix, or the unrotated factor-structure matrix. Thus the transformation matrix, whose transpose is used to find the new unit-length reference axes, may be applied to the initial reference-structure matrix to find the new reference-structure matrix. The matrix $\mathbf{V}_{R(1)}$ is given in Table 10.3.
5. Once the new unit-length reference axes in $\mathbf{V}_{(1)}$, along with the new reference-structure matrix $\mathbf{V}_{R(1)}$, have been determined, we are ready to prepare a new set of plots. These new plots are prepared as in step 1 with a plot of the variables for each pair of reference axes, using the respective coefficients in the matrix $\mathbf{V}_{R(1)}$ as coordinates of the variables. The plots for the new reference axes are illustrated in Figures 10.5 through 10.7 and are based on Table 10.3.
6. As in step 2, the next step is to examine the new plots to find even better positions for the hyperplanes tentatively selected in connection with certain variables in the previous iteration. For example, in

TABLE 10.3

Matrices in Computations for First Iteration of Rotation

$$\mathbf{L}'_{01} = \mathbf{D}_{01}^{-1}\mathbf{N}'_{01}$$

$$\begin{bmatrix} .502 & .000 & .865 \\ .551 & -.835 & .000 \\ .000 & .652 & -.758 \end{bmatrix} = \begin{bmatrix} .865 & .000 & .000 \\ .000 & .830 & .000 \\ .000 & .000 & .760 \end{bmatrix} \begin{bmatrix} .580 & .000 & 1.000 \\ .660 & -1.000 & .000 \\ .000 & .860 & -1.000 \end{bmatrix}$$

$$\mathbf{V}_{R(1)} = \mathbf{V}_{R(0)}\mathbf{L}_{01}$$

$$\begin{bmatrix} .186 & .002 & .495 \\ .117 & .001 & .378 \\ .005 & .280 & .426 \\ .752 & .414 & -.319 \\ .515 & .022 & -.004 \\ .635 & .219 & -.231 \\ .260 & .917 & -.339 \\ .189 & .768 & -.072 \\ .548 & .691 & -.347 \\ .425 & .397 & .044 \end{bmatrix}$$

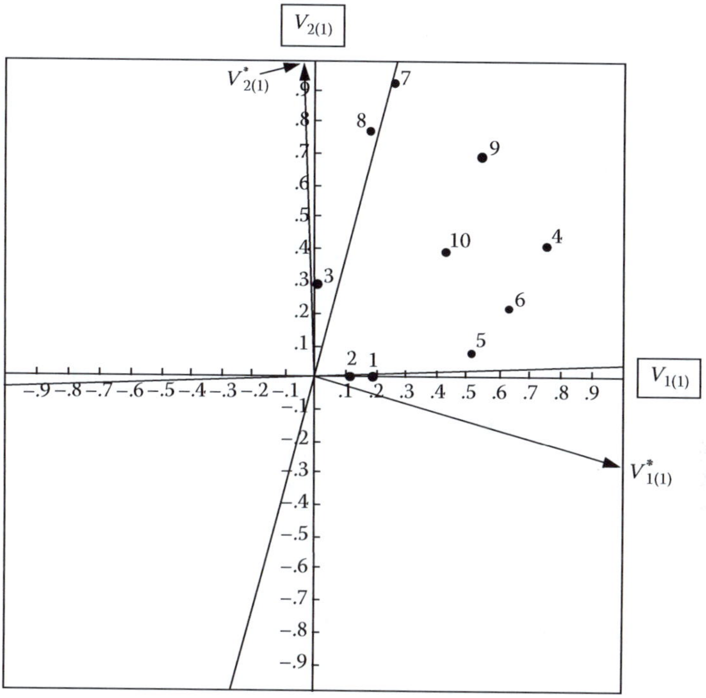

FIGURE 10.5
New plot of reference vectors $V_{1(1)}$ and $V_{2(1)}$ and resulting reference vectors $V^*_{2(1)}$ (aligned so that its coordinate hyperplane contains variables 1 and 2) and $V^*_{1(1)}$ (aligned so that its coordinate hyperplane contains variables 7 and 8).

the plot for reference axes $V_{1(1)}$ and $V_{2(1)}$ illustrated in Figure 10.5 we see the possibility for adjusting two of the hyperplanes: Consider a hyperplane passing between variables 7 and 8 and through the origin. This hyperplane will approximate the final simple-structure hyperplane containing variables 3, 7, and 8 to an even greater degree than before. Similarly, some improvement can be made by passing a hyperplane through variables 1, 2, and 5.

No new hyperplane can be found in the plot for reference axes $V_{1(1)}$ and $V_{3(1)}$, illustrated in Figure 10.6. (If one asks why not pass a hyperplane through points 1 and 2, the answer is that these points have already been taken care of by a new hyperplane in the plot for axes $V_{1(1)}$ and $V_{2(1)}$ in Figure 10.5.

In the plot for reference axes $V_{2(1)}$ and $V_{3(1)}$ illustrated in Figure 10.7, we also find a location for a new position for the third hyperplane. Recall that in the preceding set of plots in Figure 10.4, we located a hyperplane that ran parallel to a trace of points consisting of

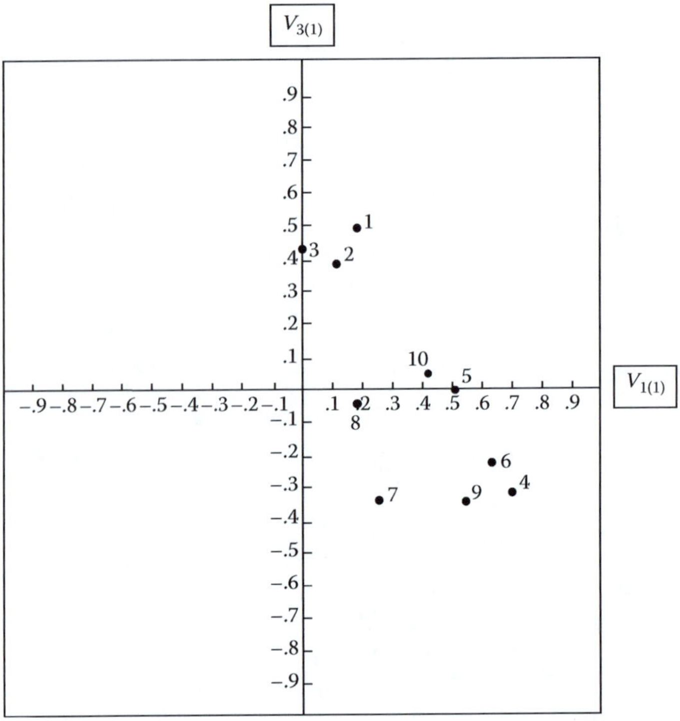

FIGURE 10.6
Plot of reference vectors $V_{1(1)}$ and $V_{3(1)}$ suggests no adjustments.

variables 4, 6, 7, and 9. This was to make it possible to pick up a hyperplane for these points.

In this new plot, we see these points suggesting a streak, and we choose to pass a new position for the third hyperplane through point 7. (We now expect variable 7 to be common to two of the final-solution hyperplanes and at the intersection of these two hyperplanes, effectively collinear with a factor, and so we pass the new hyperplane through variable 7).

7. Next, as in step 3, we draw vectors at right angles to the hyperplanes and extend these vectors from the origin again to the frame lines running parallel with the axes at unit distance from the origin. Where the new reference axes arrows intersect the frame lines, we note their coordinates, for they will define a new set of provisional reference axes, $V^*_{1(2)}$, $V^*_{2(2)}$, and $V^*_{3(2)}$, which do not have unit length. In this case, we may write equations for these new vectors in terms of the present unit-length reference axes, that is, $V^*_{1(2)} = 1.00V_{1(1)} - .26V_{2(1)} + 0V_{3(1)}$,

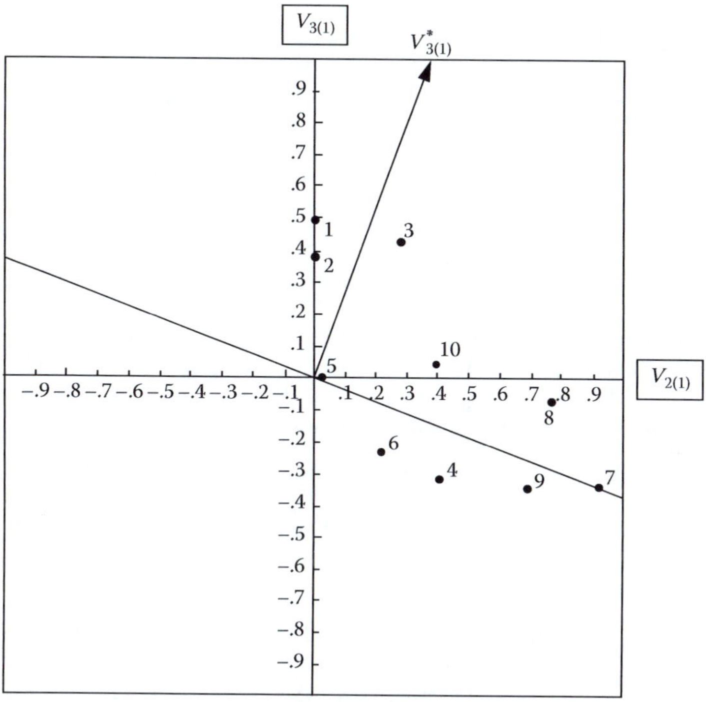

FIGURE 10.7
Plot of reference vectors $V_{2(1)}$ and $V_{3(1)}$ suggests a new $V^*_{3(1)}$ orthogonal (eventually) to variables 4, 6, 7, and 9.

$V^*_{2(2)} = -.04V_{1(1)} + 1.00V_{2(1)} + 0V_{3(1)}$, $V^*_{3(2)} = 0V_{1(1)} + .38V_{2(1)} + 1.00V_{3(1)}$. These equations may be expressed collectively by matrix equations as $\mathbf{V}^*_{(2)} = \mathbf{N}'_{12}\mathbf{V}_{(1)}$, where $\mathbf{N}'_{12}$ is the matrix of direction numbers transforming the axes in $\mathbf{V}_{(1)}$ into the provisional new reference axes in $\mathbf{V}^*_{(2)}$, that is,

$$\begin{bmatrix} V^*_{1(2)} \\ V^*_{2(2)} \\ V^*_{3(2)} \end{bmatrix} = \begin{bmatrix} 1.00 & -.26 & .0 \\ -.04 & 1.00 & .0 \\ .00 & .38 & 1.00 \end{bmatrix} \begin{bmatrix} V_{1(1)} \\ V_{2(1)} \\ V_{3(1)} \end{bmatrix}; \quad \begin{bmatrix} .373 & .226 & .900 \\ .536 & -.843 & -.035 \\ .245 & .392 & -.887 \end{bmatrix}$$

$$\mathbf{V}^*_{(2)} = \mathbf{N}'_{12}\,\mathbf{V}_{(1)}; \quad \mathbf{L}'_{02}$$

8. We now must find a new set of unit-length reference axes, $V_{1(2)}$, $V_{2(2)}$, and $V_{3(2)}$ from $V^*_{1(2)}$, $V^*_{2(2)}$, and $V^*_{3(2)}$ by normalizing the latter vectors to unit length. Let

$$\mathbf{D}_{12}^2 = \left[\text{diag}[E(\mathbf{V}_{(2)}^{*}\mathbf{V}_{(2)}^{*\prime})]\right] = \left[\text{diag}[\mathbf{N}_{12}'E(\mathbf{V}_{(1)}\mathbf{V}_{(1)}')\mathbf{N}_{12}]\right]$$

$$= \left[\text{diag}\,\mathbf{N}_{12}'\mathbf{L}_{01}'E(\mathbf{V}_{(0)}\mathbf{V}_{(0)}')\mathbf{L}_{01}\mathbf{N}_{12}\right] = \left[\text{diag}\,\mathbf{N}_{12}'\mathbf{L}_{01}'\mathbf{L}_{01}\mathbf{N}_{12}\right]$$

The diagonal matrix, $\mathbf{D}_{12}^2$, is based on the diagonal elements of the matrix $\mathbf{N}_{12}'\mathbf{L}_{01}'\mathbf{L}_{01}\mathbf{N}_{12}$. We may then write $\mathbf{V}_{(2)} = \mathbf{D}_{12}^{-1}\mathbf{V}_{(1)}^{*}$ or $\mathbf{V}_{(2)} = \mathbf{D}_{12}^{-1}\mathbf{N}_{12}'\mathbf{V}_{(1)}$. But we may substitute an earlier derived expression for $\mathbf{V}_{(1)}$ into the last equation to obtain $\mathbf{V}_{(2)} = \mathbf{D}_{12}^{-1}\mathbf{N}_{12}'\mathbf{L}_{01}'\mathbf{V}_{(0)}$ which, in words, states that $\mathbf{V}_{(2)}$ may be obtained from the initial reference axes $\mathbf{V}_{(0)}$ by a linear transformation. Let the matrix that transforms $\mathbf{V}_{(0)}$ into $\mathbf{V}_{(2)}$ be designated by $\mathbf{L}_{02}'$; then $\mathbf{L}_{02}' = \mathbf{D}_{12}^{-1}\mathbf{N}_{12}'\mathbf{L}_{01}'$; $\mathbf{L}_{02}'$ may be obtained from the previous transformation matrix $\mathbf{L}_{01}'$ and the matrix of direction numbers $\mathbf{N}_{12}' \cdot \mathbf{L}_{02}'$ is shown above.

9. At this point, we are ready to obtain a new reference-structure matrix $\mathbf{V}_{(2)}$. By definition $\mathbf{V}_{(2)} = E(\mathbf{Y}\mathbf{V}_{(2)}')$, but by developments in step 8 we now may write this as $\mathbf{V}_{(2)} = E(\mathbf{Y}\mathbf{V}_{(0)}\mathbf{L}_{02})$. Having found $\mathbf{V}_{(2)}$, we are again ready to plot.

The reader may by now see a pattern developing in these steps. We shall now describe these steps as an iterative procedure in which we shall use the subscript (*i*) to stand for the iteration number:

1. Using the reference-structure matrix, $\mathbf{V}_{R(i)}$ is obtained as a result of the *i*th iteration (beginning with $i=0$). Plot the points representing the projections of the variables onto planes defined by pairs of reference vectors.
2. Examine the plots to determine locations for improved estimates of the simple-structure hyperplanes.
3. Draw vectors at right angles to the hyperplanes and extended from the origin to a frame line parallel with one of the reference vectors and at unit distance from the origin. These vectors will be the new, provisional reference vectors in $\mathbf{V}_{(i+1)}^{*}$. Use the coordinates of these vectors to determine a matrix of direction numbers, $\mathbf{N}_{i,i+1}'$, such that $\mathbf{V}_{(i+1)}^{*} = \mathbf{N}_{i,i+1}'\mathbf{V}_{(i)}$.
4. Find the diagonal matrix $\mathbf{D}_{i,i+1}^2 = \left[\text{diag}\,\mathbf{N}_{i,i+1}'\mathbf{L}_{0,i}'\mathbf{L}_{0,i}\mathbf{N}_{i,i+1}\right]$. Then the new transformation matrix, $\mathbf{L}_{0,i+1}'$, for transforming the initial reference vectors in $\mathbf{V}_{(0)}$ into the new unit-length reference vectors, $\mathbf{V}_{(i+1)}$, is given by the equation $\mathbf{L}_{0,i+1}' = \mathbf{D}_{i,i+1}^{-1}\mathbf{N}_{i,i+1}'\mathbf{L}_{0,i}'$.
5. Finally, the new reference-structure matrix, $\mathbf{V}_{R(i+1)}$, is given by the equation $\mathbf{V}_{R(i+1)} = \mathbf{V}_{R(0)}\mathbf{L}_{0,i+1}$.
6. At this point, return to step 1 and continue to the next iteration. If, however, one is satisfied that no improvement can be made on the

reference axes after $\mathbf{V}_{R(i+1)}$ is found, then one takes $\mathbf{V}_{R(i+1)}$ to be the reference-structure matrix $\mathbf{V}_R$ of the simple-structure solution and the iterations are brought to an end.

7. Then using $\mathbf{V}_R$, one must compute the factor-pattern matrix $\mathbf{\Lambda}$ of the simple-structure solution as follows: (a) Let $\mathbf{L}' = \mathbf{L}'_{0,i+1}$. Then the reference axes of the simple-structure solution are given by $\mathbf{V} = \mathbf{L}'\mathbf{V}_{(0)}$ whereas the matrix of correlations among the reference axes of the simple-structure solution is given by $\mathbf{R}_{VV} = E(\mathbf{V}\mathbf{V}') = \mathbf{L}'\mathbf{L}$. (b) Invert the matrix $\mathbf{R}_{VV}$ and obtain the diagonal matrix $\mathbf{D}_V^2 = \left[\text{diag } \mathbf{R}_{VV}^{-1}\right]$. (c) Then $\mathbf{\Lambda} = \mathbf{V}_R\mathbf{D}_V$. (d) Finally, the matrix of correlations among the primary factors is given by $\mathbf{R}_{XX} = \mathbf{D}_V^{-1}\mathbf{R}_{VV}^{-1}\mathbf{D}_V^{-1} = \mathbf{D}_V^{-1}(\mathbf{L}'\mathbf{L})\mathbf{D}_V^{-1}$.

At this point, we will skip over the next several iterations in the example problem to show what the final solution looked like. After five iterations in which clusters of variables moved into alignment with the reference axes as they became coordinate hyperplanes orthogonal to the complementary reference vectors, the following matrices were obtained (Tables 10.4 and 10.5):

In Figures 10.8 through 10.10, we see that three clusters of variables fall in coordinate hyperplanes. In Figure 10.8, variables 3, 7, and 8 are in a hyperplane orthogonal to $V_{1(5)}$, while variables 1, 2, and 5 are orthogonal to $V_{2(5)}$. In Figure 10.9, variables 3, 7, and 8 remain orthogonal to $V_{1(5)}$, but now,

TABLE 10.4

Matrices in Computations for Fifth Iteration of Rotation

$$\mathbf{L}'_{05} = \mathbf{D}_{45}^{-1}\mathbf{N}'_{45}\mathbf{L}'_{04}$$

$$\begin{bmatrix} .410 & .305 & .860 \\ .539 & -.842 & -.039 \\ .407 & .556 & .725 \end{bmatrix} = \begin{bmatrix} .999 & .00 & .00 \\ .00 & .999 & .00 \\ .00 & .00 & 1.006 \end{bmatrix} \begin{bmatrix} 1.00 & -.025 & .00 \\ .00 & 1.00 & .00 \\ .02 & .00 & 1.00 \end{bmatrix} \begin{bmatrix} .424 & .284 & .860 \\ .539 & -.842 & -.039 \\ .396 & .547 & -.738 \end{bmatrix}$$

$$\mathbf{V}_{R(5)} = \mathbf{V}_{R(0)}\mathbf{L}_{05}$$

$$\begin{bmatrix} .269 & -.001 & .749 \\ .179 & -.001 & .561 \\ -.013 & .287 & .701 \\ .682 & .389 & .005 \\ .564 & .002 & .191 \\ .616 & .195 & .004 \\ .002 & .917 & .003 \\ -.003 & .770 & .282 \\ .380 & .677 & .005 \\ .372 & .386 & .377 \end{bmatrix}$$

TABLE 10.5

Correlations among Reference Axes, Primary Factors, and Factor-Pattern Matrix

$$\mathbf{R}_{VV} = \mathbf{L}'\mathbf{L} = \mathbf{L}'_{05}\mathbf{L}_{05}$$

$$\begin{bmatrix} 1.000 & -.069 & -.287 \\ -.069 & 1.000 & -.220 \\ -.287 & -.220 & 1.000 \end{bmatrix}$$

$$\mathbf{R}_{XX} = \left[\text{diag } \mathbf{R}_{VV}^{-1}\right]^{-\frac{1}{2}} \mathbf{R}_{VV}^{-1} \left[\text{diag } \mathbf{R}_{VV}^{-1}\right]^{-\frac{1}{2}}$$

$$\begin{bmatrix} 1.000 & .142 & .311 \\ .142 & 1.000 & .251 \\ .311 & .251 & 1.000 \end{bmatrix}$$

$$\mathbf{\Lambda} = \mathbf{V}_R \left[\text{diag } \mathbf{R}_{VV}^{-1}\right]^{1/2}$$

$$\begin{bmatrix} .284 & -.001 & .808 \\ .188 & -.001 & .606 \\ -.014 & .297 & .756 \\ .719 & .402 & .005 \\ .594 & .002 & .206 \\ .649 & .202 & .004 \\ .002 & .949 & .004 \\ -.003 & .798 & .304 \\ .401 & .701 & .005 \\ .392 & .400 & .407 \end{bmatrix}$$

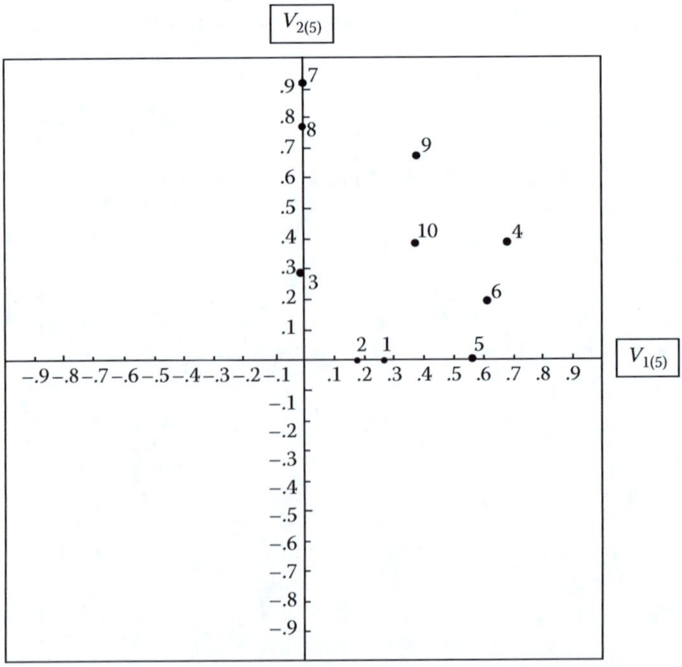

FIGURE 10.8
Skipping to the result after five iterations, we see in the plot of $V_{1(5)}$ and $V_{2(5)}$ that $V_{1(5)}$ is orthogonal to variables 3, 7, and 8, and $V_{2(5)}$ is orthogonal to variables 1, 2, and 5.

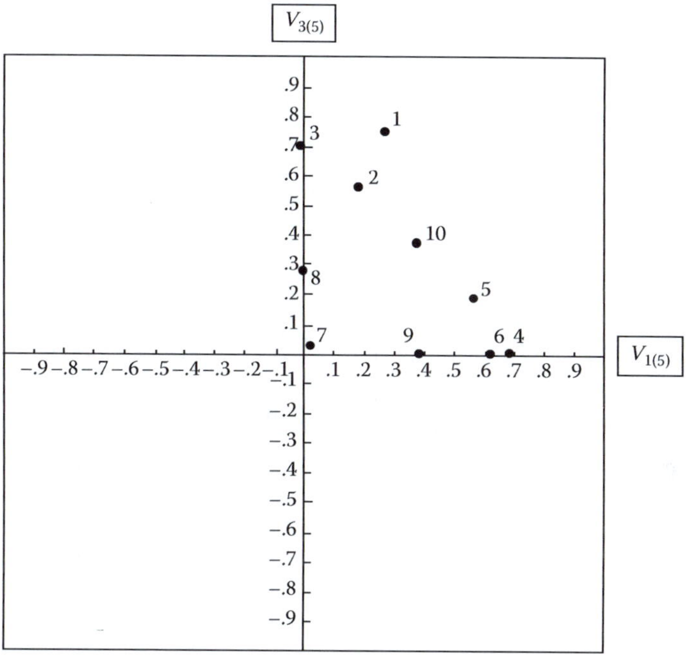

FIGURE 10.9
After five iterations, $V_{1(5)}$ is orthogonal to variables 3, 7, and 8, while $V_{3(5)}$ is orthogonal to variables 4, 6, 7, and 9.

variables 4, 6, 7, and 9 fall in another hyperplane orthogonal to $V_{3(5)}$. In Figure 10.10, variables 1, 2, and 5 remain orthogonal to $V_{2(5)}$ while variables 4, 6, 7, and 9 remain orthogonal to $V_{3(5)}$.

Once the final plots are made and the final reference-structure matrix calculated, the final step is to compute the correlations, $\mathbf{R}_{VV}$, among the reference axes, and from these in turn the correlations, $\mathbf{R}_{XX}$, among the primary factors. Taking the square roots of the diagonal elements of $\mathbf{R}_{VV}^{-1}$, we are then able to rescale the columns of the reference-structure matrix by multiplying them by these respective square roots to obtain the factor-pattern matrix, $\mathbf{\Lambda}$.

The problem used to illustrate graphical oblique rotation was artificially contrived in such a way that there was a perfect simple structure to be found. However, in rotating factors obtained from empirical data the variables do not always lie exactly within the hyperplanes but may vary around them in a way suggestive of random error due to sampling and other influences. Thus, in empirical studies the hyperplanes will be relatively indistinct, but this indistinctness may be minimized to some extent by careful selection of variables to overdetermine the potential factors in the domain studied as well as by the use of large samples.

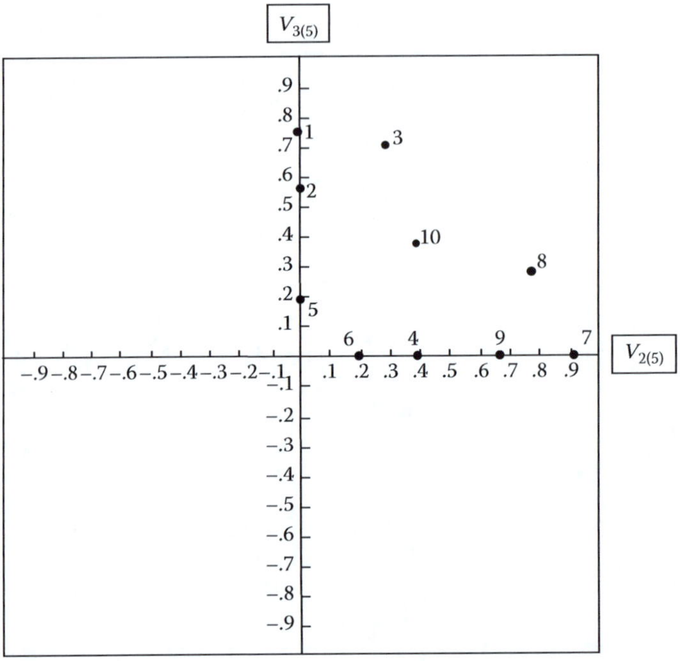

FIGURE 10.10
After five iterations $V_{2(5)}$ is orthogonal to variables 1, 2, and 5 while $V_{3(5)}$ is orthogonal to variables 4, 6, 7, and 9.

The accuracy of graphical rotation remains limited to the scale of the plots and the accuracy with which points are plotted by the researcher. The procedure described, however, is self-correcting, since the reference-structure matrix at any stage is a linear transformation of the unrotated reference-structure matrix.

Graphical rotation is a tedious activity. Each iteration involves rotation of $r(r-1)/2$ pairs of reference axes, and this number of pairs increases dramatically with the size of r. The process might be made less onerous and faster if one could work with computer-generated graphs and a graphical interface that allows one to define the hyperplanes by mouse, and then calculates the next reference-structure matrix automatically. But this presumes that graphical rotation is the optimal method, and researchers were not convinced that that was the case. The next developments in the history of factor analysis involved the search for computer algorithms whereby the rotation process could be made entirely automatic. But this depended on finding a way to represent the simple-structure criteria of L.L. Thurstone mathematically and to rotate to optimize these criteria.

11

Orthogonal Analytic Rotation

11.1 Introduction

Although Thurstone's five criteria for a simple-structure solution, discussed in Section 10.2, appear to be consistent with the idea that each variable should be accounted for by fewer than the total number of common factors, many factor analysts have felt that these criteria are not rigorous and that their implementation is all too subjective.

But the intrinsic value of objectivity and conceptual rigor has not been the only reason motivating factor analysts to develop a rigorous, objective criterion for the simple structure. In fact, objectivity and rigor per se appear less important than simply mechanizing the factor-rotation process. Rotating a moderate number of factors by a graphical, hand-plotted procedure is frequently a month-long enterprise involving considerable drudgery. It is not surprising that, when high-speed computers were first developed, factor analysts expended considerable effort in computerizing the rotation process, and this involved defining, objectively and mathematically, criteria for simple structure so that the computer could be instructed to do the rotation.

In discussing the objective methods of factor rotation—which normally are called "analytic" methods as opposed to intuitive, subjective methods—we shall need to distinguish between two major classes of rotation: orthogonal rotation and oblique rotation. The efforts to find analytic criteria for orthogonal rotation were the first to be successful, and for the student they are initially the easiest to understand. Accordingly, we will deal with analytic orthogonal rotation first.

But I need to interject here that orthogonal rotation is not really rotation to the simple structure in Thurstone's sense, since he presumed that the factors would be correlated with one another. Forcing the factors to be mutually orthogonal introduces an artifact that prevents the method from achieving a true simple structure. In nature, the presence of other unmeasured common causes of the common factors themselves will cause the factors to be correlated. Restriction of range can also introduce correlations between factors that otherwise are uncorrelated. So, forcing the factors to be orthogonal does not yield the most accurate rendition of what is in the world.

Having said that, I will still discuss the methods of orthogonal analytic rotation. An understanding of the history of factor analysis demands it, and furthermore, concepts learned in developing the algorithms of orthogonal rotation carried over into the development of algorithms for oblique rotation. Orthogonal rotations can also serve as intermediate solutions, which will be further modified to become oblique rotations to simple structure. So, there will be a use for some of the orthogonal solutions.

Before discussing particular methods of orthogonal analytic rotation, we must consider algebraically what the problem of defining an analytic criterion involves. Let **A** be an $n \times r$ unrotated-factor-structure matrix obtained in connection with an orthogonal-factor solution. Then the solution to the orthogonal-analytic-rotation problem involves finding an $r \times r$ orthonormal-transformation matrix, **T**, such that the resulting $n \times r$ transformed matrix, $\mathbf{B} = \mathbf{AT}$, has certain properties. Note that $\mathbf{TT}' = \mathbf{TT}' = \mathbf{I}$, an identity matrix of order r. The question is: "What properties must the matrix **B** have in order to be an orthogonal simple-structure solution?"

11.2 Quartimax Criterion

In the 1950s, several factor analysts using different approaches arrived at what is now known, due to Neuhaus and Wrigley (1954), as the quartimax criterion. Carroll (1953) was the first of these. He proposed that the matrix **B** would represent a simple-structure solution if the value of

$$q_1 = \sum_{j<k=1}^{r} \sum_{i=1}^{n} b_{ij}^2 b_{ik}^2$$

is a minimum. A certain parallel exists between this function when it is a minimum and Thurstone's third criterion, that for every pair of columns of the reference-structure matrix $\mathbf{V}_R$ several zero entries in one column should correspond to nonzero entries in the other; when q_1 is a minimum, it means that the product of any two coefficients in a given row of the matrix **B** will tend to be small, suggesting that either both are small (near zero) or one is large and the other near zero.

Very shortly after Carroll's paper appeared, Neuhaus and Wrigley (1954), Saunders (1953), and Ferguson (1954) proposed their own criteria which, as we shall soon see, are mathematically equivalent to Carroll's criterion. Neuhaus and Wrigley suggested that the matrix **B** would be the most interpretable if the variance of the $n \times r$ squared loadings of the matrix **B** was a maximum, that is,

$$q_2 = \frac{nr \sum_{i=1}^{n} \sum_{j=1}^{r} (b_{ij}^2)^2 - \left(\sum_{i=1}^{n} \sum_{j=1}^{r} b_{ij}^2 \right)^2}{n^2 r^2} = \text{Maximum}$$

This criterion ensures that there will be many large and many small coefficients in **B**. Saunders reasoned in almost the same way, but instead of maximizing the variance of squared loadings, he wished to maximize the kurtosis of all loadings and their reflections (corresponding values with opposite signs), using as criterion

$$q_3 = \frac{nr\sum_{i=1}^{n}\sum_{j=1}^{r} b_{ij}^4}{\left(\sum_{i=1}^{n}\sum_{j=1}^{r} b_{ij}^2\right)^2} = \text{Maximum}$$

Ferguson wished simply to maximize the sum of the coefficients each raised to the fourth power, using for his rationale certain developments in information theory:

$$q_4 = \sum_{i=1}^{n}\sum_{j=1}^{r} b_{ij}^4 = \text{Maximum}$$

The reason all four of these criteria lead to the same solution for the matrix **B** is as follows: First, consider that for any variable *i* the sum of its squared loadings in a factor-structure matrix for orthogonal factors (the communality of variable *i*) is invariant under an orthonormal transformation of that matrix, i.e.,

$$\text{Constant} = \sum_{j=1}^{r} b_{ij}^2 = \sum_{j=1}^{r} a_{ij}^2 = h_i^2 = \text{Communality of variable } i$$

As proof of this assertion, consider any arbitrary $r \times r$ orthonormal-transformation matrix **T** applied to the matrix **A**, that is,

$$\mathbf{B} = \mathbf{AT} \tag{11.1}$$

Then the sum of squared loadings for each row of either **A** or **B** is given by the diagonal element of the matrix [diag **AA′**] or [diag **BB′**], respectively. Note that **AA′** = **BB′** because, after substituting from Equation 11.1, we have **AA′** = **ATT′A′** = **BB′**, made possible by the orthonormal properties of the matrix **T**. As a consequence the diagonal elements of the matrices **AA′** and **BB′** must be the same. This completes the proof.

If the communality of a variable is invariant, so is the square of its communality, which we may written as

$$\text{Constant} = (h_i^2)^2 = \left(\sum_{j=1}^{r} b_{ij}^2\right)^2 = \sum_{j=1}^{r} b_{ij}^4 + 2\sum_{j<k=1}^{r} b_{ij}^2 b_{ik}^2$$

Likewise, the sum of squared communalities across all tests must be constant, i.e.,

$$\text{Constant} = \sum_{i=1}^{n}\sum_{j=1}^{r} b_{ij}^{4} + 2\sum_{i=1}^{n}\sum_{j<k=1}^{r} b_{ij}^{2}b_{ik}^{2}$$

which we may note is equivalent to

$$\text{Constant} = q_4 + 2q_1$$

That is, the quartimax criterion of Ferguson plus twice the criterion of Carroll is a constant. Thus, maximizing q_4 is equivalent to minimizing q_1.

Maximizing Neuhaus and Wrigley's q_2 or Saunders q_3 is also equivalent to maximizing Ferguson's q_4, because in q_2 and q_3 appears the term

$$\left(\sum_{i=1}^{n}\sum_{j=1}^{r} b_{ij}^{2}\right)^{2}$$

which is equivalent to the square of the sum of the communalities, a constant. In q_2, this term divided by nr is subtracted from nr times the term $\sum_{i=1}^{n}\sum_{j=1}^{r} b_{ij}^{4}$ whereas in q_3 it is divided into the latter term. Either subtracting, multiplying, or dividing by a constant does not change the point at which an expression is a maximum; since in this case the expression to achieve a maximum is equivalent to Ferguson's q_4 criterion, maximizing q_2 and q_3 is equivalent to maximizing q_4. Because of its mathematical simplicity, we will use Ferguson's q_4 criterion as the quartimax criterion in subsequent discussion and analysis.

To find simultaneously all the coefficients of an orthonormal-transformation matrix **T** which when postmultiplied by **A** produces a **B** matrix for which the criterion q_4 is a maximum is a difficult mathematical problem to solve. Thus most attempts to do quartimax rotation have concentrated not upon finding a transformation which simultaneously rotates all factors but on an iterative procedure which for each pair of different factors rotates the pair to maximize the criterion for that pair. No proof has ever been offered that such an iterative procedure will always converge, but with empirical structure matrices it always has. Thus we will discuss pairwise quartimax factor rotation.

Let a_{ij} and a_{ik} be the factor-structure coefficients for the ith variable on the jth and kth factors, respectively, of the unrotated-factor-structure matrix **A**. Let b_{ij} and b_{ik} be the factor-structure coefficients of the ith variable on the same factors j and k in the rotated-factor matrix **B** after only factors j and k have been rotated. From a commonly known theorem of analytic geometry we know that the coordinates b_{ij} and b_{ik} of the ith variable on factors j and k after a rigid orthogonal rotation of the axes through an angle θ can be expressed in terms of the variables' coordinates on the original axes as

$$b_{ij} = a_{ij}\cos\theta - a_{ik}\sin\theta$$

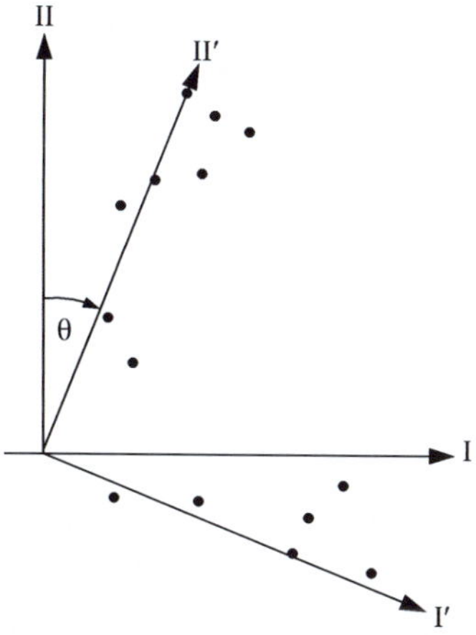

FIGURE 11.1
Graphical illustration of rotation of factors I and II through an angle, *q*, to obtain factors I′ and II′, representing the improved simple-structure solution.

$$b_{ik} = a_{ij} \sin\theta + a_{ik} \cos\theta \tag{11.2}$$

For a two-factor rotation, the quartimax criterion, q_4, to be maximized may be expressed as

$$Q(\theta) = \sum_{i=1}^{n} (b_{ij}^4 + b_{ik}^4) = \text{Maximum} \tag{11.3}$$

That is, by varying the angle θ through which the factors *j* and *k* are rotated, one wishes to maximize $Q(\theta)$; this involves summing over variables the sums for the respective variables of the factor-structure coefficients (raised to the fourth power) on the resulting rotated factors. As an illustration, consider Figure 11.1 which in the case of the plot of two factors shows an orthogonal rotation of the two factors through an angle θ; this results in an improved simple-structure solution.

Now, the ultimate objective is to find the b_{ij}'s and b_{ik}'s corresponding to the quartimax solution. Historically, one way to approach the solution of this problem was to substitute for b_{ij} and b_{ik} in Equation 11.3 the expressions in Equation 11.2 and then to take the derivative of $Q(\theta)$ with respect to θ, to set the derivative equal to zero, and to solve for θ. The angle θ which maximized $Q(\theta)$ was then a function of the initial a_{ij}'s and a_{ik}'s. Once θ was determined, the b_{ij}'s and b_{ik}'s could be determined by using Equation 11.2. This was the manner in which the quartimax solution for two-factor rotation was obtained.

However, Saunders (1962) suggested that the quartimax solution may be obtained not by solving for the angle θ directly but by solving for the value of tan θ, which, as θ varies, varies between $-\infty$ and $+\infty$, from which one could very easily (in a computer program) obtain cos θ and sin θ needed to carry out the transformation in Equation 11.2. To illustrate, let

$$b_{ij}^{*} = \sec\theta\, b_{ij} = \frac{b_{ij}}{\cos\theta} = a_{ij} - a_{ik}\tan\theta$$

$$b_{ik}^{*} = \sec\theta\, b_{ik} = \frac{b_{ik}}{\cos\theta} = a_{ij}\tan\theta + a_{ik} \tag{11.4}$$

and

$$Q^{*} = \sum_{i=1}^{n}(b_{ij}^{*4} + b_{ik}^{*4}) = \sec^{4}\theta \sum_{i=1}^{n}(b_{ij}^{4} + b_{ik}^{4}) \tag{11.5}$$

Note that Q in Equation 11.3 is equal to $\sec^{-4}\theta Q^{*}$. Now, let $x = \tan\theta$; then by a trigonometric identity, $\sec^{4}\theta = (1 + \tan^{2}\theta)^{2} = (1 + x^{2})^{2}$ and $\sec^{-4}\theta = (1 + x^{2})^{-2}$. If we now treat Q as a function of x instead of θ, we may consider maximizing $Q(x)$ that is

$$\text{Maximum } Q(x) = \text{Maximum}(1 + x^{2})^{-2}\, Q^{*}(x)$$

where

$$Q^{*}(x) = \left(\sum_{i=1}^{n} a_{ij}^{4} + \sum_{i=1}^{n} a_{ij}^{4}\right)x^{4} + 4\left[\sum_{i=1}^{n}(a_{ij}a_{ik}(a_{ij}^{2} - a_{ik}^{2}))\right]x^{3}$$
$$+ 12\left(\sum_{i=1}^{n} a_{ij}^{2}a_{ik}^{2}\right)x^{2} + 4\left[\sum_{i=1}^{n}(a_{ij}a_{ik}(a_{ik}^{2} - a_{ij}^{2}))\right]x$$
$$+ \sum_{i=1}^{n} a_{ij}^{4} + \sum_{i=1}^{n} a_{ik}^{4}$$

To find the value of x which maximizes $Q(x)$, take the derivative of $Q(x)$ with respect to x:

$$\frac{dQ(x)}{dx} = \frac{d[(1 + x^{2})^{-2}]Q^{*}(x)}{dx}$$
$$= -4x(1 + x^{2})^{-3}Q^{*}(x) + \frac{(1 + x^{2})^{-2}dQ^{*}(x)}{dx} \tag{11.6}$$

Set the derivative in Equation 11.6 equal to zero and multiply both sides of this resulting equation by $(1 + x^{2})^{3}$ to obtain (after rearranging):

$$\frac{(1 + x^{2})dQ^{*}(x)}{dx} - 4xQ^{*}(x) = 0 \tag{11.7}$$

The algebraic equivalents to $dQ^*(x)/dx$ and $Q^*(x)$ expressed in terms of the a_{ij}'s and a_{ik}'s may then be substituted into Equation 11.7 and the resulting equation simplified to the following quartic equation:

$$x^4 + \beta x^3 - 6x^2 - \beta x + 1 = 0 \tag{11.8}$$

with

$$\beta = \frac{6\sum_{i=1}^{n} a_{ij}^2 a_{ik}^2 - \left(\sum_{i=1}^{n} a_{ij}^4 + \sum_{i=1}^{n} a_{ik}^4\right)}{\sum_{i=1}^{n} a_{ij}^3 a_{ik} - \sum_{i=1}^{n} a_{ij} a_{ik}^3} \tag{11.9}$$

The problem now remains to solve for x in Equation 11.8. Since Equation 11.8 is a quartic equation, it must have four roots. According to Descartes' rule of signs, the number of positive real roots of a polynomial equation

$$b_0 x^n + b_1 x^{n-1} + b_2 x^{n-2} + \cdots + b_{n-1} x + b_n = 0$$

with real coefficients is never greater than the number of variations of sign in the sequence of its coefficients, $b_0, b_1, b_2, \ldots, b_n$, and if it is less then it is less by an even integer. Thus, in Equation 11.8 there are two variations in sign, and we may provisionally suppose the equation has either two positive roots or no positive roots. If we consider the equation

$$\frac{dQ^*(x)}{dx} = x^4 - \beta x^3 - 6x^2 + \beta x + 1 = 0$$

we find that the number of positive roots of this equation equals the number of negative roots of Equation 11.8. In the latter equation, there are two variations in sign and so at most two positive roots or no positive roots; hence we conclude that in Equation 11.8 there are at most two negative roots or no negative roots. If we evaluate Equation 11.8 when $x = -1$, $x = 0$, and $x = 1$, we obtain -4, 1, and -4, respectively, which tells us that the equation must cross the x-axis between -1 and 0 and again between 0 and +1; this immediately establishes that there is at least one negative root and one positive root and therefore that there must be two negative roots and two positive roots, by Descartes' rule of signs. Thus we need not concern ourselves with the possibility of imaginary roots for this equation.

To find the roots of Equation 11.8, we may use the Newton–Raphson method. According to this method, if $y(x) = 0$ is an equation whose roots are to be found, then, beginning with an initial trial value x_0 (say $x_0 = 0$) for a root to this equation, we may use the following iterative algorithm to derive successively better approximations to a root of the equation:

$$x_{(i+1)} = x_{(i)} - \frac{y(x_{(i)})}{y'(x_{(i)})}$$

where

$x_{(i+1)}$ and $x_{(i)}$ are the values approximating the root on the $(i + 1)$st and ith iterations, respectively

$y(x_{(i)})$ is the value of $y(x)$ evaluated at $x_{(i)}$

$y'(x_{(i)})$ is the derivative of $y(x)$ with respect to x evaluated at $x_{(i)}$

In the case of Equation 11.8, we have the following iterative equation:

$$x_{(i+1)} = x_{(i)} - \frac{x_{(i)}^4 + \beta x_{(i)}^3 - 6x_{(i)}^2 - \beta x + 1}{4x_{(i)}^3 + 3\beta x_{(i)}^2 - 12x_{(i)} - \beta} \tag{11.10}$$

The iterative procedure is carried out until $|x_{(i+1)} - x_{(i)}| < \varepsilon$, where ε could be, for example, $.0001\times(|x_{(i+1)}|)$, a tolerance level relative to the magnitude of the estimate of the root. After m iterations, the final approximate value, $x_{(m)}$, is taken as an estimate of the root.

Before continuing, let us consider what the four roots of Equation 11.8 represent in terms of rotations. Figure 11.2 shows the plot of a set of variables in the plane identified by factors j and k (solid lines). The position represents the quartimax solution in this plane is given by axes j' and k'. Axes j'' and k'' are axes for which the quartimax criterion is a "minimum," which on intuitive grounds we may accept as being 45° away from axes j' and k' which maximize the criterion. Angle θ is the angle of rotation smaller than 45° of the original axes j and k to a position which maximizes the quartimax criterion. Angle γ is an angle smaller than 45° which would rotate to minimize the criterion, and η is an angle greater than 45° which would rotate to minimize the criterion.

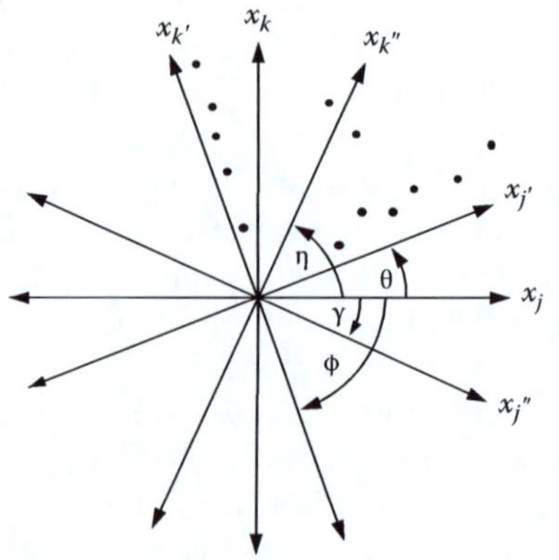

FIGURE 11.2
Plot of set of variables in plane identified by factors x_j and x_k. Position representing quartimax solution in this plane is given by axes $x_{j'}$ and $x_{k'}$, whereas axes $x_{j'}$ and $x_{k'}$ represent axes for which quartimax criterion is minimized.

By the Newton–Raphson procedure a solution $x_{(m)}$ to Equation 11.8 may represent the tangent of any one of these four angles θ, ϕ, γ, or η. What we would like to find is the tangent of the angle involving the smallest rotation to achieve a maximization of the criterion. This may be done regardless of the value of $x_{(m)}$ produced by the Newton–Raphson procedure.

Let p be the denominator of β in Equation 11.9; $r = (1 + x_{(m)}^2)$; $A = x_{(m)}^4 + \beta x_{(m)}^3 - 6x_{(m)}^2 - \beta x_{(m)} + 1$; and $B = 4x_{(m)}^3 + 3\beta x_{(m)}^2 - 12x_{(m)} - \beta$. Then the sign of $D = p(A + rB)$ is the sign of the second derivative of Q at $x_{(m)}$. If D is negative, then $x_{(m)}$ represents a rotation to a maximum, whereas if it is positive, $x_{(m)}$ represents a rotation to a minimum.

Suppose, for the time being, the second derivative is positive so that $x_{(m)}$ represents a rotation to a minimum. How can we find another root $x'_{(m)}$ which represents a rotation to a maximum without going through the Newton–Raphson process again? First, we test to see whether $x_{(m)}$ is positive or negative. If it is positive, we subtract 45° from the angle of rotation represented by $x_{(m)}$, but if it is negative, we add 45° to the angle of rotation. These additions to the angles will involve the following modifications of $x_{(m)}$ in order to obtain $x'_{(m)}$: (1) if $x_{(m)}$ is positive, then $x'_{(m)} = (x_{(m)} + 1)/(1 - x_{(m)})$; (2) if $x_{(m)}$ is negative, then $x'_{(m)} = (x_{(m)} + 1)/(1 - x_{(m)})$.

If we have found immediately the tangent of an angle of rotation maximizing the criterion, that is, $x'_{(m)} = x_{(m)}$, or if we have modified a solution yielding a minimization, to obtain a solution $x'_{(m)}$ yielding a maximization, we next must determine whether $x'_{(m)}$ represents the smallest angle of rotation maximizing the criterion. This is readily done, for if the absolute value of $x'_{(m)}$ is greater than 1 (representing the tangent of an angle greater than 45°), then the solution corresponding to the smaller angle of rotation is 90° away from the solution obtained from $x'_{(m)}$ and the tangent of the smaller angle, θ, is then $\tan\theta = \tan(\phi - 90°) = x_\theta = -1/x'_{(m)}$. Once x_θ is found, the sine and cosine of this angle, θ, may be found by

$$\cos\theta = (1 + x_\theta^2)^{-1/2}$$

$$\sin\theta = \frac{x_\theta}{(1 + x_\theta^2)^{1/2}}$$

Otherwise, $\tan\theta = x_\theta = x'_{(m)}$.

However, as one takes in turn each pair of columns j and k, $j = 1,\ldots,r$ and $j < k = 1,\ldots,r$ of the matrix **A** and determines the angle of rotation, which will yield a maximization of the pairwise quartimax criterion in Equation 11.3, it will be found that for some pairs of the factors the angle of rotation required is very small. One need not carry out the rotation if the angle is less than, say 0.5°, which has the tangent value of .0087. Thus, if $|x_\theta| < .0087$, the rotation according to Equation 11.2 is not carried out.

For each pairwise rotation, we may consider the matrix **A** to be successively transformed into a matrix **B** for which the total quartimax criterion q_4

is a maximum. As these transformations are carried out by successively pairing each column j with each other column k to the right of it in matrix $\mathbf{A}$, the coefficients of column j (as well as those of column k) will change. In the case of some pairs of columns we need not make any changes because the angle of rotation is negligible. But the question remains when to stop the process of rotation. One criterion for stopping rotation is to stop when successively all $r(r-1)/2$ pairs of columns yield pairwise rotations through negligible angles of rotation, i.e., through angles less than 0.5°. At this point the solution should be very close to the true quartimax solution.

11.3 Varimax Criterion

In an important paper, Kaiser (1958) noted that in connection with the quartimax method of rotation one might define the simplicity of the factorial composition of the ith observed variable to be

$$q_i = \frac{r\sum_{j=1}^{r}(a_{ij}^2)^2 - \left(\sum_{j=1}^{r}a_{ij}^2\right)^2}{r^2}$$

that is, the variance of the squared loadings across the factors for variable i. The variance will be large when there are large as well as near-zero loadings on the factors, indicating that a variable is accounted for by a smaller number than the total number, r, of factors. The variance will be a minimum when the squared loadings are near-uniform magnitude across the factors. Kaiser then showed that maximizing

$$q = \sum_{i=1}^{n} q_i = \sum_{i=1}^{n}\frac{\sum_{j=1}^{r}(a_{ij}^2)^2 - \left(\sum_{j=1}^{r}a_{ij}^2\right)^2}{r^2}$$

the sum of the q_i across the n variables would be equivalent to maximizing the quartimax criterion, q_4, because constant terms (e.g., communalities) would vanish when differentiated. Kaiser thus concluded that the aim of the quartimax criterion is to simplify the rows or variables of the factor-structure matrix but is unconcerned with simplifying the columns or the factors. This implies that the quartimax criterion would frequently yield a general factor. This is because there is no reason why, under the criterion q_i for the simplicity of the ith variable, a large loading could not appear on the same factor for each variable. This tendency to produce a general factor by the quartimax criterion is especially enhanced when the first, principal component dominates in the account of the variance of the variables.

Kaiser (1958) thus suggested that a modification of the quartimax criterion would be to define the simplicity of a factor as

$$v_j^* = \frac{n\sum_{i=1}^{n}(a_{ij}^2)^2 - \left(\sum_{i=1}^{n} a_{ij}^2\right)^2}{n^2} \tag{11.11}$$

that is, the variance of the squared loadings on the factor is an index of the factor's simplicity, and the criterion for all factors would be the maximization of

$$v^* = \sum_{j=1}^{r} v_j^* = \sum_{j=1}^{r} \frac{n\sum_{i=1}^{n}(a_{ij}^2)^2 - \left(\sum_{i=1}^{n} a_{ij}^2\right)^2}{n^2} \tag{11.12}$$

the sum of variances of squared loadings in columns across all columns.

Thus, Kaiser's criterion in Equation 11.12 serves to maximize the interpretability of the factors by simplifying them so that each factor has only a minimum number of variables with large loadings on it.

Kaiser originally designated his criterion v^* in Equation 11.12 as the "raw varimax" criterion, for it applied to the "raw" loadings themselves. In applications of the raw varimax criterion, Kaiser discovered that the criterion produced solutions that were only moderately suggestive of simple-structure solutions. However, Saunders suggested, in a private communication to Kaiser, that he might improve the varimax method if he applied it to the normalized loadings of the variables. Normalizing the loadings can be accomplished by making each variable a unit-length vector in common-factor space, that is by dividing the raw loadings of each variable by the square root of the respective variable's communality. As a consequence the normalized varimax criterion became

$$v = \sum_{j=1}^{r} \frac{n\sum_{i=1}^{n}(a_{ij}^2/h_i^2)^2 - \left(\sum_{i=1}^{n} a_{ij}^2/h_i^2\right)^2}{n^2} \tag{11.13}$$

where h_i^2 is the communality of the ith test. Thus the matrix acted upon by the normalized varimax method of rotation is the matrix $\mathbf{H}^{-1}\mathbf{A}$, where $\mathbf{H}^2$ is an $n \times n$ diagonal matrix of communalities, and $\mathbf{A}$ is the initial $n \times r$ unrotated-factor-structure matrix. The resulting matrix, $\mathbf{B}^* = \mathbf{H}^{-1}\mathbf{A}\mathbf{T}_\upsilon$, where $\mathbf{T}_\upsilon$ is an $r \times r$ orthonormal-transformation matrix determined by the normalized varimax criterion, is then modified into the "original metric" of the variables by premultiplying it by the diagonal matrix of the square roots of the communalities, that is, $\mathbf{B} = \mathbf{H}\mathbf{B}^*$, with $\mathbf{B}$ the resulting rotated-factor-loading matrix to be interpreted. Whereas the raw varimax criterion was biased by inequalities in the communalities of variables and did not always produce a "clean" simple-structure solution, the normalized varimax procedure has performed extremely well, giving solutions considered to be very good orthogonal simple structure. In fact, the normalized varimax criterion is today perhaps the most used orthogonal-rotation procedure for attaining an approximation to simple structure.

Computing a varimax rotation is little different from computing a quartimax rotation. Factors are typically rotated in pairs in an iterative procedure which eventually converges to a point which maximizes the value of v in Equation 11.13. Kaiser (1958) originally worked out equations which for any pair of factors would find the angle of rotation which would maximize the sum of the variances of the squared normalized loadings in the two corresponding columns of the normalized factor-structure matrix $\mathbf{H}^{-1}\mathbf{A}$. Saunders' (1962) method of solving for the tangent of the angle instead of the angle itself, however, leads to an algorithm which can be used with minor modifications in not just varimax rotations but in quartimax and other orthogonal rotations. In fact, the same quartic equation

$$x^4 + \beta x^3 - 6x^2 - \beta x + 1 = 0 \tag{11.8}$$

must be solved for x in varimax rotation as in quartimax rotation; however, in the case of varimax rotation

$$\beta = \frac{\left(\sum_{i=1}^{n} a_{ij}^4 - 6\sum_{i=1}^{n} a_{ij}^2 a_{ik}^2 + \sum_{i=1}^{n} a_{ik}^4\right) + K\left[4\left(\sum_{i=1}^{n} a_{ij}a_{ik}\right)^2 - \left(\sum_{i=1}^{n} a_{ik}^2 - \sum_{i=1}^{n} a_{ij}^2\right)^2\right]\Big/n}{\left(\sum_{i=1}^{n} a_{ij}^3 a_{ik} - \sum_{i=1}^{n} a_{ij}a_{ik}^3\right) + K\left[\sum_{i=1}^{n} a_{ij}a_{ik}\left(\sum_{i=1}^{n} a_{ik}^2 - \sum_{i=1}^{n} a_{ij}^2\right)\right]\Big/n} \tag{11.14}$$

where the coefficients a_{ij} and a_{ik} are elements of the jth and kth columns of the normalized factor-structure matrix, $\mathbf{H}^{-1}\mathbf{A}$. In varimax rotation, the constant K has the value of 1 whereas in quartimax K is zero. In quartimax, the rows of $\mathbf{A}$ are not normalized.

11.4 Transvarimax Methods

Saunders (1962), in seeking to integrate the quartimax and varimax methods into a computer program for oblimax oblique analytic rotation, discovered that because of the restrictions in a single-plane orthogonal rotation, the tangent of a desirable angle of rotation would always be a root of an equation of the form of Equation 11.8, with the coefficient β in this equation varying according to the method of orthogonal rotation involved. He then found that Equation 11.14 would determine β in a manner general to both the quartimax and varimax methods, with $K = 0$ in the quartimax method and $K = 1$ in the varimax method. You may see this if you compare Equation 11.9 to

Equation 11.14. Saunders noted that the varimax method of rotation involved the essential terms of a quartimax rotation as well as terms not in quartimax rotation and so he wondered whether an improved method of rotation might be found by weighting more heavily the terms unique to the varimax method. He thus investigated the consequences of using values for K greater than 1 with data for which he felt there was room for improvement over the solution obtained by varimax rotation. He called such solutions "trans-varimax solutions." Saunders found that satisfactory orthogonal simple structure was available over a considerable range of values for K as soon as some minimum value, dictated by the data, was exceeded. In some cases, if the simple structure was sufficiently clear, the minimum requirement for K could be less than 1 or even less than zero. However, he found that K could not become infinite without entirely dropping the fourth-degree moments from Equation 11.14, and so there had to be an optimum K for most sets of data.

Saunders sought to determine the optimum value for K. After some experimentation, he found the most satisfactory results occurred when it reached a value of K equal to about one-half the number of factors rotated, that is, when

$$K = \frac{r}{2} \tag{11.15}$$

Saunders decided to call a solution in which K was $r/2$ the "equamax" solution, because it gave increased weight to the second-degree moments in Equation 11.14 as the number of factors rotated increased. Saunders claimed that this process may be rationalized intuitively as compensating for the decreased proportion of variables expected to load on any single factor by equalizing the contribution of the second-degree terms.

Henry Kaiser in a personal communication to the author indicated that in his experience the larger the value of K in Equation 11.14 the more nearly equal will be the column sums of squares of rotated-factor loadings. Indeed, when $K = +\infty$ the column sums of squares are equal; on the other hand, when $K = -\infty$ the solution is equivalent to the principal-axes solution (provided that the unrotated solution is a principal-axes solution except for a linear transformation).

11.4.1 Parsimax

Subsequently, Crawford (1967) sought to combine the row-simplifying features of quartimax rotation with the column-simplifying features of varimax rotation in an orthogonal-rotation method known as the "parsimax" (i.e., maximum parsimony) method. The parsimax method produces results that for some data are very similar to equamax solutions and thus casts some light on the reason why equamax may produce acceptable simple-structure solutions.

Crawford considers that an index of row simplicity is Carroll's q_1 criterion, that is, $q_1 = \sum_{i=1}^{n}\sum_{j=1}^{r}\sum_{k=1}^{r} a_{ij}^2 a_{ik}^2, j \neq k$. When q_1 is a minimum, each variable is accounted for by the fewest number of common factors; hence one has maximum parsimony in using the factors to account for the common variance of the variables. An analogous criterion, which we shall designate as p_1 may be used as an index of column simplicity, that is, $p_1 = \sum_{j=1}^{r}\sum_{i=1}^{n}\sum_{h=1}^{n} a_{ij}^2 a_{hj}^2, h \neq i$. When p_1 is a minimum, each factor is determined by fewer than the total number of variables, and one has, in effect, maximum parsimony in determining the factors from the variables. These two criteria may then be combined in the parsimax criterion

$$p_c = K_1 q_1 + K_2 p_1 \tag{11.16}$$

with K_1 and K_2 weights chosen so that q_1 and p_1 always have the same weight regardless of the number of factors rotated. Since there are $nr(n-1)$ terms in p_1 and $nr(r-1)$ terms in q_1, we must have $K_1 = n-1$ and $K_2 = r-1$ to make their contributions relatively equal.

Crawford then shows that minimizing p_c in Equation 11.16 is essentially equivalent to using the transvarimax method of Saunders, with the value of K in Equation 11.14 set at $K = n(r-1)/(n+r-2)$. When the number of variables is relatively small, say on the order of 25, and the number of factors is between one-fourth and one-third the number of variables, then the value of K as determined by the parsimax method will not differ greatly from the value of $r/2$ used for K by the equamax equation; as a result, the two methods will produce similar results. However, as the number of variables increases, the value of K in the parsimax method increases much more rapidly than it does in the equamax method; thus results produced by these two methods should not always be equivalent.

In evaluating the orthogonal varimax and transvarimax methods, Kaiser (personal communication) indicates that the only systematic fault of the varimax method is that in studies where the factors are underdetermined, the varimax method tends not to distribute the variables sufficiently across the factors. Thus, there will be a few "general" factors and several underdetermined factors that are difficult to interpret. Equamax, on the other hand, tends to distribute the variables more nearly equally across the different factors, and so the factors will appear more interpretable. However, Kaiser claims that it is possible to construct cases for which equamax will behave disastrously, e.g., perform a 45° rotation on two pure uncorrelated clusters. Thus, he concludes that although varimax may give at times a moderately distorted view of things it will never result in a catastrophe, as might the equamax method. Crawford emphasized the desirability of knowing the "correct" number of factors to rotate in the parsimax method but favors parsimax to equamax because of parsimax's more explicit rationale. He does, however, recommend that the varimax method be used when there is doubt about the number of factors to rotate.

TABLE 11.1

Quartimax Rotation for 24 Psychological Tests

	1	2	3	4
1	.594	.190	.376	.070
2	.380	.068	.244	.042
3	.472	.010	.314	.014
4	.454	.072	.366	−.014
5	−.018	.130	.809	−.040
6	−.004	.031	.817	.053
7	−.047	.066	.858	−.093
8	.195	.204	.662	−.034
9	−.024	−.058	.860	.097
10	−.125	.702	.234	.113
11	.008	.615	.307	.231
12	.190	.691	.171	−.011
13	.318	.566	.354	−.084
14	−.027	.193	.327	.426
15	.102	.116	.254	.448
16	.370	.142	.285	.364
17	.017	.240	.287	.571
18	.303	.323	.220	.467
19	.190	.180	.279	.321
20	.344	.095	.522	.139
21	.346	.388	.347	.146
22	.288	.039	.528	.259
23	.436	.190	.554	.088
24	.097	.434	.488	.198

Source: Holzinger, K.J. and Harman, H.H., *Factor Analysis*, University of Chicago Press, Chicago, IL, 1941.

Tables 11.1 through 11.4 give examples of factor-structure matrices for a classic study of 24 psychological tests (Holzinger and Harman, 1941), illustrating the quartimax, varimax, equamax, and parsimax methods of rotation. The initial solution was a centroid solution. Input to the rotation program was Kaiser's normalized varimax solution for this study (Kaiser, 1958).

11.5 Simultaneous Orthogonal Varimax and Parsimax

The computing algorithms involving single-plane rotations of pairs of factors in orthogonal-rotation schemes, though convergent, are generally very time-consuming and, today, take up a lion's share of the time used in computing a

TABLE 11.2

Normalized Varimax Rotation (Kaiser, 1958) for 24 Psychological Tests

	1	2	3	4
1	.140	.190	.670	.170
2	.100	.070	.430	.100
3	.150	.020	.540	.080
4	.200	.090	.540	.070
5	.750	.210	.220	.130
6	.750	.100	.230	.210
7	.820	.160	.210	.080
8	.540	.260	.380	.120
9	.800	.010	.220	.250
10	.150	.700	−.060	.240
11	.170	.600	.080	.360
12	.020	.690	.230	.110
13	.180	.590	.410	.060
14	.220	.160	.040	.500
15	.120	.070	.140	.500
16	.080	.100	.410	.430
17	.140	.180	.060	.640
18	.000	.260	.320	.540
19	.130	.150	.240	.390
20	.350	.110	.470	.250
21	.150	.380	.420	.260
22	.360	.040	.410	.360
23	.350	.210	.570	.220
24	.097	.434	.488	.198

factor analysis. Thus, some factor analysts have tried to discover algorithms that directly determine the transformation matrix needed to transform the initial factor-structure matrix into the final factor-structure matrix in the hope that such algorithms might be more efficient than the single-plane-rotation algorithms. The result is that two researchers have been successful by arriving independently at essentially the same algorithm for a simultaneous rotation of all factors in the varimax method. The first solution was given by Horst (1965, pp. 428–429) in his text on factor analysis. The second solution was found by Sherin (1966) and published in *Psychometrika* in 1966. (The present author has learned privately that neither of these authors knew of the other's work.)

To avoid some duplication of discussion, we will discuss the Horst–Sherin algorithm for simultaneous rotation of all factors in the general case which includes both quartimax and varimax as special cases along with equamax and parsimax.

TABLE 11.3

Unnormalized Equamax Rotation for 24 Psychological Tests

	1	2	3	4
1	.193	.203	.654	.165
2	.134	.078	.419	.099
3	.193	.028	.526	.081
4	.242	.098	.522	.067
5	.765	.225	.159	.103
6	.769	.119	.168	.188
7	.833	.173	.145	.054
8	.567	.273	.335	.098
9	.820	.032	.154	.231
10	.142	.712	−.077	.202
11	.177	.619	.060	.328
12	.032	.696	.224	.079
13	.205	.596	.392	.032
14	.232	.186	.016	.487
15	.142	.096	.124	.494
16	.120	.123	.397	.427
17	.157	.212	.040	.628
18	.034	.287	.311	.530
19	.156	.171	.224	.382
20	.390	.128	.438	.241
21	.184	.396	.402	.243
22	.399	.063	.376	.353
23	.396	.227	.538	.207
24	.359	.460	.187	.313

Recall Kaiser's (1958) observation that a criterion for row simplicity for a final factor-structure matrix, $\mathbf{V}_R$, might be

$$Q = \sum_{i=1}^{n}\left[\sum_{j=1}^{r} v_{ij}^4 - \frac{1}{r}\left(\sum_{j=1}^{r} v_{ij}^2\right)^2\right]$$

the sum across variables of the sums of squared deviations of squared factor loadings around the average squared factor loading for each variable. In the orthogonal case this corresponds to the quartimax criterion. On the other hand, a criterion for column simplicity in the final factor-structure matrix, **V**, is

$$V = \sum_{j=1}^{r}\left[\sum_{i=1}^{n} v_{ij}^4 - \frac{1}{n}\left(\sum_{i=1}^{n} v_{ij}^2\right)^2\right] \tag{11.17}$$

TABLE 11.4

Unnormalized Parsimax Rotation for 24 Psychological Tests

	1	2	3	4
1	.161	.196	.656	.202
2	.115	.073	.420	.123
3	.171	.023	.529	.112
4	.218	.094	.529	.097
5	.753	.241	.188	.114
6	.759	.135	.191	.200
7	.823	.191	.179	.064
8	.547	.281	.357	.118
9	.813	.049	.175	.243
10	.128	.719	−.069	.194
11	.159	.624	.060	.328
12	.006	.693	.234	.088
13	.175	.594	.410	.052
14	.226	.195	.001	.487
15	.133	.100	.102	.501
16	.101	.121	.379	.449
17	.150	.220	.014	.629
18	.014	.286	.287	.546
19	.142	.173	.211	.394
20	.369	.130	.442	.266
21	.157	.394	.403	.264
22	.381	.067	.373	.376
23	.368	.227	.545	.238
24	.339	.468	.192	.322

the sum across factors of the sums of squared deviations of squared factor loadings around the average squared factor loading for each factor. In the orthogonal case, V corresponds to the raw varimax criterion. A formulation of the various criteria can then be expressed by the general equation

$$P = K_1 Q + K_2 V \tag{11.18}$$

where

P denotes a general rotation criterion

K_1 and K_2 are weights

Recalling the definitions for Q and V and carrying out some algebraic simplification, we may rewrite Equation 11.18 as

$$P = C_3 \sum_{i=1}^{n} \sum_{j=1}^{r} v_{ij}^4 - C_1 \sum_{i=1}^{n} \left(\sum_{j=1}^{r} v_{ij}^2 \right)^2 - C_2 \sum_{j=1}^{r} \left(\sum_{i=1}^{n} v_{ij}^2 \right)^2 \tag{11.19}$$

where

$C_3 = K_1 + K_2$
$C_1 = K_1/r$
$C_2 = K_2/n$

Actually, the criterion in Equation 11.19 may be made even more general by trying other values for the weights C_1, C_2, and C_3. For example, if $C_3 = 1$, and $C_1 = C_2 = 0$ we have the quartimax criterion. If $C_3 = 1$, $C_1 = 0$, and $C_2 = 1/n$, we have the raw varimax criterion. If $C_3 = 1$, $C_1 = 0$, and $C_2 = (r/2)/n$ we have the equamax criterion, and if $C_3 = 1$, $C_1 = 0$, and $C_2 = (r - 1)/(n + r - 2)$ we have Crawford's (1967) orthogonal parsimax criterion. Under orthogonal transformations the term involving C_1 will always be invariant and thus not contribute to the maximization of the criterion, but we retain this term in the equation for generalization later on to the case of transformations.

Now, let **A** be the initial, unrotated, rowwise normalized factor-structure matrix and **T** be an $r \times r$ orthonormal-transformation matrix such that

$$\mathbf{V} = \mathbf{AT}$$

or

$$\mathbf{V} = \left\| v_{ij} \right\| = \left\| \sum_{m=1}^{r} a_{im} t_{mj} \right\| \tag{11.20}$$

which indicates that the matrix **V** is a function of the matrix **T**.

We wish to maximize P as a function of the matrix $\mathbf{T} = \left\| t_{qs} \right\|$ under the restraints that

$$u_{jh} = \sum_{k=1}^{r} t_{kj} t_{kh} = \begin{cases} 1, & j = h \\ 0, & j \neq h \end{cases}, \quad j, h = 1, \ldots, r \tag{11.21}$$

which requires the matrix **T** to be orthonormal. Observe that $u_{jh} = u_{hj}$ and so there are only $r(r + 1)/2$ independent restraints. For each restraint we may form an equation. Of these equations r will have the form

$$u_{jj} - 1 = \sum_{k=1}^{r} t_{kj}^2 - 1 = 0, \quad j = 1, \ldots, r \tag{11.22a}$$

and $r(r - 1)$ of the equations will be of the form

$$u_{jh} = \sum_{k=1}^{r} t_{kj} t_{kh} = 0, \quad j \neq h = 1, \ldots, r \tag{11.22b}$$

Let $\theta = \left\| \theta_{jh} \right\|$, with each element θ_{jh} corresponding to the Lagrange multiplier $2\gamma_{jh}$ of an equation $u_{jh} - \delta_{jh} = 0$. δ_{jh} is the Kronecker delta, where $\delta_{jh} = 1$ if $j = h$, and $\delta_{jh} = 0$ if $j \neq h$. Keep in mind that the matrix θ is symmetric because $u_{jh} = u_{hj}$.

The symmetry of θ will be used later on to eliminate θ and thereby obtain a matrix equation for **T**. According to the method of Lagrange for maximizing a function subject to restraints, a necessary condition that P in Equation 11.19 be a maximum subject to the restraints in Equation 11.21 is

$$\frac{\partial P}{\partial t_{qs}} = \frac{\partial}{\partial t_{qs}} \sum_{j-1}^{r} \sum_{j=1}^{r} 2\theta_{jh} \left(\sum_{k=1}^{r} t_{kj} t_{kh} - \delta_{jh} \right) = 0 \tag{11.23}$$

where δ_{jh} is the Kronecker delta.

As background for the partial differentiation of P with respect to t_{qs} for each q and s

$$\frac{\partial v_{ij}}{\partial t_{qs}} = 0 \quad \text{if } j \neq s$$

Moreover, taking account of Equation 11.20, we have

$$\frac{\partial v_{is}}{\partial t_{qs}} = \frac{\partial \sum_{k=1}^{r} a_{im} t_{ms}}{\partial t_{qs}} = a_{iq}$$

$$\frac{\partial P}{\partial t_{qs}} = 4C_3 \sum_{i=1}^{n} v_{is}^3 a_{iq} - 4C_1 \sum_{i=1}^{n} \left(\sum_{j=1}^{r} v_{ij}^2 \right) v_{is} a_{iq} - 4C_2 \sum_{i=1}^{n} v_{is}^2 \left(\sum_{i=1}^{n} v_{is} a_{iq} \right) \tag{11.24}$$

On the other hand,

$$\frac{\partial}{\partial t_{qs}} \sum_{j=1}^{r} \sum_{h=1}^{r} 2\theta_{jh} \left(\sum_{k=1}^{r} t_{kj} t_{kh} - \delta_{jh} \right) = 4 \sum_{j=1}^{r} \theta_{js} t_{qj} \tag{11.25}$$

Substituting Equations 11.24 and 11.25 in Equation 11.23 and dividing both sides of the equation by 4, we have

$$C_3 \sum_{i=1}^{n} v_{is}^3 a_{iq} - C_1 \sum_{i=1}^{n} \left(\sum_{j=1}^{r} v_{ij}^2 \right) v_{is} a_{iq} - C_2 \sum_{i=1}^{n} v_{is}^2 \left(\sum_{i=1}^{n} v_{is} a_{iq} \right) - \sum_{j=1}^{r} \theta_{js} t_{qj} = 0 \tag{11.26}$$

Now, following a notation used by Sherin (1966), define

$$\mathbf{W} = \| w_{ij} \| = C_3 \| v_{ij}{}^3 \|$$

$$\mathbf{D} = C_2[\text{diag } \mathbf{V'V}] \tag{11.27}$$

the $r \times r$ diagonal matrix having as diagonal elements

$$d_{jj} = C_2 \left(\sum_{i=1}^{n} v_{ij}^2 \right) \tag{11.28}$$

and

$$\mathbf{H} = C_1[\text{diag } \mathbf{VV'}]$$

the $n \times n$ diagonal matrix having as diagonal elements

$$h_{ii} = C_1\left(\sum_{j=1}^{r} v_{ij}^2\right) \tag{11.29}$$

Substituting Equations 11.27 through 11.29 in Equation 11.26, we have

$$\sum_{i=1}^{n} w_{is}a_{iq} - \sum_{i=1}^{n} h_{ii}v_{is}a_{iq} - d_{ss}\sum_{i=1}^{n} v_{is}a_{iq} - \sum_{j=1}^{r} \theta_{js}t_{qj} = 0 \tag{11.30}$$

which may be rewritten in matrix form as

$$\mathbf{A'W} - \mathbf{A'HV} - \mathbf{A'VD} - \mathbf{T\theta} = \mathbf{0} \tag{11.31}$$

Recall that $\mathbf{T}$ is an orthonormal matrix, where $\mathbf{T'} = \mathbf{T}^{-1}$; then use this fact to obtain

$$\boldsymbol{\theta} = \mathbf{T'A'(W - HV - VD)} \tag{11.32}$$

However, if we substitute the expression for $\boldsymbol{\theta}$ obtained by Equation 11.32 in Equation 10.32 to eliminate $\boldsymbol{\theta}$ we obtain a trivial result. But we can avoid the trivial result if we recall that $\boldsymbol{\theta}$ is a symmetric matrix, that is, $\boldsymbol{\theta'} = \boldsymbol{\theta}$, and write

$$\boldsymbol{\theta} = \boldsymbol{\theta'} = \mathbf{(W - HV - VD)'AT} \tag{11.33}$$

We may substitute the expression in Equation 11.33 for $\boldsymbol{\theta}$ in Equation 11.31 to obtain

$$\mathbf{A'(W - HV - VD)} = \mathbf{T(W - HV - VD)'AT} \tag{11.34}$$

Postmultiply both sides of Equation 11.33 by $\mathbf{T'}$ to obtain Equation 11.35

$$\mathbf{A'(W - HV - VD)T'} = \mathbf{T(W - HV - VD)'A} \tag{11.35}$$

We are now in a position to consider an iterative computing algorithm to find the matrix $\mathbf{T}$. Let $\mathbf{G} = \mathbf{A'(W - HV - VD)}$ so that Equation 11.35 becomes

$$\mathbf{GT'} = \mathbf{TG'} \tag{11.36}$$

Post-multiply both sides of Equation 11.36 by $\mathbf{TG'}$; this gives

$$\mathbf{GG'} = \mathbf{TG'TG'} = [\mathbf{TG'}]^2 \tag{11.37}$$

Take the square root of both sides (by an eigenvector–eigenvalue decomposition of **GG′**), obtaining

$$[\mathbf{GG}']^{1/2} = \mathbf{QD}_{GG}^{1/2}\mathbf{Q}' = \mathbf{TG}' \tag{11.38}$$

where

Q is the eigenvector matrix

$\mathbf{D}_{GG}$ the eigenvalue matrix of **GG′**

Now, left-multiply by **T′**, right-multiply by $[\mathbf{GG}']^{-1/2}$, and transpose both sides to obtain

$$\mathbf{T} = [\mathbf{GG}']^{-1/2}\mathbf{G} \tag{11.39}$$

The matrix **T** in Equation 11.38 is orthogonal, as may be verified by multiplying it by its transpose and carrying out the necessary matrix algebra.

We may use Equation 11.39 as the basis for an iterative computing algorithm. Let $\mathbf{T}_{(0)} = \mathbf{I}$ be the initial value of the orthonormal-transformation matrix where $\mathbf{T}_{(g)}$ is the value of the transformation matrix in the gth iteration. Then define

$$\mathbf{V}_{(g)} = \mathbf{AT}_{(g)}\mathbf{V}_{(9)} = \mathbf{AT}_{(9)}$$

$$\mathbf{W}_{(g)} = \left\|v_{ij}^3\right\|_{(g)} C_3$$

$$\mathbf{H}_{(g)} = [\text{diag}\mathbf{V}_{(g)}\mathbf{V}'_{(g)}]C_1$$

$$\mathbf{D}_{(g)} = [\text{diag}\mathbf{V}'_{(g)}\mathbf{V}_{(g)}]C_2$$

$$\mathbf{G}_{(g)} = \mathbf{A}'(\mathbf{W}_{(g)} - \mathbf{H}_{(g)}\mathbf{V}_{(g)} - \mathbf{V}_{(g)}\mathbf{D}_{(g)})$$

$$\mathbf{T}_{(g+1)} = [\mathbf{G}_{(g)}\mathbf{G}'_{(g)}]^{-1/2}\mathbf{G}_{(g)} \tag{11.40}$$

The matrix $\mathbf{T}_{(g+1)}$ in Equation 11.40 will always be orthonormal.

Horst (1965, p. 428) points out in connection with the varimax method ($C_3 = 1$, $C_2 = 1/n$) that the above algorithm is self-correcting and does not accumulate round-off error. He does not report much experience with the method but does note that with small matrices the Kaiser method of single-plane rotations is faster, whereas with larger matrices the simultaneous transformation method appears faster. Sherin (1966) uses a slight modification of the above algorithm in which $\mathbf{V}_{(g-1)}$ is substituted for **A** in each step so that $\mathbf{V}_{(g)} = \mathbf{V}_{(g-1)}\mathbf{T}_{(g)}$, etc. In this case the method converges to the proper **V** matrix and the final transformation matrix **T** is the identity matrix. Sherin's method, however, may accumulate round-off error. Sherin reports that his version of the algorithm converged slowly and the rate of convergence decreased as it neared the final solution. The Kaiser method of single-plane rotations appeared faster, according to Sherin.

11.5.1 Gradient Projection Algorithm

The simultaneous varimax method just described has similarities to a more general algorithm, proposed by Jennrich (2002, 2004), known as the "gradient projection algorithm" (GPA). The GPA algorithm can be readily adapted to both orthogonal and oblique rotation and many different rotation criteria. GPA is iterative and has a loop with two major steps. Given a loading matrix **A** for orthogonal factors that is to be rotated, let **T** be an initial $r \times r$ transformation matrix, and **V** be the resulting rotated matrix $\mathbf{V} = \mathbf{AT}$. In the orthogonal case **T** must be an orthogonal matrix, implying that $\mathbf{T'T} = \mathbf{I}$. A criterion of rotation to be optimized, $Q(\mathbf{V})$ must be specified as a function of the elements of the matrix **V**. $Q(\mathbf{V})$ can further be regarded as a function of **T**, in that we seek to optimize $Q(\mathbf{AT})$. Using the theory of differentials, Jennrich (2004) shows that the $r \times r$ gradient matrix **G** in the orthogonal transformation case is equal to

$$\mathbf{G} = \mathbf{A'G}_q \tag{11.41}$$

where
- $\mathbf{A'}$ is the $r \times n$ transpose of the original unrotated matrix
- $\mathbf{G}_q$ is an $n \times r$ matrix representing the gradient of Q at **V**
- $\mathbf{G}_q$ may be obtained by obtaining the partial derivative of Q with respect to each element in **V**

In the case of varimax rotation the criterion for Q is given as

$$Q(\mathbf{V}) = -\text{tr}[(\mathbf{V}^{[2]} - \mathbf{NV}^{[2]})'(\mathbf{V}^{[2]}) - \mathbf{NV}^{[2]})]/4$$

where $\mathbf{V}^{[2]} = [v_{ij}^2]$ is the matrix obtained from **V** by replacing each of its elements by the square of the respective element. The matrix **N** is an $n \times n$ matrix of 1s, each of which is divided by n. That is,

$$\mathbf{N} = (1/n)(\mathbf{1}_n\mathbf{1}'_n)$$

where $\mathbf{1}_n$ is an $n \times 1$ sum vector of 1s.

On the other hand, for the varimax criterion, $\mathbf{G}_q$ is given as

$$\mathbf{G}_q = -\mathbf{V} \cdot (\mathbf{V}^{[2]} - \mathbf{NV}^{[2]})$$

where $\cdot$ denotes an element-by-corresponding-element multiplication of the elements of the first matrix times the elements of the second to produce a matrix with the same dimensions.

The gradient projection algorithm then proceeds as follows. Let α be some number initially equal to 1, but to be increased or decreased during the iterations, serving to determine the degree to which the gradient elements in **G** are to be added to a current **T** to get a new **T**. More about this shortly.

The two major steps of the gradient projection algorithm are as follows:

1. Compute $Q(\mathbf{V})$, $\mathbf{G}_q$.
   Compute $\mathbf{G} = \mathbf{A}'\mathbf{G}_q$, $\mathbf{M} = \mathbf{T}'\mathbf{G}$, $\mathbf{S} = (1/2)(\mathbf{M} + \mathbf{M}')$

   $$\mathbf{G}_p = \mathbf{G} - \mathbf{TS}$$

   Compute $\mathbf{W} = \mathbf{T} - \alpha\mathbf{G}_p$, find singular value decomposition of

   $$\mathbf{W} \quad \text{as} \quad \mathbf{W} = \mathbf{PDQ}'$$

2. Replace **T** with **PQ**′. Replace **V** with **AT**.
   Go to 1.

Jennrich (2002) has discussed various methods of accelerating and detecting convergence with the GPA algorithm. We will not discuss them here, but will do so in the next chapter.

Bernaards and Jennrich (2005) have posted programs in several different programming languages for both orthogonal and oblique rotations using the GPA algorithm according to most of the rotation criteria at http://www.stat.ucla.edu/research

12

Oblique Analytic Rotation

12.1 General

Major developments in the methodology of rotation have occurred since 1965. First, the paper of Jennrich and Sampson (1966) showed that it was not necessary or desirable to rotate first to a reference-structure matrix and then convert that to a factor-pattern matrix. One could apply the rotational criteria directly to the determination of a factor-pattern matrix, and this would more easily avoid a collapse of the factor space into fewer dimensions. Second, while planar rotations in two dimensions continued to dominate the approach to rotation through to the 1990s, since 2000, a number of papers by Jennrich (2001, 2002, 2004, 2006) have demonstrated a unified approach to all rotational methods using simultaneous rotations of all factors via $r \times r$ transformation matrices **T**. Furthermore, Jennrich has shown, through a compact but easily mastered notation, that these approaches can be quite easy to program and implement. Thus I regard Jennrich's contributions to be the wave of the future, and in this chapter I will not dwell on rotations to reference-structure matrices, and will illustrate the planar rotation approach with only the method of direct oblimin. I will describe the criteria of the various popular methods of rotation and then show how they may be applied to obtain factor-pattern matrices by Jennrich's general approach using gradient-projection algorithms (GPAs).

12.1.1 Distinctness of the Criteria in Oblique Rotation

In our previous discussion of analytic orthogonal rotation, we emphasized how most of the criteria for rotation represented attempts to express mathematically the essence of Thurstone's five criteria for simple structure. Recall, for example, that Carroll proposed minimizing the criterion

$$q_1 = \sum_{j<k=1}^{r} \sum_{i=1}^{n} b_{ij}^2 b_{ik}^2 \quad j \neq k \tag{12.1}$$

which was an index of the degree to which factor loadings for any given variable differed in magnitude on different factors. Neuhaus and Wrigley (1954), on the other hand, sought to maximize the variance of squared loadings in the rotated reference-structure matrix by maximizing the function

$$q_2 = \frac{nr\sum_{i=1}^{n}\sum_{j=1}^{r}(b_{ij}^2)^2 - \left(\sum_{i=1}^{n}\sum_{j=1}^{r} b_{ij}^2\right)^2}{n^2r^2} = \text{Maximum} \tag{12.2}$$

Saunders, in turn, wanted to maximize the kurtosis of the loadings and their reflections by maximizing the criterion

$$q_3 = \frac{nr\sum_{i=1}^{n}\sum_{j=1}^{r} b_{ij}^4}{\left(\sum_{i=1}^{n}\sum_{j=1}^{r} b_{ij}^2\right)^2} = \text{Maximum} \tag{12.3}$$

and Ferguson (1954) simply sought to maximize the function

$$q_4 = \sum_{i=1}^{n}\sum_{j=1}^{r} b_{ij}^4 = \text{Maximum}$$

In the case of orthogonal rotations, each of these criteria led to the same solution because of the invariance of the sum, $\sum_{j=1}^{r} v_{ij}^2 = h_i^2$, under orthogonal transformations. These criteria are not equivalent when, to consider oblique transformations, we relax the restraint that transformations must be orthogonal, because the sums of squared elements in the rows are not invariant under oblique transformations. Thus, in the case of oblique transformation, there may be several different ways of defining criteria for the simple structure.

12.2 Oblimin Family

A criterion for oblique rotation that has led to a family of criteria that have been quite popular is $q_1 = \sum_{j=1}^{r}\sum_{k=1}^{r}\left(\sum_{i=1}^{n}\lambda_{ij}^2\lambda_{ik}^2\right) j \neq k$ (now applied to directly obtain a pattern matrix). Carroll (1953) and others used this criterion quite successfully for oblique rotation with its name coming to be called "quartimin." In the same vein, a criterion related both logically and mathematically to Carroll's q_1 was also proposed by Kaiser (1958) as the oblique counterpart to his varimax method, involving the minimization across all pairs of reference axes of the sum of covariances of squared factor loadings (summing for each pair across variables), that is, minimizing

$$k = \sum_{j=1}^{r}\sum_{k=1}^{r}\left[\sum_{i=1}^{n}\lambda_{ij}^{2}\lambda_{ik}^{2} - \frac{1}{n}\left(\sum_{i=1}^{n}\lambda_{ij}^{2}\right)\left(\sum_{i=1}^{n}\lambda_{ik}^{2}\right)\right] \quad j \neq k \tag{12.4}$$

That this is related to Carroll's criterion, q_1, in Chapter 11 should be obvious, for we may rewrite Equation 12.4 as

$$k = q_1 - \frac{1}{n}\sum_{j=1}^{r}\sum_{k=1}^{r}\left(\sum_{i=1}^{n}\lambda_{ij}^{2}\right)\left(\sum_{i=1}^{n}\lambda_{ik}^{2}\right) \quad j \neq k$$

Carroll (1957) noted this relationship between q_1 and Kaiser's criterion k (which Carroll named the "covarimin" criterion) and proposed a new criterion B which was a weighted combination of q_1 and k, that is,

$$B = \alpha q_1 + \beta k \tag{12.5}$$

Because k contains q_1, we may rewrite Equation 11.10 as

$$B = (\alpha + \beta)q_1 - \frac{\beta}{n}\sum_{j=1}^{r}\sum_{k=1}^{r}\left(\sum_{i=1}^{n}\lambda_{ij}^{2}\right)\left(\sum_{i=1}^{n}\lambda_{ik}^{2}\right) \quad j \neq k \tag{12.6}$$

But we may divide both sides of Equation 12.11 by $\alpha + \beta$ and, letting $\gamma = \beta/(\alpha + \beta)$, rewrite our criterion finally as

$$B = \sum_{j=1}^{r}\sum_{k=1}^{r}\left[\sum_{i=1}^{n}\lambda_{ij}^{2}\lambda_{ik}^{2} - \frac{\gamma}{n}\left(\sum_{i=1}^{n}\lambda_{ij}^{2}\right)\left(\sum_{j=1}^{n}\lambda_{ik}^{2}\right)\right] \quad j \neq k \tag{12.7}$$

When B is minimized, we should have a simple-structure solution. Experience has shown that when $\gamma = 0$ and B is equivalent to Carroll's original quartimin criterion, q_1, the resulting solution produces reference axes that are too oblique, whereas when letting B be Kaiser's covarimin criterion, k, by setting $\gamma = 1$, the resulting reference axes tend to be too nearly orthogonal. Thus one would suppose that setting γ equal to a value between 0 and 1 would affect a desirable compromise. Carroll (1957) reported that when $\gamma = .5$ the results were the most satisfactory, and he designated the criterion when $\gamma = .5$ the "biquartimin" criterion. Unfortunately, this choice for the value of γ is based only on the fact that most of the time letting γ equal .5 results in good simple structure, whereas in reality for each different set of data there may be an optimum but different value for γ which for those data will produce the "best" simple-structure solution. As yet, however, no a priori rationale has been provided for choosing an optimum value for γ with each set of data.

12.2.1 Direct Oblimin by Planar Rotations

A radical departure from the usual approaches to oblique rotation, proposed by Jennrich and Sampson (1966) and named the "direct oblimin" method by Harman (1967), involved directly finding the primary-factor-pattern matrix, $\mathbf{\Lambda}$. This method differs from the usual practice of first finding the reference-structure matrix, $\mathbf{V}_R$, and then finding $\mathbf{\Lambda}$ by the equation $\mathbf{\Lambda}=\mathbf{V}_R\mathbf{D}_V$, where $\mathbf{D}_V^2=[\text{diag }\mathbf{R}_{VV}^{-1}]$, and $\mathbf{R}_{VV}$ is the matrix of correlations among the reference axes. $\mathbf{\Lambda}$ may also exhibit a simple structure because the zeros in $\mathbf{\Lambda}$ correspond to zeros in $\mathbf{V}_R$.

Jennrich and Sampson suggest that if one can maximize or minimize a simplicity function $f(\mathbf{V}_R)$, for example, the oblimin function, in connection with the reference-structure matrix, $\mathbf{V}_R=\mathbf{AL}$, under the restraint that $[\text{diag }\mathbf{L'L}]=\mathbf{I}$, one can similarly maximize or minimize the same simplicity function $f(\mathbf{\Lambda})$ in connection with the primary-factor-pattern matrix, $\mathbf{\Lambda}=\mathbf{A}(\mathbf{T}')^{-1}$. Here $\mathbf{T}$ is the $r\times r$ transformation matrix in the equation $\mathbf{X}=\mathbf{T'X}_{(0)}$ used to transform the initial primary axes, designated by the random vector, $\mathbf{X}_{(0)}$, into the final primary axes, designated by $\mathbf{X}$, with $\mathbf{T}$ satisfying the restraint that $[\text{diag }\mathbf{T'T}]=\mathbf{I}$.

To illustrate their point, Jennrich and Sampson developed an algorithm for applying Carroll's oblimin criterion, q_1, to the primary-factor-pattern matrix, $\mathbf{\Lambda}$. Their algorithm proceeds by a sequence of elementary single-plane rotations. To describe this algorithm, let $X_1,\ldots,X_r$ represent the primary factors at an intermediate step, let $\mathbf{\Lambda}$ be the corresponding factor-pattern matrix, and $\mathbf{R}_{XX}$ the corresponding matrix of correlations among the primary factors. If one chooses two factors, say X_s and X_h, then a simple rotation consists of rotating X_s to a new position X_s^*, in the plane of X_s and X_h in such a way that the resulting factor-pattern matrix $\mathbf{\Lambda}^*$ minimizes the oblimin function applied to $(n-r)\times r$. The rotated factor

$$X_s^* = t_sX_s + t_hX_h \tag{12.8}$$

is a simple linear combination of just the vectors X_s and X_h, with the weights t_s and t_h chosen so that X_s^* has unit length (or unit standard deviations). The unit-length requirement for X_s^* implies that

$$t_s^2 + 2t_st_hr_{sh} + t_h^2 = 1 \tag{12.9}$$

since the length of X_s^* is given by

$$\begin{aligned}|X_s^*| &= (X_s^{*\prime}X_s^*)^{1/2} = [(t_sX_s + t_hX_h)'(t_sX_s + t_hX_h)]^{1/2}\\ &= (t_s^2X_s'X_s + t_st_hX_s'X_h + t_st_hX_h'X_s + t_h^2X_h'X_h)^{1/2}\\ &= (t_s^2 + 2t_st_hr_{sh} + t_h^2)^{1/2}\end{aligned}$$

with $X_s'X_s=X_h'X_h=1$ (because they represent correlations of the variables with themselves), implying the original factors have unit length (standard deviations) and $X'_s X_h = r_{sh}$, the correlation between X_s and X_h. Having worked thus far with transformations of reference-structure loadings, we must now consider the effect of single-plane transformations of the primary axes on the primary-factor-pattern loadings. As background, suppose we have a vector

$$\mathbf{y} = a_1\mathbf{x}_1 + a_2\mathbf{x}_2 + \cdots + a_r\mathbf{x}_r \tag{12.10}$$

that is a linear combination of some basis vectors $\mathbf{x}_1,\ldots,\mathbf{x}_r$, with $a_1,\ldots,a_r$ the weights applied to these vectors. Let us now consider the effect of modifying the set of basis vectors by replacing the first basis vector, $\mathbf{x}_1$, with a new basis vector $\mathbf{x}_1^* = w_1\mathbf{x}_1 + w_2\mathbf{x}_2$, which is a linear combination of just vectors $\mathbf{x}_1$ and $\mathbf{x}_2$. (The vector $\mathbf{x}_1^*$ evidently lies in the plane defined by $\mathbf{x}_1$ and $\mathbf{x}_2$.) We may describe the vector $\mathbf{y}$ as a linear combination of the new set of basis vectors, that is,

$$\mathbf{y} = b_1\mathbf{x}_1^* + b_2\mathbf{x}_2 + \cdots + b_r\mathbf{x}_r \tag{12.11}$$

and the question is, "What is the relationship between the weights $a_1,\ldots,a_r$ and the weights $b_1,\ldots,b_r$?"

Consider that $\mathbf{x}_1 = (1/w_1)(\mathbf{x}_1^* - w_2\mathbf{x}_2)$ by rearranging the equation for $\mathbf{x}_1^*$. Or we may rewrite this, by removing parentheses, as

$$\mathbf{x}_1 = \frac{1}{w_1}\mathbf{x}_1^* - w_2\mathbf{x}_2 \tag{12.12}$$

Substituting Equation 12.17 for $\mathbf{x}_1$ in Equation 12.15 we have, after collecting terms,

$$\mathbf{y} = \frac{a_1}{w_1}\mathbf{x}_1^* + \left(a_2 - \frac{w_1}{w_2}a_1\right)\mathbf{x}_2 + \cdots + a_r\mathbf{x}_r \tag{12.13}$$

This equation expresses $\mathbf{y}$ in terms of the new set of basis vectors, and so, evidently,

$$b_1 = \frac{a_1}{w_1} \qquad b_2 = a_2 - \frac{w_2}{w_1}a_1 \quad \text{and} \quad b_k = a_k \qquad k = 3,\ldots,r \tag{12.14}$$

Let us return now to the question of what the coefficients of $\boldsymbol{\Lambda}^*$ will be after the change of basis is carried out by replacing primary factor X_s by a new factor X_s^* defined as in Equation 12.8. If we denote by λ_{ij} and λ_{ij}^* the current and subsequent factor-pattern loadings, respectively, for the ith variable on the jth factor, then using the results of Equation 12.14, we have

$$\lambda_{is}^* = \frac{\lambda_{is}}{t_s} \quad \lambda_{ih}^* = \left(\lambda_{ih} - \frac{t_h}{t_s}\lambda_{is}\right) \quad \lambda_{ik}^* = \lambda_{ik} \quad s,h \neq k \tag{12.15}$$

From Equation 12.15, we can conclude that the change in one primary factor produces a change in two columns of the resulting factor-pattern matrix Λ^*. Thus if we denote by $B_{s,h}$ the oblimin criterion applied to the factor-pattern matrix Λ^*, after replacing primary factor X_s by a new factor X_s^* lying in the plane defined by primary factors X_s and X_h, we have

$$\begin{aligned} B_{s,h} &= \sum_{j=1}^{r}\sum_{k=1}^{r}\left(\sum_{i=1}^{n}\lambda_{ij}^{*2}\lambda_{ik}^{*2}\right) \\ &= \sum_{i=1}^{n}\lambda_{is}^{*2}\lambda_{ih}^{*2} + \sum_{i=1}^{n}\left(\lambda_{is}^{*2} + \lambda_{ih}^{*2}\right)\sum_{i=1}^{n}\lambda_{ik}^{*2} + K \quad s,h \neq k \end{aligned} \tag{12.16}$$

where K is a constant term unaffected by the change of factor X_s to X_s^*. Let

$$w_i = \sum_{k=1}^{r}\lambda_{ik}^{*2} \quad s,h \neq k$$

Then our aim is to minimize the function

$$g = \sum_{i=1}^{n}\left(\frac{\lambda_{is}}{t_s}\right)^2\left(\lambda_{ih} - \frac{t_h}{t_s}\lambda_{is}\right)^2 + \sum_{i=1}^{n}\left[\left(\frac{\lambda_{is}}{t_s}\right)^2 + \left(\lambda_{ih} - \frac{t_h}{t_s}\lambda_{is}\right)^2\right]w_i \tag{12.17}$$

under the unit-length restraint in Equation 12.9.

Let us use the change of variable

$$y = \frac{1}{t_s} \quad x = \frac{t_h}{t_s}$$

Then, after a fair amount of algebraic manipulation, we obtain from Equations 12.9 and 12.17

$$y^2 = 1 + 2r_{sh}x + x^2 \tag{12.18}$$

and

$$g = a_4x^4 + a_3x^3 + a_2x^2 + a_1x + a_0 \tag{12.19}$$

where

$$a_0 = \sum_{i=1}^{n} w_i(\lambda_{is}^2 + \lambda_{ih}^2) + \sum_{i=1}^{n}\lambda_{is}^2\lambda_{ih}^2$$

$$a_1 = 2r_{sh}\sum_{i=1}^{n}(\lambda_{is}^2(w_i + \lambda_{ih}^2) - 2\sum_{i=1}^{r}\lambda_{is}\lambda_{ih}(w_i + \lambda_{is}^2)$$

$$a_2 = \sum_{i=1}^{n}\lambda_{is}^2(\lambda_{is}^2 - 4r_{sh}\lambda_{is}\lambda_{ih} + \lambda_{ih}^2 + w_i)$$

$$a_3 = 2\sum_{i=1}^{n}\lambda_{is}^2(r_{sh}\lambda_{is}^2 - \lambda_{is}\lambda_{ih})$$

$$a_4 = \sum_{i=1}^{n}\lambda_{is}^4 \qquad (12.20)$$

We need concern ourselves only with minimizing Equation 12.19 without restrictions. Equation 12.19 will be a minimum when its derivative with respect to x is zero. Hence we need only to find roots to the equation

$$\frac{dg}{dx} = 4a_4x^3 + 3a_3x^2 + 2a_2x + a_1 = 0 \qquad (12.21)$$

which may be done readily by Newton–Raphson approximation. A good initial trial value in the Newton–Raphson procedure is $x=0$, especially in later rotations when the process has nearly converged. Whether a root of Equation 12.21 minimizes Equation 12.19 or not may be tested by examining the sign of

$$\frac{d^2g}{dx^2} = 12a_4x^2 + 6a_3x + 2a_2 \qquad (12.22)$$

at the root. If Equation 12.22 is positive, then the root corresponds to a minimum of Equation 12.19.

When the minimizing root x has been found, then one finds the corresponding value for y by taking the square root of Equation 12.18 after substituting into it the value for x which minimizes g. (The sign of the square root is arbitrary.) Then

$$t_s = \frac{1}{y} \quad t_h = t_s x \qquad (12.23)$$

and one obtains the new loadings for the matrix $\boldsymbol{\Lambda}^*$ by using Equations 12.23 and 12.15. The new coefficients in the matrix $\mathbf{R}_{XX}^* = [r_{jk}^*]$ of correlations among factors is given by

$$r_{sj}^* = t_s r_{sj} + t_h r_{hj} \quad j \neq s$$

$$r_{jk}^* = r_{jk} \quad j,k \neq s \tag{12.24}$$

When the rotations have converged, $\mathbf{\Lambda}$ contains the rotated primary-factor-pattern matrix, and $\mathbf{R}_{XX}$ contains the correlations among the primary factors.

Jennrich and Sampson (1966) made some interesting comparisons between their direct oblimin method and the indirect oblimin methods. They point out that the pure oblimin method, using the value of zero for the coefficient γ in Equation 12.7, will produce perfect simple-structure solutions when they exist, as is the case in artificial problems such as the one used to illustrate graphical rotation, earlier in this chapter. The disadvantage with the indirect oblimin procedure is that with some data the reference axes tend to become so correlated that the solution tends to become singular and lose a distinct factor in the process. Avoiding singularity, of course, was the motive for the biquartimin and covarimin methods of Carroll and Kaiser, respectively.

However, Jennrich and Sampson (1966) illustrate in their discussion of the direct oblimin method how indirect methods will not always recover a perfect simple structure when it exists. Actually, they were not sure whether one could always find a set of linearly independent reference axes whose reference-structure matrix, $\mathbf{V}_R$, would minimize the oblimin criterion, but they present a proof that with any data there always is a factor-pattern matrix, $\mathbf{\Lambda}$, which minimizes the oblimin criterion and at the same time retains the linear independence of the primary axes. In the opinion of Jennrich and Sampson, the difficulty with indirect oblimin methods is that the elements of the reference-structure matrix must never exceed 1 in absolute magnitude, whereas in the direct oblimin method the elements of the factor-pattern matrix are free to range, potentially, between minus and plus infinity in magnitude.

We will come back to the oblimin family farther on in connection with Jennrich's GPA.

12.3 Harris–Kaiser Oblique Transformations

In most factor analyses, the researcher initially obtains an $n \times r$ factor-pattern matrix, $\mathbf{A}$, corresponding to some orthogonal solution (e.g., principal-axes solution). Then he transforms $\mathbf{A}$ into either a reference-structure matrix, $\mathbf{V}_R = \mathbf{AL}$, or a factor-pattern matrix, $\mathbf{\Lambda} = \mathbf{AM}$, representing a simple-structure solution, where $\mathbf{L}$ and $\mathbf{M}$ are $r \times r$ transformation matrices. Harris and Kaiser (1964) suggest that a promising line of attack on the problem of oblique transformations of factors involves regarding the transformation matrices, $\mathbf{L}$ and $\mathbf{M}$, as themselves composed of simple orthonormal and diagonal transformation matrices. For example, an arbitrary transformation matrix, $\mathbf{M}$, may be regarded as the product

$$\mathbf{M} = \mathbf{T}_2\mathbf{D}_2\mathbf{T}_1\mathbf{D}_1 \tag{12.25}$$

where

$\mathbf{T}_2$ and $\mathbf{T}_1$ are orthonormal $r \times r$ matrices
$\mathbf{D}_2$ and $\mathbf{D}_1$ are $r \times r$ diagonal matrices

The algebraic basis for the composition of **M** as in Equation 12.25 is a theorem originally advanced (but not rigorously proved) by Eckart and Young (1936) but later proved by Johnson (1963). This theorem is also known as the theorem of the "singular-value decomposition" of a matrix. This theorem states that, for any real matrix **A**, two orthogonal matrices **P** and **Q** can be found for which **P′AQ** is a real diagonal matrix **D** with no negative elements. This theorem, which represents one of the few original contributions of psychologists to the theory of matrices, may be applied in many situations in the theory of factor analysis and multivariate statistics. We will first outline a proof of this theorem given by Johnson (1963) and then will discuss its application to Equation 12.25. Let **A** be an arbitrary real, $n \times r$ matrix, having rank r, where $n \geq r$. Consider that the matrix **AA′** may be diagonalized by the $n \times n$ orthonormal matrix **P** containing the eigenvectors of **AA′** so that

$$\mathbf{P}'\mathbf{A}\mathbf{A}'\mathbf{P} = \begin{bmatrix} \mathbf{D} & \mathbf{0} \\ \mathbf{0} & \mathbf{0} \end{bmatrix} \tag{12.26}$$

where **D** is an $r \times r$ diagonal matrix containing the nonzero eigenvalues of **AA′**. Now, let

$$\mathbf{P}'\mathbf{A} = \begin{bmatrix} \mathbf{G} \\ \mathbf{H} \end{bmatrix} \tag{12.27}$$

where

G is an $r \times r$ matrix
H is an $(n-r) \times r$ matrix

Evidently, from Equation 12.33,

$$\mathbf{P}'\mathbf{A}\mathbf{A}'\mathbf{P} = \begin{bmatrix} \mathbf{G} \\ \mathbf{H} \end{bmatrix} \begin{bmatrix} \mathbf{G}' & \mathbf{H}' \end{bmatrix} = \begin{bmatrix} \mathbf{G}\mathbf{G}' & \mathbf{G}\mathbf{H}' \\ \mathbf{H}\mathbf{G}' & \mathbf{H}\mathbf{H}' \end{bmatrix} \tag{12.28}$$

Using Equation 12.28, we can now conclude that $\mathbf{G}\mathbf{G}' = \mathbf{D}$ and $\mathbf{H}\mathbf{H}' = \mathbf{0}$. That $\mathbf{H}\mathbf{H}' = \mathbf{0}$ implies that $\mathbf{H} = \mathbf{0}$, because the diagonal elements of **HH′** are the sums of squares of the row elements of **H**. On the other hand, since $\mathbf{G}\mathbf{G}' = \mathbf{D}$ is a diagonal matrix, **G** must itself be a matrix composed of the diagonal matrix $\mathbf{D}^{1/2}$ and some unknown $r \times r$ orthonormal matrix **Q′**, that is, $\mathbf{G} = \mathbf{D}^{1/2}\mathbf{Q}'$, or

$$\mathbf{P'A} = \begin{bmatrix} \mathbf{D}^{1/2}\mathbf{Q'} \\ \mathbf{0} \end{bmatrix} = \begin{bmatrix} \mathbf{D}^{1/2} \\ \mathbf{0} \end{bmatrix} \mathbf{Q'} \tag{12.29}$$

Thus, if we postmultiply Equation 12.29 by **Q**, we obtain

$$\mathbf{P'AQ} = \begin{bmatrix} \mathbf{D}^{1/2} \\ \mathbf{0} \end{bmatrix} \tag{12.30}$$

This shows that any real matrix **A** can be diagonalized by premultiplication and postmultiplication by appropriately chosen orthogonal matrices.

If the matrix **P** can be identified as the matrix of eigenvectors of **AA′**, the matrix **Q** can be identified as the matrix of eigenvectors of **A′A**. For proof, obtain from Equation 12.30

$$(\mathbf{P'AQ})'(\mathbf{P'AQ}) = \mathbf{Q'A'PP'AQ} = \begin{bmatrix} \mathbf{D}^{1/2} & \mathbf{0} \end{bmatrix} \begin{bmatrix} \mathbf{D}^{1/2} \\ \mathbf{0} \end{bmatrix} = \mathbf{D}$$

But $\mathbf{PP'} = \mathbf{I}$ (since **P** is an orthonormal matrix), and so we have

$$\mathbf{Q'A'AQ} = \mathbf{D}$$

which shows that **Q**, being an orthonormal matrix in the first place, must also be the matrix of eigenvectors of **A′A**, since **Q** diagonalizes **A′A**. The matrix **D** evidently contains the nonzero eigenvalues of **AA′** and the eigenvalues of **A′A**. When **A** is a square matrix, **P**. **Q** and **D** are all square matrices of the same order as **A**.

On the other hand, by the "singular-value decomposition" of **A**, we shall mean

$$\mathbf{A}_{n\times r} = \mathbf{P}_1\mathbf{D}_r^{1/2}\mathbf{Q}' \quad n > r \tag{12.31a}$$

where

- $\mathbf{P}_1$ is an $n \times r$ matrix whose columns consist of just the first r eigenvectors in **P** corresponding to the r eigenvalues in **D** of $\mathbf{AA'} = \mathbf{PDP'}$
- $\mathbf{D}_r^{1/2}$ is the $r \times r$ diagonal matrix consisting of the square roots of the first r eigenvalues of **D**
- **Q** is the $r \times r$ matrix of the first r eigenvectors corresponding to the eigenvalues of **A′A**

Generalizing, we may also indicate that

$$\mathbf{A}_{n\times r} = \mathbf{P}\,\mathbf{D}_r^{1/2}\mathbf{Q}_1' \quad n < r \tag{12.31b}$$

where $\mathbf{Q}_1$ consists of the first r eigenvectors of $\mathbf{A'A}$. And

$$\mathbf{A}_{n\times r} = \mathbf{PD}_r^{1/2}\mathbf{Q}' \quad n = r \tag{12.31c}$$

Returning now to Equation 12.25, let us rewrite Equation 12.25 as

$$\mathbf{M} = \mathbf{M}^*\mathbf{D}_1 \tag{12.32}$$

Equation 12.32 is the form in which we most frequently consider the oblique-transformation problem, for we seek a transformation $\mathbf{M}^*$ of $\mathbf{A}$ which, except for normalization of the resulting factors, represents a transformation of the unrotated factors into a desirable configuration as far as their direction in space is concerned. But using the theorem of Eckart and Young (1936), we can regard any arbitrary $\mathbf{M}^*$ as $\mathbf{T}_2\mathbf{D}_2\mathbf{T}_1$, identifying $\mathbf{T}_2$, $\mathbf{T}_1$, and $\mathbf{D}_2$ with the matrices $\mathbf{P}$, $\mathbf{Q}'$, and $\mathbf{D}$ of the theorem. The diagonal matrix, $\mathbf{D}_1$, in Equations 12.25 and 12.32 serves only to make the corresponding transformed factors have unit length, that is, we can identify $\mathbf{D}_1$ as $\mathbf{D}_1 = [\text{diag}(\mathbf{M}^{*\prime}\mathbf{M})^{-1}]^{1/2}$. Thus we can regard any transformation matrix, $\mathbf{M}$, in terms of a series of orthonormal and length-adjusting transformations of the factors.

Harris and Kaiser (1964) consider three classes of transformations. Class I is the set of all possible orthogonal transformations, where

$$\mathbf{M} = \mathbf{T}_2(\mathbf{I})\mathbf{T}_1(\mathbf{I}) \tag{12.33}$$

with $\mathbf{D}_2 = \mathbf{D}_1 = \mathbf{I}$. The remaining two classes involve oblique transformations. In class II, one lets

$$\mathbf{M} = (\mathbf{I})\mathbf{D}_2\mathbf{T}_1\mathbf{D}_1 \tag{12.34}$$

where $\mathbf{T}_2$ is set equal to the identity matrix. In this class, the transformation involves an initial rescaling of the columns of $\mathbf{A}$ followed by an orthogonal transformation, and then a rescaling again of the columns of the result. Harris and Kaiser (1964) suggest that when $\mathbf{\Lambda} = \mathbf{AD}^{1/2}$ is a principal-axes solution, with $\mathbf{A}$ the $n \times r$ matrix of retained eigenvectors and $\mathbf{D}$ the $r \times r$ diagonal of retained eigenvalues of the reproduced matrix $\mathbf{R}^* = \mathbf{\Lambda\Lambda}'$, $\mathbf{D}_2$ in Equation 12.31 be set equal to $\mathbf{D}^{-1/2}$. Then the resulting pattern matrix $\mathbf{\Lambda M} = \mathbf{AT}_1\mathbf{D}_1$ may be obtained by letting $\mathbf{T}_1$ be one of the orthogonal-simple-structure transformations corresponding to a rotation of $\mathbf{A}$ (not $\mathbf{AD}^{1/2}$). The matrix $\mathbf{D}_1$ only normalizes the resulting factors.

Harris and Kaiser recommend that the researcher use this procedure whenever he believes that each factor will be determined by independent, nonoverlapping clusters of the observed variables. On the other hand, if the researcher suspects that there are clear hyperplanes to be found, but the factors are determined by overlapping clusters of variables, Harris and Kaiser recommend choosing $\mathbf{D}_2$ to be some other power of $\mathbf{D}$, say $\mathbf{D}_2 = \mathbf{D}^{-1/2}$. In this case, $\mathbf{\Lambda M} = \mathbf{AD}^{1/2}\mathbf{T}_1\mathbf{D}_1$. $\mathbf{T}_1$ may be one of the "orthomax" solutions for simple structure, but precisely which orthomax method applied to $\mathbf{A}$ leads to the optimum oblique transformation in this case has not been reported by Harris and Kaiser in the literature.

The third class of oblique solutions places no restrictions on the matrices $\mathbf{T}_2$, $\mathbf{T}_1$, $\mathbf{D}_2$, and $\mathbf{D}_1$ in Equation 12.31 and so represents the general case, where class II solutions are a special case in which $\mathbf{T}_1 = \mathbf{I}$. Rationales for optimally choosing $\mathbf{T}_2$, $\mathbf{T}_1$, and $\mathbf{D}_2$ have not yet been advanced except for those representing class II solutions as given by Harris and Kaiser (1964).

12.4 Weighted Oblique Rotation

We have already indicated how Kaiser's varimax method was improved by first weighting the variables by the reciprocals of the square roots of their communalities. Instead of rotating the principal-axes pattern $\mathbf{A}$, we rotated $\mathbf{H}^{-1}\mathbf{A}$, where $\mathbf{H}^2 = \text{diag}(\mathbf{AA}')$. Then when the final pattern was found as $\mathbf{\Lambda}^* = \mathbf{AT}^{-1}$, this was rescaled back to the original metric by $\mathbf{\Lambda} = \mathbf{H\Lambda}^*$. There are other situations in which certain kinds of reweightings of the variables might improve the solutions.

Some data sets present problems for rotation schemes, such as those we have discussed up to now. One kind of data set is one where a number of variables are dependent on more than one factor, and particularly on all of them. The classic case in the factor analysis literature is Thurstone's Box Problem (Thurstone, 1947) where Thurstone took a sizeable number of boxes of varying shapes and took various measurements, such as length, width, height (x, y, and z) and made up several additional measurements representing varying functions of these basic dimensions (e.g., xy, xy^2, $\sqrt{x^2+y^2}$, $\sqrt{x^2+z^2}$, $\sqrt{y^2+z^2}$, xz, x/y, x/z, y/x, $\sqrt{x^2+y^2+z^2}$, xyz, etc.) Although one would expect the basic dimensions of x, y, and z would be found as the primary axes factors, many analytic methods of rotation failed to recover them as the primary axes, instead lining up with variables representing pairwise and triple combinations of the factors such as $\sqrt{x^2+y^2}$, $\sqrt{x^2+z^2}$, and $\sqrt{y^2+z^2}$.

In Table 12.1, we show the original unrotated pattern matrix for the Box Problem variables. Also shown is the weight vector derived from the unrotated pattern according to the method of Cureton and Mulaik (1975). The weight vector is subsequently converted to a diagonal matrix.

Cureton and Mulaik (1975) diagnosed the problem in these cases as due to the overdetermination of false hyperplanes by numerous variables representing combinations of two or more factors. What was needed was to identify the "bounding hyperplanes" at the effective boundaries of the configuration of test vectors, which might be relatively underrepresented in the

TABLE 12.1

Unrotated Principal-Axes Factor Pattern for Thurstone's Invariant Box Problem as Given by Cureton and Mulaik (1975) and the Resulting Weight Vector

Variable	Unrotated Pattern			w
x	.629	−.494	.579	.986
y	.751	.602	.125	.817
z	.765	−.230	−.572	.808
xy	.866	.131	.459	.544
xz	.873	−.473	−.042	.535
yz	.906	.250	−.323	.412
x^2y	.824	−.149	.528	.678
xy^2	.859	.358	.306	.540
x^2z	.812	−.518	.203	.698
xz^2	.951	−.441	−.254	.524
y^2z	.876	.406	−.185	.490
yz^2	.885	.095	−.431	.476
x/y	.102	−.936	.322	.068
y/x	.102	.936	−.322	.068
x/z	.081	−.163	.969	.044
z/x	.081	.163	−.969	.044
y/z	.006	.810	.582	.001
z/y	−.006	−.810	−.582	.001
$2x + 2y$	.852	.223	.420	.553
$2x + 2z$	.861	−.483	−.094	.569
$2y + 2z$	.912	.248	−.304	.385
$(x^2 + y^2)^{1/2}$	.847	.218	.405	.534
$(x^2 + z^2)^{1/2}$	.845	−.456	−.106	.547
$(y^2 + z^2)^{1/2}$	.902	.246	−.272	.353
xyz	.987	−.026	.043	.008
$(x^2 + y^2 + z^2)^{1/2}$	.965	.057	−.028	.013

set of tests analyzed. The solution they came up with was to identify those variables that were at the boundaries, and during the rotation process to give them greater weight, while giving smaller weights to those variables toward the center of the configuration of test vectors.

To implement their solution, they first took the initial, unrotated $n \times r$ principal-axes factor matrix, $\mathbf{\Lambda}_0$, and normalized its rows, forming a new matrix

$$\mathbf{G} = \mathbf{H}^{-1}\mathbf{\Lambda}_0 = \left\| \frac{\lambda_{ij}}{h_i} \right\| \qquad (12.35)$$

where $\mathbf{H}^2 = [\text{diag}\ \mathbf{\Lambda}_0\mathbf{\Lambda}_0']$ is the diagonal matrix of communalities. This made each variable be represented by a unit length vector in common factor space. Next, they reflected each variable having a negative loading on the first-principal-axis factor by multiplying it by −1. The effect was to make the negative loadings on the first factor become positive, while reversing the signs of loadings on other factors. The resulting matrix was called **A**.

Normalization and reflection have no effect on the determination of the primary axes or the reference axes.

They next reasoned that if the first factor in **A** dominated the other factors in the size of loadings and amounts of variance accounted for, then the primary axes of an orthogonal approximation to simple structure would lie roughly symmetrically about the first axis of **A**. The angle which each simple structure primary axis would have with the first-principal-axes factor in **A** would have a cosine of approximately $(1/r)^{1/2}$. It would be exactly $(1/r)^{1/2}$ if the symmetry about the first axis of **A** is exact (Landahl, 1938). If the configuration of the correlated factors is acute (high positive intercorrelations among factors), the oblique primary axes will make smaller angles with the first axis of **A** (and have larger cosines with the first axis). If the configuration is obtuse, the angles will be greater and have smaller cosines with the first axis.

They argued that we should give the greatest weight to those test vectors that are closest to the primary axes. These would be those whose loading a_{i1} on the first principal axis is closest to $(1/r)^{1/2}$. (This assumes that among the variables, each factor will be represented by at least one variable that is dependent on just this factor.)

Any other test vectors that are not as close to the primary axis factors but still close to the bounding hyperplanes will have large angles with the first principal axis of **A**. Their loadings a_{i1} will not greatly differ from $(1/r)^{1/2}$. Such test vectors should receive intermediate weights.

Near zero weights should be given to test vectors with high loadings near unity on the first principal axis of **A**. Near zero weights should also be given to any test vector orthogonal (or nearly so) to the first principal axis of **A**. Such test vectors as the latter will contribute little to determining the principal axes.

A weighting function with the desired properties was

$$w_i = \cos^2\left(\frac{\cos^{-1}(1/r)^{1/2} - \cos^{-1} a_{i1}}{\cos^{-1}(1/r)^{1/2}} \times 90°\right) + .001 \quad \text{if } a_{i1} \geq (1/r)^{1/2}$$

$$w_i = \cos^2\left(\frac{\cos^{-1}(1/r)^{1/2} - \cos^{-1} a_{i1}}{90° - \cos^{-1}(1/r)^{1/2}} \times 90°\right) + .001 \quad \text{if } a_{i1} < (1/r)^{1/2} \qquad (12.36)$$

The value of .001 is added to prevent any weight from being exactly zero, which would prevent obtaining its reciprocal.

Now, a weighting function would involve weights that range continuously from 0 to 1, with 0 for test vectors almost collinear with the first-principal-axis factor, through to unity for test vectors lying close to the primary axes (with loadings a_{i1} approximately $(1/r)^{1/2}$ in value). Then as the test vectors approached having zero loadings on the first principal axis, the weights given them would approach 0 also.

Each variable is given a weight w_i according to the scheme in Table 12.2. Each of these weights is then inserted as the respective diagonal element in the ith row of a diagonal matrix **W**. Then the matrix **WA** is submitted to the rotation algorithm to yield a final solution $\mathbf{\Lambda}_p^* = \mathbf{WA}(\mathbf{T}')^{-1}$ where **T** represents the transformation carrying **WA** through to its final simple-structure solution. However, the weights must be stripped back out to put the variables back into their original metric. So, the final loading matrix should be $\mathbf{\Lambda}_p = \mathbf{W}^{-1}\mathbf{\Lambda}_p^*$.

This weighting scheme can be applied to any rotation method. Cureton and Mulaik reported successfully recovering the primary axes of x, y, and z of Thurstone's box problem with a weighted Promax rotation.

TABLE 12.2

Weighted, Normalized Varimax versus Raw Normalized Varimax Applied to Thurstone's Box Problem

Variable	Normalized Varimax			w	Weighted Normalized Varimax		
x	**.781**	.428	−.426	.986	**.983**	.076	−.062
y	.134	**.825**	.493	.817	.169	**.941**	−.168
z	−.352	.784	**−.476**	.808	.194	.183	**−.945**
xy	.577	.796	.107	.544	.699	.690	−.114
xz	.228	.762	−.595	.535	.719	.139	−.672
yz	−.190	.975	.023	.412	.167	.685	−.701
x^2y	.695	.691	−.138	.678	.871	.455	−.116
xy^2	.380	.856	.288	.540	.475	.841	−.162
x^2z	.461	.658	−.569	.698	.862	.106	−.462
xz^2	.033	.876	−.628	.524	.612	.178	−.870
y^2z	−.097	.955	.211	.490	.151	.812	−.533
yz^2	−.264	.941	−.149	.476	.172	.533	−.815
x/y	.529	−.129	−.833	.068	.769	−.627	−.079
y/x	−.490	.327	.802	.068	−.655	.748	−.039
x/z	.979	−.099	.069	.044	.740	.072	.647
z/x	−.948	.256	−.094	.044	−.650	.024	−.741
y/z	.384	.074	.917	.001	−.074	.734	.671
z/y	−.384	−.074	−.917	.001	.074	−.734	−.671
$2x + 2y$	.517	.806	.188	.553	.616	.748	−.109
$2x + 2z$	.178	.757	−.616	.569	.685	.116	−.708
$2y + 2z$	−.170	.978	.025	.385	.183	.690	−.690
$(x^2 + y^2)^{1/2}$	.503	.802	.180	.534	.607	.739	−.119
$(x^2 + z^2)^{1/2}$	.158	.748	−.591	.547	.654	.126	−.700
$(y^2 + z^2)^{1/2}$	−.141	.963	.032	.353	.199	.687	−.660
xyz	.235	.946	−.164	.008	.592	.573	−.546
$(x^2 + y^2 + z^2)^{1/2}$	.145	.951	−.098	.013	.490	.614	−.564

Note: How normalized varimax fails to identify the true factors associated respectively with the variables x, y, and z but the weighted solution recovers these quite well. Bold numbers emphasize the largest loadings of x, y, and z, respectively.

In Table 12.2, we show the unweighted normalized varimax solution for the Box Problem and, for comparison, the weighted, normalized varimax solution, which recovers the factors corresponding to x, y, and z. The unweighted solution does not align with these variables.

In Table 12.3, we show the raw quartimin that fails to yield factors corresponding to x, y, and z, and the weighted quartimin that succeeds at this.

TABLE 12.3

Raw Quartimin versus Weighted and Normalized Quartimin Applied to the Thurstone Box Problem

Variable	Raw Quartimin			w	Weighted Quartimin		
x	.470	−.296	.763	.986	−.020	−.009	**.997**
y	.784	.596	.011	.817	.026	**.981**	−.067
z	.824	−.415	−.400	.808	**.990**	.035	−.065
xy	.790	.254	.483	.544	−.036	.674	.555
xz	.816	−.479	.189	.535	.657	−.013	.560
yz	.973	.117	−.294	.412	.648	.620	−.142
x^2y	.707	.011	.629	.678	−.011	.405	.785
xy^2	.833	.421	.268	.540	.011	.849	.277
x^2z	.710	−.442	.430	.698	.422	−.028	.768
xz^2	.932	−.517	−.015	.524	.876	.006	.392
y^2z	.937	.309	−.212	.490	.447	.786	−.146
yz^2	.954	−.064	−.352	.476	.796	.437	−.132
x/y	−.054	−.787	.600	.068	.123	−.756	.921
y/x	.255	.783	−.579	.068	−.021	.860	−.856
x/z	−.100	.153	.981	.044	−.776	.119	.915
z/x	.259	−.156	−.965	.044	.857	−.036	−.864
y/z	−.004	.954	.311	.001	−.841	.908	−.074
z/y	.004	−.954	−.311	.001	.841	−.908	.074
$2x + 2y$	.793	.329	.417	.553	−.044	.745	.457
$2x + 2z$	.812	−.504	.141	.569	.702	−.041	.521
$2y + 2z$	.976	.121	−.275	.385	.634	.626	−.124
$(x^2 + y^2)^{1/2}$	.790	.320	.404	.534	−.030	.734	.447
$(x^2 + z^2)^{1/2}$	.801	−.482	.120	.547	.695	−.026	.488
$(y^2 + z^2)^{1/2}$	.960	.129	−.245	.353	.601	.627	−.100
xyz	.962	−.030	.146	.008	.463	.488	.364
$(x^2 + y^2 + z^2)^{1/2}$	.961	.027	.051	.013	.484	.538	.244
	Correlations among Factors				*Correlations among Factors*		
	1.000	.031	.094		1.000	.354	.344
	.031	1.000	−.018		.354	1.000	.328
	.094	.018	1.000		.344	.328	1.000

Note: How raw quartimin fails to recover the "true" factors but weighted quartimin succeeds. Bold numbers emphasize the largest loadings of x, y, and z, respectively.

However, I should note that the weighted quartimin did not always yield the expected factors. When I applied the weighted quartimin to the centroid solution for the principal axes for the study of 24 psychological tests by Holzinger and Swineford (1939), the weighted solution recovered factors very similar to those found by other methods of rotation, including the raw quartimin. However, when I applied the weighted quartimin to the unrotated maximum-likelihood solution for the pattern for the 24 psychological tests, the resulting loadings were not of a clean simple structure as found with other methods. The only difference was the input matrices. I tried several random starts to see if there was a local minimum causing the discrepancy, but the same solution resulted in each case. The unweighted quartimin successfully recovered an expected pattern similar to those found by other methods.

12.5 Oblique Procrustean Transformations

Let **A** and **B** be $n \times r$ matrices with $n \geq r$. Consider now the class of $r \times r$ linear-transformation matrices **T** in the equation

$$\mathbf{B} = \mathbf{AT} + \mathbf{E} \tag{12.37}$$

We wish to find that particular transformation matrix **T** such that

$$\text{tr}(\mathbf{E}'\mathbf{E}) = \text{tr}(\mathbf{B} - \mathbf{AT})'(\mathbf{B} - \mathbf{AT}) \quad \text{is a minimum} \tag{12.38}$$

With no restriction placed upon **T**, the solution for **T** is readily found: Take the partial derivative of Equation 12.36 with respect to **T** and set this equal to 0; in other words, find that

$$\frac{\partial}{\partial \mathbf{T}} \text{tr}(\mathbf{E}'\mathbf{E}) = -2\mathbf{A}'\mathbf{B} + 2\mathbf{A}'\mathbf{AT} = \mathbf{0} \tag{12.39}$$

Equation 12.37 may be rewritten as

$$\mathbf{A}'\mathbf{AT} = \mathbf{A}'\mathbf{B} \tag{12.40}$$

or

$$\mathbf{T} = (\mathbf{A}'\mathbf{A})^{-1}\mathbf{A}'\mathbf{B} \tag{12.41}$$

The transformation matrix **T** does not generally produce new basis vectors with unit lengths. Because of this feature it is not an appropriate transformation to use in directly transforming an unrotated-factor-structure matrix into a reference-structure matrix, although, because computing such a transformation by Equation 12.39 is so simple, it may be used as the first step in computing an approximation of such a minimizing transformation. For example, if **A** is an unrotated-factor-structure matrix and **B** is a hypothetical reference-structure matrix, we can compute the transformation matrix **T** as in Equation 12.39 and then normalize **T** so that diag $\mathbf{T}'\mathbf{T}=\mathbf{I}$. To normalize **T** we first find the matrix $\mathbf{D}_T^{-1}=\text{diag } \mathbf{T}'\mathbf{T}$ and then find the normalized transformation matrix $\mathbf{N}=\mathbf{T}\mathbf{D}_T^{-1}$. The transformed matrix **AN** is a plausible approximation to **B**, although **AN** is not the *least-squares* linear approximation of **B** under the restraint that diag $\mathbf{T}'\mathbf{T}=\mathbf{I}$. Nonetheless, **T** represents a "procrustean transformation" of **A** to approximate a target matrix **B**. The name was given originally by Hurley and Cattell (1962), being based on the Greek legend in which a highway bandit, Procrustes, by name, would tie his victims to an iron bed and stretch their legs or cut them off to make them fit its length. The term "procrustean" has come to refer to any inflexible procedure or action taken to force a result to conform to some preconceived criterion or idea.

12.5.1 Promax Oblique Rotation

Hendrickson and White (1964) devised an ingenious application of the unrestricted oblique procrustean transformation in Equation 12.39 to the problem of "blind" oblique transformation of factors to simple structure. Hendrickson and white begin with the assumption that an orthogonal-simple-structure solution, such as varimax, has been obtained and that it is fairly close to the optimum oblique-simple-structure solution. Therefore one can use functions of the elements of the orthogonal simple-structure factor-pattern matrix to construct a hypothetical oblique factor-pattern matrix to be approximated by an unrestricted procrustean transformation of the orthogonal factor-pattern matrix.

By $\mathbf{\Lambda}_V$ let us mean the $n\times r$ orthogonal-simple-structure factor-pattern matrix obtained by varimax rotation of the original orthogonal unrotated principal-axes factors. Consider now an $n\times r$ matrix **B** with elements

$$b_{ij}=\frac{a_{ij}^{m+1}}{a_{ij}} \quad |a_{ij}|>0 \tag{12.42}$$

where m is some power to which the element $a_{ij}=\lambda_{ij}$ is raised. (First raising the element to the $m+1$st power then dividing by a_{ij} makes the resulting element b_{ij} have the same sign as the original a_{ij}.) Hendrickson and White recommended that $m=4$. Now, the purpose of raising the elements to the mth power was to make elements closer to zero approach zero much more rapidly than those distant from zero.

In the Promax method the matrix $\mathbf{\Lambda}_V$ is to be rotated and the matrix **B** of elements as given above is the "target matrix." A transformation matrix $\mathbf{T}_1$ is then obtained by least squares according to Equation 12.39:

$$\mathbf{T}_1 = (\mathbf{\Lambda}_V'\mathbf{\Lambda}_V)^{-1}\mathbf{\Lambda}_V'\mathbf{B} \tag{12.43}$$

However, beginning with unit-length orthogonal factors in connection with $\mathbf{\Lambda}_V$, the resulting matrix $\mathbf{\Lambda}_V\mathbf{T}_1$ will not be a factor-pattern matrix that minimizes the least-squares criterion given in Equation 12.38. Nevertheless, we may still obtain an excellent approximation by rescaling the columns of $\mathbf{T}_1$ by

$$\mathbf{T} = \mathbf{T}_1\mathbf{D}_1$$

where $\mathbf{D}_1^2 = [\text{diag}(\mathbf{T}_1'\mathbf{T}_1)^{-1}]$. Note that

$$[\text{diag}(\mathbf{T}'\mathbf{T})^{-1}] = [\text{diag}\,\mathbf{D}_1^{-1}(\mathbf{T}_1'\mathbf{T}_1)^{-1}\mathbf{D}_1^{-1})] = \mathbf{I}$$

which satisfies the proper normalization of **T**. The desired factor-pattern matrix is then given by

$$\mathbf{\Lambda}_P = \mathbf{\Lambda}_V\mathbf{T} = \mathbf{\Lambda}_V\mathbf{T}_1\mathbf{D}_1 \tag{12.44}$$

The Promax method of oblique rotation became quite popular up to 2000 because it is easy to program for computers, does not require much computer time to compute, and produces results frequently as good as those produced by more sophisticated methods of oblique rotation. However, it should be replaced by a direct procrustean rotation to a factor-pattern matrix that minimizes the least-squares criterion, which we discuss next. A simple procedure for obtaining this will be discussed in connection with the GPA of Jennrich (2002).

12.5.2 Rotation to a Factor-Pattern Matrix Approximating a Given Target Matrix

Let **A** and **B** be $n \times r$ matrices with $n \geq r$. Consider now the class of $r \times r$ linear-transformation matrices **T** in the equation $\mathbf{B} = \mathbf{A}(\mathbf{T}')^{-1} + \mathbf{E}$. We wish to find that particular least-squares transformation matrix **T** such that $\text{tr}(\mathbf{E}'\mathbf{E}) = \text{tr}[\mathbf{B} - \mathbf{A}(\mathbf{T}')^{-1}][\mathbf{B} - \mathbf{A}\,(\mathbf{T}')^{-1}]$ is a minimum under the constraint that $\text{diag}\,(\mathbf{T}'\mathbf{T}) = \mathbf{I}$.

Mulaik (1972) showed how the single-plane rotation method used by Jennrich and Sampson (1966) to find the direct oblimin solution could be applied to this problem. We will forego discussion of it here, since it can be replaced by a special application of the much simpler general GPA of Jennrich (2002), which we will turn to shortly.

12.5.3 Promaj

Trendafilov (1994) sought to improve on the Promax method by developing a superior target matrix. The method he used was "vector majorization"

(Marshall and Olkin, 1979). According to Trendafilov (1994) "The vector **X** is said to majorize another vector **Y**... if the inequalities

$$\sum_{i=1}^{k} X_{[i]} \le \sum_{i=1}^{k} Y_{[i]}$$

hold for each $k=1, 2, \ldots, n-1$, where $X_{[1]} \le X_{[2]} \le, \ldots, \le X_{[n]}$ denote the components of **X** in increasing order and the equality

$$\sum_{i=1}^{n} X_{[i]} = \sum_{i=1}^{n} Y_{[i]}$$

Holds" (p. 387).

More specifically for the purposes of oblique rotation, let **A** be a normalized varimax factor-pattern solution for orthogonal factors. We will seek to create a target matrix, **B**, from the elements of **A** as follows: For each column $\mathbf{a}_j$ of **A**, we will create a corresponding *j*th column of **B** by majorizing $\mathbf{a}_j$. To do this for column *j* let $\bar{a}_j^2 = (1/n)\Sigma_{i=1}^{n} a_{ij}^2$ be the average of the squared elements of column $\mathbf{a}_j$. Now order the squared elements from $\mathbf{a}_j$ in ascending order, that is, such that $a_{[1]j}^2 \le a_{[2]j}^2 \le \cdots \le a_{[n-k]j}^2 \le \bar{a}_j^2 \le a_{[n-k+1]}^2 \le \cdots a_{[n]}^2$. Note that we have inserted in the ordering of the elements of $\mathbf{a}_j$ the average squared element in $\mathbf{a}_j$: $\bar{\mathrm{a}}_j^2$. We do this to make the next step more perspicuous.

In Table 12.4, we show both the Promax and the corresponding Promaj solutions from the unrotated maximum-likelihood estimation of the pattern matrix of the 24 psychological tests of Holzinger and Swineford (1939).

More specifically for purposes of oblique rotation, let **A** be a normalized varimax factor-pattern solution for orthogonal factors. We will seek to create a target matrix **B** from the elements of **A** as follows: For each column $\mathbf{a}_j$ of **A**, we will create a corresponding *j*th column of **B** by majorizing $\mathbf{a}_j$. To do this for column *j* let $\bar{a}_j^2 = (1/n)\Sigma_{i=1}^{n} a_{ij}^2$ be the average of the squared elements of column $\mathbf{a}_j$. Now order the squared elements from $\mathbf{a}_j$ in ascending order, that is, such that $a_{[1]j}^2 \le a_{[2]j}^2 \le \cdots \le a_{[n-k]j}^2 \le \bar{a}_j^2 \le a_{[n-k+1]}^2 \le \cdots a_{[n]}^2$. Note that we have inserted in the ordering of the elements of $\mathbf{a}_j$ the average squared element in $\mathbf{a}_j$: $\bar{\mathbf{a}}_j^2$. We do this to make the next step more perspicuous. Now let $c_j = a^2_{[n-k]j}$, that is, let c_j equal the squared element immediately less than or equal to $\bar{\mathbf{a}}_j^2$ in the ordering. We now are ready to construct the *j*th column $\mathbf{b}_j$ of **B** by the elements

$$b_{ij} \leftarrow \mathrm{sign}(a_{ij})\sqrt{\max(a_{ij}^2 - c_j, 0)} \tag{12.45}$$

We then proceed the same as we did with Promax rotation, using the majorized target matrix **B**. In contrast with Promax, the effect of majorizing is to place zeros in the target matrix whenever a squared loading, a_{ij}^2, is less than or equal to c_j in the respective column. Promax simply raised each element to the fourth power, retaining the sign of the original element. Thus

TABLE 12.4

Promax and Promaj Solutions from the Maximum-Likelihood Solution (See Table 12.6) of the 24 Psychological Tests of Holzinger and Swineford (1939)

Variable	Promax				Promaj			
Visual perception	.044	.075	**.671**	.097	−.013	.116	**.695**	.041
Cubes	.050	.010	**.423**	.060	.016	.034	**.441**	.024
Paper form board	.067	−.121	**.563**	.084	.025	−.094	**.595**	.040
Flags	.163	.010	**.505**	.022	.127	.035	**.524**	−.025
General information	**.725**	.159	.039	.030	**.731**	.141	.037	−.003
Paragraph comprehension	**.764**	−.008	.040	.140	**.768**	−.033	.055	.110
Sentence completion	**.812**	.097	.058	−.050	**.825**	.076	.058	−.087
Word classification	**.523**	.174	.234	.020	**.509**	.176	.236	−.021
Word meaning	**.811**	−.036	.032	.133	**.818**	−.064	.048	.102
Addition	.066	**.879**	−.206	.026	.052	**.894**	−.268	.024
Code	.070	**.487**	.008	.294	.035	**.499**	−.009	.284
Counting dots	−.111	**.719**	.188	−.034	−.149	**.759**	.146	−.059
Straight-curved capitals	.068	**.477**	**.408**	−.038	.026	**.516**	.390	−.083
Word recognition	.137	.025	−.100	**.561**	.109	.010	−.077	**.569**
Number recognition	.052	.013	−.009	**.532**	.018	.006	.016	**.536**
Figure recognition	−.040	−.042	.310	**.536**	−.099	−.028	**.349**	**.518**
Object—number	.055	.170	−.085	**.567**	.019	.164	−.072	**.574**
Number—figure	−.101	.273	.204	**.424**	−.156	.294	.213	**.410**
Figure—word	.070	.099	.149	.342	.034	.105	.165	**.329**
Deduction	.308	.026	.299	.246	.275	.032	**.322**	.212
Numerical puzzles	.058	.382	.321	.133	.014	**.411**	**.314**	.099
Problem reasoning	.295	.031	.298	.247	.262	.038	**.320**	.213
Series completion	.276	.150	**.417**	.155	.235	.169	**.431**	.109
Arithmetic problems	.279	**.464**	.032	.196	.254	**.470**	.015	.176
	Correlations among Factors				*Correlations among Factors*			
	1.000	.243	.328	.297	1.000	.337	.405	.382
	.243	1.000	.250	.306	.337	1.000	.303	.363
	.328	.250	1.000	.316	.405	.303	1.000	.365
	.297	.306	.316	1.000	.382	.363	.365	1.000

Note: Bold numbers emphasize the highest loadings in a column.

an element would never be made to be exactly zero in Promax, unless it was already zero. In Promaj the elements a_{ij}^2 in each column greater than c_j are made to equal $\text{sign}(a_{ij})\sqrt{a_{ij}^2 - c_j}$.

12.5.4 Promin

A limitation of both Promax and Promaj noted by Lorenzo-Seva (1999) is that both of these methods depend on a prior normalized varimax solution from which is constructed a completely specified target matrix **B**. In discussing the weighted oblique rotation method of Cureton and Mulaik (1975), we already noted that there are some data sets, such as given in the Box Problem (Thurstone, 1947) for which the ordinary normalized varimax and other rotational methods fail to recover the optimal simple structure. Lorenzo-Seva (1999) seeks to correct this by recommending an algorithm he calls Promin, which begins with a weighted normalized varimax solution from which one constructs a partially specified target for procrustean rotation. The key to this method is the manner in which 0 elements are determined in a partially specified target matrix.

Let **A** be the weighted, normalized orthogonal varimax rotation of an original, unrotated orthogonal principal-axes pattern matrix. For each column j of **A** compute the mean m_j and standard deviation s_j of the respective column "squared" elements. Then one builds the target matrix as follows:

$$\text{If } a_{ij}^2 < m_j + s_j/4 \quad \text{then } b_{ij} = 0$$

$$\text{If } a_{ij}^2 \geq m_j + s_j/4 \quad \text{then } b_{ij} \text{ is unspecified} \qquad (12.46)$$

Lorenzo-Seva (1999) argues that using the standard deviation along with the mean compensates for cases where there may be many variables, only a few of which load on a given factor. In such cases, the mean will be low, and using the mean alone some low nonzero loadings may be made into unspecified cases, which will degrade the solution. If, as is usually the case, there are a few variables with loadings near 1.00 on the factor, this will increase the standard deviation of the loadings. So only if a squared loading is less than the mean plus 1/4 times the standard deviation will it be given a zero value, but otherwise it will be left unspecified. Leaving elements unspecified gives a greater freedom for the algorithm to find solutions that minimize the lack of fit to the specified zeros.

The best algorithm for finding the Promin solution is given by Jennrich (2002): Construct the $n \times r$ partially specified target matrix **B**. Then let **W** be an $n \times r$ matrix with a 1 in every position corresponding to a specified (zero) element in **B**. Insert a zero in **W** in every position corresponding to an unspecified element in **B**. Define now $\tilde{\mathbf{B}} = \mathbf{W} \cdot \mathbf{B}$, where $\cdot$ denotes element-by-element multiplication, that is, $\mathbf{C} = \mathbf{A} \cdot \mathbf{B}$ denotes $c_{ij} = a_{ij} b_{ij}$ for all i and j. Programs like SPSS MATRIX easily handle operations like this. Then the function to minimize is given as

$$Q(\mathbf{\Lambda}) = \text{tr}[(\mathbf{W} \cdot \mathbf{\Lambda} - \tilde{\mathbf{B}})'(\mathbf{W} \cdot \mathbf{\Lambda} - \tilde{\mathbf{B}})] \qquad (12.47)$$

$\tilde{\mathbf{B}}$ in this case becomes essentially a null matrix and could be omitted.

The Promin solution for 24 psychological tests is shown in Table 12.5. The original solution was a maximum-likelihood solution rotated to a weighted,

TABLE 12.5

Promin Solutions for 24 Psychological Tests

Variables		**Promin**		
Visual perception	−.042	.081	**.715**	.042
Cubes	−.002	.013	**.454**	.023
Paper form board	.000	−.120	**.607**	.035
Flags	.108	.014	**.551**	−.037
General information	**.733**	.128	.075	−.032
Paragraph comprehension	**.766**	−.055	.076	.087
Sentence completion	**.830**	.070	.108	−.131
Word classification	**.504**	.158	.274	−.044
Word meaning	**.817**	−.084	.071	.075
Addition	.062	**.883**	−.249	.047
Code	.026	**.463**	−.025	.324
Counting dots	−.153	**.745**	.175	−.042
Straight-curved capitals	.014	**.497**	**.430**	−.085
Word recognition	.093	−.039	−.142	**.623**
Number recognition	−.001	−.043	−.046	**.590**
Figure recognition	−.131	−.087	.293	**.572**
Object—number	.003	.111	−.136	**.636**
Number—figure	−.179	.244	.174	**.464**
Figure—word	.016	.066	.136	**.362**
Deduction	.256	−.005	**.320**	.220
Numerical puzzles	−.001	**.381**	**.326**	.115
Problem reasoning	.244	.001	**.318**	.222
Series completion	.216	.135	**.448**	.111
Arithmetic problems	.249	**.440**	.020	.194
		Correlations among Factors		
	1.000	.334	.427	.476
	.334	1.000	.314	.429
	.427	.314	1.000	.501
	.476	.429	.501	1.000

Note: Bold numbers emphasize highest loadings in a column.

normalized varimax pattern matrix. Discrepancies between these solutions and those reported by other authors may be due to the initial input matrix. Some solutions use the original centroid solution, others the principal-components solution, and the one here, the maximum-likelihood solution. Some may also not rotate to a proper oblique factor-pattern matrix by an inappropriate choice for constraint on the matrix **T**.

The current solutions were computed using modifications of programs created by Bernaards and Jennrich (2005) and are made available at the Web site http://www.stat.ucla.edu/research/gpa

12.6 Gradient-Projection-Algorithm Synthesis

In a series of recent papers, Jennrich (2001, 2002, 2004, 2005, 2006) has sought to synthesize and simplify the methods of rotation to simple structure. His first approach does this by using the "GPA," which allows the researcher to carry out iterations involving basically just two major steps, but requires obtaining derivatives. The second discusses how one might implement oblique rotation without obtaining derivatives. The third uses component loss functions (CLFs).

12.6.1 Gradient-Projection Algorithm

The GPA is an iterative algorithm for finding the maximum or minimum of a function under constraints placed on the independent variables of the function. It works especially well with convex functions with quadratic constraints.

To help the reader get a preliminary understanding of the algorithm, before we consider the more complex case in which constraints are placed on transformation matrices, I am going to consider finding the minimum of a quadratic function of two independent variables under a quadratic constraint placed on the independent variables. Let the function to be minimized be $z = f(x, y) = 4x^2 - 8x + y^2 + 2xy - y + 13$. A portion of the surface of this function near the minimum is shown in Figure 12.1.

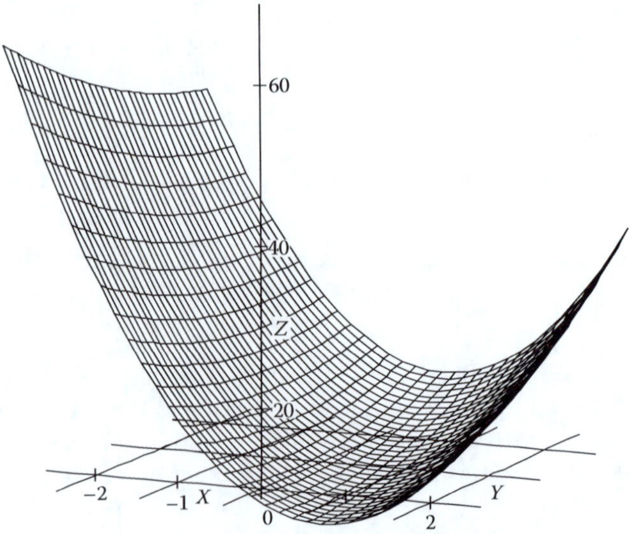

FIGURE 12.1
Surface plot of function $z = 4x^2 - 8x + y^2 + 2xy - y + 13$ used to illustrate the GPA. The aim is to minimize this function under the constraint that $y = \sqrt{4 - x^2}$.

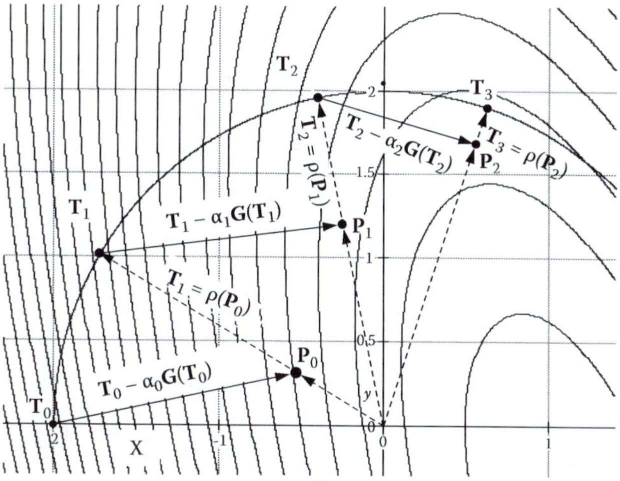

FIGURE 12.2
Illustrated is a contour plot of the quadratic function to be minimized. Superimposed on this plot is also the curve of the equation of constraint $y = \sqrt{4 - x^2}$. Beginning with an initial point $\mathbf{T}_0 = (-2,0)$ conforming to the equation of constraint, a negative gradient vector $-\mathbf{G}(\mathbf{T}_0)$ is computed, which is used to find a point $\mathbf{P}_0 = \mathbf{T}_0 - \alpha_0 \mathbf{G}(\mathbf{T}_0)$ in the direction of steepest descent from $\mathbf{T}_0$. This point is then projected onto the curve of constraint at a point $\mathbf{T}_1$ minimally distant from the point $\mathbf{P}_0$ by $\mathbf{T}_1 = \rho(\mathbf{P}_0) = (2/\sqrt{\mathbf{P}_0'\mathbf{P}_0})\,\mathbf{P}_0$. The algorithm then cycles repeatedly in this two-step manner, adjusting α_i accordingly, until convergence is reached.

Let x and y be constrained by the equation $x^2 + y^2 = 4$, which is equivalent to requiring that $y = \sqrt{4 - x^2}$ at the minimum. A contour plot of the function showing contours of points corresponding to equal value for the function in the space whose points are the coordinates (x,y) appears in Figure 12.2. Superimposed on the contour plot is the plot of the equation of constraint $y = \sqrt{4 - x^2}$ given by a bold curve. Note that for this problem the curve of constraint is a segment of a circle with a radius of 2 units. The values of x and y that minimize the function $z = f(x, y)$ under the constraint must fall on the curve $y = \sqrt{4 - x^2}$. Without the constraint the solution would likely minimize the function at some other point off the curve of constraint.

The negative gradient is a vector that points in the direction of steepest descent from some point in parameter space. The negative gradient vector is orthogonal to a straight line drawn tangent to the contour line of equal function values as it passes through the point. The coordinates of the positive gradient vector are the values of the partial derivatives of the function with respect to each of the respective independent variables evaluated at the point in parameter space. Let us denote the gradient vector by the boldface letter $\mathbf{G}(x_i, y_i)$. The values in parentheses designate arguments of a function of the coordinates of the point (x_i, y_i) used to obtain the gradient vector at that point. In this problem the gradient has just two coordinate values, so

$$\mathbf{G}(x_i, y_i) = \begin{bmatrix} \frac{\partial f}{\partial x} \\ \frac{\partial f}{\partial y} \end{bmatrix}_{(x_i, y_i)}$$

The partial derivatives of the function above are as follows:

$$\frac{\partial f}{\partial x} = 8x + 2y - 8$$

$$\frac{\partial f}{\partial y} = 2x + 2y - 1$$

The values of each of these partial derivatives at the point (x_i, y_i) are given by substituting x_i and y_i for x and y in the expressions shown for the partial derivatives. So, if $x = -2$ and $y = 0$ then

$$\mathbf{G}(-2, 0) = \begin{bmatrix} -24 \\ -5 \end{bmatrix}$$

Since the negative gradient points in the most downhill direction from the point (–2, 0), we need to go in that direction to find a better solution for the minimizing point. By $\mathbf{T}_0 = \begin{bmatrix} x_0 \\ y_0 \end{bmatrix}$ we shall mean the vector of coordinates of an initial point that satisfies the equation of constraint. From $\mathbf{T}_0$ we will go in the direction of the negative gradient to a new point, with coordinates also expressed as a 2×1 vector $\mathbf{P}_0 = \mathbf{T}_0 - \alpha_0 \mathbf{G}(\mathbf{T}_0)$. Because all points have coordinates expressed with respect to the origin (0,0), we can also represent $\mathbf{P}_0$ as a vector drawn as a dashed line directly from the origin to the coordinates of the point. This is shown in Figure 12.2.

By α_0 we shall mean the quantity $\alpha_0 = \beta_0 / |\mathbf{T}_0|$ where β_0 is some scalar quantity used to expand ($\beta_0 > 1$) or shrink ($\beta_0 < 1$) the length of the gradient vector, and $|\mathbf{T}_0|$ is the length of the raw gradient vector, used to normalize the gradient vector $\mathbf{G}(\mathbf{T}_0)$ to unit length. We will suppose that initially for $\mathbf{T}_0$ that $\beta_0 = 1.5$, since we want to make large initial steps. We may diminish or increase this value during later iterations

The next task is to find the point $\mathbf{T}_1$ on the curve of constraint $y = \sqrt{4 - x^2}$ that is minimally distant from $\mathbf{P}_0 = \mathbf{T}_0 - \alpha_0 \mathbf{G}(\mathbf{T}_0)$. $\mathbf{T}_1$ represents the projection of the point $\mathbf{P}_0 = \mathbf{T}_0 - \alpha_0 \mathbf{G}(\mathbf{T}_0)$ onto the curve $y = \sqrt{4 - x^2}$. The solution for the minimally distant point $\mathbf{T}_1$ on the curve $y = \sqrt{4 - x^2}$ is given by the matrix equation $\mathbf{T}_1 = \rho(\mathbf{P}_0) = (2/\sqrt{\mathbf{P}_0'\mathbf{P}_0})\,\mathbf{P}_0$. Note that $\sqrt{\mathbf{P}_0'\mathbf{P}_0}$ is the length of $\mathbf{P}_0$. Multiplying $\mathbf{P}_0$ by the reciprocal of its length creates a unit length vector pointed in the same direction as $\mathbf{P}_0$ from the origin (0,0). Multiplying this unit length vector by 2 creates a vector passing through $\mathbf{P}_0$ that is 2 units from the origin. Such a vector points to a point on the curve of the equation of constraint. Consider that the point on a circle that is minimally distant from a given interior point lies at the end of a line drawn from the center of the circle through the interior point to the circumference of the circle. So, the GPA for this problem begins with a point on the curve of constraint. Next, it finds a trial point in the downhill direction given by the negative gradient vector whose length has been adjusted to be in the vicinity of the curve of constraint. Then it seeks a new point that is minimally distant on the curve of constraint from the trial point. When the curve of constraint describes a segment of a circle, the new point on the curve of constraint is obtained by stretching the vector from the origin to the trial point on the curve. This completes the first cycle of the algorithm. Figure 12.2 illustrates the GPA for this problem.

In subsequent cycles we obtain the gradient vector $\mathbf{G}(\mathbf{T}_i)$ of the function $z=f(x,y)$ evaluated at the point $\mathbf{T}_i$. We determine a value β_i to shrink or extend the gradient vector and compute $\alpha_i = \beta_i/|\mathbf{T}_i|$. Then we find the point $\mathbf{P}_i=\mathbf{T}_i-\alpha_i\mathbf{G}(\mathbf{T}_i)$. The coordinates of this point are then used to find the coordinates of $\mathbf{T}_{i+1}$ by computing $\mathbf{T}_{i+1}=\rho(\mathbf{P}_i)=\left(2/\sqrt{\mathbf{P}_i'\mathbf{P}_i}\right)\mathbf{P}_i$. The key to attaining successful convergence is adjusting the size of β_i at each cycle. We will not discuss how this is done now, but will consider it as we discuss Jennrich's use of this algorithm to find transformation matrices to rotate a factor-pattern matrix to optimize some criterion for simple structure.

12.6.2 Jennrich's Use of the GPA

I am indebted to Robert Jennrich (personal communication) for much of the following description, including the wording, of the GPA.

An $n \times r$ factor-pattern matrix $\mathbf{\Lambda}$, is a rotation of an initial $n \times r$ loading matrix $\mathbf{A}$ if and only if

$$\mathbf{\Lambda} = \mathbf{A}(\mathbf{T}')^{-1} \tag{12.48}$$

for some $r \times r$ matrix $\mathbf{T}$ subject to the constraint that diag($\mathbf{T}'\mathbf{T}$)=$\mathbf{I}$. Let $\mathbf{T}$ stand for the manifold of $r \times r$ nonsingular matrices with columns of length one that satisfy this constraint. Next, let $\mathbf{Q}(\mathbf{\Lambda})$ be a rotation criterion evaluated at $\mathbf{\Lambda}$. We seek to minimize (maximize)

$$f(\mathbf{T}) = Q(\mathbf{A}(\mathbf{T}')^{-1})$$

subject to the constraint that the columns of $\mathbf{T}$ have unit length.

Jennrich (2002) gave a GPA for doing this. He showed that the gradient $\mathbf{G}$ of f at $\mathbf{T}$ is

$$\mathbf{G} = -\left(\mathbf{\Lambda}'\frac{\partial Q}{\partial \mathbf{\Lambda}}\mathbf{T}^{-1}\right)' = -(\mathbf{\Lambda}'\mathbf{G}_q\mathbf{T}^{-1})' \tag{12.49}$$

Each rotational criterion Q will have its own expression for $\mathbf{G}_q$. Jennrich (2002) gives expressions for $\mathbf{G}_q$ for each of the most popular oblique rotation criteria.

Next, for any nonsingular $r \times r$ matrix $\mathbf{M}$, the projection of $\mathbf{M}$ onto T is given by

$$\rho(\mathbf{M}) = \mathbf{M}[\text{diag}(\mathbf{M}'\mathbf{M})]^{-1/2} \tag{12.50}$$

The projection of $\mathbf{M}$ (which initially may not have columns of unit length) onto the manifold T is written as $\rho(\mathbf{M})$ and is simply the matrix $\mathbf{M}$ whose columns have been normalized to unit length.

Now, according to Jennrich (2002) a step of the GPA is as follows: The algorithm begins with a transformation matrix $\mathbf{T}$ on the manifold T. The gradient $\mathbf{G}$ is computed using Equations 12.48 and 12.49. Then $\mathbf{T}$ is moved in the direction of the negative gradient a distance given by the step value α to $\mathbf{T} - \alpha\mathbf{G}$. From there, using the projection function in Equation 12.50, it is projected back onto $\tilde{\mathbf{T}}$ on the manifold T. The algorithm continues by replacing $\mathbf{T}$ by $\tilde{\mathbf{T}}$ in each iteration, until the algorithm converges.

If the gradient $\mathbf{G} \neq \mathbf{0}$, moving $\mathbf{T}$ by a small amount in the negative gradient direction will decrease $f(\mathbf{T})$. Thus for a sufficiently small α $f(\mathbf{T} - \alpha\mathbf{G}) < f(\mathbf{T})$. However, $\mathbf{T} - \alpha\mathbf{G}$ will likely not be on T. On the other hand, $\tilde{\mathbf{T}}$ is on T and Jennrich (2002) has shown that if $\mathbf{T}$ is not a stationary point of f restricted to T, it will be the case that $f(\tilde{\mathbf{T}}) < f(\mathbf{T})$, if α is sufficiently small. Thus if α is sufficiently small, the GPA algorithm is a strictly decreasing algorithm.

But a well-defined algorithm needs a method for choosing and modifying α, and a stopping rule.

Jennrich recommends modifying α as one goes along during the iteration process. If $f(\tilde{\mathbf{T}}) \geq f(\mathbf{T})$, then one might replace α by $\alpha/2$, recompute $\tilde{\mathbf{T}}$, and if necessary continue this until $f(\tilde{\mathbf{T}}) < f(\mathbf{T})$. This halving of α is guaranteed to succeed in arriving at $f(\tilde{\mathbf{T}}) < f(\mathbf{T})$ within a finite number of halvings and yield an algorithm that is strictly decreasing. Jennrich, however, recommends using a slightly stronger requirement, that $f(\tilde{\mathbf{T}}) < f(\mathbf{T}) - .5s^2\alpha$, (where s is a quantity we will describe shortly), because with this modification, the algorithm must converge to a stationary point of f restricted to T.

Jennrich has shown that $\mathbf{T}$ is a stationary point of f restricted to T, if and only if

$$s = \|\mathbf{G} - \mathbf{T}\,\text{diag}(\mathbf{T}'\mathbf{G})\|$$

is zero. By $\|\mathbf{U}\| = (\mathbf{U},\mathbf{U})^{1/2} = [\text{tr}(\mathbf{U}'\mathbf{U})]^{1/2}$ he means the "Froebenius norm" of $\mathbf{U}$. He recommends using $s < 10^{-6}$ as a stopping rule. He also uses the notation $(\mathbf{U},\mathbf{V}) \equiv \text{tr}(\mathbf{U}'\mathbf{V})$.

12.6.2.1 Gradient-Projection Algorithm

Given an initial factor-pattern matrix **A** for orthogonal factors (e.g., the unrotated principal-axes solution or a varimax solution), choose an initial **T** (e.g., **T** = **I**, or a random **T** with normalized columns) and an initial α (e.g., $\alpha = 1$). Compute $\Lambda = \mathbf{A}(\mathbf{T}')^{-1}$, $f(\mathbf{T}) = Q(\mathbf{A}(\mathbf{T}')^{-1}) = Q(\Lambda)$ and $\mathbf{G}_q(\Lambda)$.

1. Compute the gradient $\mathbf{G} = -\left(\boldsymbol{\Lambda}'\frac{\partial Q}{\partial \boldsymbol{\Lambda}}\mathbf{T}^{-1}\right)' = -(\boldsymbol{\Lambda}'\mathbf{G}_q\mathbf{T}^{-1})'$ at **T**.
2. Compute $s = \|\mathbf{G} - \mathbf{T}\,\mathrm{diag}(\mathbf{T}'\mathbf{G})\|$. If $s < 10^{-6}$, stop, print $\boldsymbol{\Lambda}$, $f(\mathbf{T})$, s, and α, and exit the program.
3. Compute $\mathbf{M} = \mathbf{T} - \alpha\mathbf{G}$ and $\tilde{\mathbf{T}} = \rho(\mathbf{M}) = \mathbf{M}[\mathrm{diag}(\mathbf{M}'\mathbf{M})]^{-1/2}$.
4. Compute $\tilde{\boldsymbol{\Lambda}} = \mathbf{A}(\tilde{\boldsymbol{\Lambda}})^{-1}$, $f(\tilde{\mathbf{T}}) = Q(\mathbf{A}(\tilde{\mathbf{T}}')^{-1}) = Q(\tilde{\boldsymbol{\Lambda}})$, $\mathbf{G}_q(\tilde{\boldsymbol{\Lambda}})$. If $f(\tilde{\mathbf{T}}) > f(\mathbf{T})$, replace α with $\alpha/2$ and go to step 3.
5. Else if $f(\tilde{\mathbf{T}}) < f(\mathbf{T}) - .5s^2\alpha$, replace **T** with $\tilde{\mathbf{T}}$, $\boldsymbol{\Lambda}$ with $\tilde{\boldsymbol{\Lambda}}$, and $f(\mathbf{T})$ with $f(\tilde{\mathbf{T}})$, then compute $\mathbf{G}_q(\boldsymbol{\Lambda})$, and go to step 1.

But Jennrich (2002) indicates that the algorithm as described is only a primitive version of it. Let us first define a new function:

$$h(\alpha) = f(\rho(\mathbf{T} - \alpha\mathbf{G})) \tag{12.51}$$

He notes that this new value for f involves a specific α. The current value is $h(0) = f(\rho(\mathbf{T}))$ and a partial step method halves α until

$$h(\alpha) < h(0)$$

He notes that a simple alternative is to repeatedly divide α by 2 until

$$h(\alpha) \le h(0) - .5s^2\alpha$$

He proves that such an α can be found in a finite number of halvings.

But to speed up slow convergence, he also begins the partial step procedure by first doubling α. If this still yields a solution where $h(\alpha) \le h(0) - .5s^2\alpha$, then one proceeds with that value, otherwise one will eventually reduce α to a proper size by further halvings.

Now, he provides the essentials for a number of rotational methods that can be implemented using the GPA.

12.6.2.2 Quartimin

The quartimin criterion was a simple version of Carroll's oblimin family of criteria. It is expressed as

$$Q(\mathbf{\Lambda}) = (\mathbf{\Lambda}^{[2]}, \mathbf{\Lambda}^{[2]}\mathbf{N})/4 = \mathrm{tr}(\mathbf{\Lambda}'^{[2]}\mathbf{\Lambda}^{[2]}\mathbf{N})/4 \tag{12.52}$$

where $\mathbf{\Lambda}^{[2]} = \mathbf{\Lambda}\cdot\mathbf{\Lambda} = [\lambda_{ij}\lambda_{ij}]$, that is, a $n \times r$ matrix $\mathbf{\Lambda}^{[2]}$ in which each element is squared, (or achieved by element-by-element multiplication of corresponding elements of $\mathbf{\Lambda}$ times those of $\mathbf{\Lambda}$). The matrix $\mathbf{N}$, on the other hand is an $r \times r$ square matrix with 1s everywhere except on the principal diagonal, which contains only 0s. The operation $\mathbf{\Lambda}^{[2]}$ squares each element in $\mathbf{\Lambda}$, and then $\mathbf{\Lambda}'^{[2]}\mathbf{\Lambda}^{[2]}$ creates an $r \times r$ symmetric matrix whose elements are the sums of products of squared elements in one column with those in either the same or another column, respectively. The sum of the products of squared elements of a given column with those of the same column are entered respectively in the principal diagonal, while those between different columns are in the off-diagonal.

Now, let us focus on the principal diagonal of the product of $\mathbf{\Lambda}'^{[2]}\mathbf{\Lambda}^{[2]}\mathbf{N}$. Each row of $\mathbf{\Lambda}'^{[2]}\mathbf{\Lambda}^{[2]}$ that is multiplied by a column of $\mathbf{N}$ ends up in the principal diagonal of $\mathbf{\Lambda}'^{[2]}\mathbf{\Lambda}^{[2]}\mathbf{N}$. Note that the 0 element in the principal diagonal of $\mathbf{N}$ pairs up with the diagonal elements of sum of squares of squared elements in $\mathbf{\Lambda}'^{[2]}\mathbf{\Lambda}^{[2]}$, and is skipped in the summing for matrix multiplication. The result is that

$$\begin{aligned} Q(\mathbf{\Lambda}) &= (\mathbf{\Lambda}^{[2]}, \mathbf{\Lambda}^{[2]}\mathbf{N})/4 = \mathrm{tr}(\mathbf{\Lambda}'^{[2]}\mathbf{\Lambda}^{[2]}\mathbf{N})/4 \\ &= \sum_{j=1}^{r}\sum_{k=1}^{r}\sum_{i=1}^{n} \lambda_{ij}^2\lambda_{ik}^2/4 \quad j \neq k \end{aligned} \tag{12.53}$$

so, this criterion is minimized to the extent that a squared element in one column pairs up with a small element in another column, over all pairs of distinct columns. That illustrates the notation, which is quite compact. But it readily lends itself to the use of routines as found, for example, in the SPSS Matrix-End Matrix language. So, it should be simple to implement these in such computer packages.

Now, Jennrich (1972) gives the gradient matrix for the quartimin criterion at Λ as

$$\mathbf{G}_q = \frac{\partial Q}{\partial \mathbf{\Lambda}} = \mathbf{\Lambda}\cdot(\mathbf{\Lambda}^{[2]}\mathbf{N}) \tag{12.54}$$

If $\mathbf{A}$ and $\mathbf{B}$ are each $n \times r$ matrices, then $\mathbf{C} = \mathbf{A}\cdot\mathbf{B}$ denotes an $n \times r$ matrix whose elements are $c_{ij} = a_{ij}b_{ij}$ (Tables 12.6 and 12.7).

12.6.2.3 Oblimin Rotation

Each of the members of Carroll's oblimin family can be obtained by an appropriate choice of the parameter γ in

TABLE 12.6

Initial Unrotated Maximum-Likelihood Pattern Matrix for Holzinger and Swineford's (1939) 24 Psychological Test Data and Quartimin Pattern Matrix with Correlations among Their Factors

Unrotated Pattern Matrix				**Quartimin Pattern Matrix**			
.553	.044	.454	−.218	.056	.026	**.687**	.069
.344	−.010	.289	−.134	.059	−.021	**.430**	.040
.377	−.111	.421	−.158	.077	−.159	**.564**	.056
.465	−.071	.298	−.198	.181	−.031	**.507**	−.010
.741	−.225	−.217	−.038	**.771**	.112	.007	−.020
.737	−.347	−.145	.060	**.807**	−.058	−.004	.088
.738	−.325	−.242	−.096	**.865**	.049	.018	−.110
.696	−.121	−.033	−.120	**.559**	.126	.219	−.024
.749	−.391	−.160	.060	**.857**	−.087	−.017	.077
.486	.617	−.378	−.014	.072	**.866**	−.165	.044
.540	.370	−.040	.138	.069	**.459**	.029	.304
.447	.572	−.040	−.191	−.109	**.699**	**.239**	−.024
.579	.307	.117	−.258	.084	**.439**	**.440**	−.054
.404	.045	.082	.427	.126	.000	−.112	**.571**
.365	.071	.162	.374	.038	−.012	−.015	**.544**
.452	.073	.419	.256	−.055	−.080	**.313**	**.540**
.438	.190	.080	.409	.041	.145	−.084	**.585**
.464	.315	.244	.181	−.116	**.242**	**.227**	**.438**
.415	.093	.174	.164	.066	.069	.152	**.342**
.602	−.091	.191	.037	**.324**	−.020	**.288**	**.218**
.561	.271	.146	−.090	.065	**.345**	**.345**	.124
.595	−.081	.193	.038	**.311**	−.014	**.288**	**.220**
.669	.000	.215	−.090	**.296**	.099	**.418**	.124
.654	.237	−.113	.056	**.294**	**.428**	.041	.187
Correlations among Factors				*Correlations among Factors*			
1.000	.000	.000	.000	1.000	.292	.404	.415
.000	1.000	.000	.000	.292	1.000	.255	.318
.000	.000	1.000	.000	.404	.255	1.000	.380
.000	.000	.000	1.000	.415	.318	.380	1.000

Note: Bold numbers emphasize the highest loadings in a column.

$$Q(\mathbf{\Lambda}) = (\mathbf{\Lambda}^{[2]}, (\mathbf{I} - \gamma\mathbf{C})\mathbf{\Lambda}^{[2]}\mathbf{N})/4 \qquad (12.55)$$

Here **C** is a matrix, all of whose elements equal $1/n$. The values for γ of 0, 0.5, 1 yield the quartimin, bi-quartimin, and covarimin criteria, respectively. In the example shown, using Carroll's recommendation of .5 produced several large loadings greater than 1.00. I tried smaller values and settled on .35 as giving the best solution.

TABLE 12.7

Biquartimin Pattern Matrix with $\gamma = .35$ Set by Trial and Error and Promax Pattern Matrix and Correlations among Their Factors

Biquartimin Pattern Matrix			
−.018	−.024	**.767**	.001
.016	−.054	**.480**	−.004
.026	−.206	**.630**	.002
.150	−.070	**.567**	−.081
.833	.081	−.025	−.082
.863	−.103	−.050	.049
.951	.019	−.006	−.195
.584	.093	.224	−.095
.920	−.133	−.065	.034
.053	**.893**	−.201	.047
.005	**.441**	−.006	.342
−.163	**.710**	.273	−.063
.040	**.425**	**.498**	−.129
.055	−.046	−.195	**.680**
−.044	−.058	−.078	**.643**
−.173	−.143	.299	**.610**
−.048	.103	−.162	**.696**
−.227	.204	.213	**.497**
.000	.031	.131	**.382**
.289	−.071	.288	.204
.005	.320	**.372**	.097
.274	−.064	.288	.207
.257	.053	**.446**	.077
.268	**.408**	.010	.187
Correlations among Factors			
1.000	.398	.573	.623
.398	1.000	.383	.456
.573	.383	1.000	.612
.623	.456	.612	1.000

Note: Data input is maximum-likelihood solution for 24 psychological tests (Holzinger and Swineford, 1939). Promax input was the normalized varimax solution. Bold numbers emphasize the highest loadings in a column.

The gradient matrix at $\mathbf{\Lambda}$ for the oblimin family is

$$\mathbf{G}_q = \mathbf{\Lambda} \cdot ((\mathbf{I} - \gamma \mathbf{C})\mathbf{\Lambda}^{[2]}\mathbf{N}) \qquad (12.56)$$

However, Jennrich (1979) recommends against using the oblimin family with any $\gamma > 0$, because there exists a manifold T of oblique rotations over which $Q(\mathbf{\Lambda})$ does not have a minimum and in fact is not bounded from below. He recommends using simply $\gamma=0$, which is quartimin.

12.6.2.4 Least-Squares Rotation to a Target Matrix

Let **B** be a target matrix and we wish to rotate an initial matrix **A** to a factor-pattern matrix **Λ** that differs minimally in a least-squares sense from **B**. Then we seek a **Λ** such that

$$Q(\mathbf{\Lambda}) = \|\mathbf{\Lambda} - \mathbf{B}\|^2 = \text{tr}(\mathbf{\Lambda} - \mathbf{B})'(\mathbf{\Lambda} - \mathbf{B}) \tag{12.57}$$

is a minimum.

The gradient matrix in this case is given by

$$\mathbf{G}_q = 2(\mathbf{\Lambda} - \mathbf{B}) \tag{12.58}$$

This routine is the heart of the Promax rotation method. In the example shown, we produce a Promax solution by rotating the initial matrix **A** to fit a target constructed from the varimax solution by obtaining the absolute value of the fifth power of each varimax element, divided by the original value of the element to recover the sign.

12.6.2.5 Least-Squares Rotation to a Partially Specified Target Pattern Matrix

Let **B** be an $n \times r$ matrix in which only a subset of its elements are specified. Since we can only fit the rotated matrix to the specified elements of **B**, then summations can only involve elements in **Λ** that match up with specified elements in **B**. We can accomplish this with an $n \times r$ matrix **W**, such that $w_{ij}=1$ if b_{ij} is specified, and 0 otherwise. Next let $\tilde{\mathbf{B}} = \mathbf{W} \cdot \mathbf{B}$ be a fully specified matrix. Then the criterion for this method is given by

$$Q(\mathbf{\Lambda}) = \|\mathbf{W} \cdot \mathbf{\Lambda} - \tilde{\mathbf{B}}\|^2 \tag{12.59}$$

while the gradient matrix is given as

$$\mathbf{G}_q = 2(\mathbf{W} \cdot \mathbf{\Lambda} - \tilde{\mathbf{\Lambda}}) \tag{12.60}$$

This routine is used by the Promin method of Lorenzo-Seva (1999).

12.6.3 Simplimax

The aim of simplimax (Kiers, 1994) is to rotate to a pattern matrix with a specified number of near-zero elements whose squared values are as small as possible, without a prespecified fixed and immutable position for them in the pattern matrix. The criterion for the simplimax method of Kiers (1994) is given by

$$Q(\mathbf{\Lambda}, \mathbf{W}) = \|\mathbf{W} \cdot \mathbf{\Lambda}\|^2 \tag{12.61}$$

where **W** must contain a specified number p of 1s, with other elements equal to 0. The 1s correspond to elements in $\mathbf{\Lambda}$ that should equal zero or be close to zero. In this case this corresponds to Browne's (1972) method of rotating to a partial target $\tilde{\mathbf{B}}$ in which $\tilde{\mathbf{B}}=\mathbf{0}$. The **W** matrix only seeks to fit certain elements of $\mathbf{\Lambda}$ to zero, which effectively seeks to minimize them in a least-squares sense. However, in simplimax, **W** may be modified during the iterations to get a better fit.

The partial gradient of Q with respect to $\mathbf{\Lambda}$ is

$$\mathbf{G}_q = 2\mathbf{W} \cdot \mathbf{\Lambda} \tag{12.62}$$

The criterion Q is to be minimized over all possible $\mathbf{\Lambda}$ and all possible **W** given p.

How should we determine p? One way would be to use some other rotational method such as the quartimin method to produce a provisional $\mathbf{\Lambda}$ matrix. We can examine this $\mathbf{\Lambda}$ to determine what is to be regarded as small, near-zero values. Count how many there are and let this be p. Set the elements of **W** corresponding to the p smallest elements of $\mathbf{\Lambda}$ to 1 and the rest to zero. This method of course has the drawback that if the initial solution is far from the simplest solution, one may be unable to escape that solution for one more optimal. This is the problem of local minima.

Kiers (1994) speculated that one might try all possible values of $p \leq nr$ and produce a "scree plot" of the values of minimum Q corresponding to each given value of p. Jennrich (2006) notes that as with scree plots used to determine the number of factors, there may be no sharp break in the curve. So this is still an open issue, although using a previous method of rotation, such as quartimin, to find the number of elements p near to zero, may have the greatest merit, albeit a certain subjectivity resides in this procedure.

In any case once determined, p should not be changed, otherwise the concept of a minimum value for Q has no specific fixed value. Still, researchers may yet investigate whether during the iterations p might be increased, for example, if more than p elements of $\mathbf{\Lambda}=\mathbf{A}(\mathbf{T}')^{-1}$ fall below the absolute value of the largest "near-zero" element of the current iteration. That is still an open question.

The initial value of **T** can be a random $r \times r$ matrix whose column sums of squares each add up to unity. In this way, if there are local minima, using several starting points may reveal their presence. Or **T** can be the $r \times r$ identity matrix (Table 12.8).

TABLE 12.8

Simplimax Pattern Matrix with $k=62$ and $k=74$ for the 24 Psychological Test Data

Simplimax Matrix $k=62$				**Simplimax Matrix $k=74$**			
.015	.091	.697	−.002	−.008	.106	.681	.062
.033	.016	.447	−.005	.020	.027	.431	.038
.042	−.124	.623	−.002	.033	−.107	.578	.067
.155	.008	.522	−.065	.138	.027	.514	−.016
.794	.097	−.027	−.051	.765	.134	.002	−.045
.830	−.090	.021	.053	.806	−.051	.000	.086
.891	.026	−.007	−.142	.864	.069	.027	−.134
.564	.140	.184	−.066	.534	.170	.212	−.049
.882	−.124	.015	.042	.858	−.082	−.008	.076
.092	.929	−.407	.051	.044	.918	−.256	−.038
.072	.504	−.068	.283	.033	.499	−.040	.273
−.119	.789	.044	−.045	−.161	.780	.170	−.095
.064	.518	.318	−.099	.026	.525	.402	−.107
.133	−.012	−.054	.552	.114	−.013	−.155	.599
.037	−.013	.045	.521	.020	−.016	−.055	.572
−.080	−.054	.397	.489	−.099	−.054	.281	.572
.045	.152	−.065	.565	.019	.145	−.143	.600
−.133	.292	.208	.401	−.163	.284	.169	.436
.058	.089	.170	.310	.037	.092	.120	.349
.315	−.005	.321	.169	.291	.015	.276	.223
.049	.408	.266	.082	.014	.412	.302	.092
.301	.002	.319	.171	.278	.021	.275	.224
.279	.138	.409	.066	.249	.158	.401	.111
.303	.462	−.059	.161	.263	.470	−.014	.149
Correlations among Factors				*Correlations among Factors*			
1.000	.414	.476	.426	1.000	.389	.448	.471
.414	1.000	.448	.376	.389	1.000	.313	.458
.476	.448	1.000	.370	.448	.313	1.000	.449
.426	.376	.370	1.000	.471	.458	.449	1.000

Kiers (1994) recommended first seeking to minimize Q for a given **T**. This can effectively be done by finding the p smallest elements of $\mathbf{\Lambda}=\mathbf{A}(\mathbf{T}')^{-1}$ and changing (if needed) **W** accordingly so that the 1s in **W** correspond to the p smallest elements of this $\mathbf{\Lambda}$.

Next, given the latest **W**, seek to minimize Q by finding a new value for **T** using the gradient-projection method. Then seek a possible new **W** given this new **T**, and so on. So, effectively the algorithm alternates back and forth between seeking to optimize **W** and then **T**.

12.7 Rotating Using Component Loss Functions

Eber (1966) developed the Maxplane method to emulate the manner in which a manual rotator would seek to minimize the hyperplane count, representing the number of variables close to the hyperplanes. A numerical value of 1 was given to a variable whose cosine with the reference vector fell within a narrow band on either side of zero. Otherwise the variable received a 0. Then the 1s were counted and the result known as the hyperplane count for the rotated solution. The aim of this method was to maximize the total hyperplane count.

Katz and Rohlf (1974) abandoned the zero-one hyperplane count for each variable in favor of a continuous function indicating the distance of the variable from the hyperplane. This made use of gradient methods possible. They then considered a two-parameter exponential family of functions, $f(w,u) = \exp[-(v_{ij}/w)^{2u}]$ where w represented the width of the hyperplane sought and u the power to which the ratio of the reference-structure loading to the hyperplane width would be raised. The aim of the function was to increase in dramatic nonlinear fashion the value of the function as the rotated factor brought the variable closer to the hyperplane. Applying this function to each reference-structure loading and maximizing their sum over all of the loadings in the matrix led to solutions very close to maxplane and superior to oblimax and varimax. Rozeboom (1991a,b) introduced his HYBALL program to emulate subjective rotation and used a four-parameter family of functions $c_i \times \psi[(\lambda_{ij} - w\lambda_{ik})/b]^2$. He too obtained very clean simple-structure solutions superior to traditional rotational methods.

A component loss function (CLF) applies the loss function to each element of $\mathbf{\Lambda}$ separately. In contrast to the direct oblimin or quartimin solution, the CLFs only concern individual elements without concern for their comparison with others. For example a CLF for an item λ_{ij} could be $h(\lambda_{ij}) = \lambda_{ij}^2$. Then a component loss criterion (CLC) for all elements in $\mathbf{\Lambda}$ could simply be the sum of all the CLF's for the individual elements, $\text{CLC} = Q(\mathbf{\Lambda}) = \sum_{i=1}^{n}\sum_{j=1}^{r} h(\lambda_{ij}) = \sum_{i=1}^{n}\sum_{j=1}^{r} \lambda_{ij}^2$, which would represent the least-squares loss function when minimized.

Jennrich (2006) considered a class of criteria involving an arbitrary CLF that was evaluated at the absolute value for each component element λ_{ij} of a factor-pattern matrix $\mathbf{\Lambda}$. In other words, the loss associated with each element λ_{ij} is to be measured by $h(\lambda_{ij}) = |\lambda_{ij}|$, the absolute value of the element. Then the sum of the loss values over all elements of $\mathbf{\Lambda}$ would be the CLC for the rotated solution, in this case $\text{CLC} = \sum_{i=1}^{n}\sum_{j=1}^{r}|\lambda_{ij}|$.

Perfect simple structure is often thought of as occurring when the factor-pattern matrix has in each row at most one nonzero value, and these nonzero elements occur in nonoverlapping sets of variables on different factors. Jennrich (2006) then considered four classes of functions: (1) a linear function; (2) a cubic, concave upward function; (3) a basic concave function;

and (4) a quadratic right constant CLF. He showed mathematically that you could always rotate with CLF functions to recover existent perfect simple structure if the CLF function is symmetric and its right half is concave and nondecreasing. He then showed that very simple forms of CLF rotation would perform much better than traditional oblique rotation schemes.

We will assume, as we have previously done in this chapter, that we seek a factor-pattern matrix $\mathbf{\Lambda}=\mathbf{A}(\mathbf{T}')^{-1}$ from an initial factor loading matrix for orthogonal factors, under the restraint that $\text{diag}(\mathbf{T}'\mathbf{T})=\mathbf{I}$. We seek further to optimize a CLF

$$Q(\mathbf{\Lambda})=\sum_{i=1}^{n}\sum_{j=1}^{r}|\lambda_{ij}| \tag{12.63}$$

called a CLC. Two simple CLFs with desirable properties proposed by Jennrich (2006) are the "linear function" defined by

$$h(\lambda_{ij})=|\lambda_{ij}| \tag{12.64}$$

and the basic concave CLF,

$$h(\lambda_{ij})=1-e^{-|\lambda_{ij}|} \tag{12.65}$$

Jennrich (2006) notes that both the simple linear function and the basic concave CLF yield essentially the same solution. That being the case, it would seem that the simple linear function is to be preferred on grounds of simplicity for working out the gradients and implementing the algorithm. Jennrich (2006) computed solutions using a derivative free method that computed approximate gradients arithmetically. He found that the simple linear CLF yielded solutions essentially the same as the basic concave, the simplimax, and the geomin solutions when applied to Thurstone's Box Problem. This is a somewhat computationally intensive and slower algorithm. I have worked out GPA versions that converge quite rapidly and yield the same solutions. These are easy to program and to implement in programs already developed by Jennrich and his colleagues at http://www.stat.ucla.edu/research/gpa in the Web.

As with all of Jennrich's programs for the GPA for factor rotation, we need to provide two quantities, $Q(\mathbf{\Lambda})$ and $\mathbf{G}_q$. $Q(\mathbf{\Lambda})$ is already given in Equation 12.60. For $\mathbf{G}_q$ we need to find expressions for the derivative of the absolute value of a variable, which is not given in most calculus textbooks. The function $h(\lambda_{ij})=|\lambda_{ij}|$ has a graph shown in Figure 12.3.

The ordinary derivative of abs(x) is given as abs(x)/x, which means that at 0 the derivative is undefined, because we would have 0/0. We may, however, for purposes of numerical analysis, and because the absolute value function is symmetric, take the symmetric derivative:

$$\frac{dy}{dx}=\lim_{\Delta x\to 0}\frac{f(x+\Delta x)-f(x-\Delta x)}{2\Delta x}$$

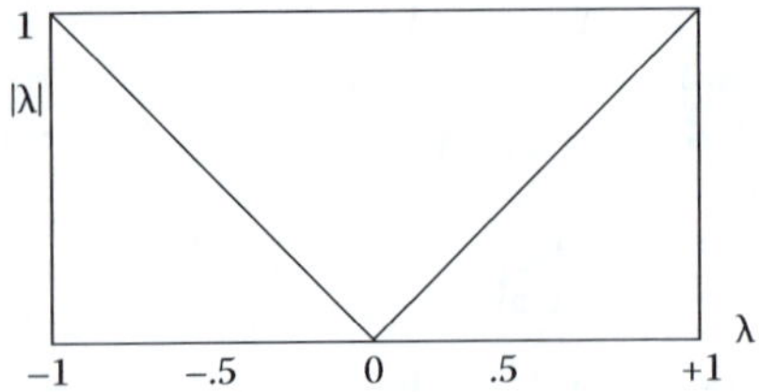

FIGURE 12.3
Graph of $h(\lambda_{ij}) = |\lambda_{ij}|$ over the interval (−1, 1). Note that the slope of the function for all values less than 0 is −1 while for those greater than 0 it is +1. The slope at 0 by a symmetric derivative is 0.

hence, when $x=0$,

$$\left.\frac{dh}{dx}\right|_0 = \lim_{\Delta x\to 0} \frac{|(0+\Delta x)|}{2\Delta x} - \frac{|-\Delta x|}{2\Delta x} = \frac{|\Delta x| - |\Delta x|}{2\Delta x} = \frac{1-1}{2} = 0$$

Thus, the matrix $\mathbf{G}_q = \partial Q(\boldsymbol{\Lambda})/\partial\boldsymbol{\Lambda}$ for the simple linear CLF function is $\mathbf{G}_q = [g_{ij}]$ where

$$g_{ij} = \begin{cases} +1 & \text{if } \lambda_{ij} > 0 \\ 0 & \text{if } \lambda_{ij} = 0 \\ -1 & \text{if } \lambda_{ij} < 0 \end{cases} \tag{12.66}$$

Using Equations 12.60 and 12.61 I have adapted Jennrich's (2002) GPA to perform this rotation, which I will call "Absolmin" from "absolute minimum." I have applied it to the 24 psychological test problem and obtained a converged solution. The solution is quite good and compares favorably with simplimax, geomin and Jennrich's simple linear with epsilon set equal to .01 (see below). However, the convergence has to be determined not by examining the slope of the criterion, since rarely will this be zero because at all values of λ_{ij} except zero, the derivative will be either −1 or + 1. Rather, convergence is detected by comparing the values of $Q(\boldsymbol{\Lambda})$ after several iterations to see if there has been no change and also examining the value of α to see if it has become very close to 0.

At this writing I have had very little experience with the "Absolmin" solution, but what I have seen has been very encouraging. Further work will have to take into account the many local minima (by using several random-oblique-transformation matrices to begin the algorithm) and to retain solutions that have the lowest value for $Q(\boldsymbol{\Lambda})$ when started from a number of different values for **T**. I have found that often one has to run the algorithm many times, starting with different random-oblique-transformation matrices, to find the solution that yields the lowest criterion value.

Random-oblique-transformation matrices can be easily made as follows, if one's programming language has a way to compute uniformly distributed random matrices. Let **M** be an $r \times r$ matrix whose elements are uniformly and independently distributed random numbers. Then one may compute a random **T** by $\mathbf{T} = \rho(\mathbf{M}) = \mathbf{M}[\text{diag}(\mathbf{M}'\mathbf{M})]^{-1/2}$ (Jennrich, personal communication). If one seeks a random-orthogonal-transformation matrix **T**, it may be obtained by application of the singular-value decomposition (see Equation 12.31) such that $\mathbf{M} = \mathbf{UDV}'$ where **U** and **V** are $r \times r$ orthogonal matrices and **D** a diagonal matrix. Then $\mathbf{T} = \mathbf{UV}'$ or $\mathbf{T} = \mathbf{U}$ or $\mathbf{T} = \mathbf{V}$ would be random $r \times r$ orthogonal matrices.

On the other hand, Jennrich (2006) was concerned with the absence of a true derivative at $\lambda_{ij} = 0$ for the simple linear algorithm. He spliced a quadratic function $a + b\lambda^2$ into the absolute value function $h(\lambda) = |\lambda|$ in the vicinity of zero (Figure 12.4). The quadratic function had to be selected in such a way that its derivative $h'(\varepsilon)$ at $\lambda = \varepsilon$ would be +1 and at $\lambda = -\varepsilon$ would be −1, where ε is a small quantity greater than zero. This could be accomplished if

$$b = \frac{h'(\varepsilon)}{2\varepsilon} \quad \text{and} \quad a = h(\varepsilon) - b\varepsilon^2$$

Then the resulting modified function is given by

$$h_\varepsilon(\lambda) = \begin{cases} h(\lambda) & |\lambda| > \varepsilon \\ a + b\lambda^2 & -\varepsilon \le \lambda \le \varepsilon \end{cases}$$

The elements of the gradient matrix $\mathbf{G}_q$ for the modified linear GPA algorithm is then $\mathbf{G}_q = [g_{ij}]$ where

$$g_{ij} = \begin{cases} +1 & \text{if } \lambda_{ij} > \varepsilon \\ \lambda_{ij}/\varepsilon & \text{if } \varepsilon \le \lambda_{ij} \le +\varepsilon \\ -1 & \text{if } \lambda_{ij} < -\varepsilon \end{cases} \tag{12.67}$$

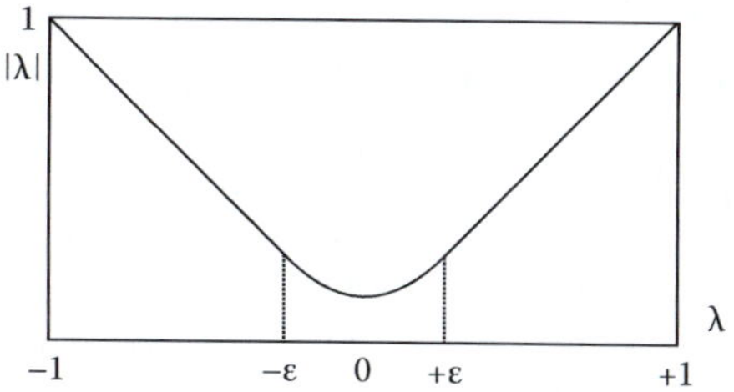

FIGURE 12.4
Magnified view of modified simple linear loss function in the vicinity of $\lambda = 0$ for a specified ε. The loss function is the quadratic function $a + b\lambda^2$ between $-\varepsilon$ and $+\varepsilon$ but is the absolute value function for all values of λ such that $|\lambda| > \varepsilon$.

TABLE 12.9

Absolmin and Modified Absolmin Solutions for the 24 Psychological Test Problem

Absolmin Pattern Matrix				**Modified Absolmin ε=.01**			
.047	.091	**.690**	.005	.082	.104	**.670**	.014
.052	.017	**.438**	.000	.071	.026	**.425**	.007
.065	−.113	**.599**	−.001	.080	−.100	**.582**	.017
.171	.008	**.509**	−.051	.192	.017	**.493**	−.047
.775	.085	−.008	−.002	**.799**	.074	−.016	−.050
.811	−.089	.030	.095	**.830**	−.097	.023	.053
.864	.015	.000	−.085	**.880**	.003	−.010	−.132
.559	.128	.197	−.028	**.588**	.123	.185	−.060
.860	−.123	.021	.086	**.876**	−.131	.014	.043
.098	**.878**	**−.308**	.077	.167	**.859**	**−.305**	.013
.092	**.483**	.002	.288	.153	**.478**	.001	.249
−.096	**.747**	.109	−.029	−.034	**.742**	.102	−.061
.085	**.490**	**.351**	−.077	.136	**.490**	**.337**	−.098
.150	.001	−.011	**.535**	.188	.002	−.007	**.512**
.059	.000	.082	**.500**	.096	.005	.084	**.485**
−.042	−.036	.418	**.464**	.000	−.023	**.411**	**.466**
.068	.156	−.008	**.547**	.118	.157	−.004	**.520**
−.097	.288	.257	**.385**	−.041	.294	.253	**.371**
.078	.093	.198	**.302**	.115	.097	.194	.290
.328	−.001	**.331**	.180	**.359**	.004	**.320**	.165
.073	**.390**	**.304**	.093	.125	**.391**	.293	.073
.315	.006	**.330**	.182	**.347**	.011	**.319**	.167
.296	.133	**.422**	.084	**.336**	.138	**.407**	.069
.312	**.439**	.001	.183	**.367**	**.431**	−.003	.137
Correlations among Factors				*Correlations among Factors*			
1.000	.353	.419	.310	1.000	.313	.396	.300
.353	1.000	.341	.275	.313	1.000	.296	.281
.419	.341	1.000	.267	.396	.296	1.000	.197
.310	.275	.267	1.000	.300	.281	.197	1.000

Note: Bold numbers emphasize the highest loadings in a column.

By varying the value of ε from large to small near zero one can make the CLC approach the least-squares criterion, on the one hand, and the simple linear Absolmin on the other. Jennrich found that setting ε at .01 yielded solutions essentially equivalent to simplimax and geomin. A proper name for this modification of the "Absolmin" criterion is the "Epsolmin" criterion.

The absolmin algorithm also succeeds in recovering the bounding hyperplanes of Thurstone's (1947) Box Problem.

However, the solutions for the Box Problem for absolmin, epsolmin, and least squares were difficult to find, due to many local minima, requiring

TABLE 12.10

Absolmin, Epsolmin, and Least-Squares Solutions to the Box Problem Are Each CLF Solutions

Variables	Absolmin			Epsolmin			Least Squares		
x	**.978**	−.019	−.055	−.063	.001	**1.001**	−.013	**.993**	−.021
y	.068	**.952**	−.011	**.935**	−.064	.072	**.925**	.089	.105
z	.000	.096	**−.952**	.109	**−.954**	.002	.069	.083	**.946**
xy	.634	.645	.000	.606	−.003	.651	.631	.655	.020
xz	.587	.023	−.677	.005	−.643	.603	.009	.652	.625
yz	−.018	.642	−.610	.642	−.646	−.016	.608	.047	.668
x^2y	.826	.383	−.044	.340	−.021	.846	.378	.847	.022
xy^2	.387	.819	−.023	.790	−.050	.398	.798	.410	.079
$x^2 z$	.776	−.008	−.465	−.037	−.420	.795	−.014	.825	.399
xz^2	.438	.057	−.876	.048	−.852	.450	.036	.519	.835
$y^2 z$	−.011	.789	−.413	.784	−.460	−.008	.756	.040	.489
yz^2	−.024	.474	−.755	.479	−.782	−.021	.441	.050	.794
x/y	.811	−.733	−.194	−.757	−.104	.828	−.711	.822	.055
y/x	−.728	.839	.092	.859	.001	−.743	.812	−.726	.051
x/z	.871	.057	.683	.007	.726	.889	.080	.823	−.731
z/x	−.804	.027	−.764	.074	−.808	−.821	.000	−.748	.814
y/z	.000	.827	.827	.803	.774	.000	.825	−.053	−.727
z/y	.000	−.827	−.827	−.803	−.774	.000	−.825	.053	.727
$2x + 2y$	.547	.713	.015	.678	.003	.562	.698	.567	.019
$2x + 2z$	.548	−.001	−.718	−.016	−.685	.562	−.015	.615	.666
$2y + 2z$	.000	.646	−.598	.646	−.634	.003	.613	.064	.655
$(x^2 + y^2)^{1/2}$	.537	.704	.003	.669	−.009	.551	.688	.557	.030
$(x^2 + z^2)^{1/2}$	.517	.013	−.708	−.001	−.677	.531	−.002	.584	.660
$(y^2 + z^2)^{1/2}$	.022	.645	−.567	.643	−.602	.025	.612	.084	.623
xyz	.451	.497	−.470	.476	−.473	.463	.475	.506	.480
$(x^2 + y^2 + z^2)^{1/2}$	.341	.549	−.481	.532	−.493	.351	.523	.397	.505
	Correlations among Factors								
	1.000	.195	−.214	1.000	−.201	.235	1.000	.169	.182
	.195	1.000	.272	−.201	1.000	−.268	.169	1.000	.205
	−.214	−.272	1.000	.235	−.268	1.000	.182	.205	1.000

Note: Bold numbers emphasize the highest loadings in a column.

many random starts to find a solution that seemed near minimum. I decided to start with the weighted varimax solution to see if that would quickly find a better minimum. In Tables 12.9 and 12.10, the absolmin solution shown was the one with the lowest $Q(\mathbf{\Lambda})$ of about 20 solutions, each begun with a random orthogonal **T**.

The results shown in Table 12.11 were all obtained directly without random searches (which should be done earlier with the weighted varimax solution).

TABLE 12.11

Absolmin, Epsolmin, and Least-Squares Solutions to the Box Problem Beginning with the Weighted Varimax Solution

Variables	Absolmin			Epsolmin			Least Squares		
x	**.992**	−.011	.010	**.994**	−.018	.011	**.994**	−.014	.020
y	.062	**.940**	−.058	.062	**.937**	−.068	.091	**.925**	−.103
z	.012	.071	**−.959**	.022	.068	**−.958**	.081	.070	**−.946**
xy	.638	.642	.006	.638	.636	−.001	.656	.630	−.020
xz	.605	.011	−.642	.613	.004	−.640	.652	.009	−.626
yz	−.015	.618	−.647	−.009	.615	−.653	.047	.609	−.667
x^2y	.835	.384	−.011	.836	.377	−.015	.848	.376	−.023
xy^2	.386	.811	−.043	.386	.807	−.051	.411	.798	−.079
x^2z	.793	−.013	−.415	.799	−.020	−.414	.824	−.015	−.400
xz^2	.455	.039	−.852	.465	.032	−.851	.518	.037	−.836
$y^2 z$	−.012	.769	−.457	−.008	.765	−.465	.040	.757	−.488
yz^2	−.018	.449	−.784	−.009	.445	−.787	.049	.442	−.793
x/y	.830	−.721	−.103	.833	−.726	−.094	.821	−.712	−.057
y/x	−.745	.824	.000	−.747	.827	−.010	−.725	.813	−.048
x/z	.872	.082	.738	.865	.078	.736	.824	.078	.729
z/x	−.805	.000	−.819	−.797	.003	−.818	−.749	.002	−.813
y/z	−.017	.837	.783	−.027	.837	.772	−.051	.824	.728
z/y	.017	−.837	−.783	.027	−.837	−.772	.051	−.824	−.728
$2x + 2y$	.549	.709	.012	.549	.704	.005	.568	.697	−.018
$2x + 2z$	.565	−.014	−.684	.574	−.020	−.682	.615	−.016	−.667
$2y + 2z$	.003	.623	−.634	.009	.619	−.640	.064	.614	−.654
$(x^2 + y^2)^{1/2}$	.538	.700	.000	.539	.694	−.007	.558	.687	−.030
$(x^2 + z^2)^{1/2}$	.534	.000	−.677	.542	−.006	−.675	.583	−.002	−.660
$(y^2 + z^2)^{1/2}$	.025	.622	−.602	.031	.618	−.607	.084	.612	−.622
xyz	.460	.484	−.469	.465	.477	−.473	.506	.475	−.480
$(x^2 + y^2 + z^2)^{1/2}$	.348	.533	−.490	.353	.527	−.494	.397	.523	−.505
				Correlations among Factors					
	1.000	.198	−.263	1.000	.204	.256	1.000	.169	−.205
	.198	1.000	−.245	.204	1.000	.236	.169	1.000	−.182
	−.263	−.245	1.000	.256	.236	1.000	−.205	−.182	1.000

Note: Bold numbers emphasize the highest loadings in a column.

12.8 Conclusions

We have noted with the Promin method, we may piece together features of each of these algorithms into a new algorithm that produces better results than does any of the component algorithms alone. Future research should involve developing programs that allow researchers to piece together the various algorithms so that the output from one may be readily made into the input into another, allowing researchers to fine tune their rotations with a given piece of data. Rozeboom (1991a,b) has developed a software package called Hyball that allows the user to choose different rotational options, and its flexibility and comprehensiveness may be the first of its kind in the area of rotation. Also work needs to be done on improving convergence and detection of convergence using the GPA. Better weighting methods may also be found.

One other observation that we may make is that there are now many good methods, but, unfortunately they do not all converge to precisely the same solution. Local minima in rotation also may be far more common than has been realized in the past, raising the question of the optimality of the solutions obtained—something that Rozeboom (personal communication) has observed as a result of running his Hyball program from different starting positions, and Jennrich (2002, 2006) has found running his GPA algorithms. The existence of local minima and many different rotational criteria all seem to do a passably good job of finding "simple structure," while converging to slightly different solutions belies to some extent Thurstone's belief in a unique, objective "simple-structure" solution. Perhaps someone in the future may yet prove Thurstone's dream of objective solutions in factor analysis to be a reality by proving one criterion to be better than all others, but until then it remains only an incomplete but still near realization. The differences clearly show up in the widely varying correlations among factors produced by different methods of rotation. Since these correlations are the basis for second-order factor analyses, the status of their second-order factor loadings as objective features of the world remains as yet problematic. Relying uncritically on these analyses for theoretical conclusions should be cautioned against.

13

Factor Scores and Factor Indeterminacy

13.1 Introduction

Up to now, our emphasis has been upon determining the $n \times r$ factor-pattern matrix, $\mathbf{\Lambda}$, either in the fundamental equation of common-factor analysis, $\mathbf{Y} = \mathbf{\Lambda X} + \mathbf{\Psi E}$, or in the fundamental equation of component analysis, $\mathbf{Y} = \mathbf{\Lambda X}$. The factor-pattern matrix is useful because it provides a basis for relating the n observed variables in $\mathbf{Y}$ to the r underlying common factors or to the n underlying components (as the case may be) in $\mathbf{X}$. But, given a particular observation on the n observed variables, represented by the $n \times 1$ observation vector $\mathbf{y}_i$, we might wish to estimate or determine a corresponding observation vector $\mathbf{x}$ which relates the ith observation to either the r common factors or the n components (as the case may be). Doing this would be desirable in a common-factor analysis because presumably from a theoretical point of view the common factors have a more fundamental importance than the observed variables and therefore we wish to relate the observations to the common factors. For example, one may construct a personality test of many self-description items, administer this test to a large number of persons, intercorrelate the items, and perform a common-factor analysis on the correlations among the items. The resulting common factors may then be interpreted as fundamental, underlying (latent) variables which influence performance across collections of items in the test. These factors may then assume importance for a theory of personality, and as a consequence one may wish to relate the people measured on the personality test to these factors, so that one could say that person i has score x_{1i} on factor 1, score x_{2i} on factor 2, and so on.

To cite another example, illustrating an application of component analysis, we might have a many-item attitude test that we administer to a large number of persons and for which we obtain estimates of the correlations among the items. We may then perform a principal-components analysis followed by an orthogonal rotation to simple structure of the first r principal components. At this point the variance of the original n variables will be largely accounted for by the smaller number of r rotated components. We may then obtain the scores of the subjects on these r components and use them as the dependent variables of a multivariate analysis of variance for testing for differences between groups of subjects in terms of attitude dimensions. The use of the r attitude components instead of the original attitude variables helps to reduce the number of dependent measures to work with, and the rotation of the components helps to provide components which are readily interpretable so that in the multivariate analysis of variance the discriminant functions may be more easily interpreted.

In this chapter, we are concerned with methods for obtaining the scores of subjects on common factors and on component variables. We first consider these methods in connection with models of component analysis, where we find that scores on the components are directly and uniquely determinable from the scores on the observed variables. Then we consider finding common-factor scores in common-factor analysis, where, because of the indeterminacy of the common-factor-analysis model, we will find that we can either imperfectly estimate the scores of subjects on the common factors by regression or construct factor scores that are not unique. We shall try to draw some conclusions from the effects of indeterminacy on the use of factor scores in common-factor analysis and on the interpretations of factors.

13.2 Scores on Component Variables

Consider the component-analysis model of full rank represented by the equation

$$\mathbf{Y} = \mathbf{\Lambda X} \tag{13.1}$$

where

$\mathbf{Y}$ is an $n \times 1$ random vector of n standardized observed variables
$\mathbf{\Lambda}$ is an $n \times n$ matrix of pattern coefficients
$\mathbf{X}$ is an $n \times 1$ random vector of n standardized component variables

We shall assume that the component variables are a linearly independent set and that the matrix $\mathbf{\Lambda}$ is nonsingular. This will imply that the variables in $\mathbf{Y}$ are linearly independent. We now can establish that

$$\mathbf{X} = \mathbf{\Lambda}^{-1}\mathbf{Y} \tag{13.2}$$

which means that the component variables are themselves linear combinations of the observed variables and are in fact completely determined from the observed variables. It is important to realize that in all models of component analysis the components are completely determined from the observed variables.

Equation 13.2 may serve as a basis for finding the scores of a subject on the component variables, given his scores on the observed variables. Let $\mathbf{y}_i$ be the $n \times 1$ observation vector of the scores of the ith subject on the n observed variables; then

$$\mathbf{x}_i = \mathbf{\Lambda}^{-1}\mathbf{y}_i \tag{13.3}$$

gives the scores of the ith subject on the n component variables. Equation 13.3 is useful, however, only when the full, nonsingular pattern matrix $\mathbf{\Lambda}$ is available.

In most models of component analysis the components are partitioned into two sets so that

$$\mathbf{Y} = \begin{bmatrix} \mathbf{\Lambda}_1 & \mathbf{\Lambda}_2 \end{bmatrix} \begin{bmatrix} \mathbf{X}_1 \\ \mathbf{X}_2 \end{bmatrix} = \mathbf{\Lambda}_1\mathbf{X}_1 + \mathbf{\Lambda}_2\mathbf{X}_2 \tag{13.4}$$

where

$\mathbf{\Lambda}_1$ and $\mathbf{\Lambda}_2$ are, respectively, $n \times r$ and $n \times (n - r)$ pattern matrices
$\mathbf{X}_1$ and $\mathbf{X}_2$ are, respectively, $r \times 1$ and $(n - r) \times 1$ random vectors of component variables

The emphasis in such a partitioned model may be only on the components in $\mathbf{X}_1$, with the remaining components in $\mathbf{X}_2$ are discarded or ignored. For example, $\mathbf{X}_1$ may represent the first r principal components of the variables in $\mathbf{Y}$. Or $\mathbf{X}_1$ may represent those components of the observed variables maximally related to a set of r common factors in a canonical-factor analysis. Or $\mathbf{X}_1$ may represent the image components of an image analysis. In general, however, to determine the components in $\mathbf{X}_1$ from the observed variables in $\mathbf{Y}$ one must use a regression procedure in which $\mathbf{X}_1$ is obtained by

$$\mathbf{X}_1 = E(\mathbf{X}_1\mathbf{Y}')E(\mathbf{Y}\mathbf{Y}')^{-1}\mathbf{Y} = \mathbf{R}_{X_1Y}\mathbf{R}_{YY}^{-1}\mathbf{Y} \tag{13.5}$$

with the consequence that

$$x_{1i} = \mathbf{R}_{X_1Y}\mathbf{R}_{YY}^{-1}\mathbf{y}_i \tag{13.6}$$

gives the scores of the ith subject on the r components in $\mathbf{X}_1$, given his scores on the n components in $\mathbf{Y}$.

To use Equation 13.6 one must determine the matrix $\mathbf{R}_{X_1Y}$. This is given by

$$\mathbf{R}_{X_1Y} = E(\mathbf{X}_1\mathbf{Y}') = E[\mathbf{X}_1(\mathbf{X}_1'\ \mathbf{\Lambda}_1' + \mathbf{X}_2'\mathbf{\Lambda}_2') = \mathbf{R}_{X_1X_1}\mathbf{\Lambda}_1' + \mathbf{R}_{X_1X_2}\mathbf{\Lambda}_2' \tag{13.7a}$$

However, if $\mathbf{R}_{X1X2} = E(\mathbf{X}_1\mathbf{X}_2') = \mathbf{0}$, then

$$\mathbf{R}_{X_1Y} = \mathbf{R}_{X_1X_1}\mathbf{\Lambda}_1' \tag{13.7b}$$

and one need know only $\mathbf{R}_{X1X1}$ and $\mathbf{\Lambda}_1'$.

When $\mathbf{X}_1$ represents an orthogonal transformation of the first r principal components, then the transposed structure matrix is equal to $\mathbf{\Lambda}_1' = \mathbf{R}_{X1Y} = \mathbf{T}'\mathbf{D}_r^{1/2}\mathbf{A}_r'$, where $\mathbf{A}_r$ is the $n \times r$ matrix whose r columns are the first r eigenvectors of $\mathbf{R}_{YY}$, $\mathbf{D}_r$ is the $r \times r$ diagonal matrix of the first r eigenvalues of $\mathbf{R}_{YY}$ and $\mathbf{T}$ an orthonormal-transformation matrix. Noting that $\mathbf{R}_{YY}^{-1} = \mathbf{AD}^{-1}\mathbf{A}'$, we may in this case write Equation 13.6 as

$$\mathbf{x}_{1i} = \mathbf{T}'\mathbf{D}_r^{1/2}\mathbf{A}_r'\mathbf{AD}^{-1}\mathbf{A}'^{-1}\mathbf{y}_i = \mathbf{T}'\mathbf{D}_r^{1/2}\left[\mathbf{I}_r\ \mathbf{0}\right]\mathbf{D}^{-1}\mathbf{A}'\mathbf{y}_i$$

$$= \mathbf{T}'\mathbf{D}_r^{1/2}[\mathbf{D}_r^{-1}\ \mathbf{0}]\mathbf{A}'\mathbf{y}_i = \mathbf{T}'\mathbf{D}_r^{-1/2}\mathbf{A}_r'\mathbf{y}_i \qquad (13.8)$$

Equation 13.8 is a useful canonical form for factor scores obtained from principal-components analyses; however, to use it, one must know $\mathbf{T}$.

Frequently we know only the rotated matrix $\mathbf{\Lambda}_1 = \mathbf{A}_r\mathbf{D}_r^{1/2}\mathbf{T} = \mathbf{R}_{YX1}$ (here we assume $\mathbf{T}$ is orthonormal). It would seem that we then have to go through the process of computing $\mathbf{R}_{YY}^{-1}$ to apply Equation 13.6 to this situation but, in fact, computing is not necessary. There are two ways to avoid computing $\mathbf{R}_{YY}^{-1}$ when we have an orthogonal transformation of the principal-components factor-structure matrix: One of these ways computes the factor scores as

$$\mathbf{x}_{1i} = (\mathbf{\Lambda}_1'\ \mathbf{\Lambda}_1)^{-1}\ \mathbf{\Lambda}_1'\mathbf{y} \qquad (13.9)$$

which we can show is equivalent to Equation 13.8 because

$$(\mathbf{\Lambda}_1'\mathbf{\Lambda}_1)^{-1}\mathbf{\Lambda}_1' = (\mathbf{T}'\mathbf{D}_r^{1/2}\mathbf{A}_r'\mathbf{A}_r\mathbf{D}_r^{1/2})^{-1}\mathbf{T}\mathbf{D}_r^{1/2}\mathbf{A}_r'$$

$$= (\mathbf{T}'\mathbf{D}_r\mathbf{T})^{-1}\mathbf{T}'\mathbf{D}_r^{1/2}\mathbf{A}_r'$$

$$= (\mathbf{T}'\mathbf{D}_r^{-1}\mathbf{T})\mathbf{T}'\mathbf{D}_r^{1/2}\mathbf{A}_r'$$

$$= \mathbf{T}'\mathbf{D}_r^{-1.2}\mathbf{A}_r'$$

since $\mathbf{A}_r'\mathbf{A}_r = \mathbf{I}_r$ and $\mathbf{TT}' = \mathbf{I}_r$.

The other way computes a generalized inverse $\mathbf{R}_{YY}^{+}$ pertaining to just the retained variance of $\mathbf{R}_{YY}$ from the retained eigenvectors and eigenvalues and stores this before computing the orthogonal rotation $\mathbf{\Lambda}_1 = \mathbf{A}_r\mathbf{D}_r^{1/2}\mathbf{T}$ of the principal-components factor-structure matrix, that is,

$$\mathbf{R}_{YY}^{+} = \mathbf{A}_r\mathbf{D}_r^{-1}\mathbf{A}_r'$$

and then, for oblique rotated $\mathbf{\Lambda}_1 = \mathbf{A}_r\mathbf{D}_r^{1/2}(\mathbf{T}')^{-1}$

$$\mathbf{x}_{1i} = \mathbf{R}_{X_1X_1}\mathbf{\Lambda}'_1\mathbf{R}_{YY}^{+}\mathbf{y}_i$$

$$= \mathbf{T}^{-1}\mathbf{D}_r^{1/2}\mathbf{A}_r'\mathbf{A}_r\mathbf{D}_r^{-1}\mathbf{A}_r'\mathbf{y}_i$$

$$= \mathbf{T}^{-1}\mathbf{D}_r^{-1/2}\mathbf{A}_r'\mathbf{y}_i \qquad (13.10)$$

The last method is especially useful in constructing rotated principal-components-analysis computer programs with factor-score output, because $\mathbf{R}_{YY}^{+}$ is very easy to construct, given that one already has $\mathbf{A}_r$ and $\mathbf{D}_r$. Although in some situations computing a generalized inverse in this way has severe round-off problems because the smallest roots retained may be quite small, in most component analyses the retained roots are equal to or greater than 1, which means that the reciprocals of these roots will not be greatly affected by round-off errors in the estimation of the roots.

13.2.1 Component Scores in Canonical-Component Analysis and Image Analysis

13.2.1.1 Canonical-Component Analysis

Recall from Chapter 9 that the aim of canonical-component analysis is to find those components (linear combinations) of the observed variables that are maximally correlated with corresponding linear combinations of the common factors. The method by which a canonical-component analysis is carried out is the method of weighted principal components, which rescales the observed variables by the reciprocals of the square roots of their unique variances and then finds the principal components of these rescaled variables. In practice, the matrix analyzed by the method of canonical component analysis is the matrix $\mathbf{\Psi}^{-1}\mathbf{R}_{YY}\mathbf{\Psi}^{-1}$, where $\mathbf{\Psi}^2$ is the $n \times n$ diagonal matrix of unique variances for the n observed variables and $\mathbf{R}_{YY} = E(\mathbf{YY}')$. Note that $\mathbf{\Psi}^{-1}\mathbf{R}_{YY}\mathbf{\Psi}^{-1} = E(\mathbf{\Psi}^{-1}\mathbf{YY\Psi}^{-1})$. The products of the principal-axes analysis of $\mathbf{\Psi}^{-1}\mathbf{R}_{YY}\mathbf{\Psi}^{-1}$ are the $n \times n$ matrix $\mathbf{A}$ of eigenvectors and the $n \times n$ diagonal matrix $[\gamma_j]$ of eigenvalues of the matrix $\mathbf{\Psi}^{-1}\mathbf{R}_{YY}\mathbf{\Psi}^{-1}$. The full set of n components extracted from $\mathbf{\Psi}^{-1}\mathbf{R}_{YY}\mathbf{\Psi}^{-1}$ in this way is given by Equation 9.32:

$$\tilde{\mathbf{X}} = \left[\gamma_j\right]^{-1/2}\mathbf{A}'\mathbf{\Psi}^{-1}\mathbf{Y} \tag{9.32}$$

where $\tilde{\mathbf{X}}$ is the $n \times 1$ random vector of normalized principal components of $\mathbf{\Psi}^{-1}\mathbf{Y}$.

In canonical-component analysis, however, not all components are retained; rather, only the first r components corresponding to eigenvalues greater than 1 of the matrix $\mathbf{\Psi}^{-1}\mathbf{R}_{YY}\mathbf{\Psi}^{-1}$ correspond to the common factors and therefore are retained. Consequently let $\tilde{\mathbf{X}}_1$ be an $r \times 1$ random vector of canonical components associated with the r common factors; then we may write

$$\tilde{\mathbf{x}}_{1i} = \left[\gamma_j\right]_r^{-1/2}\mathbf{A}_r'\mathbf{\Psi}^{-1}\mathbf{y}_i \tag{13.11}$$

as the equation for finding the scores of the ith subject on the r canonical components, given his or her scores in $\mathbf{y}_i$ on the n observed variables, with $[\gamma_j]$ r the $r \times r$ diagonal matrix of the first r eigenvalues greater than 1 and $\mathbf{A}_r$ the $n \times r$ matrix of corresponding eigenvectors of the matrix $\mathbf{\Psi}^{-1}\mathbf{R}_{YY}\mathbf{\Psi}^{-1}$. If the r

canonical components are subjected to an orthonormal transformation with $\mathbf{T}$, the $r \times r$ orthonormal-transformation matrix, then

$$\tilde{\mathbf{x}}_{1i} = \mathbf{T}'\left[\gamma_j\right]_r^{-1/2} \mathbf{A}_r'\mathbf{\Psi}^{-1}\mathbf{y}_i \tag{13.12}$$

However, frequently after an "orthogonal" rotation one does not know the transformation matrix $\mathbf{T}$ separately from the rotated matrix $\mathbf{\Lambda}_r = \mathbf{\Psi}\mathbf{A}_r[\gamma_j]_r^{-1/2}\mathbf{T}$, which represents the correlations between the observed variables and the rotated canonical components. In this situation one may apply Equation 13.6 to find the scores of subject i on the r rotated components, given his scores on the n observed variables. However, it is possible to avoid computing the inverse matrix $\mathbf{R}_{YY}^{-1}$ separately by computing

$$\mathbf{R}_{YY}^{+} = \mathbf{\Psi}^{-1}\mathbf{A}_r[\gamma_j]_r^{-1}\mathbf{A}_r'\mathbf{\Psi}^{-1}$$

Then one can obtain the component scores for subject i with

$$\mathbf{x}_{1i} = \mathbf{\Lambda}_r'\mathbf{R}_{YY}^{+}\mathbf{y}_i \tag{13.13}$$

13.2.1.2 Image Analysis

An image analysis is carried out in a manner similar to that of canonical-component analysis. In place of the matrix $\mathbf{\Psi}^2$ of unique variances in canonical-component analysis one substitutes the matrix $\mathbf{S}^2 = [\text{diag}\mathbf{R}_{YY}^{-1}]^{-1}$ and then carries out a principal-axes factoring of the matrix $\mathbf{S}^{-1}\mathbf{R}_{YY}\mathbf{S}^{-1}$ to obtain the $n \times n$ matrix $\mathbf{A}$ of eigenvectors and the $n \times n$ diagonal matrix $[\gamma_j]$ of eigenvalues of the matrix $\mathbf{S}^{-1}\mathbf{R}_{YY}\mathbf{S}^{-1}$. As in canonical-component analysis, one retains only those image components corresponding to eigenvalues greater than 1 of the matrix $\mathbf{S}^{-1}\mathbf{R}_{YY}\mathbf{S}^{-1}$.

In image analysis we may encounter either one of two matrices from which the image-component scores may be derived. First, we may be given a matrix $\mathbf{\Lambda}_r = \mathbf{S}\mathbf{A}_r[\gamma_j]_r^{-1/2}\mathbf{T}$, which represents the correlations between the observed variables and the retained normalized image components subjected to an orthonormal transformation by the $r \times r$ orthonormal matrix $\mathbf{T}$. Since $\mathbf{R}_{YY}^{-1}$ should already be available from computing $\mathbf{S}^2$, we can in this case obtain the component scores by applying Equation 13.6, $\mathbf{x}_{1i} = \mathbf{\Lambda}_r'\mathbf{R}_{YY}^{-1}\mathbf{y}_i$. On the other hand, we may be given the matrix $\mathbf{\Lambda}_g = \mathbf{S}\mathbf{A}_r[(\gamma_j - 1)^2/\gamma_j]_r^{1/2}\mathbf{T}$, which represents the matrix of covariances between the images of the observed variables and the rotated normalized image components. In this case the objective is to find an equation equivalent to $\tilde{\mathbf{X}}_{1i} = \mathbf{T}'\mathbf{D}_r^{-1/2}\mathbf{A}_r'\mathbf{S}^{-1}\mathbf{y}_i$, which is analogous to Equation 13.11 found in canonical-components analysis. This may be obtained by the equation

$$\tilde{\mathbf{x}}_{1i} = \mathbf{\Lambda}_g'\mathbf{S}^{-1}\mathbf{A}_r'[(\gamma_j - 1)^{-1}]_r\mathbf{A}_r'\mathbf{S}^{-1}\mathbf{y}_i \tag{13.14}$$

which is equivalent to

$$\tilde{\mathbf{x}}_{1i} = \mathbf{T}'\left[\frac{\gamma_j - 1}{\gamma_j^{1/2}}\right]_r \mathbf{S}\mathbf{S}^{-1}\mathbf{A}_r'[(\gamma_j - 1)^{-1}]_r \mathbf{A}_r'\mathbf{S}^{-1}\mathbf{y}_i$$

where we assume that all the eigenvalues γ_j, $j = 1, \ldots, r$, are greater than 1.

13.3 Indeterminacy of Common-Factor Scores

In common-factor analyses the scores on the common factors are most often estimated using regression (Thurstone, 1935) by the equation

$$\hat{\mathbf{x}}_i = \mathbf{R}_{XX}\mathbf{\Lambda}\mathbf{R}_{YY}^{-1}\mathbf{y}_i = \mathbf{R}_{XY}\mathbf{R}_{YY}^{-1}\mathbf{y}_i \qquad (13.15)$$

where

$\hat{\mathbf{x}}_i$ is a random vector of the ith subject's estimated scores on r common factors
$\mathbf{R}_{XX}$ is the matrix of correlations among the factors
$\mathbf{\Lambda}$ the factor-pattern matrix
$\mathbf{R}_{XY}$ an $r \times n$ matrix of correlations between the r common factors and the n observed variables
$\mathbf{R}_{YY}^{-1}$ the inverse of the correlation matrix among the n observed variables
$\mathbf{y}_i$ a random vector of the ith subject's standardized scores on the n observed variables

The squared multiple correlations between the observed variables and the factors is given by

$$\mathbf{P}^2 = [\text{diag}(\mathbf{R}_{XY}\mathbf{R}_{YY}^{-1}\mathbf{R}_{YX})] \qquad (13.16)$$

$\mathbf{P}^2$ (rho squared) is an $r \times r$ diagonal matrix in whose principal diagonal are the squared multiple correlations ρ_j^2 for predicting each factor respectively. Usually these squared multiple correlations are less than unity, which is why it is said that the common factors are indeterminate. Squared multiple correlations for estimating the factors should always be reported when obtaining factor scores.

Squared multiple correlations less than unity is the main contrast between common-factor analysis and components analysis: factor scores in component analysis are mathematically determinate with squared multiple correlations necessarily equal to unity; in common-factor analysis factor scores are indeterminate with squared multiple correlations less than unity. But there is more to factor indeterminacy than the inability to perfectly determine the values of a common-factor variable from the observed variables. The same holds for determining the values of the unique-factor variables.

E.B. Wilson (1928a,b, 1929) was the first to appreciate the implications of factor indeterminacy in reviews of Spearman's (1927) *The Abilities of Man*, where Spearman had put forth his theories of the general intelligence factor *g*. Wilson, a Yale mathematician and statistician who had edited into a text (Gibbs and Wilson, 1901) the lecture notes of J. Willard Gibbs on *Vector Analysis*, a book on the differential and integral calculus of vectors, was quite adept at thinking in terms of vectors and vector spaces. He quickly saw the full implications of the indeterminacy of *g* flowing from the fact that one could in principal not perfectly estimate the scores on the common factor from the observed variables. And it concerned more than the problem of getting exact scores for the factor. It concerned the fact that there was no unique variable that could function as *g*. So, which was it?

To understand this indeterminacy better, consider that the common-factor model postulates the existence of r common factors and n unique factors from which the n observed variables are derived as linear combinations. This means that, if variables are treated as vectors in a vector space, the common-factor model postulates a vector space of $r + n$ dimensions in which the r common factors and the n unique factors are together a set of basis vectors. In this space the n observed variables are embedded as linear combinations of the common and unique factors as given by the fundamental equation of the common-factor model: $\mathbf{Y} = \mathbf{\Lambda X} + \mathbf{\Psi E}$. On the other hand, the n observed variables can also serve as a set of basis vectors, and all linear combinations of these n variables lie within the space spanned by the n variables. The common and unique factors lie partly outside this space and consequently cannot be simple linear combinations of the observed variables. Factor scores obtained from regression of the common factors onto the observed variables, in effect, are examples of such linear combinations of the observed variables.

When the squared multiple correlation ρ_j^2 for predicting factor j is less than unity, it means that the common-factor variable consists of two components, a part estimable from the observed variables and a part that is not estimable. Hence there is an indeterminacy in determining the common and unique factors from within the common-factor model.

But indeterminacy is not confined to just the $(r + n)$-dimensional space encompassed by the model. Implicit in the use of the common-factor model as a framework for formulating substantive theory about relations between variables is the existence of an even larger space of variables and "factors" in which the observed variables are embedded. This is the space of all possible variables which may be defined logically on a population of interest with respect to a domain of attributes. When a common factor (or even a unique factor) is isolated in a factor analysis, the problem is to decide with which of the variables in the larger space this common factor can be identified. This decision is not easy in the light of indeterminacy and is especially difficult when variables are selected in a haphazard way for factor analysis, without any idea of what common factors one should expect to find.

In the latter case the common factors may be only poorly represented among the variables, and no linear combination of the observed variables will relate very strongly to the common factors. When that occurs, factor scores will be next to useless as intermediate measures of the common factors.

13.3.1 Geometry of Correlational Indeterminacy

As background for the meaning of indeterminacy for the relationship of factor scores to common factors, consider the following form of indeterminacy which always exists in correlation problems. Suppose we have a variable X_1 and we know that another variable, X_2, correlates .80 with X_1. If we know where X_1 is located in total variable space, where might we locate X_2 in this space? Consider Figure 13.1 in which we have circumscribed a cone with its apex at the origin around the vector X_1. The angle between the locus of this cone and the vector X_2 is θ, and the cosine of θ is equivalent to the correlation between any vector in the surface of this cone and the vector X_1. Thus the vector X_2 will be found somewhere in the locus of this cone, but not knowing anything else other than the fact that X_1 correlates .80 with X_2, we will be unable to determine where within the locus of this cone X_2 will be found.

Consider another situation. Suppose both X_2 and X_3 correlate .80 with X_1. To what extent are X_2 and X_3 correlated? With only this much information, all we know is that both X_2 and X_3 occupy the locus of the same cone circumscribed around X_1. The maximum possible correlation between X_2 and X_3 is 1, which would occur if the two variables were collinear. But what is the "minimum possible" correlation between X_2 and X_3? Variables X_2 and X_3 are minimally correlated when they occupy positions in the locus of the cone directly opposite one another across the cone, as illustrated in Figure 13.2. When minimally correlated, the angle between X_2 and X_3 is 2θ. The correlation between them is thus $\cos 2\theta$, and our problem is to find the value of $\cos 2\theta$, given $\cos \theta = \rho_{12} = \rho_{13}$. From the trigonometric identities we have $\cos 2\theta = 2\cos^2 \theta - 1$. Hence the minimum correlation between X_2 and X_3 is $\rho^2_{23(\min)} = 2\rho^2_{12} - 1$ Thus in the present example, given that $\rho_{12} = \rho_{13} = .80$, the minimum possible correlation for ρ_{23} is $2(.64) - 1 = .28$. If $\rho_{12} = \rho_{13} = .707$, the minimum possible value for ρ_{23} is 0, and if $\rho_{12} = \rho_{13} < .707$ then the minimum possible value for ρ_{23} is less than 0 or negative.

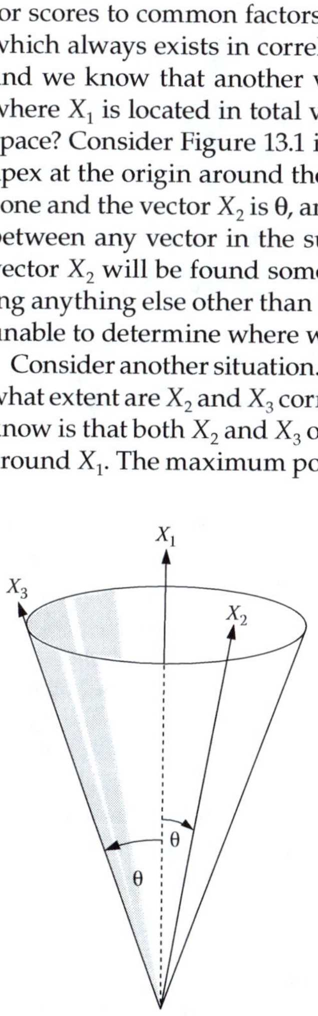

FIGURE 13.1
Cone representing locus of all vectors having angle θ with vector X_1.

The fact that two variables correlate moderately and equally well with a third variable is no guarantee that the first two variables are even moderately correlated. This principle is often invoked by psychometricians to discourage those who find an intelligence test moderately correlated with an aptitude test, and the aptitude test in turn moderately correlated with a job-performance rating, from necessarily inferring that the intelligence test is thus moderately related to the job-performance rating. In fact, it is quite likely that in this situation intelligence could be negatively correlated with the job-performance rating. In other words, inferences from correlation coefficients are not transitive across variables.

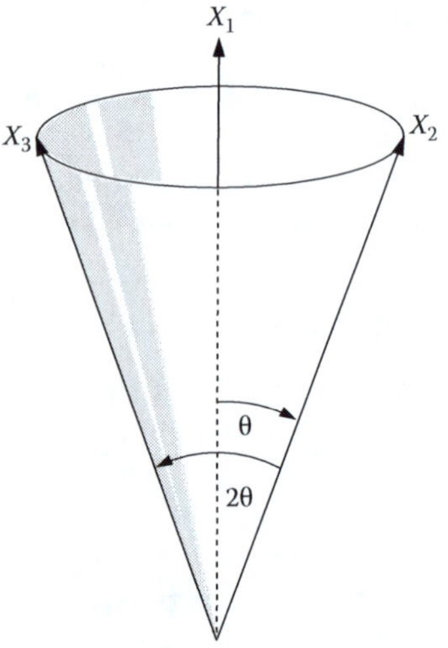

FIGURE 13.2
Graphical representation of two vectors X_2 and X_3 equally correlated with vector X_1 but minimally intercorrelated with each other.

The principles of indeterminacy in bivariate correlation also generalize to multiple correlation. Suppose, again, we have three variables, X_1, X_2, and X_3 (represented vectorially by unit-length vectors), and we know the correlations of X_1 with X_2 and X_3, respectively, and between X_2 and X_3. Where is X_1 located in space with respect to X_2 and X_3? Consider Figure 13.3 which illustrates geometrically the situation described. The vectors X_2 and X_3 define a plane which passes through them. All linear combinations of X_2 and X_3 are found within this plane. The variable X_1, however, does not lie within this plane but in the three-dimensional space occupied by the plane. The orthogonal projection of X_1 onto the plane defined by X_2 and X_3 falls on that linear combination of the variables X_2 and X_3 which represents the least-squares estimate of X_1 from X_2 and X_3 by multiple correlation. Actually, as one can observe, there are two possible vectors X_1 and X_1^* in the three-dimensional space which project themselves onto the same least-squares estimate and have, therefore, the same correlations with X_2 and X_3. Thus, if two variables are fixed absolutely within a three-dimensional space, there are only two other variables in that three-dimensional space which each have the same pattern of correlations with the first two variables, and the latter two variables have a correlation of $2R^2_{1\cdot23} - 1$ with one another, where $R^2_{1\cdot23}$ is the squared multiple correlation of X_1 onto X_2 and X_3 and is the equivalent to cos θ in Figure 13.3.

At this point, we might feel that the ambiguity is not as great as in the bivariate case illustrated in Figure 13.1, since our uncertainty in 3-space is reduced to two vectors each having the same pattern of correlations with two other vectors. This reduction in uncertainty is only illusory, however, if we assume that our uncertainty is confined not simply to the location of X_1,

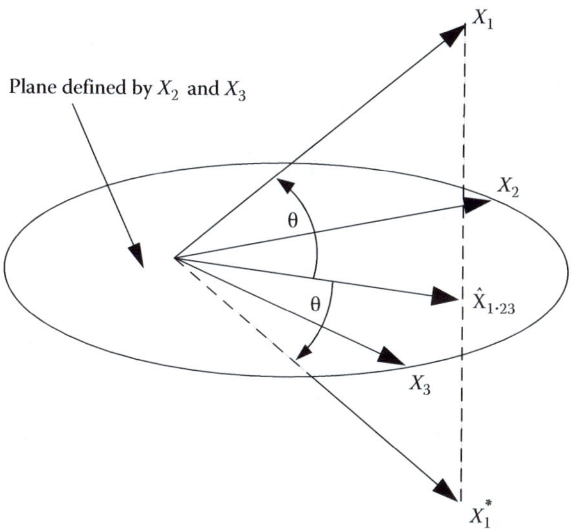

FIGURE 13.3
Representation of regression of a vector X_1 onto X_2 and X_3. The estimator, $\hat{X}_{1\cdot23}$, is in the plane defined by X_2 and X_3. Both X_1 and X_1^* have same multiple correlation with X_2 and X_3.

X_2, and X_3 with respect to the third dimension but to an additional fourth dimension as well. Uncertainty regarding the additional fourth dimension gives us freedom with which to locate X_1 relative to variables X_2 and X_3. In effect, we can locate a hypercone in the four-dimensional space revolving around the two-dimensional plane such that any vector found within this hypercone's surface has the same respective angles with respect to the vectors within the two-dimensional plane. Moreover, the minimum possible cosine of angle between two arbitrary vectors picked from the locus of this hypercone is $\cos 2\theta = 2\cos^2\theta - 1$, where θ is the angle of the surface of the hypercone with respect to the two-dimensional plane. This result generalizes to higher-dimensional problems as well.

In common-factor analysis, we know both the correlations among the observed variables and the correlations between the observed variables and the common factors. Therefore, it is possible to estimate the common factors by a multiple-regression procedure (given earlier as Equation 13.15), that is,

$$\hat{\mathbf{X}} = \mathbf{R}_{XX}\mathbf{\Lambda}'\mathbf{R}_{YY}^{-1}\mathbf{Y} \tag{13.15}$$

where

$\hat{\mathbf{X}}$ is the estimated random vector of common-factor variables based upon the observed variables in $\mathbf{Y}$

$\mathbf{R}_{XX}$ is the $r \times r$ matrix of correlations among the common factors

$\mathbf{\Lambda}$ is the $n \times r$ factor-pattern matrix

$\mathbf{R}_{YY}$ the matrix of correlations among the observed variables

The matrix of covariances among the estimator variables in $\hat{\mathbf{X}}$ is given as $\boldsymbol{\Sigma}_{\hat{X}\hat{X}} = \mathbf{R}_{XX}\boldsymbol{\Lambda}'\mathbf{R}_{YY}^{-1}\boldsymbol{\Lambda}\mathbf{R}_{XX}$. The diagonal matrix containing multiple correlations of the observed variables with the respective common factors is then given by the matrix $\mathbf{D}_P^2 = \text{diag}\left[\mathbf{R}_{XX}\boldsymbol{\Lambda}'\mathbf{R}_{YY}^{-1}\boldsymbol{\Lambda}\mathbf{R}_{XX}\right]$. The $r \times r$ matrix of correlations among the estimates of the common factors is then

$$\mathbf{R}_{\hat{X}\hat{X}} = \mathbf{D}_P^{-1}\mathbf{R}_{XX}\boldsymbol{\Lambda}'\mathbf{R}_{YY}^{-1}\boldsymbol{\Lambda}\mathbf{R}_{XX}\mathbf{D}_P^{-1} \tag{13.17}$$

In general, $\mathbf{R}_{\hat{X}\hat{X}} \neq \mathbf{R}_{XX}$.

The squared multiple-correlation coefficients given by the diagonal elements of $\mathbf{D}_P^2 = \mathbf{P}$ (rho) give an indication of the degree to which the common factors are indeterminate. This does not mean simply that the estimators of these common factors may stand in poor stead of the common factors in terms of their scores. The smaller the squared multiple-correlation coefficients, the more the common factors are free mathematically to be any one of many possible variables, some of which are quite different from one another, if not even contradictory.

13.4 Further History of Factor Indeterminacy

While Mulaik and McDonald (1978) gave some of the history of factor indeterminacy, a more detailed history was given by Steiger (1979). We cannot go into the detail covered by these articles, but will briefly cover the main points. Mulaik (2005) also covers history after 1970.

After Wilson (1928a,b, 1929) pointed out the indeterminacy of Spearman's g factor, indicating that the g factor itself was not uniquely determined by the factor analysis, Spearman (1929) seemingly misunderstood or ignored the nature of Wilson's critique, which pointed out the *nonuniqueness* for the meaning of Spearman's g factor. Spearman suggested that the indeterminacy could be easily eliminated by adding a variable that was perfectly correlated with g to the set of observed variables (Steiger, 1979). He seemed unaware that that could be done, with different additional variables, in the case of each distinct alternative g variable. Still the innumerable possible g's would remain.

In the 1930s, the issue of factor indeterminacy was considered by a number of British authors. Piaggio (1931, 1933) showed that if the number of variables satisfying the single-common-factor model could be increased without bound (and each additional variable has a nonzero absolute correlation with the factor), then the multiple correlation of the tests with the common factor would approach unity in the limit, which would make the common factor determinate (Mulaik, 2005). Piaggio (1935) was the first to demonstrate how one might construct variables that behaved exactly as the common factor in question. He further noted how these constructed factor variables were not unique because they could be made to vary by simply varying one component of the equation used to construct the common-factor variable. (We will show this shortly.)

A related approach to "overcoming" indeterminacy, following Spearman's use of an additional variable that correlates perfectly with the common factor in question, was to consider adding variables to the analysis that each pertained to the original factor, until the squared multiple correlation approached unity. Steiger (1979) called this the "indeterminacy in the limit" solution. Many at that time thought this resolved the matter.

A third approach to indeterminacy was offered by G. Thomson (1935), who showed that it was possible within the space of $n + r$ dimensions spanned by the common and unique factors to find infinitely many transformations of the common and unique factors that would each function as the common and unique factors of the variables.

Let the common-factor model equation be expressed as

$$\mathbf{Y} = [\boldsymbol{\Lambda}\, \boldsymbol{\Psi}] \begin{bmatrix} \mathbf{X} \\ \mathbf{E} \end{bmatrix}$$

where the left-most matrix on the right is an $n \times (n + r)$ matrix of factor pattern loadings and unique-factor loadings for the n variables, while the right-most matrix is an $(n + r) \times 1$ vector of common- and unique-factor variables in that order. Then presume that the common and unique factors are all mutually orthogonal standard-score variables, that is,

$$E\left(\begin{bmatrix} \mathbf{X} \\ \mathbf{E} \end{bmatrix} [\mathbf{X}'\ \mathbf{E}'] \right) = \mathbf{I}_{n+r}$$

Now, let $\mathbf{T}$ be an orthogonal, $(n + r) \times (n + r)$ transformation matrix, such that

$$\mathbf{TT}' = \mathbf{I}_{n+r} = \mathbf{T}'\mathbf{T}$$

and

$$[\boldsymbol{\Lambda}\ \boldsymbol{\Psi}]\mathbf{T} = [\boldsymbol{\Lambda}\ \boldsymbol{\Psi}] \tag{13.18}$$

Then

$$\mathbf{Y} = [\boldsymbol{\Lambda}\ \boldsymbol{\Psi}]\mathbf{TT}' \begin{bmatrix} \mathbf{X} \\ \mathbf{E} \end{bmatrix} = [\boldsymbol{\Lambda}\ \boldsymbol{\Psi}]\mathbf{T}' \begin{bmatrix} \mathbf{X} \\ \mathbf{E} \end{bmatrix}$$

$$= [\boldsymbol{\Lambda}\ \boldsymbol{\Psi}] \begin{bmatrix} \mathbf{X}^* \\ \mathbf{E}^* \end{bmatrix} \tag{13.19}$$

where $\mathbf{X}^*$ and $\mathbf{E}^*$ represent alternative common- and unique-factor variables obtained by orthogonal transformation of the initial orthogonal common- and unique-factor variables.

The matrix **T** is known as the "orthogonal right unit" of $[\mathbf{\Lambda}\ \mathbf{\Psi}]$ (Schönemann, 1971). It is not unique. A solution (Schönemann, 1971) for the orthogonal right unit of $[\mathbf{\Lambda}\ \mathbf{\Psi}]$ is given by means of its singular value composition (see Equation 12.31):

$$[\mathbf{\Lambda}\ \mathbf{\Psi}] = \mathbf{P}_{n\times n}\mathbf{D}^{1/2}_{n\times n}\mathbf{Q}'_{1(n\times(r+n))} \tag{13.20}$$

Here **P** (Roman "*P*") is the $n \times n$ eigenvector matrix of

$$\mathbf{R}_{YY} = [\mathbf{\Lambda}\ \mathbf{\Psi}]\begin{bmatrix}\mathbf{\Lambda}' \\ \mathbf{\Psi}\end{bmatrix} = \mathbf{PDP}'$$

D is the eigenvalue matrix of $\mathbf{R}_{YY}$, and

$$\mathbf{Q} = [\mathbf{Q}_1\ \mathbf{Q}_2]$$

is the $(n + r) \times (n + r)$ eigenvector matrix of the $(n + r) \times (n + r)$ matrix

$$\mathbf{M} = \begin{bmatrix}\mathbf{\Lambda}' \\ \mathbf{\Psi}\end{bmatrix}[\mathbf{\Lambda}\ \mathbf{\Psi}]$$

where

$$\mathbf{Q}'\begin{bmatrix}\mathbf{\Lambda}' \\ \mathbf{\Psi}\end{bmatrix}[\mathbf{\Lambda}\ \mathbf{\Psi}]\mathbf{Q} = \begin{bmatrix}\mathbf{D} & \mathbf{0} \\ \mathbf{0} & \mathbf{0}\end{bmatrix}$$

$\mathbf{Q}_1$ is an $(n + r) \times n$ matrix consisting of the first n eigenvectors of **M** and $\mathbf{Q}_2$ is an $(n + r) \times r$ matrix consisting of the r last eigenvectors of **M** orthogonal with $\mathbf{Q}_1$. Schönemann (1971) then gives **T** as

$$\mathbf{T} = \mathbf{Q}_1\mathbf{Q}'_1 + \mathbf{Q}_2\mathbf{S}\mathbf{Q}'_2 \tag{13.21}$$

where **S** is an arbitrary $r \times r$ orthogonal matrix, implying $\mathbf{SS}' = \mathbf{I}$.

To show that **T** is orthogonal, [amusingly, Schönemann (1971, p. 23) says "**T** … is clearly orthogonal" but does not work it out]:

$$\begin{aligned}\mathbf{TT}' &= (\mathbf{Q}_1\mathbf{Q}'_1 + \mathbf{Q}_2\mathbf{S}\mathbf{Q}'_2)(\mathbf{Q}_1\mathbf{Q}'_1 + \mathbf{Q}_2\mathbf{S}'\mathbf{Q}'_2) \\ &= \mathbf{Q}_1\mathbf{Q}'_1\mathbf{Q}_1\mathbf{Q}'_1 + \mathbf{Q}_1\mathbf{Q}'_1\mathbf{Q}_2\mathbf{S}'\mathbf{Q}'_2 + \mathbf{Q}_2\mathbf{S}\mathbf{Q}'_2\mathbf{Q}_1\mathbf{Q}'_1 + \mathbf{Q}_2\mathbf{S}\mathbf{Q}'_2\mathbf{Q}_2\mathbf{S}'\mathbf{Q}'_2 \\ &= \mathbf{Q}_1\mathbf{I}_n\mathbf{Q}'_1 + \mathbf{0} + \mathbf{0} + \mathbf{Q}_2\mathbf{I}_r\mathbf{Q}'_2 = \mathbf{Q}_1\mathbf{Q}'_1 + \mathbf{Q}_2\mathbf{Q}'_2\end{aligned}$$

Well, so far we have not yet shown that **T** is clearly orthogonal. Nevertheless, we still can do so, perhaps, not so clearly.

Consider that we can write $\mathbf{I}_{n+r} = \mathbf{QQ}'$. But let us now partition **Q** so that

$$\mathbf{I}_{n+r} = [\mathbf{Q}_1 \ \mathbf{Q}_2]\begin{bmatrix}\mathbf{Q}_1' \\ \mathbf{Q}_2'\end{bmatrix} = \mathbf{Q}_1\mathbf{Q}_1' + \mathbf{Q}_2\mathbf{Q}_2'$$

It is now clear that $\mathbf{TT}' = \mathbf{I}_{n+r}$, which establishes that **T** is orthonormal.

Next, we must show that Equation 13.18 holds:

$$\begin{aligned}[\boldsymbol{\Lambda} \ \boldsymbol{\Psi}]\mathbf{T} &= \mathbf{PD}_n^{1/2}\mathbf{Q}_1'(\mathbf{Q}_1\mathbf{Q}_1' + \mathbf{Q}_2\mathbf{S}\mathbf{Q}_2') \\ &= \mathbf{PD}_n^{1/2}\mathbf{Q}_1'\mathbf{Q}_1\mathbf{Q}_1' + \mathbf{PD}_n^{1/2}\mathbf{Q}_1'\mathbf{Q}_2\mathbf{S}\mathbf{Q}_2' \\ &= \mathbf{PD}_n^{1/2}\mathbf{Q}_1' = [\boldsymbol{\Lambda} \ \boldsymbol{\Psi}]\end{aligned}$$

$\mathbf{T}'$ is a nonunique $(n + r) \times (n + r)$ orthogonal transformation matrix by which different solutions for the common- and unique-factor variables may be found that are consistent with the fundamental equation of factor analysis. The nonuniqueness of **T** is produced by the arbitrary orthonormal transformation matrix **S** which implies that there are countless solutions for $\begin{bmatrix}\mathbf{X}^* \\ \mathbf{E}^*\end{bmatrix}$.

Steiger (1979) notes that factor analysis entered a new era under L. L. Thurstone and his students from 1941 to 1951. My observation is that in this decade the Thurstonians paid little attention to the earlier works on factor analysis by the British. Steiger (1979) observes "Unfortunately, factor indeterminacy and a related problem which Wilson had uncovered (the 'unidentifiability of $\mathbf{U}^2$ [$\boldsymbol{\Psi}^2$]') were either forgotten or ignored, to the extent that even the major texts (e.g., Holzinger and Harman, 1941; Thurstone, 1947) failed to mention their existence."

Following the Thurstonian era, the period from 1952 to 1969 involved a dramatic acceleration of developments of the statistical theory and computational methodology, eventually involving the early mainframe computers. With factor analysis now an almost effortless undertaking, many researchers came to use it blindly, almost as an automatic generator of empirical truths. Steiger (1979) notes in this period one major development pertinent to the issue of factor indeterminacy: Louis Guttman (1955) published an extensive discussion of factor indeterminacy and its meaning. In his paper (1), he gave formulas for constructing "exact" factor scores, but showed how they were not unique; (2) he introduced the concept of a universe of tests and showed how two tests in that universe could correlate with the observed variables of a study with the same correlations as a common factor of the study, and yet be minimally correlated $2\rho^2 - 1$ with one another (where ρ^2 is the squared multiple correlation between the observed variables and the common factor). If ρ^2 were less than .5, then the two minimally correlated alternative tests could be negatively correlated, implying that one contains components that contradict the other. Guttman regarded $2\rho^2 - 1$ as the index of the degree of factor indeterminacy. Guttman argued that this showed that the observed variables did not uniquely identify the nature or meaning of the factors and was a serious problem for the method. (3) Guttman (1955) argued that the use of second-order factor analysis should be discouraged because such factors were highly indeterminate. (4) Guttman then developed a theory of how indeterminacy might be reduced in the limit as the number of variables increased without bound. In conclusion, Guttman argued that the usual practice of interpreting factors according to the variables that have high loadings was "illogical." When there were an infinity of distinct variables that all could be the common factor, no interpretation has any special license and "the Spearman-Thurstone approach may have to be discarded for lack of determinacy of its factor scores" (Guttman, 1955, p. 79).

Steiger (1979) notes that Guttman (1955) was generally ignored by his contemporaries. Some have said that some of Guttman's negative feelings toward common-factor analysis may have reflected a degree of personal animus between him and Thurstone. Nevertheless, he raised some serious issues that were as yet unresolved.

13.4.1 Factor Indeterminacy from 1970 to 1980

Schönemann (1971) derived new demonstrations of factor indeterminacy and how alternative sets of common factors might be constructed for a given set of observed variables and a given factor-pattern matrix. Mulaik (1972) devoted a chapter to the question of factor scores and factor indeterminacy. In that chapter, he described Guttman's index $2\rho^2 - 1$ of indeterminacy and showed how it could relate generally to the indeterminacy in identifying a variable from only knowing its correlation with a given, known variable. Much of the earlier portion of this chapter is based on that (Mulaik, 1972) chapter.

But Schönemann and Wang (1972) reignited the controversy over factor indeterminacy. They presented formulas for constructing common- and unique-factor "scores" that would be "exact" and not merely estimates. However, these were not unique. These equations are as follows:

$$\mathbf{X} = \mathbf{\Phi}_{XX}\mathbf{\Lambda}'\mathbf{\Sigma}_{YY}^{-1}\mathbf{Y} + \mathbf{\Pi s} \tag{13.22}$$

and

$$\mathbf{E} = \mathbf{\Psi}\mathbf{\Sigma}_{YY}^{-1}\mathbf{Y} - \mathbf{\Psi}^{-1}\mathbf{\Lambda\Pi s} \tag{13.23}$$

Here $\mathbf{X}$ is an $r \times 1$ common-factor random vector and $\mathbf{E}$ an $n \times 1$ unique-factor random vector. $\mathbf{\Phi}_{XX}$ is the $r \times r$ matrix of correlations among the common factors. $\mathbf{\Lambda}$ is the $n \times r$ factor-pattern matrix. $\mathbf{\Sigma}_{YY}$ is the $n \times n$ matrix of covariances among the observed variables in **Y**. is an $r \times r$ matrix defined as any gram factor of the matrix

$$\mathbf{\Pi\Pi}' = (\mathbf{\Phi}_{XX} - \mathbf{\Phi}_{XX}\mathbf{\Lambda}'\mathbf{\Sigma}_{YY}^{-1}\mathbf{\Lambda\Phi}_{XX}) \tag{13.24}$$

s is an $r \times 1$ random vector with the properties that $E(\mathbf{s}) = 0$, var(**s**) = **I** and cov(**Y**,**s**) = **0**, which means **s** is any vector of mutually uncorrelated standard score variables that are also all uncorrelated with **Y**. Mulaik (1976a,b) further showed that an unlimited number of random vectors **s** could be constructed as transformations of

$$\mathbf{s} = \mathbf{T}_1[\mathbf{I}_r : \mathbf{0}]\mathbf{T}_2\boldsymbol{\sigma}$$

where

$\mathbf{T}_1$ is an arbitrary $r \times r$ orthonormal transformation matrix
$\mathbf{I}_r$ is an $r \times r$ identity matrix
0 is an $r \times (v - r)$ null matrix
$\mathbf{T}_2$ is an $v \times v$ arbitrary orthonormal transformation matrix
$\boldsymbol{\sigma}$ a $v \times 1$ random vector, with the properties $E(\boldsymbol{\sigma}) = \mathbf{0}$, var($\boldsymbol{\sigma}$) = $\mathbf{I}_s$ and cov (**Y**,$\boldsymbol{\sigma}$) = **0**, with $v \geq n + 1$

While one may be skeptical that the variables **X** and **E** in Equations 13.22 and 13.23 behave exactly as common and unique factors, proof comes from substituting the expressions for **X** and **E** in Equations 13.22 and 13.23, respectively, into $\mathbf{Y} = (\mathbf{\Lambda X} + \mathbf{\Psi E})$ and showing that the right hand side reduces to **Y**. (Mulaik (2005) gives the proof explicitly.)

Schönemann and Wang (1972) demonstrated that they could construct variables that would have all the properties of common and unique factors by use of Equations 13.22 and 13.23 and letting **s** be any $r \times 1$ vector of random independent orthonormal variables generated by random number generators in a computer. They presumed that with probability 1 random numbers generated in a computer would be correlated zero with any empirical variables **Y**. They also showed how one could simultaneously obtain Guttman's minimum correlation between alternative constructions for each of the common and unique factors by

$$\mathbf{\Gamma} = [\text{diag}(2\mathbf{\Phi}_{XX}\mathbf{\Lambda}'\mathbf{\Sigma}_{YY}^{-1}\mathbf{\Lambda\Phi}_{XX} - \mathbf{I})] \tag{13.25}$$

for the common factors, and

$$\mathbf{U} = [\text{diag}(2\mathbf{\Psi\Sigma}_{YY}^{-1}\mathbf{\Psi} - \mathbf{I})] \tag{13.26}$$

for the unique factors.

Schönemann and Wang (1972) also performed reanalyses by maximum-likelihood factor analysis of 13 studies in the literature in which they demonstrated a number of studies in which the smallest minimum correlation between alternative factors was negative, sometimes on the order of –.466.

Provoked by Schönemann (1971) and Schönemann and Wang (1972), and after reviewing Guttman (1955), McDonald (1974) concluded that Guttman's minimum-correlation index of factor indeterminacy was "inconsistent with the foundations of factor analysis in probability theory" (McDonald, 1974, p. 203). McDonald seemed to have the idea that Guttman (1955) was arguing that a common-factor variable had simultaneously more than one value for any subject's observations on the observed variables. And this made no sense. Framing it like that, it does make no sense.

But actually Guttman and McDonald were looking at the issue from within two different frames. McDonald was looking at the problem where the only variables in consideration were the observed variables and the common- and unique-factor variables. The common and unique factors had specified relations with the observed variables given by the model equation. In fact, one might consider the common and unique factors to be causes of the observed variables. It was not logical to presume that for each common or unique factor, the factor had more than one score for each observed subject. But Guttman had introduced the concept of a universe of variables in which the common-factor model was to be embedded. Each subject would have values on each of the infinitely many variables in the universe of variables. The researcher could not uniquely determine, from the observed variables' correlations with the common factors given by the analysis, which variables in the universe were the common and unique factors. So, it was quite possible mathematically for a subject to have scores on many different variables in the universe that would have the same correlations with the observed variables as the common and unique factors had in the factor analysis. So, which variables are the common factors? Or does it make sense to even ask this in this way?

McDonald (1974) also argued that the usual squared multiple correlation was the more appropriate index of indeterminacy in factor analysis. This has already been given here simultaneously for each common factor as the diagonal elements of the matrix $\mathbf{P}^2$ in Equation 13.16. These measure the determinacy of prediction of a common factor from the observed variables. As a measure of the degree of indeterminacy one has the error of estimate given by $\mathbf{I} - \mathbf{P}^2$. McDonald (1974) also questioned the likelihood of encountering two solutions for a factor that would be as minimally correlated as Guttman's index suggests. He felt they would approach having a measure of zero in probability, but he had no proof.

Mulaik (1976b) was an attempt to clarify Guttman's meaning of factor indeterminacy, which McDonald (1974) seemed not to grasp. I reviewed my account in Mulaik (1972). I also introduced a construction equation like Equation 13.22. But I also noted that rarely do researchers in practice seem to have strong disagreements about the interpretation of factors. Researchers would regard a factor as something in the world that they identified as common to the variables having the factor in common. After all, the interpretations of factors are not based on the scores on the common factors, which are usually unavailable, but on the factor-pattern matrix $\mathbf{\Lambda}$ or the factor structure matrix $\mathbf{R}_{YX} = \mathbf{\Lambda\Phi}$, which remain the same for all alternative solutions for a given set of common factors. How many natural variables in the world can one identify as having the pattern of relations between factors and variables given by the coefficients of the factor pattern or the factor structure matrices? If for convenience we set $\mathbf{T}_1 = \mathbf{I}$, and there were differences of interpretation, these would correspond to different versions for the common factors obtained by different rotations of $\mathbf{\Lambda}$ by $\mathbf{T}_2$. That suggested to me that one might develop a distribution for the different interpretations of a factor. The varying factor interpretations would be mapped onto points on the surface of a hypersphere generated by rotation of $\mathbf{T}_2$. Now an interpretation a of a factor, ξ_{ja}, would be given by a constructed factor variable

$$\xi_{ja} = \mathbf{v}'_j \mathbf{\Sigma}_{YY}^{-1} \mathbf{Y} + (1-\rho^2)^{1/2} s_{ja}$$

where

$\mathbf{v}_j$ is the jth column of $\mathbf{R}_{YX} = \mathbf{\Lambda\Phi}_{XX}$, the factor structure matrix, corresponding to factor j

ρ^2 is the squared multiple correlation for predicting factor j from the variables in $\mathbf{Y}$

s_{ja} is the component in vector $\mathbf{s}_a$ obtained by a given transformation $\mathbf{T}_{2a}$ of $\boldsymbol{\sigma}$ that corresponds to factor j

Since factor j is the same factor in the following discussion, we will drop it from the notation.

If ξ_a and ξ_b represent two alternative interpretations for the factor, then Mulaik gave the correlation between them as

$$\rho(\xi_a, \xi_b) = \rho^2 + (1-\rho^2)\rho(s_a, s_b) \tag{13.27}$$

where ρ^2 is the squared multiple correlation for predicting the factor from the observed variables in $\mathbf{Y}$. $\rho(s_a, s_b)$ is the correlation between the respective components in ξ_a and ξ_b that are due to distinct components s_a and s_b that are uncorrelated with $\mathbf{Y}$. The variation in correlation between two constructed factors is due simply to variation in the correlation between their undetermined parts s_a and s_b.

Mulaik then showed that

$$E[\rho(\xi_a, \xi_b)] = \rho^2 + (1-\rho^2)E[\rho(s_a, s_b)] \tag{13.28}$$

while

$$\text{var}(\rho(\xi_a, \xi_b)) = (1-\rho^2)^2 \,\text{var}(\rho(s_a, s_b)) \tag{13.29}$$

Mulaik (1976a,b) then made the possibly unrealistic assumption that all distinct interpretations of the factor were uniformly distributed on the surface of the v-dimensional hypersphere made by orthogonal rotations of $\boldsymbol{\sigma}$ by $\mathbf{T}_2$. Using the gamma distribution Mulaik was able to show that $E[\rho(s_a, s_b)] = 0$, and $\text{var}[\rho(s_a, s_b)] = 1/v$. These results implied that $E[\rho(\xi_a, \xi_b)] = \rho^2$ and $\text{var}[\rho(\xi_a, \xi_b)] = (1 - \rho^2)^2/v$. So, as the dimensionality of $\boldsymbol{\sigma}$ increased without bound, the expected correlation between alternative interpretations for a factor converged in probability to the squared multiple correlation for predicting the factor from the observed variables. In a sense one of McDonald's (1974) contentions was vindicated.

However, the assumption of a uniform distribution for all possible alternative solutions is unrealistic. Researchers will focus on some features of the world with greater likelihood than others in their interpretations. In this case, Mulaik (1976a,b) then showed that

$$E[\rho(\xi_a, \xi_b)] \geq \rho^2 \tag{13.30}$$

In other words, the squared multiple correlation is a lower bound to the average correlation between alternative solutions for a factor. There may well be some degree of agreement among researchers as to how to interpret a factor.

What Mulaik (1976a,b) concluded was that Guttman's minimum possible correlation and the squared multiple correlation simply represented different but complementary aspects of factor indeterminacy.

In April of 1976, before Mulaik (1976b) was published, I presented a paper (Mulaik, 1976a) at the Psychometric Society meetings at Bell Labs, in Murray Hill, New Jersey based much on the paper to be published. The paper was included in a symposium between myself, James Steiger (student of Peter Schönemann) and Roderick P. McDonald on factor indeterminacy. At the conclusion of the symposium, I remember particularly Darrell Bock's pointing out that the indeterminacy problem was not a fundamental problem because one could always increase the number of variables and raise the squared multiple correlation. This was Spearman's (1929, 1933) answer to Wilson (1928a,b). I felt this argument still failed to see that this solution would not work relative to the original set of variables.

I set about writing a paper showing why including more variables would not solve the initial indeterminacy, and I sent a draft to McDonald. This led to a long exchange by correspondence, in which at each turn, McDonald would disagree, rewrite my paper, and send it back to me. Finally, after an exchange of counter- and counter-counterexamples, I finally produced one where he saw the point of Guttman's factor indeterminacy. At this point, I invited him to be the second author of the paper in question (since he had written half of it anyway), and this was submitted by us and published as Mulaik and McDonald (1978) in *Psychometrika*.

The aim of this paper was to show that simply increasing the number of variables to infinity would not eliminate the initial factor indeterminacy.

Suppose, we said, we begin with a set of variables in a random vector $\boldsymbol{\eta}$ of variables satisfying the single common-factor model $\boldsymbol{\eta} = \boldsymbol{\lambda}\xi + \boldsymbol{\Psi}\boldsymbol{\varepsilon}$, where $\boldsymbol{\lambda}$ is an $n \times 1$ column of factor pattern loadings on a general (g-) factor.

Now suppose there are two researchers and the first researcher introduces k_1 new variables in random vector $\mathbf{v}_1$, which he/she now includes in a new model with the original variables in $\boldsymbol{\eta}$. The resulting model would be

$$\begin{bmatrix} \boldsymbol{\eta} \\ \mathbf{v}_1 \end{bmatrix} = \begin{bmatrix} \boldsymbol{\lambda} & \mathbf{P}_0 \\ \mathbf{c}_1 & \mathbf{P}_1 \end{bmatrix} \begin{bmatrix} \xi_0 \\ \xi_1 \end{bmatrix} + \begin{bmatrix} \boldsymbol{\Psi}_0^2 & 0 \\ 0 & \boldsymbol{\Psi}_1^2 \end{bmatrix} \tag{13.31}$$

Here $\mathbf{c}_1$ contains the factor loadings of variables in $\mathbf{v}_1$ on the first g factor. Any elements in $\mathbf{c}_1$ must range between -1 and $+1$. Additional common factors in ξ_1 may enter with the inclusion of the variables in $\mathbf{v}_1$. $\mathbf{P}_0$ and $\mathbf{P}_1$ are loadings of $\boldsymbol{\eta}$ and $\mathbf{v}_1$, respectively on the r_1 additional common factors. Mulaik and McDonald (1978) argued that this model would satisfy the "g-factor law" if there were no Heywood cases (Heywood, 1931) where row sums of squares of unrotated factor pattern loadings for orthogonal factors exceeded unity, and the additional proviso (Mulaik, 2005),

$$\begin{bmatrix} \boldsymbol{\Sigma}_{00} & \boldsymbol{\Sigma}_{01} \\ \boldsymbol{\Sigma}_{10} & \boldsymbol{\Sigma}_{11} \end{bmatrix} - \begin{bmatrix} \boldsymbol{\lambda} \\ \mathbf{c}_1 \end{bmatrix} [\boldsymbol{\lambda}' \quad \mathbf{c}_1'] \tag{13.32}$$

is nonnegative definite, held. The g-factor law simply asserted that there was a common general factor in the augmented set of variables consistent with the one in the initial set of variables.

We then asserted that any unit variance variable, X, would be a general factor g of $[\boldsymbol{\eta}, \mathbf{v}_1]$ if and only if $E(\boldsymbol{\eta}X) = \boldsymbol{\lambda}$ and $E(\mathbf{v}_1 X) = \mathbf{c}_1$. Furthermore, X would be a one-factor common factor of $\boldsymbol{\eta}$. But, we said, not every one-factor $X = \xi$ would be consistent with the g-factor law for $[\boldsymbol{\eta}, \mathbf{v}_1]$. This follows from the constraint asserted by Guttman (1955) that the correlation between any two alternative solutions for the general factor of $\boldsymbol{\eta}$, X_1, and X_2 would be correlated in the range of $2\rho_0^2 - 1 \le \rho_{12} \le 1$, where ρ_0^2 is the squared multiple correlation for predicting the common factor from $\boldsymbol{\eta}$. We will show how shortly.

This constraint is why the indeterminacy of the factors of the original set of variables in $\boldsymbol{\eta}$ is unavoidable. Additional variables do not reduce the original indeterminacy. They only reduce the indeterminacy for the augmented set of variables.

Consider, for example, that we have two alternative solutions, X_1^* and X_2^*, for the g-factor of $[\boldsymbol{\eta}, \mathbf{v}_1]$. Their correlation ρ_{12}^* must satisfy the constraint $2\rho_{(0+1)}^2 - 1 \le \rho_{12}^* \le 1$, where $\rho_{(0+1)}^2$ is the squared multiple correlation for predicting the general factor from the variables in $[\boldsymbol{\eta}, \mathbf{v}_1]$. But there can exist variables X_1 and X_2 that would be alternative solutions for the general factor of $\boldsymbol{\eta}$ such that at least one of which could not be a g-factor of $[\boldsymbol{\eta}, \mathbf{v}_1]$, because their correlation would be outside the constrained range. Now, we know that $\rho_0^2 \le \rho_{0+1}^2$, because a multiple correlation for predicting a criterion from an augmented set of predictor variables is always greater than or equal to the multiple correlation for predicting the criterion from a subset of the predictor variables. Suppose now, they are minimally correlated $2\rho_0^2 - 1 < 2\rho_{0+1}{}^2 - 1$, implying $\rho_0^2 < \rho_{0+1}^2$, then their correlation would lie outside the range for correlations between alternative solutions for the *g*-factor of $[\boldsymbol{\eta}, \mathbf{v}_1]$. In one case they could both be outside the range for *g*-factors of $[\boldsymbol{\eta}, \mathbf{v}_1]$, while being *g*-factors of $\boldsymbol{\eta}$. But if, say, X_2 is both a *g*-factor of $[\boldsymbol{\eta}, \mathbf{v}_1]$ and of $\boldsymbol{\eta}$, X_1 cannot be, because their correlation $2\rho_0^2 - 1 < 2\rho_{0+1}^2 - 1$ would be outside the range of solutions for the *g*-factor of $[\boldsymbol{\eta}, \mathbf{v}_1]$: $2\rho_{(0+1)}^2 - 1 \leq \rho_{12}^* \leq 1$. So, the range of possible alternative solutions for the *g*-factor of $[\boldsymbol{\eta}, \mathbf{v}_1]$ is reduced from that of $\boldsymbol{\eta}$.

Now, the two researchers could construct two augmented sets of variables $[\boldsymbol{\eta}, \mathbf{v}_1]$ and $[\boldsymbol{\eta}, \mathbf{v}_2]$ that each, separately, satisfied the *g*-factor law. But there could be a *g*-factor variable X_1 of $[\boldsymbol{\eta}, \mathbf{v}_1]$ that could not be a *g*-factor variable of $[\boldsymbol{\eta}, \mathbf{v}_2]$, and, similarly, a *g*-factor variable X_2 of $[\boldsymbol{\eta}, \mathbf{v}_2]$ that could not be a *g*-factor variable of $[\boldsymbol{\eta}, \mathbf{v}_1]$.

These two researchers also might find that if they were to merge the two augmented sets of variables together into $[\boldsymbol{\eta}, \mathbf{v}_1, \mathbf{v}_2]$, the doubly augmented set might not satisfy the *g*-factor law at all or even the common-factor model. If that occurred, they would learn that they had different conceptions of the common factor. That could be useful to know, because they are probably focusing on different things that still happen to be conformable by the *g*-factor law to $\boldsymbol{\eta}$ initially.

Suppose, further that they were able to augment $\boldsymbol{\eta}$ with an additional infinite set of variables in $\mathbf{v}_1$ so that $[\boldsymbol{\eta}, \mathbf{v}_1]$ conformed to the *g*-factor law, and the resulting squared multiple correlation were unity, and X_1 was the variable determined to be the unique *g*-factor for this set of variables. Similarly, suppose they augmented $\boldsymbol{\eta}$ with a second infinite set of variables in $\mathbf{v}_2$, with resulting squared multiple correlation of unity, and X_2 was the determinate solution for the *g*-factor of $[\boldsymbol{\eta}, \mathbf{v}_2]$. Then if $\rho(X_1, X_2) < 1$ (meaning they are not identical) and they were to combine all sets of variables together into the infinite vector $[\boldsymbol{\eta}, \mathbf{v}_1, \mathbf{v}_2]$, this doubly augmented set of variables could not satisfy the *g*-factor law and quite possibly even the common-factor model. X_1 and X_2 would be alternative solutions to the *g*-factor of $\boldsymbol{\eta}$, but they could not both be the *g*-factor of $[\boldsymbol{\eta}, \mathbf{v}_1, \mathbf{v}_2]$. The reason for this is that if the *g*-factor law held for the infinite vector $[\boldsymbol{\eta}, \mathbf{v}_1, \mathbf{v}_2]$, then the squared multiple correlation would be $\rho_{(0+1+2)}^2 = 1$ and the correlation ρ_{12} between X_1 and X_2 could not be less than 1, by the inequality $2\rho_{(0+1+2)}^2 - 1 \leq \rho_{12} \leq 1$. Yet the following inequality would hold: $2\rho_0^2 - 1 \leq \rho_{12} \leq 1$ with respect to the original set of variables in $\boldsymbol{\eta}$, which made distinct alternative solutions for X_1 and X_2 possible relative to $\boldsymbol{\eta}$.

This has practical consequences. If $\rho(X_1, X_2)$ is small enough, it might not be possible for two researchers to detect that they have different conceptions for the common factor for even a moderately large number of variables, and in some cases they would have to have an enormous number of variables to detect the difference. This result applies not just to exploratory factor analysis, but to confirmatory factor analysis as well.

So, different infinite sets of variables chosen to augment the original finite set of variables η while continuing to satisfy the g-factor law would determine the different solutions X_g originally for the g-factor of η. And these common factors, X_g, could be as minimally correlated as the alternative solutions were for the g-factor of η.

Consequently, it should be clear that augmenting the original set of variables with additional variables that continue to conform to the original common factor of the variables in η does not remove the original indeterminacy.

There is no need to assume that the world contains only one variable that conforms to the g-factor law with respect to η. So the resulting reduction in indeterminacy applies only to the augmented set of variables and the g-factor common to them.

This also suggests that there is some leeway in what additional variables will fit with a g-factor model. If the initial squared multiple correlation for predicting the g-factor in η is low, then there is considerable freedom in finding variables that when added with η will satisfy the g-factor law. Yet different sets of additional variables need not represent the same factor. Suppose, for example, we regard η to be a set of marker variables for a common factor, and we include it with some additional variables $\mathbf{v}_1$ in one study and find the resulting variables satisfy the g-factor law. Then we include η with a different set of variables $\mathbf{v}_2$ and these satisfy the g-factor law. It would be wrong to conclude then that this is conclusive evidence that the two studies reveal the same factor. Even combining the two additional sets with η and satisfying the g-factor law, will not prove that the common factor of $[\eta, \mathbf{v}_1, \mathbf{v}_2]$ is the same factor one thinks it is for $[\eta, \mathbf{v}_1]$ or someone else thinks it is for $[\eta, \mathbf{v}_2]$. The common factor of $[\eta, \mathbf{v}_1, \mathbf{v}_2]$ may be a third alternative solution for the common factor of η that is consistent with both $[\eta, \mathbf{v}_1]$ and $[\eta, \mathbf{v}_2]$, and this will be possible if $\rho^2_{(0+1+2)} < 1$.

Suppose then researcher 1 adds variables $\mathbf{v}_1^*$ (further consistent with his/her conception of the common factor) to $[\eta, \mathbf{v}_1]$ to obtain $[\eta, \mathbf{v}_1, \mathbf{v}_1^*]$ and this satisfies the g-factor law and confirms researcher 1s conception of his/her factor. And suppose researcher 2 similarly adds $\mathbf{v}_2^*$ to $[\eta, \mathbf{v}_2]$ to obtain $[\eta, \mathbf{v}_2, \mathbf{v}_2^*]$, and this also satisfies the g-factor law. But we now can gain evidence that possibly disconfirms the view that they are thinking of the same factor if $[\eta, \mathbf{v}_1, \mathbf{v}_1^*, \mathbf{v}_2, \mathbf{v}_2^*]$ does not satisfy the g-factor law or even the common-factor model. And similarly we would have had disconfirming evidence if $[\eta, \mathbf{v}_1, \mathbf{v}_2]$ did not satisfy the g-factor law or the common-factor model.

This points out the need to get the squared multiple correlation for predicting the common factor of the marker set η to be as high as possible. Optimally one should try to achieve this with as few variables as possible, by initially choosing indicators of it that have high loadings on the factor, say, up in the .90s. The squared multiple correlation is always greater than the largest squared correlation of a single predictor with the criterion. So, if the squared correlation of a single predictor with the common factor is very high, this will assure that the squared multiple correlation is high and indeterminacy will be greatly reduced.

It also points out that variables offered by one researcher as indicators of a given common factor should be merged and analyzed with those of other researchers to determine if they refer to the same interpretation for the common factor.

Of course, one must also be aware of the possibility of multicollinearity if the correlations among the predictors of the common factor approach 1. Achieving a singular correlation matrix for the observed variables will not be desirable.

So, factor indeterminacy is not just about scores on the factor, but also about the meaning of factors, (especially when indicators of them are placed among other variables), which can vary considerably without detection if the squared multiple correlation for predicting them is low. The meaning of a factor can change when different variables are added to an original set of variables.

13.4.1.1 "Infinite Domain" Position

On the other hand, Mulaik and McDonald (1978) established the following theorem:

> **Theorem 6.** Let $\mathbf{v} = [V_1, \ldots]$ be an infinite domain of variables having a determinate factor space with r common factors (r not necessarily finite). Then the following is true: (a) if $[\eta, \mathbf{v}_1]$ and $[\eta, \mathbf{v}_2]$ are each selections of, respectively, $p + k_1$ and $p + k_2$ variables chosen from the domain such that each conforms to the g-factor law, then $[\eta, \mathbf{v}_1, \mathbf{v}_2]$ conforms to the g-factor law. (b) In the limit, if k_1 and k_2 increase without bound, the g-factors, respectively of $[\eta, \mathbf{v}_1]$ and $[\eta, \mathbf{v}_2]$ and the g-factor of $[\eta, \mathbf{v}_1, \mathbf{v}_2]$ are each determinate and are the same variable. (c) In the limit if k_1 and k_2 increase without bound, $\mathbf{v}_1$ and $\mathbf{v}_2$ need not share any variables in common, nor need the set $[\eta, \mathbf{v}_1, \mathbf{v}_2]$ contain every variable in the domain having the specified factor in common. (d) The set of all possible components of $\mathbf{v}_1$ and all possible components of $\mathbf{v}_2$ are [is] the same (Mulaik and McDonald, 1978, p. 187).

The above theorem represents what most factor analysts naively assume is the case about the variables they are selecting for their studies—that they all come from the same, determinate infinite domain satisfying the common-factor model. Some commentators (Steiger, 1994) on the Mulaik and McDonald (1978) paper have not read this paper very closely and have misinterpreted this theorem as being their basic position, calling it the "infinite domain position," when in fact it is not their position. Theorem 6 from their paper is conditioned on the assumption that there is an infinite determinate factor domain and that variables are sampled from it. From that assumption, one can derive the conclusions of the theorem. But there must be some empirical way to ascertain in advance that the variables are from such a domain independent of doing a factor analysis. For example, there would need to be some set of rules by which one would be able to determine by examining their external characteristics independent of factor analysis that they all belong to the same domain. But the conditions under which the assumption would be reasonable in practice are stringent and never established in a purely exploratory factor analysis. The problem is not that we are simply dealing with infinities, but rather that we must come up with reliable generating rules for generating large numbers of, if not infinitely many, variables that would represent a determinate domain. Generally such rules might arise as inductive generalizations from the external in-the-world characteristics of a set of variables satisfying the *g*-factor law. The rule might then be chosen arbitrarily to be the norm for what one means by the *g*-factor. When there are indeterminacies there is room for the arbitrary establishment of norms, which does not rule out the possibility of other alternative norms. But by a chosen rule one can proceed to select variables that satisfy the rule and then bring them together with the original variables on which the rule was based and analyze them together to see if they satisfy the *g*-factor law. If they do, there is still no proof that the *g*-factor of the current set of variables is the same variable as a *g*-factor that one started with. Actually, the last statement is incoherent, because there is no way to know what the original *g*-factor variable was: it was one of infinitely many. The value of passing the test is that one has a rule that now applies to selecting a much larger set of variables for which the *g*-factor is much more determinate. Whatever the alternative conceptions are, they are now known to be much closer to one another than before. The test, however, can falsify the assumption that the augmented set of variables has the same common factor, if the *g*-factor law does not apply to them. So, the value of the test is in the possibility of falsifying a hypothesis that the rule defines how to select variables having *a* (not *the*) common *g*-factor not just for the original set of variables on which the rule was based, but also for the augmented set of variables.

There are already efforts being directed to the solution of the problem of finding rules for defining domains of variables in the area of computerized adaptive testing, where the computer generates at random the items to present to testees (Embretson and Reise, 2000). But at present these work only for geometric figures with a limited set of attributes to vary. A comparable solution for generating verbal cognitive-ability, personality, or attitude items, say, has yet to be found. The computer needs clear definitions for the attributes of items that are to be varied at random.

While exploratory factor analytic researchers for the past 100 years may have presumed they were sampling variables from infinite domains with a fixed set of common factors, it is doubtful whether more than a few of them had clear concepts of these domains and the factors that were to be expected in them. Treated as an exploratory technique, a researcher begins a factor analysis usually without clear concepts of what factors might be found in a selection of variables, and so variables selected are not chosen with any attempt at rigor in assuring that they are necessarily indicators of a set of factors in question. More often the variables are a set available at hand and one wants to know "What are the common factors of these variables?"

Factor indeterminacy also reveals the folly of believing that factor analysis itself is going to give you *the* answer. Many interpretations are possible–just as many interpretations are possible of a Rorschach card or of what one sees in the clouds. Factor analysis used this way is just a mathematical way of organizing information about relations among variables, an abductive, hypothesis-formulating method when applied to experience, and it is up to the researcher's experience, skill, and knowledge of the world to determine what useful meaning to give to the factors of a factor analysis. So, researchers, given constraints imposed by the world, give meaning to factors. The factor analysis alone cannot do this.

Factor analytic results in the form of outputs from computer programs when combined with experience of the world can provoke or stimulate the formulation of hypotheses about the factors. Such hypotheses are grounded not simply in the mathematics but in the experienced world by objectively applicable rules of how to go on in the same way in terms of a conceived and hypothesized invariant in the world. So, the researcher's hypotheses will have to be tested against the world in future studies with additional variables, using other techniques such as confirmatory factor analysis and structural equation modeling. But even there the researcher's concept of the invariant factor, being based on variables studied so far, may itself be still sufficiently underdetermined and thus free to drift along certain lines as a result of the indeterminacy, as the concept is subtly modified while being generalized to additional variables in the world.

This is not as serious as it sounds. And the theorems of Mulaik and McDonald (1978) show us why. The end result, as more and more variables from the world are brought in and shown to be in conformity to a given concept of the factor (which itself may be evolving to a more specific and determinate concept in accord with variables in the world that remain consistent to the common-factor model to which it is applied) is something more determinate and thus objective. (Objective, determinate knowledge of the world is the ideal goal of science.) And the concept of the common factor of the augmented set of variables need not be the initial concept one had for the original set of variables. (Scientists can change their concepts in accordance with additional data.)

Another advantage of reduced indeterminacy resulting from basing a common-factor concept on increasing numbers of variables, is that the concept can furthermore be differentiated from other concepts of the common factor. This can be done by noting differences in the sets of variables $[\boldsymbol{\eta}, \mathbf{v}_1]$ and $[\boldsymbol{\eta}, \mathbf{v}_2]$ to which they each may be separately applied and conform to the *g*-factor law, while the joint set $[\boldsymbol{\eta}, \mathbf{v}_1, \mathbf{v}_2]$ does not so conform. The original indeterminacy surrounding $\boldsymbol{\eta}$ is not eliminated, but the indeterminacy is reduced and may even be eliminated for a world-constrained concept relative to a sufficiently augmented set $[\boldsymbol{\eta}, \mathbf{v}_1]$ of variables. Thus initially there is no need to assume that there is one and only one infinite domain from which variables are being sampled. But as time goes on, continuing to seek sets of variables that continue to satisfy the *g*-factor law leads to sets of variables in the world that give rise to more determinate factors. And the sets of variables associated with them, when carefully examined for their substantive content, can suggest a more rigorously defined concept for a factor and criteria by which to select other variables representative of it. So, a single infinite domain might be the result of such a process, but not an initial presupposition of it. And there may result many infinite domains for different concepts, that nevertheless correspond to a common factor of some nuclear set of variables $\boldsymbol{\eta}$.

If one pauses to think about it, science advances generally in the same way. Scientists may initially have many equally acceptable interpretations for what unites a given set of phenomena. But with additional data derived from more observations and experiments, some of these interpretations will be ruled out, while others survive and new ones are formulated to encompass all that is known up to that point. In this way scientific concepts tend to converge to solutions that encompass broader and broader sets of phenomena and at the same time become increasingly better anchored in experience.

13.4.2 Researchers with Well-Defined Concepts of Their Domains

On the other hand, there have been researchers who did have more or less clear concepts of what factors they were to find and what a domain of variables spanned by those factors would consist of. To some extent I believe Spearman and Thurstone were each researchers of this kind.

Spearman (1904, 1927) did not discover *g* using factor analysis. He did not use the centroid method or know about eigenvectors and eigenvalues. He presumed from a conceptual analysis of the mental operations he believed were involved in performing on certain tests that they would have a mental factor in common. He used certain tests of "tetrad differences," each involving correlations among four variables, to determine whether they had but one common factor. He was doing confirmatory factor analysis. More traditional concepts of factor analysis entered in later when he sought to determine how much his *g* entered into various tests. There the concept of a factor loading came into play.

But reading Spearman's (1927), I also have the impression that his concept of *g*, which was based on conceptual analyses of different mental tests to find what is common to them, tended to vary across a number of concepts that I regard as distinct. His *g* involved mental energy invested in various mental operations, analogous to the way Freud said libido was invested in various defense mechanisms. Chief among the mental operations were the degree to which one apprehended one's own experience, the degree to which one was able to educe relations, and the degree to which one was able to educe correlates of relations. But it also included the ability to discern identity, similarities, attributions, causality, and constitution. And the list went on and on. What was lacking was a clear idea of what is *not g*. Nevertheless, because he had all these conceptions of what was involved in g derived not from factor analysis but from conceptual analyses of the mental operations involved in various tests, he felt g was identified in the world and not indeterminate. Setting up a common-factor model with a single general factor and using the tetrad differences tests was only a way of testing his model.

Anyone who reads Thurstone's (1951) research report, *An Analysis of Mechanical Aptitude*, will realize that Thurstone was not doing purely exploratory research. He devoted months to combing the literature to find tests that would measure what he thought were the factors of a given domain, such as, in this case, mechanical aptitude. And when he did not find such tests, he spent another series of months constructing and trying new tests that would be measures of his expected factors. He further demanded at least four tests for each anticipated factor, to "overdetermine" the factor. He even defined a domain of tests as all those tests that have a given set of factors in common (Thurstone, 1947). So, it would be absurd to think that Thurstone had no idea of what factors (as variables in the world) were to be found when he did a factor analysis. And so, for Thurstone, indeterminacy made no sense, if by that it meant that he would be uncertain as to which alternative variables in the world were the factors. He already had a definite idea. In effect, Thurstone was doing confirmatory factor analyses, using the weaker methodology of exploratory factor analysis to test his hypotheses in an era before confirmatory factor analysis had been invented. Perhaps it is for this reason that Thurstone tended to ignore Guttman (1955). Perhaps that Thurstone did not respond in print to Guttman (1955) is also due to the fact that Thurstone died in September of 1955 and had little opportunity to write anything in response. But it would also partly explain why Thurstone and his students tended to ignore Wilson (1928a,b).

The Thurstonians would have been more concerned about it had they known about the problem which Mulaik and McDonald (1978) demonstrated where two researchers are unable to detect that they have different conceptions of the factors because their studies did not have sufficient numbers of variables to reveal that their two selections of additional variables do not yield a common-factor model when combined with the original variables. That is a real and practical consequence of indeterminacy for those doing confirmatory studies.

Thus, indeterminacy is also present in a confirmatory factor analysis designed to test a hypothesis about what variables are effects of what common factors (see Vittadini, 1989). In formulating his or her hypothesis the researcher is not lacking in concepts of what interpretation to give to a factor and is not waiting on the output from the computer to make that interpretation. Where factor indeterminacy enters into confirmatory factor analysis is as a basis for the usual indeterminacy in scientific settings when, *after* subjecting a prior hypothesis to the test against empirical data, one finds that one's prior interpretation of the data *conforms* to the data. While there is support for the hypothesis, there also remains a certain indeterminacy or uncertainty regarding possible alternative explanations that would fit as well or better these same data (Garrison, 1986). Factor indeterminacy is one source of this kind of indeterminacy in the case of the use of latent variable models. But these can only be revealed by further studies with more observations, and, in this case, more variables.

One final paper of note between 1970 and 1980 also appeared in 1978. Williams (1978) demonstrated necessary and sufficient conditions for a common-factor model to exist with unique and determinate common-factor scores. His argument has parallels with those in Mulaik and McDonald (1978), especially with their Theorem 6 quoted previously. However, unlike Mulaik and McDonald (1978) he did not allow the addition of variables to the core set in η to bring in additional factors, which made his demonstrations a bit artificial. He also did not treat the problem of defining a way of generating an infinite sequence of variables that in practice would converge to a determinate common-factor model.

13.4.2.1 Factor Indeterminacy from 1980 to 2000

Bartholomew (1981) presented a new approach to the factor indeterminacy issue. He set forth the distinction between a statistical model, which he characterized as the "posterior distribution position" and a mathematical model, which he ascribed to the proponents of the factor indeterminacy issue. He felt that adherents of the mathematical model did not understand the nature and role of a random variable. At this point he revealed that his approach was to take a Bayesian position on indeterminacy. The issue, as he saw it, was to characterize the indeterminacy and uncertainty before and after the data is collected and seen. In Bartholomew (1996a) he repeated his argument by saying that

> Once the model is specified in random variable terms the question of what can be said about the latent variable X, ineluctably follows. As soon as **Y** has been observed, we can no longer treat it as a random variable. The act of observation converts it into a vector of real numbers. X is still not observed and cannot be observed. The relevant distribution for making statements about X is then $f_X(x|\ \mathbf{Y} = y)$, the posterior density of x. An obvious predictor of X is $E(X|\mathbf{Y} = \mathbf{y})$, though others are possible.
>
> (Bartholomew, 1996a, p. 553)

Bartholomew's concept of **Y** as not a random variable once one sees some scores on it, strikes me as strange. In a game of craps one may have seen the results of previous tosses, but that does not take away from the randomness of the outcomes of future tosses. In any case, for Bartholomew the regression estimator of the common factor (or something similar) is all we have to work with in attempting to reduce uncertainty about the scores on the latent variables.

But he then went on to argue that the mathematical model only expresses relations between real numbers. There is no need, he said, to indicate any probability distribution. In fact, he said, it would be misleading to enter them into the discussion. The formulas for constructing determinate factor scores are merely mathematical constructs. But this does not mean that they describe random variables. In fact there are no random variables in this approach.

It is interesting that Schönemann (1996a) agrees with Bartholomew, that there are no random variables in the mathematics of his position regarding latent common factors and factor indeterminacy.

> Strictly speaking, the factors of the factor model cannot be random variables anyway. Random variables are usually defined as maps of probability spaces into the (sometimes product) space of the reals, and maps as many-one relations. Now, while the relation from the sample space to the test space is many-one, and thus a map, the relation from the test space to the factor space is, in view of the indeterminacy, many-many and thus not a map. Hence the composite relation from the sample space to the factor space is not a map either, and factors cannot be random variables by the conventional definitions. This is just another way of saying that the homomorphism from the vector space of the factors to that of the tests is not invertible. However, compared with other, more urgent problems of the factor model, this minor technicality hardly seems worth worrying about.
>
> (Schönemann, 1996a, pp. 573–574)

One wonders then why alternative solutions for the factors not being random variables would be an issue for Bartholomew (1996a) in the case of factor indeterminacy but not in the case of the factors of the common-factor model itself. His formula for the posterior expected value of X given $\mathbf{Y} = \boldsymbol{y}$ presumes X is a random variable. But if Schönemann is right, it isn't.

It strikes me as tunnel vision to persist in focusing just on relations between the observed variables and the common factors once we have seen that the observed variables and common factors will have to be embedded in a context of many more variables as research proceeds.

The indeterminacy that Bartholemew allows in $f_X(x|\mathbf{Y} = y)$ may still apply to any number of distinct variables X in the world. Furthermore, anticipating the consequences of indeterminacy does not need actual scores on factors. The results of some of Mulaik and McDonald's (1978) theorems do not require actually obtaining scores on factors, but can be interpreted simply as predicting possible failures of the common-factor model to fit when certain sets of variables that had originally fit the model are combined. In this, mathematics has practical value.

By citing Bartholomew (1996a) and Schönemann (1996a), I have already revealed the existence of a debate over the factor indeterminacy issue that occurred in "multivariate behavioral research." This to me represents the last major treatment of factor indeterminacy in the twentieth century. The initial protagonist in the debate was Michael Maraun (1996a,b,c) who initiated the debate with a target article, which combined contemporary philosophical views taken from Wittgenstein with the "alternative solution position" (ASP) of factor indeterminacy. Others commenting on his paper were James Steiger (1996a,b), representing ASP, David Bartholomew (1996a, 1996b), representing the posterior distribution position (PDP), William Rozeboom (1996a,b), Peter Schönemann (1996a,b) (ASP), Stanley Mulaik (1996a,b) (ASP), and R.P. Mcdonald (1996a,b) (ASP).

In my opinion, nothing really new in the way of mathematical results was produced in this debate, but it was informative, if not entertaining, to get the sharply conflicting views of the participants in the debate as they debated both with Maraun and among themselves. Of interest would be the different philosophical points of view that were expressed. It would be an excellent topic for a graduate seminar to review these papers, but it would take up too many more pages here with nothing of major importance we have not already covered. I will therefore leave the history at that. For further reactions of my own to this history, see Mulaik (2005).

13.5 Other Estimators of Common Factors

Thurstone (1935) recommended the least-squares regression method of Equation 13.15 for estimating the common factors from the observed variables. Although from a validity point of view this method of estimation is as good as any, it has two drawbacks: First, with one exception, even though the common factors may be uncorrelated, the method generally produces correlated estimates of the common factors. Second, the estimates are normally not univocal; that is, even when the common factors are themselves uncorrelated, the estimator of one common factor will also correlate with other common factors. Generally, it is not possible to obtain both orthogonal and univocal estimators of the factors at the same time. Heerman (1963), however, investigated estimators that would be either orthogonal or univocal, but orthogonality among common factors is today not regarded as essential as in Heerman's day, and thus his procedures are less often used now. In addition, the variances of the estimates from the regression procedure are normally not unity (because they equal the squared multiple correlations) when the common factors have unit variance, although this is not a serious problem because the estimates can always be normalized after they have been obtained. (Normalization, however, reduces accuracy of estimation.) Nevertheless, the regression least-squares estimators by definition have the best fit in a least-squares sense to the common factors, of any linear combination of the observed variables. There is, however, another procedure also called "least squares" (Horst 1965).

13.5.1 Least Squares

Suppose $\mathbf{X}_0$ is a "true" common-factor variable and $\hat{\mathbf{X}}_L$ a least-squares estimator that minimizes

$$\begin{aligned} LS &= \mathrm{tr}[E\{(\mathbf{Y} - \mathbf{\Lambda}\hat{\mathbf{X}}_L)(\mathbf{Y} - \mathbf{\Lambda}\hat{\mathbf{X}}_L)'\}] \\ &= \mathrm{tr}[E\{((\mathbf{\Lambda}\mathbf{X}_0 + \mathbf{\Psi}\mathbf{E}) - \mathbf{\Lambda}\hat{\mathbf{X}}_L)((\mathbf{\Lambda}\mathbf{X}_0 + \mathbf{\Psi}\mathbf{E}) - \mathbf{\Lambda}\hat{\mathbf{X}}_L)'\}] \\ &= \mathrm{tr}[E\{(\mathbf{\Lambda}(\mathbf{X}_0 - \hat{\mathbf{X}}_L) + \mathbf{\Psi}\mathbf{E})(\mathbf{\Lambda}(\mathbf{X}_0 - \hat{\mathbf{X}}_L) + \mathbf{\Psi}\mathbf{E})'\}] \end{aligned}$$

When $\mathbf{X}_0 = \hat{\mathbf{X}}_L$, this reduces to $\mathrm{tr}(\mathbf{\Psi}^2)$. So, effectively the least-squares method here described tries to reduce the variance in $\mathbf{Y}$ due to the difference between the true and estimated common factors to a minimum. The first of the above equations is used to derive the estimator.

$$\begin{aligned} LS &= \mathrm{tr}[E\{(\mathbf{Y} - \mathbf{\Lambda}\hat{\mathbf{X}}_L)(\mathbf{Y} - \mathbf{\Lambda}\hat{\mathbf{X}}_L)'\}] \\ &= \mathrm{tr}[E\{(\mathbf{Y} - \mathbf{\Lambda}\hat{\mathbf{X}}_L)(\mathbf{Y}' - \hat{\mathbf{X}}_L'\mathbf{\Lambda}')\}] \\ &= \mathrm{tr}\, E(\mathbf{Y}\mathbf{Y}') - \mathrm{tr}\, E(\mathbf{Y}\mathbf{X}_L'\mathbf{\Lambda}') - \mathrm{tr}\, E(\mathbf{\Lambda}\mathbf{X}_L\mathbf{Y}') + \mathrm{tr}\, E(\mathbf{\Lambda}\mathbf{X}_L\mathbf{X}_L'\mathbf{\Lambda}') \end{aligned} \tag{13.33}$$

From Equations 4.20 and 4.21, the last equation can be rewritten as

$$LS = \mathrm{tr}\, E(\mathbf{Y}\mathbf{Y}') - 2\,\mathrm{tr}\, E(\mathbf{Y}'\mathbf{\Lambda}\hat{\mathbf{X}}_L) + \mathrm{tr}\, E(\hat{\mathbf{X}}_L'\mathbf{\Lambda}'\mathbf{\Lambda}\hat{\mathbf{X}}_L) \tag{13.34}$$

We now take the partial derivative of Equation 13.34 with respect to $\hat{\mathbf{X}}_L$ using Equations 4.16 and 4.17:

$$\frac{\partial LS}{\partial \hat{\mathbf{X}}_L} = \mathbf{0} - 2\mathbf{\Lambda}'\mathbf{Y} + 2\mathbf{\Lambda}'\mathbf{\Lambda}\hat{\mathbf{X}}_L \tag{13.35}$$

Setting Equation 13.35 equal to zero and solving for $\hat{\mathbf{X}}_L$, we obtain

$$2\mathbf{\Lambda}'\mathbf{\Lambda}\hat{\mathbf{X}}_L = 2\mathbf{\Lambda}'\mathbf{Y}$$

or

$$\hat{\mathbf{X}}_L = (\mathbf{\Lambda}'\mathbf{\Lambda})^{-1}\mathbf{\Lambda}'\mathbf{Y} \tag{13.36}$$

The covariance matrix among the estimated factor scores is given by

$$E(\hat{\mathbf{X}}_L\hat{\mathbf{X}}_L') = (\mathbf{\Lambda}'\mathbf{\Lambda})^{-1}\mathbf{\Lambda}'\mathbf{\Sigma}_{YY}\mathbf{\Lambda}(\mathbf{\Lambda}'\mathbf{\Lambda})^{-1}$$

The matrix of covariances between the "true" common-factors and the least-squares estimators is

$$\begin{aligned} E(\mathbf{X}_0\hat{\mathbf{X}}_L') &= E[\mathbf{X}_0\mathbf{Y}'\boldsymbol{\Lambda}(\boldsymbol{\Lambda}'\boldsymbol{\Lambda})^{-1}] \\ &= E(\mathbf{X}_0(\mathbf{X}_0'\boldsymbol{\Lambda}' + \mathbf{E}'\boldsymbol{\Psi})\boldsymbol{\Lambda}(\boldsymbol{\Lambda}'\boldsymbol{\Lambda})^{-1} \\ &= E(\mathbf{X}_0\mathbf{X}_0')\boldsymbol{\Lambda}'\boldsymbol{\Lambda}(\boldsymbol{\Lambda}'\boldsymbol{\Lambda})^{-1} = \boldsymbol{\Lambda}_{X_0X_0} \end{aligned} \tag{13.37}$$

Assuming that the common factors are normalized to have unit variances, the correlation matrix is

$$\mathbf{R}_{X_0\hat{X}_L} = \mathbf{R}_{X_0X_0}[\mathrm{diag}(\boldsymbol{\Lambda}'\boldsymbol{\Lambda})^{-1}\boldsymbol{\Lambda}'\boldsymbol{\Sigma}_{Y_0Y_0}\boldsymbol{\Lambda}(\boldsymbol{\Lambda}'\boldsymbol{\Lambda})^{-1}]^{-1/2} \tag{13.38}$$

13.5.2 Bartlett's Method

Bartlett's (1937) method of estimating common factors approaches the estimation problem differently from the least-square techniques.

In this method, we treat each $n \times 1$ observation vector, $\mathbf{y}_i$, separately and seek to minimize the sum of the squares of the ith individual's estimated unique-factor scores across the n unique factors after estimating his common-factor scores. That is to say, let $\boldsymbol{\Lambda}$ be an $n \times r$ factor-pattern matrix and $\boldsymbol{\Psi}^2$ the $n \times n$ diagonal matrix of unique-factor variances obtained from a factor analysis of the variables in $\mathbf{Y}$; then we may write

$$\mathbf{y}_i = \boldsymbol{\Lambda}\mathbf{x}_i + \boldsymbol{\Psi}\mathbf{e}_i \tag{13.39}$$

for the ith individual, where $\mathbf{y}_i$, is as defined above, $\mathbf{x}_i$ is the $r \times 1$ vector of the ith individual's common-factor scores (to be estimated), and $\mathbf{e}_i$, is the $n \times 1$ vector of his unique-factor scores (to be estimated as a consequence of estimating $\mathbf{x}_i$). Equation 13.19 may be rewritten as

$$\mathbf{y}_i - \boldsymbol{\Lambda}\mathbf{x}_i = \boldsymbol{\Psi}\mathbf{e}_i$$

or

$$\mathbf{e}_i = \boldsymbol{\Psi}^{-1}\mathbf{y}_i - \boldsymbol{\Psi}^{-1}\boldsymbol{\Lambda}\mathbf{x}_i \tag{13.40}$$

With $\boldsymbol{\Psi}$ and $\boldsymbol{\Lambda}$ given, Bartlett's method seeks to find values for the elements of $\mathbf{x}_i$ such that

$$\mathrm{tr}(\mathbf{e}_i\mathbf{e}_i') = \mathrm{tr}[(\boldsymbol{\Psi}^{-1}\mathbf{y}_i - \boldsymbol{\Psi}^{-1}\boldsymbol{\Lambda}\mathbf{x}_i)(\boldsymbol{\Psi}^{-1}\mathbf{y}_i - \boldsymbol{\Psi}^{-1}\boldsymbol{\Lambda}\mathbf{x}_i)'] \tag{13.41}$$

is a minimum.

It is possible to show, after some manipulation, that Equation 13.41 is equivalent to

$$\mathrm{tr}(\mathbf{e}_i\mathbf{e}_i') = \mathrm{tr}(\boldsymbol{\Psi}^{-1}\mathbf{y}_i\mathbf{y}_i'\boldsymbol{\Psi}^{-1}) - 2\,\mathrm{tr}(\boldsymbol{\Psi}^{-1}\boldsymbol{\Lambda}\mathbf{x}_i\mathbf{y}_i'\boldsymbol{\Psi}^{-1}) + \mathrm{tr}(\boldsymbol{\Psi}^{-1}\boldsymbol{\Lambda}\mathbf{x}_i\mathbf{x}_i'\boldsymbol{\Lambda}'\boldsymbol{\Psi}^{-1}) \quad (13.42)$$

A necessary condition that $\mathrm{tr}(\mathbf{e}_i\mathbf{e}_i')$ is a minimum is that its partial derivatives with respect to the elements of $\mathbf{x}_i$ are all equal to zero. Taking in turn the partial derivatives of each of the three expressions on the right-hand side of Equation 13.42 we have

$$\frac{\partial\,\mathrm{tr}(\boldsymbol{\Psi}^{-1}\mathbf{y}_i\mathbf{y}_i'\boldsymbol{\Psi}^{-1})}{\partial\mathbf{x}_i} = 0$$

$$\frac{\partial\,\mathrm{tr}(\boldsymbol{\Psi}^{-1}\boldsymbol{\Lambda}\mathbf{x}_i\mathbf{y}_i'\boldsymbol{\Psi}^{-1})}{\partial\mathbf{x}_i} = \frac{\partial\,\mathrm{tr}(\mathbf{y}'\boldsymbol{\Psi}^{-2}\boldsymbol{\Lambda}\mathbf{x}_i)}{\partial\mathbf{x}_i} = \boldsymbol{\Lambda}'\boldsymbol{\Psi}^{-2}\mathbf{y}_i$$

by Equations 4.19 and 4.16, and similarly by Equations 4.19 and 4.17

$$\frac{\partial\,\mathrm{tr}(\boldsymbol{\Psi}^{-1}\boldsymbol{\Lambda}\mathbf{x}_i\mathbf{x}_i'\boldsymbol{\Lambda}'\boldsymbol{\Psi}^{-1})}{\partial\mathbf{x}_i} = \frac{\partial\,\mathrm{tr}(\mathbf{x}'\boldsymbol{\Lambda}'\boldsymbol{\Psi}^{-2}\boldsymbol{\Lambda}\mathbf{x}_i)}{\partial\mathbf{x}_i} = 2\boldsymbol{\Lambda}'\boldsymbol{\Psi}^{-2}\boldsymbol{\Lambda}\mathbf{x}_i$$

Thus

$$\frac{\partial\,\mathrm{tr}(\mathbf{e}_i\mathbf{e}_i')}{\partial\mathbf{x}_\mathrm{i}} = -2\boldsymbol{\Lambda}'\boldsymbol{\Psi}^{-2}\mathbf{y}_i + 2(\boldsymbol{\Lambda}'\boldsymbol{\Psi}^{-2}\boldsymbol{\Lambda})\mathbf{x}_i \quad (13.43)$$

Solving for $\mathbf{x}_i$ in Equation 13.43, we obtain

$$\hat{\mathbf{x}}_i = (\hat{\boldsymbol{\Lambda}}'\hat{\boldsymbol{\Psi}}^{-2}\hat{\boldsymbol{\Lambda}})^{-1}\hat{\boldsymbol{\Lambda}}'\hat{\boldsymbol{\Psi}}^{-2}\mathbf{y}_i \quad (13.44)$$

which is Bartlett's equation for the vector of estimates of the common factors for the *i*th individual (observation).

By $\hat{\mathbf{X}}_\mathrm{B} = (\boldsymbol{\Lambda}\boldsymbol{\Psi}^{-2}\boldsymbol{\Lambda})^{-1}\boldsymbol{\Lambda}\boldsymbol{\Psi}^{-2}\mathbf{Y}$, let us mean the $r \times 1$ random vector of estimates of the common factors obtained from the observed variables by Bartlett's method. Then, it can be shown that

$$E(\hat{\mathbf{X}}_\mathrm{B}\hat{\mathbf{X}}_\mathrm{B}') = (\boldsymbol{\Lambda}'\boldsymbol{\Psi}^{-2}\boldsymbol{\Lambda})^{-1}\boldsymbol{\Lambda}'\boldsymbol{\Psi}^{-2}\mathbf{R}_{YY}\boldsymbol{\Psi}^{-2}\boldsymbol{\Lambda}(\boldsymbol{\Lambda}'\boldsymbol{\Psi}^{-2}\boldsymbol{\Lambda})^{-1} \quad (13.45)$$

If the matrices $\boldsymbol{\Lambda}$ and $\boldsymbol{\Psi}$ are such that $\mathbf{R}_{YY} = (\boldsymbol{\Lambda}\boldsymbol{\Lambda}' + \boldsymbol{\Psi}^2)$, it can be shown that

$$E(\hat{\mathbf{X}}_\mathrm{B}\hat{\mathbf{X}}_\mathrm{B}') = \mathbf{I} + (\boldsymbol{\Lambda}'\boldsymbol{\Psi}^{-2}\boldsymbol{\Lambda})^{-1} \quad (13.46)$$

What then is the covariance matrix between the "true" factors $\mathbf{X}$ and the Bartlett estimators? Assuming again that $\mathbf{R}_{YY} = \boldsymbol{\Lambda}\boldsymbol{\Lambda}' + \boldsymbol{\Psi}^2$, and $E(\mathbf{X}\mathbf{E}') = \mathbf{0}$, we have

$$
\begin{aligned}
E(\mathbf{X}\hat{\mathbf{X}}'_B) &= E(\mathbf{X}\mathbf{Y}')\boldsymbol{\Psi}^{-2}\boldsymbol{\Lambda}(\boldsymbol{\Lambda}'\boldsymbol{\Psi}^{-2}\boldsymbol{\Lambda})^{-1} \\
&= E[\mathbf{X}(\mathbf{X}'\boldsymbol{\Lambda}' + \mathbf{E}'\boldsymbol{\Psi})]\boldsymbol{\Psi}^{-2}\boldsymbol{\Lambda}(\boldsymbol{\Lambda}'\boldsymbol{\Psi}^{-2}\boldsymbol{\Lambda})^{-1} \\
&= E(\mathbf{X}\mathbf{X}'\boldsymbol{\Lambda}')\boldsymbol{\Psi}^{-2}\boldsymbol{\Lambda}(\boldsymbol{\Lambda}'\boldsymbol{\Psi}^{-2}\boldsymbol{\Lambda})^{-1} \\
&= \mathbf{R}_{XX}(\boldsymbol{\Lambda}'\boldsymbol{\Psi}^{-2}\boldsymbol{\Lambda})(\boldsymbol{\Lambda}'\boldsymbol{\Psi}^{-2}\boldsymbol{\Lambda})^{-1} = \mathbf{R}_{XX} \qquad (13.47)
\end{aligned}
$$

On the other hand the matrix of correlations between variables in **X** and those in $\hat{\mathbf{X}}_B$ is given by

$$
\mathbf{R}_{X\hat{X}_B} = \mathbf{R}_{XX}[\text{diag}(\mathbf{I} + \boldsymbol{\Lambda}'\boldsymbol{\Psi}^{-2}\boldsymbol{\Lambda})]^{-1/2}
$$

Since by fiat the diagonal elements of $\mathbf{R}_{XX}$ are unities, this means that the correlations of estimators with corresponding factors are to be found in [diag $(\mathbf{I} + \boldsymbol{\Lambda}'\boldsymbol{\Psi}^{-2}\boldsymbol{\Lambda})]^{-1/2}$.

The Bartlett estimators of the common factors are maximum-likelihood estimators of the common factors when $\boldsymbol{\Lambda}$ and $\boldsymbol{\Psi}^2$ are known and **Y** is distributed according to the multivariate normal distribution.

13.5.3 Evaluation of Estimation Methods

In the 1960s, orthogonal rotations were routinely desired and employed, although researchers did not fully appreciate the artifactuality of such solutions. Consequently, it was desired that the factor score estimators of individual factors be high and univocal, the latter being the condition in which the factor score estimator of a factor only has nonzero correlations with its intended factor and zero correlations with the rest. In reviews of methods for estimating common-factor scores, Harris (1967) and McDonald and Burr (1967) conclude that Bartlett estimators are generally superior to other estimators of the common-factor scores when one prefers univocality to the other properties of estimators. However, McDonald and Burr (1967) point out that least-squares regression estimators using Equation 13.15 appear superior to other methods in general when validity (correlation with the common factors) is a prime consideration. Horst (1969) points out that the regression estimators are also free of scale of the original variables, a property he considers important.

In general, the regression estimators and the Bartlett estimators yield distinct solutions for the estimates of the common-factor scores. However, there is one case in which they both yield, except for scale, identical as well as simultaneously orthogonal and univocal estimates of the common-factor scores. This is the case when $\boldsymbol{\Lambda} = \boldsymbol{\Psi}\mathbf{A}_r[\gamma_j - 1]_r^{1/2}$, the solution for the common-factor-structure matrix in canonical-factor analysis. In this case, the regression estimators are equivalent to

$$\hat{\mathbf{X}} = [\gamma_j - 1]_r^{1/2}[\gamma_j]_r^{-1}\mathbf{A}_r'\mathbf{\Psi}^{-1}\mathbf{Y} \tag{13.48}$$

and the Bartlett estimators are

$$\hat{\mathbf{X}} = [\gamma_j - 1]_r^{-1/2}\mathbf{A}_r'\mathbf{\Psi}^{-1}\mathbf{Y} \tag{13.49}$$

Since the two solutions differ only in terms of the premultiplying diagonal matrices, they therefore differ only in scale. When these two sets of estimators are normalized they yield scores equivalent with the normalized canonical components as obtained by Equation 13.11. Harris (1967) and McDonald and Burr (1967) both point out this property of these estimators of the common factors. Harris, however, stresses that, if in the case of the canonical-factor solution both the canonical factors and their estimators are subjected to the same orthonormal transformation, the regression estimators and the normalized canonical components (as estimators) lose their univocal properties whereas the Bartlett estimators do not. Here again this seems to be a desirable feature of the Bartlett estimator for orthogonal solutions.

But with oblique rotations considered to be the proper way to obtain simple structure solutions, orthogonal solutions are less used, and the accuracy of estimation is paramount. In this regard, the regression estimator is optimal.

14

Factorial Invariance

14.1 Introduction

When used as a scientific tool, factor analysis has as its goal the decomposition of variables into fundamental elements, the factors, which may then be used to explain the interrelationships among variables. The extent to which the factors "really" are fundamental, however, depends largely on the extent to which the same factors may be found under widely varying conditions. For example, if different sets of variables representing the same domain of behavior are factor-analyzed, will the same fundamental factors be found? Or if differently selected experimental populations are studied factor-analytically with the same set of variables, will the same fundamental factors be found, indicating that regardless of how the cases are selected for study the same factors operate on the variables in the individual cases studied? These two questions are basic to what is known as the problem of factorial invariance, which we will consider in this chapter.

14.2 Invariance under Selection of Variables

That the same factors should be found in different sets of variables sampled from a given domain of variables is an idea originating with L.L. Thurstone (1947). To understand the basis for this idea, we must understand Thurstone's conception of factor analysis as a tool for the scientific study of human behavior. Thurstone, aside from being one of the great pioneers in the field of quantitative psychology, was also an erstwhile engineer and amateur inventor. Thus, it is not unreasonable to characterize his approach to the study of human behavior as the approach of the engineer to the study of the inner workings of the classic "black box"—a device which cannot be opened for inspection but which can be understood in terms of the relationships between inputs to the box and outputs from the box. For Thurstone, the human being was a "black box," and the basis for human behavior—locked up inside the human being—was not available for direct observation. But the human being's behavior could be understood by searching for those functionally independent unities in behavior which presumably have structural and functional counterparts inside the human being.

To carry out this search, Thurstone chose common-factor analysis. He used psychological tests to determine standardized input conditions, and the measured performances of people on these tests served as outputs to be correlated. If two tests were correlated (in terms of performances on them), this implied to some extent that the same internal mechanisms were involved in the performances on these tests. If the factors obtained from the factor analysis of intercorrelations among these test performances were identified with the functionally independent mechanisms within the human being, the saturations of the test performances on these factors would indicate the degree to which these mechanisms entered into the performances.

However, a major difficulty involved with this identification of the factors with internal mechanisms inside the human being was that the factors, serving as frames of reference among the performance variables, could not be regarded as being locatable in any absolute way with respect to the original variables. This, of course, is the problem of rotational indeterminacy in factor analysis. Presumably two researchers could factor-analyze the same variables and come up with entirely different sets of factors—with correspondingly different hypothesized functional mechanisms within the human being—for explaining the correlations among variables. All they would have to use would be different criteria of factor rotation or extraction. Thurstone sought to overcome this indeterminacy by having researchers in an exploratory factor analysis of a domain of behavior choose a particular set of factors in the domain as a standard reference frame to be referred to in future factor analyses of the domain. In particular, he recommended that the standard reference frame should consist of factors having an oblique simple structure.

This standard was chosen for several reasons. One reason, of course, was that factors in an oblique simple structure would be the least ambiguous to interpret. But not the least of the reasons was that the primary factors of an oblique simple structure would be invariant across factor analyses of different sets of variables sampled from the particular domain of interest.

Thurstone's assertion that the primary factors of an oblique simple structure are invariant across studies with different variables but having the same common factors is based upon a little appreciated stipulation: The variables in these different studies must be selected with care so that enough of them lie close to or within the coordinate hyperplanes defining the primary factors in question to determine these same factors in the different studies.

Figure 14.1 is an illustration, using a three-dimensional example, of how the coordinate hyperplanes (the intersections of which are the primary factors) may be defined by different sets of variables. Consider the plane passing through both primary factors, X_1 and X_2. In that plane are found the vectors representing the variables, Y_1, Y_2, Y_3, Y_4, and Y_5. If different selections

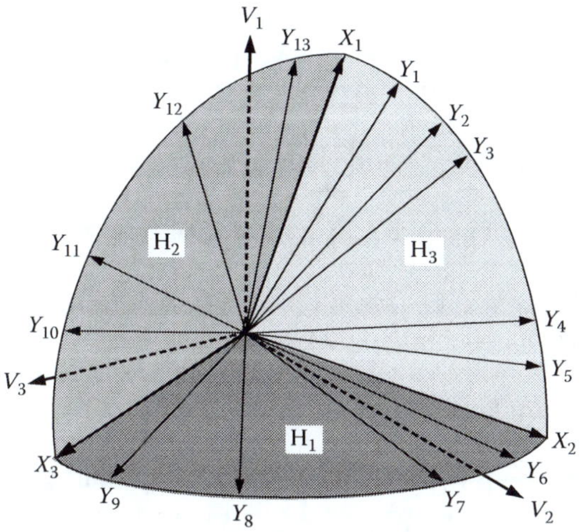

FIGURE 14.1
In this illustration axes X_1, X_2, and X_3 represent factors defined by the three planes. Note that any pair of variables, Y_1, Y_2, Y_3, Y_4, and Y_5, may define a plane passing through factors X_1 and X_2.

of these five variables were used in different studies, as long as two of them (in general, one less than the number of common factors) were present in a study, they would still define the same plane common to primary factors X_1 and X_2. If this were true of the variables defining the other planes, then the same primary factors would be found. Of course, to be able to ensure that variables which lie within these hyperplanes are chosen, one must have a good understanding of the nature of the primary factors.

In other words, one cannot expect the same primary factors to be found if the variables chosen in the different studies represent random samples from the universe of all logically possible variables measuring a given domain of behavior. For it is logically possible that the universe of variables may be represented by the points on the surface of an r-dimensional hypersphere, with r the number of common factors in the universe. No simple structure would exist in such a universe. But certain samples taken from this universe might emphasize certain directions in it and thus determine an oblique simple structure. If all such samples emphasized the same directions, the primary factors in these samples would be invariant, and these samples would have a common frame of reference for describing the performance in the domain. The common frame of reference, like a common language, would make possible a consistent account of the domain of behavior from researcher to researcher.

It should be clear by now that Thurstone did not believe in an indiscriminate selection of tests for a factor-analytic study, especially for those factor-analytic studies representing the tests of hypotheses regarding the nature of the primary factors found in some initial exploratory study. Thurstone took great care to understand the behavioral processes represented by primary factors so that tests could be constructed measuring specified combinations of these processes for inclusion in subsequent hypothesis-testing factor analyses. The validity of interpretations given to the primary factors were always tested in the construct-validity sense by constructing tests which should contain specified portions of the primary factors and relate in specified ways to other tests.

Thurstone's approach to factor analysis thus stands in contrast to the common practice of collecting data on some haphazardly selected variables, of feeding these data into a computer to perform a factor analysis, and of expecting the computer to do one's thinking for him by finding those hypothetical constructs that explain the relationships among the variables. From Thurstone's point of view, the experimenter always had to do the explaining, with the results of a factor analysis serving to generate new as well as to test old hypotheses. In any case, the invariance of the primary factors depended upon the variables selected for factor analysis. And this invariance could not be maintained without some deliberate effort on the part of the researcher.

14.3 Invariance under Selection of Experimental Populations

Up to now, we have considered the invariance of results obtained across factor analyses using different variables. Implicit in this discussion was the idea that the experimental population used in these different factor-analytic studies was the same. But what happens if one uses the same variables with differently selected experimental populations? Can one identify the same factors? Do the relationships among these factors remain the same? Are correlations of the variables with these factors invariant across populations? To obtain answers to these questions, we will have to resort to the theory of the effects of population selection upon the correlations among variables. The development of this theory is credited to Aitken (1934), Lawley (1943, 1944), Thomson and Lederman (1939), Thurstone (1947), and Ahmavaara (1954). More recently this theory has been amply reviewed by Gulliksen (1950, Chapters 11 through 13) and Meredith (1964a). In this section, we will rely heavily on these last two works.

14.3.1 Effect of Univariate Selection

We can best begin a discussion of the effects of selection upon factor analysis by first considering the effects of univariate selection upon bivariate correlation. Here we will be concerned with comparing the value of the correlation coefficient between two variables determined in some parent population with a corresponding correlation coefficient determined in an experimental subpopulation selected on the basis of population members' scores on one of the variables. For example, suppose an educational psychologist wishes to determine the correlation between an entrance-examination score and final grade-point average, using students who were selected for admission to a school on the basis of their scores on the entrance examination. If he obtains a low but nonzero correlation between these two variables, does this mean that the entrance examination is ineffective in selecting people who will do well in terms of grade-point average? In this case, the question concerns not the correlation between these two variables in the experimental population, but the correlation in the larger parent population of all persons who apply for admission to the school. In particular, the problem is whether one can infer a higher correlation coefficient between these two variables in the parent population than in the experimental population when one of the variables has been used in the selection of the experimental population.

Let us consider this problem on a more abstract level by designating X and Y to be two random variables whose bivariate correlation is graphically represented in Figure 14.2. The total ellipse in this figure represents a line passing through points of equal density in a parent population. We will assume

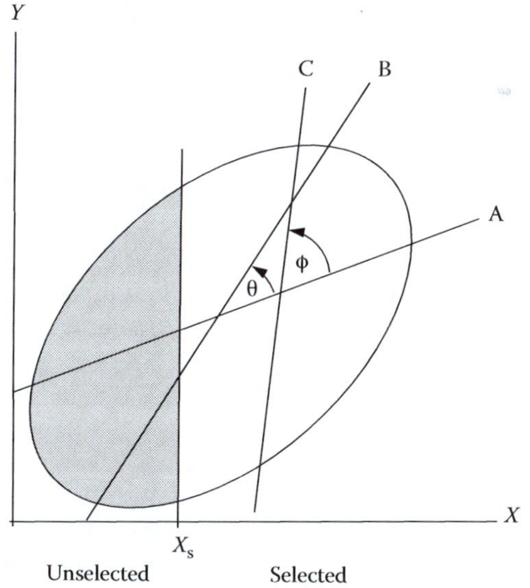

FIGURE 14.2
Representation of a bivariate distribution in parent population and selected subpopulation having values on X only greater than $X=x_s$. "A" is the line of regression of Y onto X in both parent population and subpopulation. "B" is the line of regression of X onto Y in parent population. "C" is the line of regression of X onto Y in subpopulation. θ is angle between A and B in the unselected population. φ is the angle between A and C in selected population.

that the regression of each variable on the other is strictly linear and let the straight lines A and B be the regression lines relating each of these variables to the other in the population. That is, line A passes through those points which are the expected (average) values of Y, given specified values of X in the parent population. Line B passes through points which are the expected (average) values of X, given specified values of Y in the parent population. The cosine of the angle θ determined by the intersection of the regression lines A and B is monotonically related to the correlation coefficient between these two variables in the parent population. On the other hand, the truncated right-hand side portion of the ellipse represents an experimental subpopulation of the parent population selected so that all members of the subpopulation have values of X greater than or equal to the fixed value x_s. In this subpopulation we can also determine the regression lines for relating each variable to the other by plotting lines through those points which are the expected values of a variable, given specified values of the other variable. Here we see that selection of the subpopulation on the variable X has not altered the expected values of the variable Y, given specified values of X in the subpopulation. But direct selection on X has altered the expected values of X for given values of Y in the subpopulation. Thus the regression of X on Y is not invariant under the selection on X. (The new regression line for the prediction of variable X from variable Y is represented by the line C in Figure 14.2.) We now note that the cosine of the angle Φ is less acute than the angle θ, which leads us to conclude that the correlation between X and Y in the subpopulation is closer to zero than the corresponding correlation in the parent population. As a matter of fact, if we can assume, in a parent population, that two variables are linearly related, we can conclude that the correlation between these two variables will be closer, or at least as close, to zero in any subpopulation selected on one of these variables than the corresponding correlation in the parent population.

There is one invariant property that we can infer from Figure 14.2, using the assumptions made so far. Consider the following equation for the regression of Y on the variable X in the parent population:

$$E(Y|X) = A_0 + A_1X \tag{14.1}$$

This equation states that the expected values of Y, given specified values of X, are obtained by a linear function of X, where A_0 is the expected value of Y when X is zero, and, the regression coefficient, A_1 is the slope of the regression line. Consider next a corresponding equation relating these two variables in the subpopulation [using lowercase letters to designate values in the subpopulation, a notation suggested by Gulliksen (1950)]:

$$E(y|x) = a_0 + a_1x \tag{14.2}$$

Here a_0 represents the expected value of y when x equals zero. The parameter a_1 represents the slope of the regression line in the subpopulation. Equations 14.1

and 14.2 are not necessarily equivalent. The values of y and x may be expressed as deviations from different origins than the origins of X and Y. We must assume, however, that the units of x and y equal the units of X and Y, respectively. (Standardizing these scores within the subpopulation will produce scores with different units from those in the parent population.) The values of A_0 and a_0 may or may not be equal, depending upon how the origins for the scores in the parent and subpopulations are determined. But A_1 and a_1, representing the slope of the regression line of Y on X in parent and subpopulation, respectively, will be equal. In other words, the slope, or regression coefficient, in the regression of a variable on a variable used in the selection of a subpopulation from a parent population is the same in parent and subpopulation. This is an invariant property unaffected by selection on X.

We can also make another assumption which will allow us to establish another invariant under selection: Besides linearity of regression, assume that homoscedasticity holds in the regression of Y on X. This means that the dispersion of Y for a given value of X is the same regardless of the value of X. Now we can see that selection of the subpopulation on the variable X will not affect the dispersions of the variable Y around the regression line of Y on X. Consequently we can conclude that the error of estimate in estimating Y from X is invariant under selection on X. Expressed mathematically, this invariant is given as

$$S_Y^2(1-R_{XY}^2)=s_y^2(1-r_{xy}^2) \tag{14.3}$$

where

S_Y^2 and s_y^2 are the variances of the variable Y in the parent population and subpopulation, respectively

R_{XY} and r_{xy} are the correlations between the variables X and Y in the population and subpopulation, respectively

These two invariants can be used to determine the correlation between variables X and Y in a parent population when only the variance of X in parent population and subpopulation and the correlation between X and Y in the subpopulation are known. If we write the invariance of the regression coefficients (or slope) as

$$\frac{S_Y}{S_x}R_{XY}=\frac{s_y}{s_x}r_{xy} \tag{14.4}$$

we may use this equation and Equation 14.3 to find R_{XY} given S_X^2 (parent population variance on X) and s_x^2 (subpopulation variance on X) and r_{xy} (subpopulation correlation between x and y).

Using Equation 14.4, first solve for S_Y:

$$S_Y=\frac{S_X s_y r_{xy}}{s_x R_{XY}} \tag{14.5}$$

Next, substitute Equation 14.5 for S_Y in Equation 14.3 to obtain

$$\frac{S_X^2 s_y^2 r_{xy}^2}{s_x^2 R_{XY}}(1-R_{XY}^2)=s_y^2(1-r_{xy}^2) \tag{14.6}$$

Then solve Equation 14.6 for R_{XY} to obtain

$$R_{XY}=\frac{S_X^2/s_x^2}{1-r_{xy}^2(1-S_X^2/s_x^2)} \tag{14.7}$$

Thus, under the first assumption we can determine that the correlation between two variables in a subpopulation selected on one of the variables is less than the correlation between them in the parent population. Adding the second assumption allows us to determine the exact degree of this correlation going from the subpopulation to parent population.

14.3.2 Multivariate Case

We are now in a position to generalize to the multivariate case. Let **X** be a $p\times 1$ random vector whose coordinates are p observed variables and **Y** a $q\times 1$ random vector whose coordinates are q random variables. Assume $E(\mathbf{X})=\mathbf{0}$ and $E(\mathbf{Y})=\mathbf{0}$. Let $\boldsymbol{\Sigma}_{XY}=E(\mathbf{XY}')$. We will assume that the variables in **X** are used to select some subpopulation of individuals in which relationships among and between the p and q variables are to be determined. What invariants may we establish for the multivariate case?

To make a long story short, these invariants turn out to be simply multivariate extensions of the invariants found in the bivariate case: (1) The coefficients of regression of the variables in **Y** on the variables in **X** are invariant under selection, if we assume linearity of regressions of the variables in **Y** on the variables in **X** and the same units of scale for the variables in **X** and **Y** in the parent and subpopulation. (2) The variance–covariance matrix of error scores in estimating **Y** from **X** is invariant under selection. Let us see how we might express these invariants mathematically.

Let $\mathbf{B}_{XY}$ be the matrix of regression coefficients for the regression of variables in **Y** on the variables in **X** in the population. This matrix is obtained by the equation

$$\mathbf{B}_{XY}=\boldsymbol{\Sigma}_{XX}^{-1}\boldsymbol{\Sigma}_{XY} \tag{14.8}$$

where

$\boldsymbol{\Sigma}_{XX}^{-1}$ is the inverse of the variance–covariance matrix for the variables in **X** in the parent population

$\boldsymbol{\Sigma}_{XY}$ is the matrix of covariances between the variables in **X** and in **Y** in the parent population

As a counterpart to $\mathbf{B}_{XY}$, let $\mathbf{b}_{xy}$ be the matrix of regression coefficients for predicting the variables in **Y** from the variables in **X** in the subpopulation selected on **X**. This matrix is obtained as

$$\mathbf{b}_{xy} = \boldsymbol{\sigma}_{xx}^{-1}\boldsymbol{\sigma}_{xy} \tag{14.9}$$

where

$\boldsymbol{\sigma}_{xx}^{-1}$ is the inverse of the variance–covariance matrix for the variables in **X**
$\boldsymbol{\sigma}_{xy}$ is the matrix of covariances between the variables in **X** and **Y**, respectively, in the subpopulation

Then, the first invariant is expressed as

$$\mathbf{B}_{XY} = \boldsymbol{\Sigma}_{XX}^{-1}\boldsymbol{\Sigma}_{XY} = \boldsymbol{\sigma}_{xx}^{-1}\boldsymbol{\sigma}_{xy} = \mathbf{b}_{xy} \tag{14.10}$$

Let $\boldsymbol{\Sigma}_{\mathrm{EE}}$ be the matrix of variances and covariances of the errors in estimating **Y** from the variables in **X** in the parent population. This matrix is obtained as

$$\boldsymbol{\Sigma}_{\mathrm{EE}} = \boldsymbol{\Sigma}_{YY} - \boldsymbol{\Sigma}_{YX}\boldsymbol{\Sigma}_{XX}^{-1}\boldsymbol{\Sigma}_{XY} \tag{14.11}$$

The counterpart of the matrix in Equation 14.11 in the subpopulation is σ_{ee}, obtained as in Equation 14.11 by an analogous equation. Thus, the second invariant may be written as

$$\boldsymbol{\Sigma}_{\mathrm{EE}} = \boldsymbol{\Sigma}_{YY} - \boldsymbol{\Sigma}_{YX}\boldsymbol{\Sigma}_{XX}^{-1}\boldsymbol{\Sigma}_{XY} = \boldsymbol{\sigma}_{yy} - \boldsymbol{\sigma}_{yx}\boldsymbol{\sigma}_{xx}^{-1}\boldsymbol{\sigma}_{xy} = \boldsymbol{\sigma}_{\mathrm{ee}} \tag{14.12}$$

We now have enough equations to derive something useful. Consider that it is possible to rewrite Equation 14.12 as

$$\boldsymbol{\Sigma}_{\mathrm{EE}} = \boldsymbol{\Sigma}_{YY} - \mathbf{B}'_{XY}\boldsymbol{\Sigma}_{XX}\mathbf{B}_{XY} = \boldsymbol{\sigma}_{yy} - \mathbf{b}'_{xy}\boldsymbol{\sigma}_{xx}\mathbf{b}_{xy} = \boldsymbol{\sigma}_{\mathrm{ee}} \tag{14.13}$$

Let us now take advantage of the fact that $\mathbf{B}_{XY} = \mathbf{b}_{xy}$ to solve for $\boldsymbol{\Sigma}_{yy}$ in Equation 14.13. Thus, we have

$$\boldsymbol{\sigma}_{yy} = \boldsymbol{\Sigma}_{YY} + \mathbf{B}'_{XY}(\boldsymbol{\sigma}_{xx} - \boldsymbol{\Sigma}_{XX})\mathbf{B}_{XY} \tag{14.14}$$

Thus, in Equation 14.14, we have an equation for determining the variance–covariance matrix for a set of incidentally selected variables to be observed in a selected subpopulation when we know the matrix of regression weights for predicting the incidentally selected variables and the variance–covariance matrix for the directly selected variables in the parent and subpopulation, respectively.

However, there is more to be learned from Equation 14.14. For one thing the equation applies only to variance–covariance matrices. If the matrices $\mathbf{\Sigma}_{YY}$ and $\mathbf{\Sigma}_{XX}$ are correlation matrices (variance–covariance matrices for standard scores), the matrices $\boldsymbol{\sigma}_{yy}$, and $\boldsymbol{\sigma}_{xx}$ cannot be correlation matrices but rather must be variance–covariance matrices for variables in the original metric of the standard scores for **Y** and **X**, respectively, in the parent population. In other words, the same units of scale must be used for the respective variables in parent and subpopulation. Consider also the special case where $\mathbf{\Sigma}_{YY}$ in the parent population is a diagonal matrix, indicating that the variables in **Y** are uncorrelated in the parent population. Would these same variables be uncorrelated in a subpopulation where the variables in **Y** are incidentally selected? We can see from examining Equation 14.14 that this would be possible only if the expression $\mathbf{B}'_{XY}(\sigma_{xx} - \mathbf{\Sigma}_{XX})\mathbf{B}_{XY}$ is a null or diagonal matrix. But in most instances this will never be a diagonal matrix. And as a consequence, if some incidentally selected variables (in some subpopulation) are uncorrelated in a parent population, they cannot (except very rarely) be uncorrelated in the subpopulation. These conclusions will have considerable bearing on the problem of invariance of factors in different experimental populations, the problem to which we now turn.

14.3.3 Factorial Invariance in Different Experimental Populations

In some contexts the problem arises whether two factor analyses of the same set of variables in different experimental populations will yield the same factors. In dealing with this problem we will find that we are dealing with a special case of the problem of the invariance of regression coefficients under selection. Consider, for example, that the model of common-factor analysis may be expressed by the following mathematical equation:

$$\mathbf{Y} = \mathbf{\Lambda X} + \mathbf{\Psi E} \tag{14.15}$$

where

- **Y** is an $n \times 1$ vector of n observable random variables
- **X** is an $r \times 1$ random vector of r common-factor variables
- $\mathbf{\Lambda}$ is an $n \times r$ matrix of factor-pattern or regression coefficients
- $\mathbf{\Psi}$ is an $n \times n$ diagonal matrix of unique-factor-pattern coefficients
- **E** is an $n \times 1$ random vector of n unique factors

In this equation we will explicitly assume that $E(\mathbf{Y}) = E(\mathbf{E}) = \mathbf{0}$ and $E(\mathbf{X}) = \mathbf{0}$ but will not assume that $\text{diag}[E(\mathbf{YY}')] = \mathbf{I}$ or $\text{diag}[E(\mathbf{XX}')] = \mathbf{I}$. (The reason for not using the latter assumption will become clear as we proceed.) Then the variance–covariance matrix for the scores in **Y** (say in a parent population) will be

$$\mathbf{\Sigma}_{YY} = E(\mathbf{YY}') = \mathbf{\Lambda \Phi}_{XX}\mathbf{\Lambda}' + \mathbf{\Psi}^2 \tag{14.16}$$

where

$\mathbf{\Sigma}_{YY}$ is the $n \times n$ variance–covariance matrix of observed scores in a parent population

$\mathbf{\Phi}_{XX}$ is the $r \times r$ variance–covariance matrix for the common factors

$\mathbf{\Psi}^2$ is the (diagonal) variance–covariance matrix for the unique factors

$\mathbf{\Lambda}$ and $\mathbf{Y}$ are as in Equation 14.15

Equation 14.16 is simply the fundamental equation of factor analysis.

To apply selection theory to our problem, let us suppose that there exists a set of p linearly independent variables in a $p \times 1$ random vector $\mathbf{V}$ (possibly unknown in practice) on which several subpopulations are directly selected from the parent population and that, furthermore, the regressions of the common factors and observed variables on these selection variables in the parent population are linear and demonstrate homoscedasticity. In addition, assume that the p selection variables and the n unique factors are uncorrelated. (This assumption, we should note, is a fairly strong assumption. The only case wherein one can be certain that this assumption is satisfied is where one assumes the unique variances are equivalent to the error variances of the variables. This fact provides further support for the use of true variances in the place of communalities in factor analysis.) Finally assume, for the sake of argument, that there is no direct selection on any of the variables in $\mathbf{Y}$ or $\mathbf{X}$ (although one might consider the possibility that some linear combinations of the direct selection variables may perfectly predict a common factor in $\mathbf{X}$).

To express these assumptions mathematically, let us denote by $\mathbf{V}$ the $p \times 1$ random vector of (hypothetical) variables used to select subjects in the parent population on the direct-selection variables. Furthermore, denote by $\mathbf{B}$ and $\boldsymbol{\beta}$ the matrices of regression coefficients for predicting the observed variables and common factors, respectively, from the direct-selection variables. Finally, let $\mathbf{U}$ and $\mathbf{Q}$ be residual vectors representing the parts of the total scores in $\mathbf{Y}$ and $\mathbf{X}$, respectively, not predictable from the direct-selection variables in $\mathbf{V}$. We should note, however, that the residual vector $\mathbf{U}$ consists of the unique factors $\mathbf{E}$ (since by assumption they are a portion of $\mathbf{Y}$ not predictable from $\mathbf{V}$) plus another portion $\mathbf{G}$, the common portion of $\mathbf{Y}$ not predictable from $\mathbf{V}$. The matrix $\mathbf{Q}$ represents a portion of the common factors not predictable from $\mathbf{V}$. Thus we may summarize the above assumptions with the following equations:

$$\mathbf{Y} = \mathbf{B}'\mathbf{V} + \mathbf{U} = \mathbf{B}'\mathbf{V} + \mathbf{E} + \mathbf{G} \tag{14.17}$$

$$\mathbf{X} = \boldsymbol{\beta}'\mathbf{V} + \mathbf{Q} \tag{14.18}$$

If we now substitute Equation 14.18 for the vector $\mathbf{X}$ in Equation 14.15, we may obtain the expression

$$\mathbf{Y} = \Lambda(\boldsymbol{\beta}'\mathbf{V} + \mathbf{Q}) + \mathbf{E} = \mathbf{\Lambda}\boldsymbol{\beta}'\mathbf{V} + \mathbf{\Lambda}\mathbf{Q} + \mathbf{E}$$

Evidently the expression $\mathbf{\Lambda Q}$ represents the common portion of $\mathbf{Y}$ derived from the portion of the common factors not predictable from $\mathbf{V}$ so that $\mathbf{\Lambda Q}$ must equal $\mathbf{G}$. We thus can match terms in Equation 14.17 with those in the present equation to conclude that $\mathbf{B} = \boldsymbol{\beta}\mathbf{\Lambda}'$.

Since Equations 14.17 and 14.18 represent equations for the regressions of $\mathbf{Y}$ and $\mathbf{X}$, respectively, on the direct-selection variables in $\mathbf{V}$, we may immediately establish the following equations based upon Equation 14.14 from our discussion of selection theory:

$$\boldsymbol{\sigma}_{yy} = \mathbf{\Sigma}_{YY} + \mathbf{B}'(\boldsymbol{\sigma}_{vv} - \mathbf{\Sigma}_{VV})\mathbf{B} \tag{14.19}$$

$$\boldsymbol{\phi}_{xx} = \mathbf{\Phi}_{XX} + \boldsymbol{\beta}'(\boldsymbol{\sigma}_{yy} - \mathbf{\Sigma}_{YY})\boldsymbol{\beta} \tag{14.20}$$

The expressions, $\boldsymbol{\sigma}_{yy}$, $\boldsymbol{\sigma}_{vv}$, and $\boldsymbol{\phi}_{xx}$, represent the subpopulation variance–covariance matrices corresponding to the parent population matrices $\mathbf{\Sigma}_{YY}$, $\mathbf{\Sigma}_{VV}$, and $\mathbf{\Phi}_{XX}$, respectively.

We now are in a position to demonstrate the invariance of the matrix $\mathbf{\Lambda}$, the factor-pattern matrix, in the parent population and subpopulation. First, substitute the expression in Equation 14.16 for $\mathbf{\Sigma}_{YY}$ and the expression $\boldsymbol{\beta}\mathbf{\Lambda}'$ for $\mathbf{B}$ in Equation 14.19 to obtain

$$\boldsymbol{\sigma}_{yy} = \mathbf{\Lambda}\mathbf{\Phi}_{XX}\mathbf{\Lambda}' + \mathbf{\Psi}^2 + \mathbf{\Lambda}\boldsymbol{\beta}'(\boldsymbol{\sigma}_{vv} - \mathbf{\Sigma}_{VV})\boldsymbol{\beta}\mathbf{\Lambda}' \tag{14.21}$$

Factor out the matrices $\mathbf{\Lambda}$ and $\mathbf{\Lambda}'$ in Equation 14.21 to obtain further

$$\boldsymbol{\sigma}_{yy} = \mathbf{\Lambda}[\mathbf{\Phi}_{XX} + \boldsymbol{\beta}'(\boldsymbol{\sigma}_{vv} - \mathbf{\Sigma}_{VV})\boldsymbol{\beta}]\mathbf{\Lambda}' + \mathbf{\Psi}^2 \tag{14.22}$$

which we can rewrite immediately, after substitution from Equation 14.20, as

$$\boldsymbol{\sigma}_{yy} = \mathbf{\Lambda}\boldsymbol{\phi}_{xx}\mathbf{\Lambda}' + \mathbf{\Psi}^2 \tag{14.23}$$

If we look closely at Equation 14.23, we will notice that it has the form of the fundamental equation of common-factor analysis, where $\mathbf{\Lambda}$ is a factor-pattern matrix associated with the subpopulation variance–covariance matrix $\boldsymbol{\sigma}_{yy}$ and $\boldsymbol{\sigma}_{xx}$ is the subpopulation variance–covariance matrix for the common factors (in the original metric of the parent-population common factors). The matrix $\mathbf{\Psi}^2$ is unchanged from its corresponding value in the parent population by assumption. Thus we can conclude that if $\mathbf{\Lambda}$ is a factor-pattern matrix in the parent population it will also be a factor-pattern matrix in the subpopulation. However, the factors associated with $\mathbf{\Lambda}$ in the parent population and the subpopulation will not maintain the same intercorrelations with one another. For example, if a set of factors associated with a factor-pattern matrix $\mathbf{\Lambda}$ are mutually uncorrelated in the parent population, the same factors are likely to be correlated in the subpopulation. (A symmetric conclusion is derivable going from subpopulation to parent population where uncorrelated factors in the subpopulation will be possibly correlated in the parent population.) In addition, Meredith (1964a) also demonstrates that the invariance of the factor-pattern matrix $\mathbf{\Lambda}$ is not uniquely fixed to a particular factor-pattern matrix $\mathbf{\Lambda}$, but applies to all factor-pattern matrices which are linear transformations of $\mathbf{\Lambda}$ (with corresponding linear transformations of the factors).

The invariance of the factor-pattern matrix $\mathbf{\Lambda}$ is not maintained if we consider the case where the observed scores and common-factor scores are considered to have unit variances in both parent population and subpopulation. Let $\mathbf{d}_y$ and $\mathbf{d}_x$ be diagonal matrices whose diagonal elements are the square roots of the diagonal elements of the matrices $\boldsymbol{\sigma}_{yy}$ and $\boldsymbol{\sigma}_{xx}$ respectively. Then, the fundamental equation of common-factor analysis in Equation 14.23 may be rewritten as

$$\mathbf{r}_{yy} = \mathbf{d}_y^{-1}\boldsymbol{\sigma}_{yy}\mathbf{d}_y^{-1} = \mathbf{d}_y^{-1}\mathbf{\Lambda}\mathbf{d}_x\mathbf{r}_{xx}\mathbf{d}_x\mathbf{\Lambda}'\mathbf{d}_y^{-1} + \mathbf{d}_y^{-1}\mathbf{\Psi}^2\mathbf{d}_y^{-1} \quad (14.24)$$

where $\mathbf{r}_{yy}$ and $\mathbf{r}_{xx}$ are subpopulation correlation matrices for the observed variables and common factors, respectively. Thus, if we assume that the observed variables and common factors have unit variances in both the parent population and, the subpopulation, we find that a factor-pattern matrix $\mathbf{\Lambda}$ in the parent population does not correspond in the subpopulation to the same matrix but to the matrix $\mathbf{d}_y^{-1}\mathbf{\Lambda}\mathbf{d}_x^{-1}$. (We will make more explicit the implications of this fact shortly.)

Meredith (1964a, p. 180) also shows that variances and factor-structure matrices are never invariant under selection. In this context he defines a factor-structure matrix as the matrix of covariances between observed variables and common factors (which differs slightly from the definition normally used in this book which is that a factor-structure matrix is the matrix of correlations between observed variables and common factors). In any case, if we designate this "factor-structure" matrix as $\mathbf{\Xi}$, it can be shown that

$$\mathbf{\Xi} = E(\mathbf{YX}') = E[(\mathbf{\Lambda X} + \mathbf{\Psi E})\mathbf{X}'] = \mathbf{\Lambda\Phi}_{XX} \quad (14.25)$$

in the parent population and

$$\boldsymbol{\xi} = \mathbf{\Lambda}\boldsymbol{\phi}_{xx} \quad (14.26)$$

in a subpopulation. The matrix $\mathbf{R}_{YX}$ of correlations between observed variables and common factors in the parent population is given as

$$\mathbf{R}_{YX} = \mathbf{D}_Y^{-1}\mathbf{\Lambda\Phi}_{XX}\mathbf{D}_X^{-1} \quad (14.27)$$

and in the subpopulation its counterpart, $\mathbf{r}_{yx}$, is given as

$$\mathbf{r}_{yx} = \mathbf{d}_y^{-1}\mathbf{\Lambda}\boldsymbol{\phi}_{xx}\mathbf{d}_x^{-1} \quad (14.28)$$

where $\mathbf{D}_Y$ and $\mathbf{D}_X$ are diagonal matrices whose diagonal elements are the square roots of the corresponding diagonal elements of $\mathbf{\Sigma}_{YY}$ and $\mathbf{\Phi}_{XX}$ respectively, and $\mathbf{d}_y$ and $\mathbf{d}_x$ are as defined earlier. As we can see, none of these matrices are the same in parent population and subpopulation. One problem that this suggests is that the lack of invariance of correlations between variables and factors in parent population and subpopulation may create difficulties in the stability of interpretations assigned to factors, unless care is taken to make interpretations in some standard population.

Next, let us consider the invariance of a simple-structure solution across different experimental populations. If the same metric is used in all experimental populations for observed scores and common factors, respectively, then the same factor-pattern matrix $\mathbf{\Lambda}$ is obtained, and if $\mathbf{\Lambda}$ represents the simple-structure solution, then the simple-structure solution is invariant. We can also show that, if the factor analyses in the different experimental populations are conducted after standardizing the scores of the observed and common factors to unit variances in each population, the simple-structure solution in each population will correspond to the same simple-structure solution (but the pattern coefficients not equal to zero may differ). As proof, consider that $\mathbf{d}_y^{-1}\mathbf{\Lambda}\mathbf{d}_x^{-1}$ in Equation 14.24 represents the oblique factor-pattern matrix in some (normalized) subpopulation corresponding to the oblique simple-structure-solution factor-pattern matrix $\mathbf{\Lambda}$ in the parent population. Then it is obvious that for every zero element in $\mathbf{\Lambda}$ there will be a corresponding zero element in $\mathbf{d}_y^{-1}\mathbf{\Lambda}\mathbf{d}_x^{-1}$. Since the simple structure can be defined by the zero projections in the hyperplanes, this same solution theoretically can be found independently in every subpopulation without knowledge of its nature in the parent population. The invariance in this case is in regard to the zeros and not the nonzero factor-pattern loadings.

Finally, with regard to the regressions of the common factors on the common-factor portions of the variables, Meredith (1964a, pp. 183–184) shows that the regressions are invariant under selection. On the other hand, when the common factors are regressed on the total scores of the observed variables (the practice used to find factor scores), the regressions are not invariant under selection.

14.3.4 Effects of Selection on Component Analysis

Before going on to consider the problem of matching factors across experimental populations, let us consider the effects of selection upon component analysis. Since component analysis may be conceived of as a form of common-factor analysis wherein the "common" part of the observed variables is the total-score part, then all the selection formulas obtained for the model of common-factor analysis may be modified for component analysis by simply omitting the term involving the unique variances. Thus, we have for a parent population the fundamental equation of component analysis

$$\boldsymbol{\Sigma}_{YY} = \boldsymbol{\Lambda}\boldsymbol{\Sigma}_{XX}\boldsymbol{\Lambda}' \tag{14.29}$$

where $\boldsymbol{\Lambda}$ in this case is a factor-pattern matrix (of full rank) for relating the observed variables to the component factors. Analogous to Equation 14.23 we have the corresponding equation for the subpopulation

$$\boldsymbol{\sigma}_{yy} = \boldsymbol{\Lambda}\boldsymbol{\sigma}_{xx}\boldsymbol{\Lambda}' \tag{14.30}$$

which shows that the factor-pattern matrix in component analysis is invariant under selection. On the other hand, as with common-factor analysis, matrices of covariances and correlations between observed variables and common factors are not invariant under selection.

Unlike common-factor analysis, however, the regression of component factors on the observed variables *is* invariant under selection. *Proof.* Let $\mathbf{B}_{YX}$ be the matrix of regression weights for deriving the component factors from the observed variables in the equation $\mathbf{X} = \mathbf{B}'_{YX}\mathbf{Y}$ in the parent population. Then

$$\mathbf{B}_{YX} = \boldsymbol{\Sigma}_{YY}^{-1}\boldsymbol{\Lambda}\boldsymbol{\Sigma}_{XX} = (\boldsymbol{\Lambda}\boldsymbol{\Sigma}_{XX}\boldsymbol{\Lambda}')^{-1}\boldsymbol{\Lambda}\boldsymbol{\Sigma}_{XX} = (\boldsymbol{\Lambda}')^{-1} \tag{14.31}$$

(Recall that $\boldsymbol{\Lambda}$ and $\boldsymbol{\Sigma}_{XX}$ are full rank $n \times n$ matrices). In the subpopulation, the corresponding matrix of regression weights is given as

$$\mathbf{b}_{yx} = \boldsymbol{\sigma}_{yy}^{-1}\boldsymbol{\Lambda}\boldsymbol{\sigma}_{xx} = (\boldsymbol{\Lambda}\boldsymbol{\sigma}_{xx}\boldsymbol{\Lambda}')^{-1}\boldsymbol{\Lambda}\boldsymbol{\sigma}_{xx} = (\boldsymbol{\Lambda}')^{-1} \tag{14.32}$$

which proves the invariance. The assumption is made that the same metric is used in both the parent population and the subpopulation.

14.4 Comparing Factors across Populations

As factor analysis has become a computationally more feasible methodology, factor analysts have increasingly considered problems involving several factor analyses of a given set of variables. In these problems each analysis is carried out with a different experimental population and the results of the analyses are compared across populations. For example, a factor analyst interested in the developmental aspects of intellectual-performance factors might administer the same battery of tests to different groups of children representing different age levels. A separate factor analysis of test-battery performance in each group would then be carried out and the resulting factors compared across the groups to discern the developmental pattern.

To cite another example, a factor analyst might conduct an experiment involving several different experimental conditions and then compare the factors extracted from the variables measured under those conditions. However, in doing comparative factor-analytic studies one must be able to assume that the unique variances of the variables are unaffected by selection of the experimental groups. Unless the unique variances are equated with variances of errors of measurement, this assumption may prove to be a particularly difficult one to meet exactly in practical situations. Nevertheless, procedures are available, when the assumptions can be met, for comparing factors across different populations.

14.4.1 Preliminary Requirements for Comparing Factor Analyses

In comparing factors obtained with different experimental populations, we have the problem of deciding between two alternative hypotheses: (1) the same underlying factors account for the relationships among the variables in the different experimental populations; (2) different underlying factors are needed to account for the relationships among the variables in the different experimental populations. (The second hypothesis, we should note, is equivalent to saying that the patterns of interrelationships among the variables in the different experimental populations are different.) The first hypothesis assumes the invariance of factors under selection, which means that, regardless of how one selects the experimental population for study, the same factors operate on the variables within the individual cases studied. Thus, before we reject the first hypothesis in favor of the second, we must do everything in our power to rule out that apparent differences in factors are not due simply to the effects of selection.

When two or more experimental populations are involved, there are several ways in which we can go about ruling out the effects of selection as we compare factor solutions. When our interest is directed toward showing that the factor solution for a particular experimental population is unlike any that would be found in some other experimental populations, we may regard the particular experimental population in question as a reference population and, using the principles of factorial invariance under selection, seek to identify in the other experimental populations the same factor-pattern matrix as found in the reference population. If no factor-pattern matrices can be found in these other experimental populations which are sufficiently similar to the one found in the reference population to be considered equivalent to it, then we can assume that the particular reference population is unique in some way.

On the other hand, we may have reason to believe that several experimental populations have been selected from a given parent population. Although Meredith (1964b) developed procedures for testing these hypotheses, they have been superseded by developments in confirmatory factor analysis and structural equation modeling that allow the researcher to test for the invariance of factor-pattern coefficients across groups analyzed simultaneously.

Nevertheless, certain preliminary considerations can be discussed here, which will also apply to structural equation modeling. In looking for the same factors in different populations we must make certain that the observed variables are expressed in the same metric in the different experimental populations. Meredith (1964b) recommends that the observed variables be standardized to have unit variances in some reference population or that they be standardized to have their average variances across populations be unit variances. The next consideration should be to conduct the factor analyses with variance–covariance matrices expressed in the common metric for the observed variables.

However, if the analyses are conducted with correlation matrices, the resulting factor-pattern or factor-structure matrices should be transformed into the metric of the observed variables by multiplying the factor-pattern or factor-structure coefficients corresponding to given observed variables by the standard deviations of these variables in the experimental populations (expressed in the common metric). If we have done these things, we are ready to take advantage of the fact that factor-pattern matrices (and nonsingular linear transformations of them) are invariant under selection and are the only basis for identifying the same factors in different experimental populations.

14.4.2 Inappropriate Comparisons of Factors

There are a number of instances in the literature in which factors have been compared inappropriately to test the hypothesis that the same factors are found in different experimental populations. So that the reader may avoid inappropriate comparisons of factors, we cite the practices to avoid as follows:

1. The factors compared are derived from correlation instead of variance–covariance matrices computed for samples from different experimental populations. This practice violates the principle that the analyses must be in the same metric for the factor-pattern-matrix coefficients to be comparable across populations. Using correlation coefficients in each analysis forces the variables to have unit variances in each population, thereby creating a different metric for each population. Before factors obtained from correlation matrices are compared, the factor-pattern matrices should be modified to express a common metric for the observed variables.
2. Factor-structure instead of factor-pattern coefficients are compared. This error is most frequently made when one compares orthogonal-factor solutions obtained in each experimental population. According to selection theory, if the factors are mutually orthogonal in one population, they are most likely intercorrelated in some other selected population. Thus two orthogonal-factor solutions normally do not represent the same factors across different populations.
3. Different criteria of rotation are used in obtaining the compared factors.

4. Different methods of extracting the factors are used before rotating factors. (This may not normally be a serious problem, but it may account for lack of invariance in some instances.)
5. Not enough factors are extracted to get an accurate fix on the common factors in the variables. In some cases, the selection may make some common factors in experimental populations account for very small portions of the common variance and in others account for large portions. Such factors might not be retained in some analyses and lead to noncomparable solutions.
6. The factors to be compared come from analyses in which the unique portions of the observed variables in the analyses are not independent of the selection variables. This most likely will happen when traditional communalities instead of true variances are inserted in the principal diagonal of the variance–covariance matrix. In other words, if the unique variances are identified with the error variances of the variables, one can be certain that any variables involved in the selection of the experimental populations will be uncorrelated with these unique variances. One cannot have this assurance when the unique variances contain relatively large portions of true specific variance. If the selection variables relate to these unique factors then, even though they will be uncorrelated in some reference population, because of the effects of selection, they will be correlated in other populations, thus contributing to the common variance in these populations. The effect will be a lack of invariance of factor-pattern matrices. So, it is important to evaluate the assumption that any specific true variance should not be effected by selection variables.

14.4.3 Comparing Factors from Component Analyses

All the techniques discussed earlier in this section for comparing factor-pattern matrices in common-factor analysis also apply to factor-pattern matrices obtained from component analyses. However, it is pointless to use these techniques to determine whether differences in factor-pattern matrices are due simply to selection, when working with the full-rank solution (involving as many factors as there are variables). Then we will always be able to transform exactly an $n \times n$ factor-pattern matrix obtained from a component analysis of one set of data into the $n \times n$ factor-pattern matrix obtained from a component analysis with a different set of data. Why? Because all $n \times n$, square, nonsingular matrices form a mathematical group under matrix multiplication, which means, among other things, that we can always find an $n \times n$ nonsingular transformation matrix $\mathbf{T}$ such that $\mathbf{AT}=\mathbf{B}$, when $\mathbf{A}$ and $\mathbf{B}$ are nonsingular $n \times n$ matrices, that is, $\mathbf{T}=\mathbf{A}^{-1}\mathbf{B}$. This fact is a consequence of each $n \times n$ nonsingular matrix's having an inverse. In other words, we could never disconfirm the hypothesis that the components are the same in the two groups, regardless of selection or other effects in the data. If one gives the matter a bit of thought, he will see that this is as things should be, because the components, by definition, are linear combinations of the observed variables (not just their common parts), and it is always possible to study the behavior of some arbitrarily defined linear combination of variables in different populations. In short, components are invariant, by definition, across different populations, although covariances between them are not.

One might ask: Suppose in population *A* I do a principal-component analysis, followed by a rotation of the first *r* components, and in population I also do a principal-component analysis followed by a rotation of *r* components, can I compare the $n \times r$ pattern matrices **A** and **B** obtained in these two analyses? The answer is Yes. What you will need to find is the transformation matrix **T** which will transform **A** into a matrix as much like **B** as possible. That is, we seek to see to what degree $\mathbf{AT} = \mathbf{B}$. **T** is given by the following equation:

$$\mathbf{T} = (\mathbf{A}'\mathbf{A})^{-1}\mathbf{A}'\mathbf{B}$$

But the interpretation you give to any goodness (or poorness) of fit of **AT** to **B** will differ from the interpretation you would give such fit in common-factor analysis. For goodness of fit would indicate that the first *r* principal components of the study of population *A* span a subspace which overlaps well the subspace spanned by the first *r* principal components obtained in the study of population *B*. But to the degree that you do not obtain perfect fit of **AT** to **B** you can assume that selection and other effects influenced considerably the nature of the principal components to be found in each study.

14.4.4 Contrasting Experimental Populations across Factors

In structural equation modeling, it is possible to perform confirmatory factor analyses simultaneously in several groups. In these analyses, we can begin by constraining all of the factor-pattern loadings and the variances and covariances of the factors to be equal across the groups. If these constraints are incompatible with the data, we will get a significant chi-square for the analysis, indicating that our hypothesized constraints are inappropriate. But which ones? A nested series of models can then be tested. After rejecting the constraints of equal factor pattern and equal variances and covariances among the factors across all groups, we can then free the constraints on the variances and covariances among the factors, while still constraining the factor-pattern loadings to be equal. If we obtain a nonsignificant chi-square, this means the same factor-pattern loadings apply to each group, and, by implication, the same factors. So, we can now examine the freed variances and covariances among the factors to see how they vary from one another. The differing variances and covariances among factors may provide clues as to how the selection effects of variables used explicitly or implicitly in selecting the groups, have impacts on the factors. If the factor-pattern loadings and the covariances are not the same in each group, then each group must be interpreted separately, because each has a different set of factors.

14.4.5 Limitations on Factorial Invariance

Theoretically, factorial invariance under selection of populations depends on two conditions: (1) the regressions of the factor-analyzed, observed variables on the explicit or direct-selection variables are linear; (2) the errors in the regressions of the observed variables on the explicit or direct-selection variables are homoscedastic. In practice, many variables chosen for factor analysis will not satisfy these conditions with respect to some additional set of selection variables. It is important to be aware of how departures from these conditions guaranteeing factorial invariance might come about.

Departures from the condition of linearity of the regressions of the factor-analyzed variables on the selection variables most frequently arise when the regressions of the observed variables on each other are nonlinear. When this is the case, it normally is impossible to formulate the selection variables in such a way (by appropriate nonlinear transformations) that the regressions of the observed variables on these selection variables are *all* linear. Such a situation may arise for two reasons: (1) Different levels of the observed variables may represent entirely different underlying processes. For example, Guilford (1956b, p. 387) points out that a common conclusion in the industrial-psychology literature is that the regressions of job-proficiency variables on measures of general intelligence are nonlinear. The regression curve of a job-proficiency variable regressed on general intelligence frequently rises positively in the low range of intelligence, levels off in the middle to above-average range, and then descends negativity in the high range. It seems likely that intelligence plays entirely different kinds of roles in determining performance at these different levels. One would surmise then that the job performances of individuals with different levels of intelligence have different determiners, and, hence, different factors are involved in these individuals' performances. Thus different selections of subpopulations will yield different factors for test batteries containing job-proficiency and general-intelligence variables. (2) Nonlinearities among variables may be due to differences in the scaling of variables. Lord (1953), for example, shows on theoretical grounds that for tests scored by taking the number of "right" answers on the tests—for example, aptitude and achievement tests—the regressions of the observed test scores onto the underlying abilities or traits are nonlinear although monotonically increasing. Thus, equal intervals on the underlying trait or ability continuum do not represent equal intervals on the corresponding observed-score continuum. In the same work Lord also shows that tests of the same underlying trait or ability representing different levels of difficulty are nonlinearly related to the underlying ability as well as to each other—in different ways. Guilford (1941) presents evidence to show that tests measuring a single dimension but representing different difficulty levels produce more than one factor in a factor analysis of their intercorrelations, with the resulting "extra" factors being possibly artifacts of scaling. Guilford also suggests that the nature of the tests change with differences in levels of ability of the populations studied, again pointing out a source for the lack of invariance of factors to be found with such tests.

Departures from homoscedasticity of errors in the regressions of factor-analyzed variables on direct-selection variables also arise with tests scored by totaling the number of "right" answers on the tests. Lord (1953) shows that in the regression of such total scores on the underlying-ability continuum, the variability of test-score departures from the regression line is less at the extreme ends of the underlying-ability continuum than in the middle of its range. Moreover, such a phenomenon is also present in the regressions of total scores from such tests onto total-score continua on other tests. Thus, in these cases homoscedasticity cannot be satisfied. This, however, may again represent a scaling problem.

The implications for conducting factor-analytic research to be drawn from our discussion of the basis for the lack of factorial invariance is that one should be careful how he chooses his population for study. Large, heterogeneous populations of individuals representing vastly different backgrounds and experiences should be used in factor analysis with considerable caution, for the factors to be obtained with such populations in some contexts may be largely meaningless. This is because the performances on the variables of individuals in such heterogeneous populations may be determined in widely different ways, so that different factors would be needed to account for the performances of subsets of individuals in the population. The factors obtained for the whole population, however, may represent simply an averaging of the different "real" factors. Frequently such "average" factors may be difficult to interpret. One will have evidence that this is a problem when regressions of the variables on one another in the population demonstrate nonlinearity (especially lack of monotonicity) or when different subpopulations of the population produce widely different sets of factors.

In the case of experimental studies where one is concerned in showing that different experimental conditions produce different factors in a set of variables, one should be certain that, prior to assigning individuals to the different experimental conditions, the performances of these individuals on the variables are accounted for by a single set of factors. This could be accomplished by taking random samples from populations of "factorially homogeneous" individuals and assigning these samples to the experimental conditions. However, finding such factorially homogeneous populations may be more easily said than done.

Meredith (1993) has extended the treatment of factorial invariance to the subject of measurement invariance. He too notes that unique-factor variances may not be invariant under selection. Even the assumption that common and unique factors are uncorrelated under selection may not hold up across groups. Meredith's paper should be studied carefully by the advanced student.

15

Confirmatory Factor Analysis

15.1 Introduction

Confirmatory factor analysis represents a different approach to using the common-factor model from anything presented previously in this text. Whereas earlier chapters concerned "exploratory" factor analysis, this chapter will briefly consider "confirmatory" factor analysis. Both forms of factor analysis are based on the same common-factor model. The difference between them is that whereas in exploratory factor analysis, the researcher seeks to discover, for a set of variables, common factors that account for their correlations, in confirmatory factor analysis the researcher begins with a hypothesis about what the common factors are and how they are related to the observed variables and seeks to test that hypothesis. In many respects, for the student, this may be a new way of doing statistics, because most forms of statistics taught to beginners in statistics stress estimation of parameters and then testing whether they simply differ from zero. A mean may be estimated and tested against the hypothesis that it is zero, or two means may be compared and their difference tested to see if it differs significantly from zero. Or one may compute a correlation coefficient to see if there is a relation between variables, and the estimated correlation is then tested to see if it differs significantly from zero. Only on occasion is the student shown that one can specify a hypothesis that, say, the mean is equal to 4.5, or the difference between means is 2.0, or that the correlation is .75 between two variables and test these hypotheses. In this second case, the stress is placed upon using substantive theory to provide values for these coefficients as hypotheses to test. In the first, exploratory case, a null hypothesis of zero normally does not represent a specific substantive hypothesis, since the researcher really believes there may be a nonzero value, but does not know enough to specify what it is, and if it represents a relationship, hopes that it is nonzero.

Actually this is not a very strong way of using statistics to develop substantive theory about things in the world. For example, if one believes there may be a relationship between two variables, showing that it is significantly different from zero does not lend much support for any particular substantive theory, since any number of substantive theories may assert that a nonzero correlation exists between the variables. Showing that the correlation is not zero only supports all such theories. But science seeks to develop specific theories for substantive phenomena and to rule out rival theories. And so, at some point one needs to begin testing hypotheses for specific values or specific patterns of values that are at best limited to only one theory, or at worst to only a few theories.

But a problem in testing specific values or specific patterns of values for parameters may occur to the reader. Where do these specific values or specific patterns come from? In many respects they can come from our general knowledge about things and how they are related, or more frequently, not related. In other instances, they may come from previous studies and are then to be applied to a new context as invariants.

15.1.1 Abduction, Deduction, and Induction

The researcher should realize that scientists conduct research by repeatedly cycling through three phases as the research progresses. The philosopher, logician, and physical scientist Charles Sanders Peirce (1839–1914) was perhaps the first to articulate how scientists actually function through three phases, reflecting his first-hand experience as a scientist and his deep knowledge of the history of science. He named these three phases abduction, deduction, and induction. In "abduction" the scientist is confronted with new phenomena which beg for an explanation or a theory to account for them. The scientist then proceeds to develop a hypothesis to account for all of the known phenomena in question at that point. This may involve developing a number of hypotheses and eliminating all but one by how well they are able to reproduce the known phenomena in question. Sometimes abduction is described as "inference to the best explanation." As a logician, Peirce held that the inference in abduction had the form:

> The surprising fact, *C*, is observed;
> But if *A* were true, *C* would be a matter of course,
> Hence, there is reason to suspect *A* is true.
> (CP 5.189; Forster 2001)

Peirce does not claim that *A* is now indeed the unique or true explanation of *C*. That would be to commit the fallacy of affirming the consequent. There may be many reasons for *C* instead of *A*. But the scientist has reasons to believe that *A* is the reason for *C*. Thus *A* becomes his/her hypothesis to explain *C*. Still there is a need to test the hypothesis.

The second phase in the scientific cycle according to Peirce is "deduction." Given prior knowledge and the hypothesis *A*, the scientist deduces logically some new consequence that might be observed. The inference may be of two kinds (Forster, 2001):

Necessary deduction:

All *M*s are *P*
All *S*s are *M*
Hence, all *S*s are *P*

Probabilistic deduction:

p% of *M*s are *P*
$S_1,\ldots,S_n$ is a large sample of *M*

Hence, probably and approximately *p*% of $S_1,\ldots, S_n$ are *P*.

It is essential that the consequence deduced from the hypothesis be something not in the knowledge or data previously known without the hypothesis to serve as a premise to deduce it from that knowledge.

The third phase in Peirce's scientific cycle is "induction." This is reasoning from a sample to a population. Today behavioral and social scientists know this as statistical inference based on the results of a statistical test. The researcher collects or uses data not used in formulating the hypothesis or the deduction to test the hypothesis. The data collected should not be merely another random sample from the same population from which came the data used in hypothesis formation, but data collected in a new context, sometimes with other variables not observed before. But it is not sufficient that the data conform to the hypothesis, but rather that one be able to infer that it would conform in the population from which the data is sampled.

15.1.2 Science as the Knowledge of Objects

Let me preface my remarks to follow by indicating that my philosophy of science is naturalistic and not an attempt to ground epistemology in self-evident foundational truths whether introspected or otherwise. Foundationalist philosophies were shown to be incoherent in the 1960s and 1970s. A naturalistic philosophy of science relies on cognitive science to understand how humans think and know things. It is corrigible by advances in scientific knowledge of cognitive functioning. It accepts the inevitable circularity of its position, in that to explain science, we must employ science, which is itself based on schemas in human perception and action. But science is able to modify itself without changing its basic character, because human nature changes only slowly. To understand how science arises out of human nature is our next topic.

There is another rationale for why scientists form and test hypotheses. This rationale demands for scientists throughout history less expertise of scientists in formal logic and deduction than in Peirce's logical analysis. It also does not require a formal deduction after formulating a hypothesis. The hypothesis can simply assert invariants.

Scientists generally accept that scientific knowledge is (1) acquired by the senses, (2) is unbiased, (3) is intersubjective, (4) is repeatable, (5) is falsifiable (or disconfirmable), as well as possessing other characteristics that are often blends of these five. Now, what I contend is that these characteristics of scientific knowledge all follow from an unconscious application of a metaphoric schema taken from object perception: "Science is knowledge of objects" (Mulaik, 2004, 1994).

That this may be a metaphor may not be immediately obvious. George Lakoff and Mark Johnson (1980, 1999), Lakoff (1987, 1993), Lakoff and Nuñez (2000), Lakoff and Turner (1989) have argued that abstract thought is metaphoric, with metaphors taken from embodied perception and action. Metaphors are mappings from components of a source domain to components of a target domain, so that relationships between the components in the source domain may be analogously applied to make inferences in the target domain. Here the source domain is the domain of object perception. The target domain is the domain of scientific knowledge. Aspects of the schemas of object perception are mapped onto aspects of science. By "embodied" is meant that we as humans move and observe as bodies within the world with coordinated perceptual and motor schemas adapted over millions of years of evolution that integrate what is given by the senses with muscular action that permits us most of the time to successfully move among and manipulate objects in the world.

These perceptual and motor schemas, I hold, are applied rapidly—often within milliseconds, in perception—to integrate the data coming in from sensory neurons. Or they may be applied to percepts (Jaynes, 1990/1976) received in working memory from perceptual processes within at most a few seconds of one another. In this respect, humans share much in common with most mammals and other vertebrates. But humans, I contend, also have the capability to form concepts. Concepts are formed and integrated in working memory from percepts retrieved from long term memory that have been excerpted (Jaynes, 1990/1976) from perceptual experience, sometimes hours, days, and years apart in time and/or from widely spaced locations, sometimes hundreds and thousands of miles apart in space. Or the percepts may be recorded as signs or writing and taken in through perception and then into working memory. The concept is not based on a percept that simultaneously perceives all its components in a single glance. The metaphor is a blend of schemas from perceptual and motor experience that provides the organization or structure that combines all the percepts into a synthesized whole, constituting the concept, in working memory.

When Columbus argued that the earth is round, he had not seen the earth as a whole to know this. He had a number of observations, such as seeing first the tops of ships coming toward him over the horizon, then later the hulls. Then there was the round shape of the shadow of the earth on the moon during a lunar eclipse. The existence of round bodies in the heavens, such as the moon and sun, also suggested the roundness of the earth. But the concept of a ball, a sphere, integrates all of these observations into a concept of the round earth.

Leeuenhook looks through his water-drop-lens microscope and sees tiny things moving around in some water he is focused on and calls them "little animals," assimilating them thus metaphorically to the domain of ordinary living animals seen with the naked eye. He does not see the little animals with his naked eyes.

Again, Tycho Brahe's observations of Mars, used by Kepler to determine that the orbit of Mars is an ellipse, were gathered over many years from locations in space that were often hundreds of thousands of miles apart. The metaphor of an ellipse, taken from drawings in which the ellipse was perspicuously perceived as a whole, integrated all of these observations into an orbit.

Brown (2003) cites a metaphor in chemistry in the following sentence: "[Cell] Membranes contain channels that are permeable to hydrogen ions and other positive ions." This is an example of the use of a conduit metaphor (Lakoff, 1999). A conduit guides the movement of something through space. Usually a conduit has boundaries that confine what is transported through it within the conduit. A "channel" is a particular kind of conduit, usually between bodies of water. Here hydrogen and other positive ions are confined to the channel conduits as they move from outside to the interior of a cell.

Most metaphors, according to Lakoff and Johnson, are schemas of embodied action and perception—such as, paths followed in space, conduits, locations, obstacles in paths, movement of objects—used to synthesize or integrate experience into complex concepts. So, human knowledge is conditioned on and mediated by prior schemas of embodied perception and action.

15.1.3 Objects as Invariants in the Perceptual Field

Having said that, why is science's concern with objects built around a metaphor? Most abstract scientific concepts are not based on direct percepts of objects corresponding to them. Atoms, molecules, cells, channels in cell membranes, intelligence, and cooperation are regarded as objects or attributes of objects by the scientists concerned with them, but no one sees these directly and immediately by unaided perception. The use of instruments to observe these involves complicated theories of how the instruments work to reveal the objects. These theories all work at the conceptual level and incorporate numerous metaphors. So, these are conceptual objects, not perceived objects.

What then is a perceived object, if it is the basis of science's conceptual objects? J. J. Gibson (1950) has provided what I take to be an appropriate answer here. According to Topper (1983), Gibson (1950) drew upon the mathematical concept of an invariant. In topology "…a series of transformations can be endlessly and gradually applied to a pattern without affecting its invariant properties. The retinal image of a moving observer would be an example of this principal" (Gibson, 1950, pp. 153–54). Topper (1983) paraphrases Gibson as asserting that the retinal image is a projection of a solid object onto the curved surface of the retina of the eye. As the observer or the object or both move around in space with respect to each another, the image will change. But within the image there will remain invariant properties that are analogous to those in the geometry of transformations (Topper, 1983).

But there is another aspect of object perception. While objects may be invariants in the perceptual field, they are still regarded as other to and independent of the observer. This is possible, Gibson (1950) held because the observer also simultaneously gains information in perception about the actions and position of the observer with respect to the object and its surroundings. He called this "proprioception," in contrast to the perception of objects, which he named "exteroception." As the observer moves with respect to the object, the visual image changes in regular ways. As the observer approaches the object, both it and the immediate background looms out from a point toward which the observer is moving. As the observer moves away from the object, the image of the object and objects in its immediate vicinity shrink toward a point. On the other hand, if the object approaches the subject, only the object, but not the objects in the surrounding vicinity loom, while retaining invariant relations among its components. Moving the head to the right causes the image of the object to move to the left. Thus one has information in perception that the invariant object is distinct from the observer because it remains invariant despite the regular changes to the image produced by the actions and motions of the observer.

Objects are also intersubjective, because each observer can ask another whether he/she observes the same thing and acts consistent with that view. The other's report is further evidence bearing on invariance. The observer can also move around, observe from different places to see if the same thing is perceived. The observer may close his/her eyes for a few seconds, then look again, to see if the object remains invariant, endures through time.

Thus I believe scientists unconsciously came to demand that one establish the objectivity of one's concepts because they unconsciously applied the metaphor taken from the schemas of object perception to their concepts. They did this, maybe naively, because objects they held are what are real, and the object conceived, though not directly observed, was still another object in the world. (It is a trap to fall into to believe that what we know about objects is incorrigible). Just as objects are invariants when seen from different points of view, scientific objects must be invariants when observed from different points of view in different laboratories and with different observers and means of observation. But observations are limited in time and from a few points of view, so they may not take in the whole object. Future observations from new points of view may be inconsistent with the concept formed of it so far. Hence concepts of objects are corrigible.

Scientific concepts must also be free from systematic artifacts due to the methods, conceptions, and prejudices of a specific observer or group of observers. In other words, the concept of the scientific object must not be contaminated with effects due exclusively to the observer or the observer's means of observation.

For example, when cold fusion was announced by researchers at the University of Utah in 1989, physicists all over the world scrambled to replicate the phenomenon in their laboratories. At Georgia Tech a week later, scientists there reported having conducted an experiment that yielded evidence in support of cold fusion. But a few days later, the scientists retracted their claim, saying that a neutron detector they had adapted for the experiment unexpectedly yielded inflated readings of neutrons (a sign hypothesized of cold fusion) because of the confounding influence of temperature in the experiment on the instrument. The instrument had been designed for use at much higher temperatures. Thus an artifact of observation rendered "subjective" conclusions.

15.1.4 Implications for Factor Analysis

So what does this all mean for factor analysis? Haig (2005) argues that exploratory factor analysis is an abductive method for formulating hypotheses using the common cause principle, but also to be used along with confirmatory factor analysis, which tests hypotheses. Because it uses one pattern of causation, a common cause of several effects, it is limited in the instances of causation that it can represent properly.

Confirmatory factor analysis corresponds to deduction and induction and is used when the researcher formulates and tests a hypothesis about how a set of theoretical constructs or latent variables contains the causes of another set of observed variables or indicators. The researcher may develop the theory in the abstract involving anticipated latent variables derived from general knowledge and then derive and/or construct observed indicators of the hypothesized common factors as effects of them. The resulting model with certain parameters fixed to specified values represents a hypothesis asserting invariant quantities, while other parameters are left freely estimated conditional on the fixed parameters, to fill the unknown blanks in the model. The fixed parameters may be derived in some cases deductively from theory, but frequently they will be values estimated for these parameters in other contexts, often when the observed variables have been embedded with other variables. Sometimes (actually quite often for most researchers beginning a research program) the researcher will only be able to indicate what latents are not related to what indicators, specified by fixing the respective loadings to zero.

15.2 Example of Confirmatory Factor Analysis

In a study of how personality judgments may be mediators between personality descriptions as stimuli and personality ratings as responses, Carlson and Mulaik (1993) selected 15 trait-rating scales to represent three personality-judgment dimensions. In one of three different experimental conditions each of the 280 subjects was given a randomly generated description of a person, with the descriptions varying in degree on three dimensions, Friendliness, Capability, and Outgoingness. The subjects were then asked to rate the person on each of 15 7-point bipolar personality trait-rating scales. The dimensions chosen for study were based on Osgood et al.'s (1957) "big three," evaluation, potency, and activity, as represented in different modes of the person as found by Mulaik (1963). Friendliness was the way the rater valued other persons. Intelligence was potency in the realm of mental functioning. Outgoingness was activity in the realm of interpersonal relations. Four of the scales were selected to overdetermine the Friendliness dimension: (1) friendly–unfriendly, (2) sympathetic–unsympathetic, (3) kind–unkind, (4) affectionate–unaffectionate. Four of the scales were selected to represent ability: (5) intelligent–unintelligent, (6) capable–incapable, (7) competent–incompetent, (8) smart–dumb. Four scales were selected to represent outgoingness: (9) talkative–untalkative, (10) outgoing–withdrawn, (11) gregarious–ungregarious, (12) extroverted–introverted. Two scales were selected to represent both Friendliness and Ability: (13) helpful–unhelpful, (14) cooperative–uncooperative. One final scale was selected to represent both Outgoingness and Friendliness: (15) sociable–unsociable.

The study by Carlson and Mulaik involved ratings made in three experimental conditions: (1) limited information to only the three dimensions, (2) information given on the three dimensions and on the person's cooperativeness, (3) information given on the three dimensions and on sociability. For our purposes here we will only use the results found for the first condition, ratings made from descriptions varying just on the three dimensions. The correlation matrix for this data is given in Table 8.2. [This correlation matrix was not published in Carlson and Mulaik (1993)].

Now, what differs from an exploratory factor analysis at this point is that we specify a hypothesis, based on the information given above, involving how the factor-pattern matrix will have certain loadings. Carlson and Mulaik (1993) were able to specify or fix zero loadings in certain positions of the factor-pattern matrix. They, of course, implicitly constrained correlations between the common factors and unique factors to be zero and the correlations among the unique factors to be zero. However, they did not have any knowledge of the magnitudes of certain other loadings, and so they left these loadings free. This meant the free parameters were to be estimated conditional on the fixed parameters. The values of the correlations among the common factors were also treated as free parameters, as were also the unique-factor variances.

In Table 15.1 is shown the hypothesis specified for the factor-pattern loadings, which in this case, were only 0 loadings. Free parameters are indicated by "?". Note that all unique-factor variances are free, as are also the correlations among the common factors. In this case the hypothesis only indicated what indicators were not related to what latent common factors by specifying 0 loadings. Free parameters, on the other hand, are "filler" for the model and are not part of the hypothesis, since nothing is specified for them by freeing them. Free parameters are estimated by an iterative algorithm that seeks to find estimates for the free parameters that minimize an overall discrepancy or lack-of-fit function "conditional" on the fixed parameters of the model. The discrepancy function measures the degree of lack of fit between the observed

TABLE 15.1

Hypothesized Factor-Pattern Loadings and Free Parameters for a Confirmatory Factor Analysis of the Correlations among 15 Trait-Rating Scales from the Carlson and Mulaik, 1993 Study Shown in Table 8.2

Variable	Factor 1	Factor 2	Factor 3	Unique Variances
1. Friendly	?	**0**	**0**	?
2. Sympathetic	?	**0**	**0**	?
3. Kind	?	**0**	**0**	?
4. Affectionate	?	**0**	**0**	?
5. Intelligent	**0**	?	**0**	?
6. Capable	**0**	?	**0**	?
7. Competent	**0**	?	**0**	?
8. Smart	**0**	?	**0**	?
9. Talkative	**0**	**0**	?	?
10. Outgoing	**0**	**0**	?	?
11. Gregarious	**0**	**0**	?	?
12. Extroverted	**0**	**0**	?	?
13. Helpful	?	?	**0**	?
14. Cooperative	?	?	**0**	?
15. Sociable	?	**0**	?	?
Correlations among Factors				
Factor	**1**	**2**	**3**	
1	**1.000**			
2	?	**1.000**		
3	?	?	**1.000**	

Note: Hypothesized (fixed) parameters are shown in boldface, and free parameters as question marks.

correlation matrix for the observed variables and the model's reproduced correlation matrix for these variables. In short one seeks estimates of free parameters that minimize lack of fit between the reproduced correlation matrix and the observed correlation matrix, consistent with the constraints on the fixed parameters. Within the constraints given by the fixed and constrained parameters the iterative algorithm is free to seek any values for the free parameters, which in some cases could be zero or of whatever sign. In this way, any lack of fit can then be attributed to the fixed or constrained parameters, because otherwise, without these constraints, and assuming that the model is still identified (a concept to be touched on shortly), the reproduced correlation matrix would then fit the sample correlation matrix perfectly, by mathematical necessity, but not empirical necessity.

Note also that Table 15.1 does not show the fixed zero correlations between the common and unique factors or among the unique factors. These are fundamental assumptions of most applications of the common-factor model. These fixed zero correlations among the unique factors and between the common and unique factors contribute both to making the model identified as well as overidentified. More about this will be discussed later.

The model specified in Table 15.1 was then applied to the sample correlation matrix in Table 8.2, and analyzed using Bentler's EQS program for structural equation modeling. Confirmatory factor analysis is but a special case of a structural equation model, which is a very general kind of model that allows considerable freedom in representing linear causal relationships. The result of the analysis is shown in Table 15.2.

The chi-square test for goodness of fit of the reproduced correlation matrix based on the above parameter values with the observed correlation matrix yielded a chi-square of 225.115 with .84 degrees of freedom, with a p value of less than .001, which was significant. The model did not fit to within sampling error. (A nonsignificant chi-square is what is desired). However, Bentler's CFI index of goodness of fit was .968, which is usually considered quite good as an approximation. However, the parsimony ratio of degrees of freedom divided by $n(n+1)/2$ was $84/120 = .700$, which is not indicative of a highly tested model. (A parsimony ratio of .95 or greater would be better, and combined with a high degree of fit, say, CFI > .95, this would be then a highly tested model with very good fit.) So, even if the goodness of fit is strong, as an approximation, the model awaits further development and specification of parameters to fully test it.

It is possible to evaluate the fixed parameters to determine if some are contributing significantly to the lack of fit. In Bentler's EQS program this is accomplished by use of "Lagrange multiplier tests." In LISREL, Jöreskog's structural equation program, an analogous test uses what are called "modification indices." Since the above data was analyzed using EQS, the Lagrange multiplier tests were performed on each fixed parameter, and these in turn were sorted in descending order of magnitude. The way to use the results

TABLE 15.2

Results of a Confirmatory Factor Analysis of the Model in Table 15.1 Applied to the Correlations among the 15 Personality Rating Variables in Table 8.2

	Factor 1	Factor 2	Factor 3	Unique Variances
1. Friendly	.849*	0	0	.529*
2. Sympathetic	.922*	0	0	.386*
3. Kind	.925*	0	0	.381*
4. Affectionate	.901*	0	0	.434*
5. Intelligent	0	.877*	0	.480*
6. Capable	0	.925*	0	.381*
7. Competent	0	.925*	0	.381*
8. Smart	0	.903*	0	.430*
9. Talkative	0	0	.803*	.596*
10. Outgoing	0	0	.948*	.317*
11. Gregarious	0	0	.891*	.454*
12. Extroverted	0	0	.898*	.439*
13. Helpful	.725*	.222*	0	.594*
14. Cooperative	.715*	.229*	0	.603*
15. Sociable	.172*	0	.833*	.342*

Correlations among Factors

Factor	1	2	3
1	1.000		
2	.223*	1.000	
3	.556*	.302*	1.000

Note: Estimates of free parameters are shown with asterisks following them. Coefficients are given for a solution where all variables, both observed and latent, are rescaled to have unit variances.

of the Lagrange multiplier test is to free the fixed parameter with the largest single-degree-of-freedom chi-square among the Lagrange multiplier tests, and reanalyze the resulting model. This was done several times, each time freeing the fixed parameter remaining with the largest chi-square value. We should stop if the model achieves a nonsignificant chi-square. We should also use this procedure judiciously, in that, we do not want to free too many fixed parameters, because freeing a fixed parameter results in the loss of one degree of freedom.

A degree of freedom corresponds to a condition in which the model is free to differ from the observed correlation matrix. The degrees of freedom, df, for the model are computed as

$$\text{df} = n(n+1)/2 - m$$

where
- n is the number of variables
- m the number of free parameters in the model
- $n(n+1)/2$ is the number of nonredundant elements on the diagonal and off one side of the diagonal of the correlation matrix which the model is fitted to

If no parameters are estimated, corresponding to a model with all fixed parameters, the degrees of freedom would be a maximum of $n(n+1)/2$. This would correspond to a maximally testable model for the number of variables in the correlation matrix.

If as many parameters are estimated as there are distinct elements of the correlation matrix to be fitted, the degrees of freedom will be zero and the free model parameters will always be able to be adjusted to allow the model to fit perfectly to the observed correlation matrix. Such a model, known as a saturated model, would be useless for a test, since no hypothesis has been asserted that could be disconfirmed by a possible lack of fit.

So, a degree of freedom corresponds to a condition by which the model is tested for goodness of fit (Mulaik, 2001). What we hope to attain is a model with numerousdegrees of freedom, relative to the potential number of degrees of freedom (i.e., a high parsimony ratio) that we could have, and excellent fit. Freeing up fixed parameters results in a weaker model, since fewer hypothesized constraints remain by which the model potentially could be tested, which is indicated by fewer degrees of freedom.

Nevertheless, we were able, after freeing a number of fixed parameters as indicated by the use of Lagrange multiplier tests, to find parameters to free to obtain a model that fit with a nonsignificant chi-square. The results are shown in Table 15.3.

Now a word of caution should be uttered here. The procedure of freeing fixed parameters of the model to achieve better fit will only work if the "true" model has the same structure. If there are different or more common factors than those hypothesized, the researcher may not realize this, and the resulting model may be totally misleading. Also note that the model has been reduced somewhat in terms of testability by reduction in the parsimony ratio, which was not laudably high to begin with.

How would we interpret the final model? The fact that few 0 coefficients in the factor-pattern matrix contributed to lack of fit, as indicated by the Lagrange multiplier tests, suggests that there is some support for the theory behind the three factors hypothesized. It seems that the low negative loadings of the variables "sympathetic" and "kind" on outgoingness suggest that a person with a higher degree of outgoingness will be modestly less likely to be seen as sympathetic and kind. Perhaps people who are sympathetic and kind are not seen to be as outgoing and sociable. However, we note also that there is a moderate correlation between friendliness and outgoingness. This leaves some freedom for an outgoing person to be a critic of others or

TABLE 15.3

Final Standardized Estimated Model Obtained by Freeing Eight Fixed Parameters According to Lagrange Multiplier Tests

Variables	Factor 1	Factor 2	Factor 3	Unique Variances
1. Friendly	.834*	0	0	.552*
2. Sympathetic	1.052*	0	−.210*	.360*
3. Kind	1.069*	0	−.249*	.364*
4. Affectionate	.907*	0	0	.421*
5. Intelligent	0	.868*	0	.497*
6. Capable	0	.908*	0	.418*
7. Competent	0	.949*	0	.316*
8. Smart	0	.932*	0	.363*
9. Talkative	0	0	.807*	.591*
10. Outgoing	0	0	.948*	.318*
11. Gregarious	0	0	.889*	.458*
12. Extroverted	0	0	.898*	.440*
13. Helpful	.727*	.205*	0	.596*
14. Cooperative	.717*	.207*	0	.607*
15. Sociable	.193*	0	.808*	.336*

Correlations among Factors

Factor	1	2	3
1.	**1.000**		
2.	.251*	**1.000**	
3.	.633*	.296*	**1.000**

Freed Correlations among Unique Factors

ρ(*E*3, *E*1)	.215*
ρ(*E*9, *E*1)	.181*
ρ(*E*10, *E*2)	.375*
ρ(*E*8, *E*7)	−.619*
ρ(*E*9, *E*8)	.236*
ρ(*E*14, *E*13)	.281*

Note: Degrees of freedom have been reduced from 84 to 76. Parsimony ratio is now only .63. Chi-square is 91.73 for 76 df with $p = .113$. CFI = .997.

garrulous but not as sympathetic and kind as it would seem. Mulaik and Carlson (1993) hypothesized in other parts of their study that "helpful" and "cooperative" would fall on a separate factor to some degree, since they believed ratings on these scales could be driven in part by information as to the actual helpfulness or cooperativeness of the person alone apart from information as to the person's friendliness and ability. So the correlation between unique factors E13 and E14 is not surprising.

15.3 Mathematics of Confirmatory Factor Analysis

A hypothesized model of confirmatory factor analysis has the same mathematical form as in exploratory common-factor analysis. The hypothesis is asserted by the usual factor-analysis model equation

$$\mathbf{Y} = \mathbf{\Lambda X} + \mathbf{\Psi E} \quad (15.1)$$

where

$\mathbf{Y}$ is an $n \times 1$ random vector of observed random variables
$\mathbf{\Lambda}$ is an $n \times r$ factor-pattern matrix
$\mathbf{X}$ is an $r \times 1$ random vector of latent common factors
$\mathbf{\Psi}$ is an $n \times n$ diagonal matrix of unique-factor-pattern loadings
$\mathbf{E}$ is an $n \times 1$ random vector of latent unique-factor variables

The usual assumption that common and unique factors are uncorrelated holds. However, the assumption that unique factors must always be mutually uncorrelated may be relaxed in certain circumstances.

The reproduced covariance matrix for the observed variables based on the model is given by the equation

$$\mathbf{\Sigma} = \mathbf{\Lambda \Phi \Lambda}' + \mathbf{\Theta} \quad (15.2a)$$

where $\mathbf{\Theta}$ denotes the $n \times n$ variance–covariance matrix for the unique factors. Ordinarily $\mathbf{\Theta}$ is diagonal, and Equation 15.2a then reduces to $\mathbf{\Sigma} = \mathbf{\Lambda \Phi \Lambda}' + \mathbf{\Psi}^2$. But we can allow for the unique factors to be correlated in some cases, so $\mathbf{\Theta}$ need not be regarded as always diagonal.

Now, in formulating the hypothesis, the researcher must have reason to assume that he/she has a "closed system of variables." This assumption may always be challenged by evidence to the contrary. But assuming one has a closed system of variables will be necessary for ruling out the existence of unobserved, extraneous variables. Some effort must be made to examine the validity of this assumption in the design of the study. Any extraneous variables must be identified, then excluded or controlled for, as in any experiment. Otherwise the inferences made may be highly misleading. Of course, researchers are fallible and may not always identify extraneous variables in their study. In that case one relies on the self-correcting nature of science to eventually correct such errors. But one must do one's best to reduce errors and leave as little to others to correct as possible. This is doubly important for most confirmatory factor analysis studies, because random assignment is not an available way of controlling for extraneous variables in field studies, where confirmatory factor analysis is most often applied.

15.3.1 Specifying Hypotheses

Karl Jöreskog (1969a,b, 1974) was first to recognize that in confirmatory factor analysis and in the more general structural equations modeling, researchers would typically be unable to formulate complete hypotheses involving all of the elements of the parameter matrices of the model. He thus provided in his computer program LISREL© that a researcher could specify each parameter of his model in one of three ways: (1) as a prespecified "fixed" parameter, (2) as a "constrained-equal or yoked" (my term) parameter to be equal to some other parameters yoked with it, their common value to be estimated conditional on other fixed and constrained parameters, or (3) as a "free" parameter to be estimated to minimize lack-of-fit conditional on the fixed and constrained-equal parameters of the model. Other programs, such as McDonald's COSAN program and EQS, provide for introducing equations of constraint involving sets of parameters, with the equations of constraint introducing nonlinear constraints in many cases, such as the sum of the squares of a set of parameters must equal unity.

Whether a parameter is fixed, constrained-equal, or free follows from the researcher's hypothesis. Usually fixed parameters in the factor-pattern matrix $\mathbf{\Lambda}$ enter as zeros to indicate that an observed variable is not dependent on a certain factor. But it is possible to enter nonzero fixed values also, to indicate how much one believes the observed variable changes with a unit change of a specified common factor. Variances of common factors may be fixed to unity, say, to establish a metric for the solution. In some cases the metric may be set instead by fixing one of the factor-pattern loadings in a column of $\mathbf{\Lambda}$ to unity. (One should not do both of these things for the same factor, because this will correspond to a more constrained model, perhaps not intended by the researcher, in which a variable changes one unit for a unit change of a factor with unit variance.) Elements of the factor correlation matrix, $\mathbf{\Phi}$, can also be made fixed or free or constrained equal, as also can be elements of $\mathbf{\Theta}$.

15.3.2 Identification

There is another important aspect to model specification for models with free parameters: "identification." In the broadest sense, identifying a model concerns being able to distinguish one theory from another theory having the same functional relationships between data and model parameters, when the model is applied to data (L'Esperance, 1972). In confirmatory factor analysis, the fixed and constrained-equal parameters must be specified in such a way that (a) no other similar model with the same functional relationships (with the exception of a model that differs trivially in terms of different units of measurement) can have the same fixed and constrained equal parameters, but different free parameters, and (b) still generate the same hypothetical covariance matrix, $\mathbf{\Sigma}_0$, for the observed variables (James et al., 1982). That is to say, "the hypothetical covariance matrix $\mathbf{\Sigma}_0$ associated with this set of fixed [and constrained-equal] parameters will be unique for each distinct complementary set of free parameter values" (James et al., 1982, p. 129).

The preceding condition on the fixed and free parameters of the model implies that there be a unique solution for the free parameters, given the observed variance–covariance matrix, $\mathbf{S}$, and the fixed and constrained equal parameters of the model. The equation of the fundamental theorem implies that

$$\boldsymbol{\Sigma}_0 = \boldsymbol{\Lambda\Phi\Lambda}' + \boldsymbol{\Theta} \tag{15.2b}$$

is a system of simultaneous, nonlinear equations, where the parameters $\boldsymbol{\Lambda}$, $\boldsymbol{\Phi}$, and $\boldsymbol{\Theta}$ on the right hand side of the equation determine the elements σ_{ij} of a reproduced convariance matrix $\boldsymbol{\Sigma}_0$ as distinct functional equations $\sigma_{ij} = \sigma_{ij}(\boldsymbol{\theta})$ of them.

$$\boldsymbol{\Sigma}_0 = \boldsymbol{\Sigma}(\boldsymbol{\theta})$$

where the parameters in $\boldsymbol{\Lambda}$, $\boldsymbol{\Phi}$, and $\boldsymbol{\Theta}$ have been arranged in a vector $\boldsymbol{\theta}=[\boldsymbol{\theta}^*, \hat{\boldsymbol{\theta}}]$, where $\boldsymbol{\theta}^*$ is a subvector containing fixed and constrained equal parameters, and $\hat{\boldsymbol{\theta}}$ contains free parameters. While $\boldsymbol{\Sigma}_0$ in Equation 15.2b is to be derived from the parameters on the right side of the equation, we should now consider hypothetically that we know the population variance–covariance matrix $\boldsymbol{\Sigma}$ (where $\mathbf{S}$ is the sample estimate of it) and that it is given by Equation 15.3

$$\boldsymbol{\Sigma} = \boldsymbol{\Lambda\Phi\Lambda}' + \boldsymbol{\Theta} \tag{15.3}$$

Since we do not know by hypothesis the values of the free parameters, and the model must be completed with a full set of parameters, we must determine the free parameters values on the right by solving for them in terms of the "known" elements of $\boldsymbol{\Sigma}$ and the specified and constrained-equal parameters of the matrices on the right. The free parameters are unknowns to be solved for in terms of the knowns. No solution will be possible for the free parameters if there are more free parameters than there are known parameters. Since there are $n(n+1)/2$ distinct observed parameters on the left, a "necessary" condition that the free parameters each have a unique solution is that the number of free parameters must not exceed this number. However, even if the number of free parameters is less than $n(n+1)/2$, this may not be "sufficient" in a specific case to make the model identified.

The question now before us is, given some select subset of the parameters of $\boldsymbol{\theta}$ placed into the subvector $\hat{\boldsymbol{\theta}}$ of free parameters, what constraints must we place on the remaining parameters of this system of equations so that we might in turn solve for the subset of the parameters in the subvector $\hat{\boldsymbol{\theta}}$ using the elements of $\boldsymbol{\Sigma}$ and $\boldsymbol{\theta}^*$? We will say that a parameter $\hat{\boldsymbol{\theta}}_k$ in $\hat{\boldsymbol{\theta}}$ is "identified" if there exists a subset of equations $\sigma_{ij} = \sigma_{ij}(\boldsymbol{\theta})$ that is uniquely solvable for $\hat{\boldsymbol{\theta}}_k$ given $\boldsymbol{\Sigma}$, fixed parameters $\boldsymbol{\theta}^*$, and constraints. We will say that $\hat{\boldsymbol{\theta}}$ is identified, if every element in $\hat{\boldsymbol{\theta}}$ is identified. If every element in $\hat{\boldsymbol{\theta}}$ is identified, then we say that the model is identified. On the other hand, we say that a parameter is underidentified if it is not identified. A model is underidentified if at least one of its parameters is underidentified. We will say that a parameter $\hat{\boldsymbol{\theta}}_k$ is "overidentified" if more than one distinct subset of equations $\sigma_{ij} = \sigma_{ij}(\boldsymbol{\theta})$ may be found that is solvable for $\hat{\theta}_k$. If at least one free parameter of a model is overidentified, we say the model is "overidentified." If the number of free parameters is equal to $n(n+1)/2$ (the number of distinct elements in $\mathbf{\Sigma}$) and each free parameter is identified, but none is overidentified, we say that the model is "just-identified."

Just-identified models cannot themselves be tested, because they will always perfectly reproduce the data. They can only be used in the testing of more constrained models as a basis for comparison. Based on a system of as many equations as there are unknowns that is uniquely solvable for the unknowns, there is one and only one way to solve for the unknowns of a just-identified model. When solved for in terms of the empirical covariance matrix and the constraints, the estimated parameters in turn will perfectly reproduce the empirical covariance matrix. But this is a mathematical necessity, not an empirical finding.

We cannot call a situation a test if there is no logical possibility in the situation of failing the test. Thus evaluating a just-identified model by how well it reproduces the empirical covariance matrix is not a test of the model because it is logically impossible to fit the covariance matrix other than perfectly. This might not disturb us if there was only one just-identified model that could be formulated for a given covariance matrix. But actually there are infinitely many such models. A principal component analysis model, $\mathbf{\Sigma}_{YY}=\mathbf{\Lambda\Lambda}'$, with as many principal component factors as there are observed variables, and the constraints $\mathbf{A}'\mathbf{A}=\mathbf{I}$ and $\mathbf{A}'\mathbf{\Sigma}_{YY}\mathbf{A}=\mathbf{D}$, with $\mathbf{\Lambda}=\mathbf{AD}^{1/2}$, where $\mathbf{A}$ is the eigenvector matrix and $\mathbf{D}$ the eigenvalue matrix of $\mathbf{\Sigma}_{YY}$, is a just-identified model. $\mathbf{A}'\mathbf{A}=\mathbf{I}$ specifies $n(n+1)/2$ constraints (the n diagonal and $n(n-1)/2$ off-diagonal elements of $\mathbf{I}$). $\mathbf{A}'\mathbf{\Sigma}_{YY}\mathbf{A}=\mathbf{D}$ specifies the $n(n-1)/2$ distinct off-diagonal elements of $\mathbf{D}$ to be zero. Hence, there are $n(n+1)/2+n(n-1)/2=n^2$ constraints. All models are just-identified when based on nonsingular linear transforms of the components $\mathbf{\Sigma}_{YY}=(\mathbf{\Lambda T})(\mathbf{T}^{-1}\mathbf{T}'^{-1})(\mathbf{T}'\mathbf{\Lambda}')$, where $\mathbf{T}$ is any nonsingular $n\times n$ matrix. The constraints imposed to achieve a just-identified condition may reflect simply a subjective view of the researcher.

This is especially evident when $\mathbf{\Lambda}_c=\mathbf{\Lambda T}$ is a Cholesky factor of $\mathbf{\Sigma}_{YY}$. In this case $\mathbf{\Sigma}_{YY}=\mathbf{\Lambda}_c\mathbf{I}\mathbf{\Lambda}_c'$. The matrix $\mathbf{\Lambda}_c$ is lower triangular with $n(n+1)/2$ nonzero elements on and below the diagonal. These are the elements that are estimated, while the zero elements above the diagonal of $\mathbf{\Lambda}_c$ and the off-diagonal 0s and the diagonal 1s of the identity matrix $\mathbf{I}$ are fixed. Hence, there are as many distinct elements to estimate in $\mathbf{\Lambda}$ as there are distinct elements in $\mathbf{\Sigma}_{YY}$. This implies as many knowns as there are unknowns (estimated parameters) to solve for.

Because of the relation $\mathbf{\Sigma}=\mathbf{\Sigma}(\boldsymbol{\theta})$, all solutions for an overidentified parameter $\hat{\boldsymbol{\theta}}_k$ will be consistent across these different subsets of equations that yield solutions for $\hat{\boldsymbol{\theta}}_k$. But that might not be the case if we use an empirical covariance matrix such as $\mathbf{S}$ (a sample covariance matrix)—or $\mathbf{\Sigma}$ (the corresponding population covariance matrix)—in place of $\hat{\mathbf{\Sigma}}_0$ (the reproduced covariance matrix) and seek variant solutions for an overidentified parameter with the corresponding subsets of the elements of the empirical covariance matrix.

If we obtain inconsistent (i.e., different) values for the overidentified parameter from different subsets of equations, this alerts us to the fact that the empirical covariance matrix is not consistent with the model. Thus the goal is to overidentify as many of the free parameters as is possible in order to test the model, where testing a model implies possibly disconfirming it by showing that it is inconsistent with the model. Now, we do not, in actual practice, evaluate models this way. The inconsistency in solutions for values of overidentified parameters translates itself into lack of fit between the reproduced model covariance matrix, $\hat{\boldsymbol{\Sigma}}_0$, that minimizes a discrepancy function in the parameter estimation process and the sample covariance matrix $\mathbf{S}$ (that we work with in lieu of the population covariance matrix $\boldsymbol{\Sigma}$). We use this lack of fit as evidence for the misspecification of a model.

Because the estimated values of the free and yoked parameters $\hat{\boldsymbol{\theta}}$ are those values that minimize the discrepancy function conditional on the fixed parameters $\boldsymbol{\theta}^*$ and the equality and inequality constraints, we require the free parameters to all be at least identified. Otherwise, we will be unable to distinguish between distinct models having the same functional relations fixed parameters and parameter constraints, but different values for the free and yoked parameters that nevertheless reproduce the same $\hat{\boldsymbol{\Sigma}}_0$ and consequently have the same fit to $\mathbf{S}$. By having identified models, we put the onus of any lack of fit squarely onto the fixed parameters and constraints specified by hypothesis.

15.3.3 Determining Whether Parameters and Models Are Identified

So far, we have only defined the meaning of identified parameters. We have not discussed how to determine whether a given model is identified. In theory one can establish that a parameter is identified by taking a hypothetical covariance matrix $\boldsymbol{\Sigma}=\boldsymbol{\Sigma}(\boldsymbol{\theta})$, the constraints on the model parameters, the individual model equations for determining each element of $\boldsymbol{\Sigma}$ and then finding some subset of these equations by which one can solve uniquely for a parameter in question. Except for very simple cases, such as establishing that the parameters of a single common-factor model for three variables is identified, this approach to identification can become horrendously complex and burdensome.

Nevertheless, because it is an important concept, we will illustrate this brute force approach to identification: Suppose we have two variables, Y_1 and Y_2, and hypothesize that they have a single common factor. Going to the level of individual model equations for each variable, we obtain

$$\begin{aligned} Y_1 &= \lambda_1 X + \psi_1 E \\ Y_2 &= \lambda_2 X + \psi_2 E \end{aligned} \tag{15.4}$$

where

X is the common factor
E_1 and E_2 are unique factors
λ_1 and λ_2 are common-factor-pattern coefficients
ψ_1 and ψ_2 are unique-factor-pattern coefficients

We presume (without loss of generality) that all variables have zero means, which implies, e.g., that $s^2(Y) = E(Y^2)$. We also presume that $\sigma^2(X) = 1$, and $\sigma^2(E_1) = \sigma^2(E_2) = 1$. Before going further, we need to consider simplifying our notation: by σ_i^2 we shall mean $\sigma^2(Y_i)$; by σ_{ij} we shall mean $\sigma(Y_i, Y_j)$. Now from Equation 15.4 and the usual assumptions of no correlation between common and unique factors or between distinct unique factors, we may derive the variances and covariances for these two variables:

$$\sigma_1^2 = \mathrm{E}(Y_1^2) = E(\lambda_1 X + \psi_1 E_1)^2 = \lambda_1^2 + \psi_1^2$$

By similar reasoning we get

$$\sigma_2^2 = \lambda_1^2 + \psi_2^2$$

$$\sigma_{12} = \lambda_1 \lambda_2$$

This represents a system of three equations in four unknowns, λ_1, λ_2, ψ_1, and ψ_2. There is no way we can solve uniquely for λ_1, λ_2, ψ_1, and ψ_2. These parameters are "underidentified" and the model is consequently "underidentified." On the other hand, if we fix $\lambda_2 = 1$ in addition to other constraints, then a solution for λ_1 is possible, viz. $\lambda_1 = \sigma_{12}$. But in some cases the solution for ψ_2^2 namely, $\psi_2^2 = \sigma_2^2 - 1$, may be inadmissible when $\sigma_2^2 < 1$, implying a negative variance for the unique-variance component of Y_2. Fixing both $\sigma^2(X) = 1$ and $\lambda_2 = 1$ also represents a fairly strong hypothesis, in effect, that when the variance of the common factor is unity, a unit change in X leads to a unit change in Y_2. This may not be a hypothesis one wishes to test. One can also achieve identification by constraining $\lambda_1 = \lambda_2$ In that case, $\lambda_1 = \lambda_2 = \sqrt{\sigma_{12}}$. This again is a strong hypothesis that may not represent a hypothesis one wishes to test. Furthermore if $\sigma_{12} < 0$, then λ_1 and λ_2 will take on inadmissible, imaginary values.

Let us next consider the case where we have not two but three variables, Y_1, Y_2, and Y_3. We will try to fit a single common-factor model to these three variables. We will similarly assume zero means for variables, that variances of the common factor and the unique variances are unity. Thus we have the following model equations:

$$Y_1 = \lambda_1 X + \psi_1 E_1$$

$$Y_2 = \lambda_2 X + \psi_2 E_2$$

$$Y_3 = \lambda_3 X + \psi_3 E_3 \qquad (15.5)$$

from which we can derive, using the assumption of unit variances for the common and unique factors and the usual assumptions of the common-factor model concerning correlations between common and unique factors, expressions for the variances and covariances among these observed variables as functions of the model parameters:

$$\sigma_1^2 = \lambda_1^2 + \psi_1^2$$

$$\sigma_2^2 = \lambda_2^2 + \psi_2^2$$

$$\sigma_3^2 = \lambda_3^2 + \psi_3^2$$

$$\sigma_{12} = \lambda_1\lambda_2$$

$$\sigma_{13} = \lambda_1\lambda_3$$

$$\sigma_{23} = \lambda_2\lambda_3$$

Here we have six equations in six unknowns. We can solve for each of the unknowns in terms of the observed variances and covariances: From the equation $\sigma_{12}=\lambda_1\lambda_2$, we can obtain $\lambda_2=\sigma_{12}/\lambda_1$, and from $\sigma_{13}=\lambda_1\lambda_3$, we can obtain $\lambda_3=\sigma_{13}/\lambda_1$. Substituting these two expressions into the equation $\sigma_{23}=\lambda_2\lambda_3$, we get

$$\sigma_{23} = \frac{\sigma_{13}\sigma_{12}}{\lambda_1^2}$$

or

$$\lambda_1^2 = \frac{\sigma_{13}\sigma_{12}}{\sigma_{23}},$$

hence

$$\lambda_1 = \sqrt{\frac{\sigma_{13}\sigma_{12}}{\sigma_{23}}}.$$

By similar arguments, we can obtain

$$\lambda_2 = \sqrt{\frac{\sigma_{23}\sigma_{12}}{\sigma_{13}}},$$

and

$$\lambda_3 = \sqrt{\frac{\sigma_{23}\sigma_{13}}{\sigma_{12}}}$$

Once we have solutions for the common-factor-pattern coefficients, the unique variances are readily obtained from the equations for the variances. For example, from

$$\sigma_1^2 = \lambda_1^2 + \psi_1^2$$

and the equation

$$\lambda_1^2 = \frac{\sigma_{13}\sigma_{12}}{\sigma_{23}},$$

we obtain

$$\sigma_1^2 = \frac{\sigma_{13}\sigma_{12}}{\sigma_{23}} + \psi_1^2,$$

or

$$\psi_1^2 = \sigma_1^2 = -\frac{\sigma_{13}\sigma_{12}}{\sigma_{23}}$$

Similarly

$$\psi_2^2 = \sigma_2^2 = -\frac{\sigma_{23}\sigma_{12}}{\sigma_{13}}$$

$$\psi_3^2 = \sigma_3^2 - \frac{\sigma_{23}\,\sigma_{13}}{\sigma_{12}}$$

We will now illustrate an overidentified model. Given four observed variables, $Y_1,\ldots,Y_4$ and a model in which each shares a single common factor, along with the usual assumptions of unit variances for common and unique factors and zero correlations between common and unique factors and among unique factors

$$Y_1 = \lambda_1 X + \psi_1 E_1$$

$$Y_2 = \lambda_2 X + \psi_2 E_2$$

$$Y_3 = \lambda_3 X + \psi_3 E_3$$

$$Y_4 = \lambda_4 X + \psi_4 E_4$$

we can obtain expressions for the variances and covariances among these variables in terms of the model's parameters:

$$\sigma_1^2 = \lambda_1^2 + \psi_1^2 \tag{15.6a}$$

$$\sigma_2^2 = \lambda_2^2 + \psi_2^2 \tag{15.6b}$$

$$\sigma_3^2 = \lambda_3^2 + \psi_3^2 \tag{15.6c}$$

$$\sigma_4^2 = \lambda_4^2 + \psi_4^2 \tag{15.6d}$$

$$\sigma_{12} = \lambda_1\lambda_2 \tag{15.6e}$$

$$\sigma_{13} = \lambda_1\lambda_3 \tag{15.6f}$$

$$\sigma_{14} = \lambda_1\lambda_4 \tag{15.6g}$$

$$\sigma_{23} = \lambda_2\lambda_3 \tag{15.6h}$$

$$\sigma_{24} = \lambda_2\lambda_4 \tag{15.6i}$$

$$\sigma_{34} = \lambda_3\lambda_4 \tag{15.6j}$$

Here we have 10=4(4+1)/2 equations (the number of distinct elements in the observed variables covariance matrix) in eight unknowns. We have more equations than unknowns, which is a necessary but not sufficient condition that some parameters will be overidentified.

We know from our work with the previous 3-variable case that there is a solution for λ_1: Using Equation 15.6e, f, h, we can obtain

$$\lambda_1 = \sqrt{\frac{\sigma_{13}\,\sigma_{12}}{\sigma_{23}}}$$

But we can also obtain an alternative solution using a different subset of the equations—Equation 15.6f, g, j:

$$\lambda_1 = \sqrt{\frac{\sigma_{13}\,\sigma_{14}}{\sigma_{34}}}$$

A third solution for λ_1 is obtained using Equation 15.6e, g, i:

$$\lambda_1 = \sqrt{\frac{\sigma_{12}\sigma_{14}}{\sigma_{24}}}$$

Thus λ_1 is overidentified.

In similar ways, we can find three distinct solutions for each of the other three common-factor-pattern coefficients, implying that each of them is also overidentified. Consequently, the one-factor model for four indicator variables is overidentified. It can be disconfirmed when applied to four-indicator covariance matrices that are not generated by this model.

Having only three indicators of a common factor will lead to a just-identified single-factor model, having as many unknowns to solve for as there are independent data points to determine them with. In such circumstances, as long as the covariances are all nonzero, one can always solve for the parameters, which in turn will perfectly reproduce the covariances. In these cases, there is no logical possibility of disconfirming the single-factor hypothesis, so no test of it is possible. Four indicators, on the other hand, will yield an overidentified single-factor model, with more than one subset of equations available to solve for a parameter, the solutions to which may be inconsistent when the single common-factor model is not appropriate for the empirical covariances. Consequently, researchers are encouraged, in formulating confirmatory factor analysis models, to include at least four indicators of each factor in the model. This allows one to test the hypothesis that a single common factor underlies each of the indicators of it. With only three indicators, one would always be able to fit a single factor model to them (as long as they have nonzero correlations). If one can have more than four indicators, that is even better, because that creates more overidentified conditions by which one's hypothesis of a single common factor may be tested.

Working out identification for these simple two-, three-, and four-variable cases with single common factors allows us to generalize to other situations: The following model is underidentified if $\phi_{12}=0$:

$$\mathbf{\Lambda}=\begin{bmatrix}\lambda_{11} & 0\\ \lambda_{21} & 0\\ 0 & \lambda_{32}\\ 0 & \lambda_{42}\end{bmatrix},\qquad \mathbf{\Phi}_{XX}=\begin{bmatrix}1 & \phi_{12}\\ \phi_{21} & 1\end{bmatrix},\qquad \mathbf{\psi}^2=\begin{bmatrix}\psi_1^2 & 0 & 0 & 0\\ 0 & \psi_2^2 & 0 & 0\\ 0 & 0 & \psi_3^2 & 0\\ 0 & 0 & 0 & \psi_4^2\end{bmatrix}$$

If $\phi_{12}=0$, $\sigma_{13}=\sigma_{14}=\sigma_{23}=\sigma_{24}=0$, and the estimation of λ_{11} and λ_{21} breaks down because we have only the block of variances and covariances among Y_1 and Y_2 to work with, which we know from analysis of the two indicator model is an underidentified case. A parallel breakdown in the estimation of λ_{32} and λ_{42} occurs as well. On the other hand, when $\phi_{12}\neq 0$, all of the free parameters are overidentified, with estimates of pattern coefficients λ_{11} and λ_{21} and λ_{32} and λ_{42}, respectively, made possible by covariation of variables Y_1 and Y_2 with Y_3 and Y_4. For example

$$\lambda_{11}=+\sqrt{\frac{\sigma_{41}\sigma_{21}}{\sigma_{42}}}=+\sqrt{\frac{\sigma_{31}\sigma_{21}}{\sigma_{32}}},\qquad \psi_1^2=\sigma_1^2-\frac{\sigma_{41}\sigma_{21}}{\sigma_{42}}$$

$$\lambda_{21}=+\sqrt{\frac{\sigma_{42}\sigma_{21}}{\sigma_{41}}}=+\sqrt{\frac{\sigma_{32}\sigma_{21}}{\sigma_{31}}},\qquad \psi_2^2=\sigma_2^2-\frac{\sigma_{42}\sigma_{21}}{\sigma_{41}}$$

$$\lambda_{32}=+\sqrt{\frac{\sigma_{31}\sigma_{43}}{\sigma_{41}}}=+\sqrt{\frac{\sigma_{32}\sigma_{43}}{\sigma_{42}}},\qquad \psi_3^2=\sigma_3^2-\frac{\sigma_{31}\sigma_{43}}{\sigma_{41}}$$

$$\lambda_{42}=+\sqrt{\frac{\sigma_{41}\sigma_{43}}{\sigma_{31}}}=+\sqrt{\frac{\sigma_{42}\sigma_{43}}{\sigma_{32}}},\qquad \psi_4^2=\sigma_4^2-\frac{\sigma_{41}\sigma_{43}}{\sigma_{31}}$$

$$\phi_{12} = \frac{\sigma_{31}}{\sqrt{\frac{\sigma_{41}\sigma_{21}}{\sigma_{42}}}\sqrt{\frac{\sigma_{31}\sigma_{43}}{\sigma_{41}}}} = \frac{\sigma_{41}}{\sqrt{\frac{\sigma_{31}\sigma_{21}}{\sigma_{32}}}\sqrt{\frac{\sigma_{42}\sigma_{43}}{\sigma_{32}}}}$$

$$\phi_{12} = \frac{\sigma_{32}}{\sqrt{\frac{\sigma_{42}\sigma_{21}}{\sigma_{41}}}\sqrt{\frac{\sigma_{31}\sigma_{43}}{\sigma_{41}}}} = \frac{\sigma_{42}}{\sqrt{\frac{\sigma_{32}\sigma_{21}}{\sigma_{31}}}\sqrt{\frac{\sigma_{41}\sigma_{43}}{\sigma_{31}}}}$$

The following model is identified whether ϕ_{12} is zero or not:

$$\mathbf{\Lambda} = \begin{bmatrix} \lambda_{11} & 0 \\ \lambda_{21} & 0 \\ \lambda_{31} & 0 \\ 0 & \lambda_{42} \\ 0 & \lambda_{52} \\ 0 & \lambda_{62} \end{bmatrix}, \quad \mathbf{\Phi}_{XX} = \begin{bmatrix} 1 & \phi_{12} \\ \phi_{21} & 1 \end{bmatrix}, \quad \mathbf{\Psi}^2 = \begin{bmatrix} \psi_1^2 & 0 & 0 & 0 & 0 & 0 \\ 0 & \psi_2^2 & 0 & 0 & 0 & 0 \\ 0 & 0 & \psi_3^2 & 0 & 0 & 0 \\ 0 & 0 & 0 & \psi_4^2 & 0 & 0 \\ 0 & 0 & 0 & 0 & \psi_5^2 & 0 \\ 0 & 0 & 0 & 0 & 0 & \psi_6^2 \end{bmatrix}$$

With three indicators per factor there is sufficient information within the block of variances and covariances among the indicators of a factor to find solutions for the factor-pattern coefficients. The solution does not depend on using information about relations with the second set of indicators.

15.3.4 Identification of Metrics

So far we have considered identification in the context of assuming that the common factors have unit variances. The effect of fixing the variances of latent variables to unity is to fix their "metric," i.e., their units of measurement. Because a factor-pattern coefficient indicates how many units of change will occur in an indicator variable given a unit change in the value of the common factor, the value of a factor-pattern coefficient depends on the units of measurement of both the indicator variable and the common factor. In most single sample studies, fixing the variance of the common factor arbitrarily to unity, is sufficient to establish a metric for it, and this aids the identification of other parameters involving the factor as well. But it is quite possible to fix the metric of the common factor in another way: by fixing the pattern loading of one of its indicators while freeing the variance of the common factor.

When fixing a metric for the common factor, one should fix only one parameter—a pattern loading on that factor or the variance of the factor, but not both. Otherwise one specifies a rather restricted model that may not be of interest at all. For example, if we set both the pattern loading and the factor variance to unity, this says that when, in the metric that gives it unit variance, the common factor changes one unit, the respective indicator variable changes by 1 unit. That is a very strong hypothesis, particularly in connection with the variance of the observed indicator.

There are occasions when one especially wishes to fix the pattern coefficient and not the factor variance. Suppose in a previous study one has obtained an exploratory factor analytic solution and wishes to see if the same solution will be found in a new setting. One will fix the pattern loadings on the factors to the values found in the previous study, while freeing the factor variances. Recall from the chapter on Factorial Invariance that according to Meredith (1964a, 1964b) factor-pattern coefficients are invariant under selection effects on the factors while the variances and covariances among the factors are not. So, we might expect to find changes in the factor variances and covariances in the new setting. Leaving these parameters free while fixing the pattern coefficients to values obtained from the previous study allows us to see if there are any effects on the factor variances and covariances.

We may also wish to test the hypothesis of whether there has been any change from the previous to the new setting. To do this, we can free up all the pattern loadings except r of them on each factor, with the fixed values of the pattern coefficients set to their values obtained in the previous study. Choosing the largest loading and $r-1$ lowest loadings on indicators of other factors to fix in each column while freeing the rest would be my recommendation (see also Jöreskog, 1979), although further study of this problem is required. Note that this fixes r^2 parameters. This fixes both the factor metric and the solution for the pattern loadings to what it would be in an unrestricted model rotated provisionally to the position of the previous study as defined by the fixed parameters. We leave the factor variances and covariances free. We then compare this unrestricted solution against the more restricted solution to see where any changes may have come into play. Changes may come in the form of different pattern loadings and different factor variances and covariances.

Many models studied by researchers are themselves formed by joining together several simpler models. Frequently when it is known that the simpler models are all identified by themselves, it is likely then that the parameters of the more complex model formed from them are identified. But one would like more assurances than that. Most contemporary programs for confirmatory factor analysis and linear structural equations modeling (a generalization of factor analysis) check for identification by examining the "information matrix" to see if it is positive definite.

According to Jöreskog and Sörbom (1989), "the information matrix is the probability limit of the matrix of second-order derivatives of the fit function used to estimate the model" (p. 17). Another way of putting it is that the information matrix contains estimates of the variances and covariances among the parameter estimates, and is derived from the matrix of second derivatives of the fit function with respect to the free parameters evaluated at the values for the parameters that minimize the fit function. (More about this matrix will be given in a later section.) This matrix is of the order of the number of free and distinct yoked parameters. Jöreskog and Sörbom (1989) then say, "If the model is identified, the information matrix is almost certainly positive definite. If the information matrix is singular, the model is not identified, and "the rank of the information matrix" indicates how many parameters are identified" (p. 17).

Another way of evaluating whether a model is identified, according to Jöreskog and Sörbom (1989), is to arbitrarily assign reasonable values to the free parameters and then using both fixed and assigned free parameters, generate a reproduced covariance matrix for the raw variables presumed dependent on these parameters. Next, take this reproduced covariance matrix and subject it to one's structural model, with free parameters now unknowns to be determined. If the solution does not yield the same parameters that generated the covariance matrix in the first place, then the model is likely not identified.

15.3.5 Discrepancy Functions

When measuring a model's fit to the sample variance–covariance matrix **S**, we want to regard lack of fit as due exclusively to misspecifying certain constraints on certain parameters. Thus the values for the free parameters are required to be those values for these parameters that uniquely minimize the discrepancy between the model's reproduced covariance matrix, $\hat{\mathbf{\Sigma}}_0$, and the sample covariance matrix, **S**, conditional on the explicitly constrained parameters of the model. The reason for conditioning on the constraints is that we want the estimates to be dependent on the constraints. Then any discrepancy will be due to the constraints. Conversely, if no overidentifying constraints are given, the model will be free to fit the data perfectly, but in that case nothing is hypothesized and nothing is tested. The discrepancy is measured with a discrepancy function, $F[\hat{\mathbf{\Sigma}}_0(\boldsymbol{\theta}),\mathbf{S}]$, where $\boldsymbol{\theta}$ is a vector $\boldsymbol{\theta}=(\hat{\boldsymbol{\theta}}, \boldsymbol{\theta}^*)$, with $\hat{\boldsymbol{\theta}}$ the independent free and yoked parameters and $\boldsymbol{\theta}^*$ the fixed parameters of the structural equation model.

The most frequently used discrepancy functions are the following:

Ordinary least squares

$$L = \tfrac{1}{2}\,\mathrm{tr}[(\mathbf{S} - \hat{\mathbf{\Sigma}}_0(\theta))'(\mathbf{S} - \hat{\mathbf{\Sigma}}_0(\theta))]$$

Maximum likelihood

$$F = \ln\left|\hat{\mathbf{\Sigma}}_0(\boldsymbol{\theta})\right| + \mathrm{tr}(\mathbf{S}\hat{\mathbf{\Sigma}}_0(\boldsymbol{\theta})^{-1}) - \ln|\mathbf{S}| - p$$

Generalized least squares

$$\begin{aligned} G &= \tfrac{1}{2}\,\mathrm{tr}[\mathbf{S}^{-1/2}(\mathbf{S} - \hat{\mathbf{\Sigma}}_0(\boldsymbol{\theta}))'\mathbf{S}^{-1/2}\mathbf{S}^{-1/2}(\mathbf{S} - \hat{\mathbf{\Sigma}}_0(\boldsymbol{\theta}))\mathbf{S}^{-1/2}] \\ &= \tfrac{1}{2}\,\mathrm{tr}[(\hat{\mathbf{\Sigma}}_0(\boldsymbol{\theta}) - \mathbf{S})\mathbf{S}^{-1}]^2 = \tfrac{1}{2}\,\mathrm{tr}[(\hat{\mathbf{\Sigma}}_0(\boldsymbol{\theta})\mathbf{S}^{-1} - \mathbf{I})]^2 \end{aligned}$$

where

S is the sample estimate of the unrestricted variance–covariance matrix

$\hat{\mathbf{\Sigma}}_0(\boldsymbol{\theta})$ is the reproduced variance–covariance matrix under a hypothesis, generated as a multidimensional function of the free parameters of the structural model arranged for convenience in a single vector $\boldsymbol{\theta}$, with fixed parameters and constraints on parameters implicitly in the function

$\boldsymbol{\Sigma}_{YY} = \hat{\boldsymbol{\Sigma}}_0(\boldsymbol{\theta})$ means that each element of the variance covariance matrix, e.g., σ_{ij} is a certain function $\sigma_{ij} = \sigma_{ij}(\boldsymbol{\theta})$ of the model parameters. When the values of the free parameters are chosen to minimize a discrepancy function conditional on the explicit constraints on the parameters of the model, we say that the free parameters are estimated according to that discrepancy function.

Ordinary least squares seeks to minimize the sum of squared residuals remaining in the observed covariance matrix, **S**, after subtracting from it the reproduced model covariance matrix, $\hat{\boldsymbol{\Sigma}}_0$. This method of estimation has the advantage of requiring no distributional assumptions and it is usually easier to work out the estimating equations and their implementation. Its main disadvantages are that when multivariate normality cannot be presumed and the matrix analyzed is not a covariance matrix, it lacks a distributional theory within which one can make probabilistic inferences about fit. In those cases, there is no significance test for fit. The assessment of fit is also affected by arbitrary choices for the units of measurement of the variables. Fit is usually judged according to the closeness of the approximation to **S**.

Maximum-likelihood estimation seeks simultaneously to minimize the difference between $\ln|\mathbf{S}|$ and $\ln|\hat{\boldsymbol{\Sigma}}_0|$ and between $\mathrm{tr}(\mathbf{S}\hat{\boldsymbol{\Sigma}}_0^{-1})$ and $\mathrm{tr}(\mathbf{I}) = n$. These differences will be zero if $\mathbf{S} = \hat{\boldsymbol{\Sigma}}_0$. Maximum-likelihood estimation requires the making of distributional assumptions and is usually more difficult to work with in terms of developing estimating equations and algorithms. But if one can make distributional assumptions, then one gains in the ability to perform probabilistic inferences about the degree of fit, although this may require fairly large samples. Measures of fit based on the likelihood ratio are also metric invariant.

Generalized least squares is a variant of ordinary least squares. As in ordinary least squares, one seeks to minimize the sum of squared residuals. But in generalized least squares the residuals are transformed by pre- and post-multiplying the residual matrix by the inverse of the square root of the sample covariance matrix before evaluating their sum of squares. Generalized least squares has the advantage of not requiring distributional assumptions but still allows for probabilistic inference about model fit. The complexity of its estimating equations is usually intermediate between those of ordinary least-squares and maximum-likelihood estimation. Probabilistic inference, however, may require larger samples than needed for maximum-likelihood estimation with the elliptical distribution (Browne, 1982, 1984).

There are other more esoteric discrepancy functions with such forms of estimation as asymptotic distribution-free (ADF) estimation (Browne, 1974, 1977, 1984), maximum-likelihood estimation with the elliptical distribution. We will not discuss these here.

15.3.6 Estimation by Minimizing Discrepancy Functions

Maximum-likelihood estimation proceeds by seeking to minimize the following discrepancy function

$$F_{\mathrm{ML}}(\hat{\mathbf{\Sigma}}_{YY}) = \ln\left|\hat{\mathbf{\Sigma}}_{YY}\right| + \mathrm{tr}(\hat{\mathbf{\Sigma}}_{YY}^{-1}\mathbf{S}) - \ln|\mathbf{S}| - n \tag{15.7}$$

where

$\hat{\mathbf{\Sigma}}_{YY}$ is the estimated model variance–covariance matrix among the observed variables

$\mathbf{S}$ is the sample unrestricted estimate of the population covariance matrix for the observed variables

n is the number of manifest variables

Since a function is to be minimized, differential calculus will be needed. Further, because there will be multiple independent variables for the minimization (the free parameters), partial derivatives will be required to find equations and algorithms for the minimization process. Because the iterative algorithms we will use are based on the quasi-Newton methods, only first-derivatives of the discrepancy functions with respect to the free parameters will be needed.

The derivative of the maximum-likelihood discrepancy function, F, with respect to an arbitrary parameter, θ_i, of the model is given by

$$\frac{\partial F}{\partial \theta_i} = \mathrm{tr}\left[(\mathbf{\Sigma}_{YY}^{-1} - \mathbf{\Sigma}_{YY}^{-1}\mathbf{S}\mathbf{\Sigma}_{YY}^{-1})\frac{\partial \mathbf{\Sigma}_{ZZ}}{\partial \theta_i}\right] = \mathrm{tr}\left[\mathbf{Q}\frac{\partial \mathbf{\Sigma}_{YY}}{\partial \theta_i}\right] \tag{15.8}$$

15.3.7 Derivatives of Elements of Matrices

In obtaining algorithms for the minimization of discrepancy functions over variations in the values of estimated parameters, we need to know how to obtain the derivatives of the discrepancy function with respect to the free parameters of the model. To find the first derivatives of a discrepancy function F, we will need to develop rules and notation for the taking of partial derivatives of matrices with respect to scalar values. In this regard, we will follow fairly closely the notation used by Bock and Bargmann (1966, pp. 514–515). Their rules and notation are as follows:

In the following development let the elements of the matrices involved be differentiable functions of a scalar x. Now consider the $n \times r$ matrix $\mathbf{A}$.

$$\frac{\partial \mathbf{A}}{\partial x} = \left[\frac{\partial a_{ij}}{\partial x}\right]_{(n \times r)} \quad \begin{matrix} i = 1, \ldots, n \\ j = 1, \ldots, r \end{matrix}$$

In other words, we take the partial derivative of each element in turn and place it in its respective place in a similarly sized matrix.

Let **A** be an $n \times r$ matrix and **B** a $p \times q$ matrix. Then

$$\frac{\partial(\mathbf{A}+\mathbf{B})}{\partial x} = \frac{\partial \mathbf{A}}{\partial x} + \frac{\partial \mathbf{B}}{\partial x} \quad n = p, \quad r = q \tag{15.9a}$$

$$\frac{\partial \mathbf{AB}}{\partial x} = \mathbf{A}\frac{\partial \mathbf{B}}{\partial x} + \frac{\partial \mathbf{A}}{\partial x}\mathbf{B} \quad r = p \tag{15.9b}$$

$$\frac{\partial \mathbf{A}^{-1}}{\partial x} = -\mathbf{A}^{-1}\frac{\partial \mathbf{A}}{\partial x}\mathbf{A}^{-1} \quad n = r \quad |\mathbf{A}| \neq 0 \tag{15.9c}$$

Let **C** be a constant matrix; then

$$\frac{\partial \mathrm{tr}(\mathbf{AC})}{\partial x} = \mathrm{tr}\frac{\partial \mathbf{A}}{\partial x}\mathbf{C} \tag{15.9d}$$

Given two $n \times n$ square matrices **A** and **C** with **A** nonsingular and **C** a constant matrix, the following is a useful consequence of the relationships cited above:

$$\frac{\partial\ \mathrm{tr}(\mathbf{A}^{-1}\mathbf{C})}{\partial x} = -\mathrm{tr}\left(\mathbf{A}^{-1}\frac{\partial \mathbf{A}}{\partial x}\mathbf{A}^{-1}\mathbf{C}\right) = -\mathrm{tr}\left(\frac{\partial \mathbf{A}}{\partial x}\mathbf{A}^{-1}\mathbf{C}\mathbf{A}^{-1}\right) \tag{15.10a}$$

with the latter expression on the right resulting from the invariance of the trace under cyclic permutation of the matrices. Finally [with proof given by Bock and Bargmann (1966, pp. 514–515)]

$$\frac{\partial \ln|\mathbf{A}|}{\partial x} = -\mathrm{tr}\,\mathbf{A}^{-1}\frac{\partial \mathbf{A}'}{\partial x} \tag{15.10b}$$

Next consider the derivative of an $n \times r$ matrix **A** with respect to one of its elements a_{ij}. If each of the elements of **A** is independent of the other elements

$$\frac{\partial \mathbf{A}'_{ij}}{a_{ij}} = \mathbf{1}_{ij} \quad \text{and} \quad \frac{\partial \mathbf{A}'_{ij}}{a_{ij}} = \mathbf{1}_{ji} \tag{15.11}$$

where $\mathbf{1}_{ij}$ denotes an $n \times r$ matrix with zeros in every position except the *i,j* position, which contains a 1. However, if **A** is square $n \times n$ symmetric

$$\frac{\partial \mathbf{A}}{\partial a_{ij}} = \mathbf{1}_{ij} + \mathbf{1}_{ji} - \mathbf{1}_{ij}\mathbf{1}_{ij} \tag{15.12}$$

where $\mathbf{1}_{ij}\,\mathbf{1}_{ij} = \mathbf{1}_{ij}$ when $i = j$ and $\mathbf{1}_{ij}\,\mathbf{1}_{ij} = \mathbf{0}$ when $i \neq j$, $i,j = 1,\ldots,n$.

We will also need the following result: Let **A** be an $n \times r$ matrix and **B** a $p \times q$ matrix, $\mathbf{1}_{ij}$ an $r \times p$ matrix, and $\mathbf{1}_{ts}$ a $q \times n$ matrix. Then

$$\mathrm{tr}(\mathbf{A1}_{ji}) = [\mathbf{A}]_{ij} \tag{15.13a}$$

For example

$$\mathrm{tr}(\mathbf{A1}_{24}) = [\mathbf{A}]_{42},$$

that is

$$\mathrm{tr}\begin{bmatrix} a_{11} & a_{12} & a_{13} \\ a_{21} & a_{22} & a_{23} \\ a_{31} & a_{32} & a_{33} \\ a_{41} & a_{42} & a_{43} \end{bmatrix}\begin{bmatrix} 0 & 0 & 0 & 0 \\ 0 & 0 & 0 & 1 \\ 0 & 0 & 0 & 0 \end{bmatrix}_{24} = \mathrm{tr}\begin{bmatrix} 0 & 0 & 0 & a_{12} \\ 0 & 0 & 0 & a_{22} \\ 0 & 0 & 0 & a_{32} \\ 0 & 0 & 0 & a_{42} \end{bmatrix} = [\mathbf{A}]_{42}$$

Postmultiplying the matrix $\mathbf{A}$ by the matrix $\mathbf{1}_{ji}$ has the effect of extracting the jth column of $\mathbf{A}$ and inserting it into the ith column of an $n \times n$ null matrix. This leaves only the ith element of the jth column of $\mathbf{A}$ as a diagonal element in the resulting square matrix, hence the trace of this matrix is simply this one element, the ith element of the jth column of $\mathbf{A}$, i.e., $[\mathbf{A}]_{ij}$.

Next

$$\mathrm{tr}(\mathbf{A1}_{ji}\mathbf{B1}_{ts}) = \mathrm{tr}[\mathbf{A}]_{sj}[\mathbf{B}]_{it} = a_{sj}b_{it} \tag{15.13b}$$

This result follows from the fact that $\mathbf{A1}_{ji}$ is an $n \times p$ matrix containing zeros everywhere except in its ith column, which contains the jth column $[\mathbf{A}]_j$ of $\mathbf{A}$, and $\mathbf{B1}_{ts}$ is a $p \times n$ matrix which contains zeros everywhere except in its sth column, which contains the tth column $[\mathbf{B}]_t$ of $\mathbf{B}$. The product of these two resulting matrices is an $n \times n$ matrix containing zeros everywhere except in its sth column, which contains the equivalent of $[\mathbf{A}]_j[\mathbf{B}]_{it}$. The only nonzero diagonal element of the resulting $n \times n$ matrix is in the sth row and sth column and is equal to $[\mathbf{A}]_{sj}[\mathbf{B}]_{it}$. For example

$$\begin{bmatrix} a_{11} & a_{12} & a_{13} \\ a_{21} & a_{22} & a_{23} \\ a_{31} & a_{32} & a_{33} \\ a_{41} & a_{42} & a_{43} \end{bmatrix}\begin{bmatrix} 0 & 0 & 0 & 0 & 0 \\ 0 & 0 & 0 & 1 & 0 \\ 0 & 0 & 0 & 0 & 0 \end{bmatrix}_{24} = \begin{bmatrix} 0 & 0 & 0 & a_{12} & 0 \\ 0 & 0 & 0 & a_{22} & 0 \\ 0 & 0 & 0 & a_{32} & 0 \\ 0 & 0 & 0 & a_{42} & 0 \end{bmatrix}$$

and

$$\begin{bmatrix} b_{11} & b_{12} & b_{13} \\ b_{21} & b_{22} & b_{23} \\ b_{31} & b_{32} & b_{33} \\ b_{41} & b_{42} & b_{43} \\ b_{51} & b_{52} & b_{53} \end{bmatrix}\begin{bmatrix} 0 & 1 & 0 & 0 \\ 0 & 0 & 0 & 0 \\ 0 & 0 & 0 & 0 \end{bmatrix}_{12} = \begin{bmatrix} 0 & b_{11} & 0 & 0 \\ 0 & b_{21} & 0 & 0 \\ 0 & b_{31} & 0 & 0 \\ 0 & b_{41} & 0 & 0 \\ 0 & b_{51} & 0 & 0 \end{bmatrix}$$

hence

$$\mathrm{tr}(\mathbf{A1}_{24}\mathbf{B1}_{12}) = \mathrm{tr}\left\{\begin{bmatrix} 0 & 0 & 0 & a_{12} & 0 \\ 0 & 0 & 0 & a_{22} & 0 \\ 0 & 0 & 0 & a_{32} & 0 \\ 0 & 0 & 0 & a_{42} & 0 \end{bmatrix}\begin{bmatrix} 0 & b_{11} & 0 & 0 \\ 0 & b_{21} & 0 & 0 \\ 0 & b_{31} & 0 & 0 \\ 0 & b_{41} & 0 & 0 \\ 0 & b_{51} & 0 & 0 \end{bmatrix}\right\}$$

or

$$\mathrm{tr}(\mathbf{A1}_{24}\mathbf{B1}_{12}) = \mathrm{tr}\begin{bmatrix} 0 & a_{12}b_{41} & 0 & 0 \\ 0 & a_{22}b_{41} & 0 & 0 \\ 0 & a_{32}b_{41} & 0 & 0 \\ 0 & a_{42}b_{41} & 0 & 0 \end{bmatrix} = [\mathbf{A}]_{22}[\mathbf{B}]_{41}$$

Equation 15.7 generalizes (Mulaik, 1971) to

$$\mathrm{tr}(\mathbf{A1}_{ji}\mathbf{B1}_{ts}\mathbf{C}_{pq}\cdots\mathbf{1}_{mn}\mathbf{Z1}_{uv}) = [\mathbf{A}]_{vj}[\mathbf{B}]_{it}[\mathbf{C}]_{sp}\cdots[\mathbf{Z}]_{nu} \tag{15.14}$$

Note that the right-most subscript on the right-most **1** matrix "cycles" around to the left and the subscripts are then are assigned pairwise from the left to the matrices **A**, **B**, **C**, etc.

The traces of the matrix expressions for the derivatives may be obtained by rearranging these expressions (taking advantage of the invariance of the trace under cyclic permutations) into the form of Equations 15.13 or 15.14 and simplifying the result.

There has been considerable work on the topic of working out first and second derivatives of fit functions in connection with the development of algorithms for nonlinear optimization in the realm of multivariate statistics. McDonald and Swaminathan (1973) produced one of the major advances in this area, directly in the service of developing algorithms for the analysis of covariance structures. These developments have since been expanded and systematized by mathematicians, and readers may consult Rogers (1980) and Magnus and Neudecker (1988) for integrated treatments of this topic.

15.3.8 Maximum-Likelihood Estimation in Confirmatory Factor Analysis

Jöreskog's major breakthroughs in finding parameter estimates for maximum-likelihood exploratory factor analysis, confirmatory factor analysis, and analysis of covariance structures were based on implementing a quasi-Newton algorithm known as the Fletcher-Powell algorithm (1963). Variants of this algorithm are found today in most of the leading programs that permit one to perform confirmatory factor analyses. However, the BGFS (Broyden, Fletcher, Glofarb, Shannon) method, which is similar to the Fletcher–Powell Algorithm, could be used just as well. Because these methods need only the first derivatives of the discrepancy function, we will simplify matters by only discussing how to find the first derivatives of the maximum-likelihood function. Techniques for finding both first and second derivatives (to be implemented in a Newton–Raphson algorithm) have been given by various authors (Bock and Bargman, 1966; Jöreskog, 1969a,b; Mulaik, 1971, 1972; McDonald 1976; McDonald and Swaminathan 1973; Rogers 1980; Magnus and Neudecker 1988). However, the Newton–Raphson method is primarily of theoretical rather than practical interest in confirmatory factor analysis and related techniques and we will not discuss it further here.

We seek maximum-likelihood estimates of the undetermined parameters of the model

$$\mathbf{\Sigma}_0 = \mathbf{\Lambda\Phi\Lambda}' + \mathbf{\Psi}^2 \tag{15.15}$$

where

$\mathbf{\Sigma}_0$ is an $n \times n$ variance–covariance matrix for n observed variables
$\mathbf{\Phi} = \mathbf{\Sigma}_{XX}$ is an $r \times r$ variance–covariance matrix for r latent variables (factors)
$\mathbf{\Psi}^2$ is an $n \times n$ diagonal matrix of unique-factor variances
$\mathbf{\Lambda}$ is an $n \times r$ factor-pattern or experimental-design matrix

We shall assume that we have a sample of N $(n \times 1)$ observation vectors $\mathbf{y}_1, \mathbf{y}_2, \ldots, \mathbf{y}_N$ from a multivariate normal distribution where $E(\mathbf{Y}) = \mathbf{0}$ and $E(\mathbf{YY}') = \mathbf{\Sigma}_{YY}$. The logarithm of the joint-likelihood function is given by

$$\ln L = -\frac{1}{2}Nn\ln(2\pi) - \frac{1}{2}N\ln|\Sigma_0| - \frac{1}{2}\sum_{i=1}^{N}(\mathbf{y}_i - \boldsymbol{\mu})'\mathbf{\Sigma}_0(\mathbf{y}_i - \boldsymbol{\mu}) \tag{15.16}$$

We now need to estimate simultaneously both $\boldsymbol{\mu}$ and $\mathbf{\Sigma}_{YY}$. Morrison (1967) gives the maximum-likelihood estimator for the mean vector, $\boldsymbol{\mu}$, as

$$\hat{\boldsymbol{\mu}} = \frac{1}{N}\sum_{i=1}^{N}\mathbf{y}_i \tag{15.17}$$

which we may substitute into Equation 15.16 in place of $\boldsymbol{\mu}$. Now, the sum of the outer products of the sample observation vectors corrected by the sample mean is

$$m\mathbf{S} = \sum_{i=1}^{N}\mathbf{y}_i\mathbf{y}_i' - N\hat{\boldsymbol{\mu}}\hat{\boldsymbol{\mu}}' \tag{15.18}$$

where $m = N - 1$ and $\mathbf{S}$ is an unbiased estimator of $\boldsymbol{\Sigma}_{YY}$. We may now rewrite Equation 15.15 as

$$\ln L = -K - \tfrac{1}{2} m \ln |\boldsymbol{\Sigma}_0| - \tfrac{1}{2} m \operatorname{tr}(\boldsymbol{\Sigma}_0^{-1}\mathbf{S}) \tag{15.19}$$

where K is the constant expression in Equation 15.16 and the right-most expression on the right is possible because of the invariance of the trace function under cyclic permutation of matrices.

We are now ready to seek the solution for $\boldsymbol{\Sigma}_0$, which maximizes Equation 15.16. However, Jöreskog (1969a,b) shows that the matrix $\hat{\boldsymbol{\Sigma}}_0$ that maximizes the logarithm of the likelihood function also minimizes the function

$$F(\hat{\boldsymbol{\Sigma}}_0, \mathbf{S}) = \ln |\hat{\boldsymbol{\Sigma}}_0| + \operatorname{tr}(\hat{\boldsymbol{\Sigma}}_0^{-1}\mathbf{S}) - \ln |\mathbf{S}| - n \tag{15.20}$$

which is known as a discrepancy function because it compares $\hat{\boldsymbol{\Sigma}}_0$ to $\mathbf{S}$. Note that if $\hat{\boldsymbol{\Sigma}}_0 = \mathbf{S}$, then the function F equals zero, because in this case $\ln|\hat{\boldsymbol{\Sigma}}_0| = \ln|\mathbf{S}|$ and $\operatorname{tr}(\hat{\boldsymbol{\Sigma}}_0^{-1}\mathbf{S}) = \operatorname{tr}(\mathbf{I}) = n$. Jöreskog therefore recommended that $\hat{\boldsymbol{\Sigma}}_0$ be found by minimizing this function.

Let us now obtain expressions for the first derivatives of the function in Equation 15.20 so that we may obtain the values of unspecified parameters of the model of Equation 15.2a (where $\boldsymbol{\Phi} = \boldsymbol{\Sigma}_{XX}$) which minimize this function (known as the "maximum-likelihood fit function"). Let us now consider the first partial derivatives of F with respect to an arbitrary scalar variable θ:

$$\frac{\partial F}{\partial \theta} = \frac{\partial\left[\ln|\hat{\boldsymbol{\Sigma}}_0| - \ln|S| + \operatorname{tr}(\hat{\boldsymbol{\Sigma}}_0 S) - n\right]}{\partial \theta}$$

$$= \frac{\partial \ln|\hat{\boldsymbol{\Sigma}}_0|}{\partial \theta} - \frac{\partial \ln|\mathbf{S}|}{\partial \theta} + \frac{\partial \operatorname{tr}(\hat{\boldsymbol{\Sigma}}_0^{-1}\mathbf{S})}{\partial \theta} - \frac{\partial n}{\partial \theta}$$

The second and fourth expressions on the right are zero because they are partial derivatives of constants. So, after using Equations 15.9a and b, we obtain (remembering the symmetry of $\boldsymbol{\Sigma}_{YY}$):

$$\frac{\partial F}{\partial \theta} = \operatorname{tr}\left(\hat{\boldsymbol{\Sigma}}_0^{-1}\frac{\partial \hat{\boldsymbol{\Sigma}}_0}{\partial \theta}\right) - \operatorname{tr}\left(\hat{\boldsymbol{\Sigma}}_0^{-1}\mathbf{S}\hat{\boldsymbol{\Sigma}}_0^{-1}\frac{\partial \hat{\boldsymbol{\Sigma}}_0}{\partial \theta}\right)$$

or

$$\frac{\partial F}{\partial \theta} = \operatorname{tr}\left[\left(\hat{\boldsymbol{\Sigma}}_0^{-1} - \hat{\boldsymbol{\Sigma}}_0^{-1}\mathbf{S}\hat{\boldsymbol{\Sigma}}_0^{-1}\right)\frac{\partial \hat{\boldsymbol{\Sigma}}_0}{\partial \theta}\right] \tag{15.21}$$

Next consider the following partial derivatives of the matrix $\hat{\boldsymbol{\Sigma}}_0$ taken with respect to the elements of $\boldsymbol{\Lambda}$, $\boldsymbol{\Phi}$, and $\boldsymbol{\Psi}^2$, respectively:

$$\frac{\partial \hat{\boldsymbol{\Sigma}}_0}{\partial \lambda_{ij}} = \frac{\partial(\boldsymbol{\Lambda\Phi\Lambda}' + \boldsymbol{\Psi}^2)}{\partial \lambda_{ij}} = \boldsymbol{\Lambda\Phi 1}_{ji} + \boldsymbol{1}_{ij}\boldsymbol{\Phi\Lambda}' \quad i = 1,\ldots,n, \quad j = 1,\ldots, r \tag{15.22a}$$

$$\frac{\partial \hat{\boldsymbol{\Sigma}}_0}{\partial \phi_{gh}} = \boldsymbol{\Lambda}(\boldsymbol{1}_{gh} + \boldsymbol{1}_{hg} - \boldsymbol{1}_{gh}\boldsymbol{I1}_{gh})\boldsymbol{\Lambda}' \quad g, h = 1,\ldots,r \tag{15.22b}$$

$$\frac{\partial \hat{\boldsymbol{\Sigma}}_0}{\partial \psi_{ii}^2} = \boldsymbol{1}_{ii} \quad i = 1,\ldots n \tag{15.22c}$$

Now, define $\mathbf{Q} = (\hat{\boldsymbol{\Sigma}}_0^{-1} - \hat{\boldsymbol{\Sigma}}_0^{-1}\mathbf{S}\hat{\boldsymbol{\Sigma}}_0^{-1})$ and $\mathbf{M} = \hat{\boldsymbol{\Sigma}}_0^{-1}\mathbf{S}\hat{\boldsymbol{\Sigma}}_0^{-1}$.

Then by Equations 15.8 or 15.21

$$\frac{\partial F}{\partial \lambda_{ij}} = \mathrm{tr}\left[\left(\hat{\boldsymbol{\Sigma}}_0^{-1} - \hat{\boldsymbol{\Sigma}}_0^{-1}\mathbf{S}\hat{\boldsymbol{\Sigma}}_0^{-1}\right)\frac{\partial \hat{\boldsymbol{\Sigma}}_0}{\partial \lambda_{ij}}\right]$$

$$= \mathrm{tr}\left[\left(\hat{\boldsymbol{\Sigma}}_0^{-1} - \hat{\boldsymbol{\Sigma}}_0^{-1}\mathbf{S}\hat{\boldsymbol{\Sigma}}_0^{-1}\right)\left(\boldsymbol{\Lambda\Phi 1}_{ji} + \boldsymbol{1}_{ij}\boldsymbol{\Phi\Lambda}'\right)\right]$$

by Equation 15.22a, then by substituting $\mathbf{Q}$ and distributing it

$$= \mathrm{tr}\left[\mathbf{Q}\left(\boldsymbol{\Lambda\Phi 1}_{ji} + \boldsymbol{1}_{ij}\boldsymbol{\Phi\Lambda}'\right)\right] = \mathrm{tr}\left[\mathbf{Q}\boldsymbol{\Lambda\Phi 1}_{ji} + \mathbf{Q}\boldsymbol{1}_{ij}\boldsymbol{\Phi\Lambda}'\right]$$

then, because traces of sums equal sums of traces, and by the cyclic invariance of traces

$$= \mathrm{tr}\left[\mathbf{Q}\boldsymbol{\Lambda\Phi 1}_{ji}\right] + \mathrm{tr}\left[\mathbf{Q}\boldsymbol{1}_{ij}\boldsymbol{\Phi\Lambda}'\right] = \mathrm{tr}\left[\mathbf{Q}\boldsymbol{\Lambda\Phi 1}_{ji}\right] + \mathrm{tr}\left[\boldsymbol{1}_{ij}\boldsymbol{\Phi\Lambda}'\mathbf{Q}\right]$$

but then, because of invariance of traces under transposition

$$= \mathrm{tr}\left[\mathbf{Q}\boldsymbol{\Lambda\Phi 1}_{ji}\right] + \mathrm{tr}\left[\mathbf{Q}\boldsymbol{\Lambda\Phi 1}_{ji}\right] = 2\ \mathrm{tr}\left[\mathbf{Q}\boldsymbol{\Lambda\Phi 1}_{ji}\right]$$

Applying Equation 15.13a, we get

$$\frac{\partial F}{\partial \lambda_{ij}} = 2[\mathbf{Q}\boldsymbol{\Lambda\Phi}]_{ij}, \quad i = 1,\ldots,n; \quad j = 1,\ldots,r \tag{15.23}$$

Similarly

$$\frac{\partial F}{\partial \phi_{gh}} = \text{tr}\left[\left(\hat{\boldsymbol{\Sigma}}_0^{-1} - \hat{\boldsymbol{\Sigma}}_0^{-1}\mathbf{S}\hat{\boldsymbol{\Sigma}}_0^{-1}\right)\frac{\partial \hat{\boldsymbol{\Sigma}}_0}{\partial \phi_{gh}}\right]$$

$$= \text{tr}\left[\left(\hat{\boldsymbol{\Sigma}}_0^{-1} - \hat{\boldsymbol{\Sigma}}_0^{-1}\mathbf{S}\hat{\boldsymbol{\Sigma}}_0^{-1}\right)\boldsymbol{\Lambda}\left(\mathbf{1}_{gh} + \mathbf{1}_{hg} - \mathbf{1}_{gh}\mathbf{I}\mathbf{1}_{gh}\right)\boldsymbol{\Lambda}'\right]$$

$$= \text{tr}\left[\mathbf{Q}\boldsymbol{\Lambda}\left(\mathbf{1}_{gh} + \mathbf{1}_{hg} - \mathbf{1}_{gh}\mathbf{I}\mathbf{1}_{gh}\right)\boldsymbol{\Lambda}'\right]$$

$$= \text{tr}\left(\mathbf{Q}\boldsymbol{\Lambda}\mathbf{1}_{\text{gh}}\boldsymbol{\Lambda}' + \mathbf{Q}\boldsymbol{\Lambda}\mathbf{1}_{hg}\boldsymbol{\Lambda}' - \mathbf{Q}\boldsymbol{\Lambda}\mathbf{1}_{gh}\mathbf{I}\mathbf{1}_{gh}\boldsymbol{\Lambda}'\right)$$

$$= \text{tr}\left(\mathbf{Q}\boldsymbol{\Lambda}\mathbf{1}_{\text{gh}}\boldsymbol{\Lambda}'\right) + \text{tr}\left(\mathbf{Q}\boldsymbol{\Lambda}\mathbf{1}_{\text{hg}}\boldsymbol{\Lambda}'\right) - \text{tr}\left(\mathbf{Q}\boldsymbol{\Lambda}\mathbf{1}_{\text{gh}}\mathbf{I}\mathbf{1}_{gh}\boldsymbol{\Lambda}'\right)$$

$$= \text{tr}\left(\boldsymbol{\Lambda}'\mathbf{Q}\boldsymbol{\Lambda}\mathbf{1}_{\text{gh}}\right) + \text{tr}\left(\mathbf{1}_{hg}\boldsymbol{\Lambda}'\mathbf{Q}\boldsymbol{\Lambda}\right) - \text{tr}\left(\boldsymbol{\Lambda}'\mathbf{Q}\boldsymbol{\Lambda}\mathbf{1}_{\text{gh}}\mathbf{I}\mathbf{1}_{gh}\right)$$

$$= 2\text{tr}\left(\boldsymbol{\Lambda}'\mathbf{Q}\boldsymbol{\Lambda}\mathbf{1}_{\text{gh}}\right) - \text{tr}\left(\boldsymbol{\Lambda}'\mathbf{Q}\boldsymbol{\Lambda}\mathbf{1}_{gh}\mathbf{I}\mathbf{1}_{gh}\right)$$

$$= 2[\boldsymbol{\Lambda}'\mathbf{Q}\boldsymbol{\Lambda}]_{gh} - [\boldsymbol{\Lambda}'\mathbf{Q}\boldsymbol{\Lambda}]_{hg}[\mathbf{I}]_{hg}$$

Because of the symmetry of $\boldsymbol{\Lambda}'\mathbf{Q}\boldsymbol{\Lambda}$ and $\mathbf{I}$, we may finally write this as

$$\frac{\partial F}{\partial \phi_{gh}} = (2 - [\mathbf{I}]_{gh})[\boldsymbol{\Lambda}'\mathbf{Q}\boldsymbol{\Lambda}]_{gh} \tag{15.24}$$

And finally, by similar reasoning

$$\frac{\partial F}{\partial u_{ii}^2} = \text{tr}\left[\left(\hat{\boldsymbol{\Sigma}}_0^{-1} - \hat{\boldsymbol{\Sigma}}_0^{-1}\mathbf{S}\hat{\boldsymbol{\Sigma}}_0^{-1}\right)\frac{\partial \hat{\boldsymbol{\Sigma}}_0}{\partial u_{ii}^2}\right] = \text{tr}[\mathbf{Q}\mathbf{1}_{\text{ii}}] = [\mathbf{Q}]_{ii} \tag{15.25}$$

15.3.9 Least-Squares Estimation

Much psychological data involving dichotomous or polytomous responses to questionnaires, being discrete data, obviously does not conform to the multivariate normal distribution, which is a continuous probability distribution. Nevertheless, the question for confirmatory factor analysis is to what extent does the goodness-of-fit function in Equation 15.32, when minimized, yield optimum estimates for $\boldsymbol{\Lambda}$, $\boldsymbol{\Phi}$, and $\boldsymbol{\Psi}^2$ when the distribution of the observed variables departs from the multivariate normal distribution. Howe (1955) showed that without making any distributional assumptions one can justify the goodness-of-fit function in Equation 15.20 in unrestricted common-factor analysis on the grounds that the estimators which minimize Equation 15.20, conditional on a specified number of common factors, maximize the determinant of the matrix of residual partial correlations among the observed variables after the common factors have been removed from them. However, Jöreskog (personal communication) indicates that an analogous, distribution-free justification for maximum-likelihood estimators may not exist in confirmatory factor analysis. Thus one can raise the question whether other goodness-of-fit functions can be used in confirmatory factor analysis in place of Equation 15.20. In fact, a provisional alternative to Equation 15.20 is the least-squares goodness-of-fit criterion

$$L(\mathbf{\Lambda},\mathbf{\Phi},\mathbf{\Psi}^2)=\frac{1}{2}\left[\left(\hat{\mathbf{\Sigma}}_0-S\right)'\left(\hat{\mathbf{\Sigma}}_0-S\right)\right] \tag{15.26}$$

which is one-half the sum of squared deviations between corresponding elements of $\hat{\mathbf{\Sigma}}_0$ and **S**. The least-squares criterion has been used in unrestricted common-factor analysis in connection with the minres method of factor analysis (Harman and Jones, 1966) and is known to produce solutions which frequently closely approximate the corresponding maximum-likelihood solutions. The least-squares estimates of $\mathbf{\Lambda}$, $\mathbf{\Phi}$, and $\mathbf{\Psi}^2$, however, are not scale-free, as are their maximum-likelihood estimates. In connection with an early variant of confirmatory factor analysis, known as molar correlational analysis, Marshall B. Jones (1960) speculated that a least-squares criterion might be used in fitting $\hat{\mathbf{\Sigma}}_0$ to **S**, but he did not derive the required equations. Joreskog (1969a,b) reports the derivation of the first derivatives and approximations of the second derivatives of both the least-squares and the maximum-likelihood goodness-of-fit functions for a generalization of confirmatory factor analysis known as analysis of covariance structures. He used these first derivatives with the method of Fletcher and Powell (1963) to obtain, respectively, the least-squares and the maximum-likelihood estimates of the unspecified parameter of his model. We shall now consider the first partial derivatives of the function $L(\mathbf{\Lambda}, \mathbf{\Phi}, \mathbf{\Psi}^2)$ in Equation 15.26. The first partial derivative of L with respect to an arbitrary scalar variable, θ, is equivalent to

$$\frac{\partial L}{\partial\theta}=2\frac{1}{2}\operatorname{tr}\left[\left(\hat{\mathbf{\Sigma}}_0-\mathbf{S}\right)\frac{\partial\hat{\mathbf{\Sigma}}_0}{\partial\theta}\right] \tag{15.27}$$

Letting $\mathbf{Q}=(\hat{\mathbf{\Sigma}}_0-\mathbf{S})$, and substituting the expressions for the partial derivatives of $\hat{\mathbf{\Sigma}}_0$ in Equations 15.21 and 15.22 into Equation 15.27 respectively, we obtain

$$\frac{\partial L}{\partial\lambda_{ij}}=2[\mathbf{Q\Lambda\Phi}]_{ij} \tag{15.28a}$$

$$\frac{\partial L}{\partial\phi_{gh}}=(2-[\mathbf{I}]_{gh})[\mathbf{\Lambda'Q\Lambda}]_{gh} \tag{15.28b}$$

$$\frac{\partial L}{\partial\psi_{ii}^2}=[\mathbf{Q}]_{ii} \tag{15.28c}$$

which all have the form of the partial derivatives of maximum likelihood, **Q** differing in the ordinary-least-squares case, however.

15.3.10 Generalized Least-Squares Estimation

Generalized least-squares estimation is used when the underlying distribution is unknown, the sample size is fairly large, and one wishes to use a likelihood ratio goodness-of-fit chi square test of fit. The criterion minimized is the sum of squares of the transformed residuals:

$$\begin{aligned} G &= \frac{1}{2}\mathrm{tr}\left[\mathbf{S}^{-1/2}\left(\hat{\boldsymbol{\Sigma}}_0-\mathbf{S}\right)\mathbf{S}^{-1/2}\mathbf{S}^{-1/2}\left(\hat{\boldsymbol{\Sigma}}_0-\mathbf{S}\right)\mathbf{S}^{-1/2}\right] \\ &= \frac{1}{2}\mathrm{tr}\left[\left(\hat{\boldsymbol{\Sigma}}_0-\mathbf{S}\right)\mathbf{S}^{-1}\left(\hat{\boldsymbol{\Sigma}}_0-\mathbf{S}\right)\mathbf{S}^{-1}\right] \\ &= \frac{1}{2}\mathrm{tr}\left[\left(\hat{\boldsymbol{\Sigma}}_0\mathbf{S}^{-1}-\mathbf{I}\right)^2\right] \end{aligned} \tag{15.29}$$

The partial derivative of G with respect to an arbitrary parameter θ is

$$\begin{aligned} \frac{\partial G}{\partial \theta} &= \mathrm{tr}\left[\left(\hat{\boldsymbol{\Sigma}}_0\mathbf{S}^{-1}-\mathbf{I}\right)\frac{\partial \hat{\boldsymbol{\Sigma}}_0}{\partial \theta}\mathbf{S}^{-1}\right] \\ &= \mathrm{tr}\left[\left(\mathbf{S}^{-1}\hat{\boldsymbol{\Sigma}}_0\mathbf{S}^{-1}-\mathbf{S}^{-1}\right)\frac{\partial \hat{\boldsymbol{\Sigma}}_0}{\partial \theta}\right] = \mathrm{tr}\left[\mathbf{Q}\frac{\partial \hat{\boldsymbol{\Sigma}}_0}{\partial \theta}\right] \end{aligned} \tag{15.30}$$

where $\mathbf{Q} = (\mathbf{S}^{-1}\hat{\boldsymbol{\Sigma}}_0\,\mathbf{S}^{-1} - \mathbf{S}^{-1})$. We see now that the partial derivative of the generalized least-squares criterion is very similar in form to the partial derivative of the maximum-likelihood criterion. The resulting partial derivatives should thus follow the same pattern:

$$\frac{\partial G}{\partial \lambda_{ij}} = 2[\mathbf{Q}\boldsymbol{\Lambda}\boldsymbol{\Phi}]_{ij} \tag{15.31a}$$

$$\frac{\partial G}{\partial \phi_{gh}} = (2-[\mathbf{I}]_{gh})[\boldsymbol{\Lambda}'\mathbf{Q}\boldsymbol{\Lambda}]_{gh} \tag{15.31b}$$

$$\frac{\partial G}{\partial u_{ii}^2} = [\mathbf{Q}]_{ii} \tag{15.31c}$$

15.3.11 Implementing the Quasi-Newton Algorithm

For the discussion which follows, we will assume that the researcher has prespecified more than r^2 independent parameters among the elements of the matrices $\mathbf{\Lambda}$ and $\mathbf{\Phi}$ in such a way that the model $\mathbf{\Sigma}_{YY}=\mathbf{\Lambda\Phi\Lambda}'+\mathbf{\Psi}^2$ is overidentified. Next, we will assume that we have initial starting values for the m free (unspecified) elements of the matrices $\mathbf{\Lambda}$, $\mathbf{\Phi}$, and $\mathbf{\Psi}^2$. We next place the free parameters of the matrices $\mathbf{\Lambda}$, $\mathbf{\Phi}$, and $\mathbf{\Psi}^2$ in a column vector $\hat{\boldsymbol{\theta}}$ according to the following convenient scheme: Beginning with the first and then successive columns of $\mathbf{\Lambda}$, we go down each column and place the free elements successively in the vector $\hat{\boldsymbol{\theta}}$. After these elements are inserted in $\hat{\boldsymbol{\theta}}$, we proceed to the symmetric matrix $\mathbf{\Phi}$ and from each row up to and including the diagonal we successively extract the free parameter values and place them successively in the vector $\hat{\boldsymbol{\theta}}$. Finally we take the free n diagonal elements of the matrix $\mathbf{\Psi}^2$ and make them successively the remaining elements of $\hat{\boldsymbol{\theta}}$. Let $\boldsymbol{\theta}_0$ denote the vector of free parameters with elements given the starting values, either explicitly by the researcher or obtained using the multiple group method described earlier. We then perform [as recommended by Jöreskog (1969a,b)] five steepest-descent iterations,

$$\hat{\boldsymbol{\theta}}_{k+1}=\hat{\boldsymbol{\theta}}_k-\alpha_{k+1}\nabla F(\hat{\boldsymbol{\theta}}_k)$$

where

$\hat{\boldsymbol{\theta}}_k$ denotes the vector of free parameters with values after the kth iteration

$\nabla F(\hat{\boldsymbol{\theta}}_k)$ is the gradient vector of partial derivatives of the fit function F with respect to the free parameters in the vector $\hat{\boldsymbol{\theta}}$ evaluated after the kth iteration

α_{k+1} is a value determined by line search that when multiplied times the gradient vector and the resulting vector subtracted from the vector $\hat{\boldsymbol{\theta}}_k$ "moves" it to the approximate minimum in the direction of the negative gradient in the $(k+1)$st iteration

Jöreskog notes that these steepest descent iterations are most effective at the start of the iterations, but would be very ineffective if performed much later in the iterations. After each iteration the gradient vector is reevaluated with the new values for the free parameters.

At this point we begin the quasi-Newton iterations as described earlier, given for the present case by the iterative equation

$$\hat{\boldsymbol{\theta}}_{k+1}=\hat{\boldsymbol{\theta}}_k-\alpha_{k+1}\mathbf{H}_k\nabla F(\hat{\boldsymbol{\theta}}_k)$$

where $\mathbf{H}_k$ is initially an identity matrix, which with subsequent iterations becomes equal to

$$\mathbf{H}_{k+1}=\left[\frac{\mathbf{H}_k\gamma_k\delta_k'+\delta_k\gamma_k'\mathbf{H}_k}{\delta_k'\gamma_k}\right]+\left[\frac{\gamma_k'\mathbf{H}_k\gamma_k}{\delta_k'\gamma_k}\right]\left[\frac{\delta_k\delta_k'}{\delta_k'\gamma_k}\right] \tag{15.32}$$

with $\delta_k=\hat{\boldsymbol{\theta}}_{k+1}-\hat{\boldsymbol{\theta}}_k$ and $\gamma_k=\nabla F\hat{\boldsymbol{\theta}}_{k+1})-\nabla F(\hat{\boldsymbol{\theta}}_k)$. The value for α_{k+1} is obtained by line search methods along the line in the direction of the vector $-\mathbf{H}_k\nabla F(\hat{\boldsymbol{\theta}}_k)$.

The iterations are continued until each of the partial derivatives of F with respect to each of the free parameters in $\hat{\boldsymbol{\theta}}$ falls below some very small quantity in absolute value and the differences in successive estimates of each of the parameters do not occur anywhere except after, say, the fifth decimal place.

15.3.12 Avoiding Improper Solutions

During the iterations, it is essential to avoid improper values for the parameters. For example, variances of common factors in the diagonal of $\mathbf{\Phi}$ and unique-factor variances in the diagonal of $\mathbf{\Psi}^2$ should not be negative. One approach to dealing with the case when the value of a parameter becomes negative during the iterations is to set it to zero or some very small positive quantity and proceed, hoping that at some later point in the iterations it may become positive again. Sometimes starting the iterations with different starting values also gets around this problem.

Another approach to avoiding improper estimates of variances is to reformulate the problem. Thurstone (1947) expressed the common-factor model with partitioned matrices as

$$\mathbf{Y} = [\mathbf{\Lambda} \quad \mathbf{\Psi}]\begin{bmatrix} \mathbf{X} \\ \mathbf{E} \end{bmatrix} = \mathbf{\Lambda}^* \xi$$

where

$\mathbf{\Lambda}$ is the $n \times r$ factor loading matrix

$\mathbf{\Psi}$ is the $n \times n$ diagonal matrix of "unique-factor loadings," equal to the square roots of the unique variances

$\mathbf{X}$ is the $r \times 1$ vector of common-factor random variables

$\mathbf{E}$ is the $n \times 1$ vector of unique-factor random variables

Then according to the fundamental theorem, regarding the covariances among the factors

$$\mathbf{\Sigma}_{YY} = [\mathbf{\Lambda} \quad \mathbf{\Psi}]\begin{bmatrix} \mathbf{\Phi} & \mathbf{0} \\ \mathbf{0} & \mathbf{I} \end{bmatrix}\begin{bmatrix} \mathbf{\Lambda}' \\ \mathbf{\Psi} \end{bmatrix} = \mathbf{\Lambda}^* \mathbf{\Phi}^* \mathbf{\Lambda}'$$

If the diagonal elements of the common-factor variance–covariance matrix $\mathbf{\Phi}$ are fixed to unity, and the variances of the unique factors are also fixed to unity as indicated by the identity matrix $\mathbf{I}$, then it does not matter whether estimates of the parameters in $\mathbf{\Lambda}$ or $\mathbf{\Psi}$ become negative because their contribution to the variances of the observed variables in $\mathbf{\Sigma}_{YY}$ will always be positive via squares of these terms in $\mathbf{\Lambda}$ and $\mathbf{\Psi}$. But the variances of the common factors and unique factors will never be negative as long as they are fixed to positive values.

A benefit of this alternative approach to formulating the common-factor model is that common factors and unique factors can be treated the same in deriving the partial derivatives, since the partial derivatives of the fit function with respect to elements in $\mathbf{\Psi}$ are analogous to the partial derivatives of the fit function with respect to elements of $\mathbf{\Lambda}$.

Another problem that occasionally arises is when the matrix **S** is singular. This will occur when one variable is completely linearly dependent on other variables. Its effect is that one cannot obtain estimates that depend on the inverse of the **S** matrix, which in this case does not exist. This will influence generalized least squares, which uses the inverse of **S** in the computations. There are ways of identifying the source of the singularity. It may be that a third variable is a function of two other variables, such as would occur in including total score along with subscores in the same analysis.

An analogous problem occurs when parameters are not identified. In this case, they may become linearly dependent on other estimated parameters. In estimation procedures like the Newton–Raphson method that depend on obtaining the inverse of the Hessian matrix, this will cause the method to fail, since the inverse of the Hessian matrix contains the variances and covariances among the estimates of the parameters. Most computer programs will alert the user to this condition, even those that use a quasi-Newton method, which does not have to invert the **H** matrix. The researcher should not accept solutions when this condition is reported. Sometimes a parameter becomes underidentified because of the choice for the initial start values for the parameters. Trying different start values for the estimated parameters may clear up this problem. But if a parameter is intrinsically underidentified, no start values will clear up the problem. The researcher will need to reevaluate the identification of the model's parameters and seek to correct the model accordingly.

15.3.13 Statistical Tests

The matrices $\mathbf{\Lambda}$, $\mathbf{\Phi}$, and $\mathbf{\Psi}^2$, which minimize the function F of Equation 15.20, are the maximum-likelihood estimates $\mathbf{\Lambda}$, $\mathbf{\Phi}$, and $\mathbf{\Psi}^2$, which maximize the likelihood Equation 15.16. A test of the goodness of fit of the resulting matrix $\mathbf{\Sigma}_{YY} = \mathbf{\Lambda\Phi\Lambda}' + \mathbf{\Psi}^2$ to the sample-dispersion matrix **S** is given by the likelihood ratio statistic

$$U = (N-1)\left[\ln\left|\hat{\mathbf{\Sigma}}_0\right| - \ln\left|\mathbf{S}\right|\right] + \mathrm{tr}\left(\hat{\mathbf{\Sigma}}_0^{-1}\mathbf{S}\right) - n \qquad (15.33)$$

which Jöreskog (1967, p. 457) points out is $(N-1)$ times the minimum value of the function F, minimized by the Quasi-Newton algorithm. The statistic U is approximately distributed in large samples as chi-square with

$$df = n(n+1)/2 - m$$

degrees of freedom, where m is the number of distinct estimated parameters. (Jöreskog, 1969a, 1969b) originally gave the degrees of freedom in a more complicated way, not realizing, I believe, that the degrees of freedom are simply equal to the difference between the number of distinct observed values in **S** to be fit by the model, i.e., $n(n+1)/2$, and the number of distinct estimated parameters, m, with a loss in degree of freedom for each parameter estimated. He realized this in later publications.) When a Quasi-Newton algorithm has been used, the finally obtained **H** matrix of Equation 15.20 contains a close approximation to the inverse of the matrix of second-order derivatives at the minimum. Joreskog points out (1969a,b) that, when this matrix is multiplied by $2/(N-1)$, the resulting matrix **E** is an estimate of the variance–covariance matrix for the maximum-likelihood estimates of the free parameters. However, Jöreskog indicates that his experience has been that, after q iterations have been required to achieve convergence for the estimates of the free parameters, an additional q iterations may be necessary to get an accurate estimate of the matrix of second derivatives by the matrix **H** of the quasi-Newton algorithm. If one does not wish to compute these extra iterations, one should work out the formulae for these second derivatives and compute them directly with the converged values for the estimated parameters.

Then for any estimated parameter θ_i with maximum-likelihood estimate, $\hat{\theta}_i$ and variance of estimate, ε_{ii}, then an approximate 95% confidence interval for this parameter estimate is given by the z statistic as

$$\hat{\theta}_i - 2\sqrt{\varepsilon_{ii}} < \theta_i < \hat{\theta}_i + 2\sqrt{\varepsilon_{ii}} \tag{15.34}$$

Thus, we have available not only tests of the overall model but also tests for individual parameters of the model. However, the tests of individual parameters in Equation 15.34 are not independent, and when many of them are performed, some kind of Bonferonni adjustment of the significance level for these tests is frequently recommended (cf. Larzelere and Mulaik, 1977; Hays, 1994) to reduce the number of type I errors that would occur by chance.

15.3.14 What to Do When Chi-Square Is Significant

A significant chi-square goodness-of-fit statistic indicates that probabilistically the model deviates from the data in the sample to a degree that under the null hypothesis that the model is true would be quite improbable. But because there are many, many ways in which this deviation could arise, all that we can say at this point is that somewhere there is something wrong. In fact, the specifications of the model itself may not be incorrect, but rather some other background assumptions essential to inference with the chi-square statistic may be at fault: (1) multivariate normality may not apply, as it normally would not with polytomous responses to questionnaire items or Likert rating scales. (2) Subjects may not be "causally homogenous," meaning that they vary among themselves in what latent variables govern their responses, or in the degree to which a unit change in a latent factor produces a change in response on a rating scale, or in the degree to which they have reached an equilibrium in changes induced by stimuli mediated via the latent variables at the time of measurement. (3) Responses to different items may not be independent of one another, so that responding in one way to an item determines responses on another item. Responses of different subjects may not be independent—there is collusion or common social influences among some but not all subjects in how they respond. The subject may remember what response he or she made to a previous item and respond similarly on a later item. And this may not occur with every subject or with the same items. (4) The experimental setting may introduce "random shocks" like noise from construction while subjects fill out a questionnaire, and this affects responses to several items successively for some subjects, who may be closer to the noise than others and be variously affected. It may be difficult to determine the extent to which these and other extraneous influences affect the fit of the model. Nevertheless, a significant chi-square means something is wrong, and a researcher should do his or her best to diagnose the sources of lack of fit. Many of these may still remain within the specification of the model itself and should be dealt with.

Various computer programs for structural equation modeling, which are ordinarily used to perform confirmatory factor analysis, provide diagnostics for some of these phenomena. Univariate statistics for each item, such as the mean, variance, standard deviation, skewness, and kurtosis should be examined. Byrne (2006) notes that if the data show significant nonzero univariate kurtosis, then surely multivariate normality does not apply. This will affect the trustworthiness of the chi-square statistic. Multivariate sample statistics such as Mardia's (1970, 1974) coefficient indicates deviations in kurtosis from the normal. Some programs report tests for outliers among the subjects, and these subjects might be removed and a new analysis performed.

We have already mentioned the use of Lagrange multiplier tests or modification index tests (which have a similar rationale) that might be used to discover those fixed parameters, which, if freed, would reduce the lack of fit for the most. In a common factor model, one would first look for the fixed zero loading on a factor that significantly contributes to lack of fit using these tests. The Lagrange multiplier test for freeing a single parameter is equivalent to a chi-square test with 1 degree of freedom. The best strategy is to begin by freeing the fixed factor loading with the largest, significant chi-square value and reanalyzing the model. This should be done if there is a theoretical rationale that supports a nonzero value for the loading. Then obtain the Lagrange multiplier tests on fixed zero factor loadings for this new model. Select again the single fixed-zero loading with the highest chi-square and free that and reanalyze. Do a similar sequence of tests with Lagrange multipliers on the zero covariances among the disturbances after finishing those on the factor pattern loadings.

A presumption of using the Lagrange multiplier tests or modification indices to find fixed parameters to free is that the model framework is correct, and as a consequence the model is already an approximation to a model that would fit to within sampling error. Although the prior use of theory and analysis gave the greatest support to the hypothesized model tested, there is always the logical possibility that a better model is one with a considerably different causal structure that would fit as well or better with more degrees of freedom. The researcher should consider this possibility and review what is known about the phenomena modeled to see if something was not missed in the original formulation of the hypothesis. This should especially be the case if the model fit poorly to begin with, or still fits poorly after adjustments to the model are made and the model is reanalyzed. Also note any free parameter estimates with signs opposite to those expected by your theory. This should cause a careful reexamination of your theory. Keep in mind that any parameters freed in the reformulating process should lead to a corresponding loss in degrees of freedom for the modified model when analyzed against the same sample covariance matrix.

15.3.15 Approximate Fit Indices

However, while the chi-square statistic U given in Equation 15.33 is a test of the exact null hypothesis involving the overidentifying conditions imposed on the common-factor model, its usefulness has frequently been questioned because in large samples it almost invariably leads to a rejection of the hypothesized model (Bentler and Bonett, 1980). This is because just at the point where one has a sample size large enough for U to have an approximate chi square distribution, the power of the statistic against any slightly different model also becomes very high. Because in the behavioral and social sciences we have not yet graduated beyond trying to account for large effects with our theories, we will frequently not anticipate small effects and incorporate them in our theories. Thus our models, which seek to account for large effects, will frequently be wrong in an exact sense, although they may still be very good approximations. The situation is analogous to the case in physics classes where one discusses the effects of gravity on falling bodies but fails to account for either the mass of the smaller body, the distances from the larger attracting body from which the smaller body may start to fall, or the effects of air friction, if the smaller body passes through the atmosphere. Thus simple formulas like $s=(1/2)gt^2$ may be very good approximations at the surface of the earth, but not exactly correct. In situations like this we need some way to assess the degree of approximation. Indices of approximation that vary between 0 and 1, with 0 indicating total lack of fit and 1, perfect fit, have been proposed.

The comparative fit index. Bentler and Bonett (1980) presented an index they called the NFI or "normed fit index." They argued that a norm for lack of fit for comparison purposes could be given by the chi-square of a "null model" fitted to the sample covariance matrix. The null model hypothesized that there were no nonzero covariances between the observed variables. In other words, the null model hypothesized that the covariance matrix should be a diagonal matrix with free diagonal elements. If there were any nonzero covariances between the observed variables at all, the null model would produce the worse lack of fit compared to any other model nested within it that hypothesized some sources of nonzero covariance among the variables. As a measure of the lack of fit of the null model, they used the null model's chi-square. Then from this they subtracted the chi-square of the tested model in question. The difference between the chi-squares of the null model and the tested model represented a reduction in lack of fit due to the tested model. They compared this reduction then to the chi-square of the null model, to get something analogous to a proportion of variance accounted for. Thus the NFI index was

$$\text{NFI} = \frac{(\chi^2_{\text{null}} - \chi^2_k)}{\chi^2_{\text{null}}} \tag{15.35}$$

where χ^2_k denotes the chi-square of the tested model.

Marsh et al. (1988) conducted a series of Monte Carlo studies evaluating approximate fit indices then in the literature. They reported that the NFI index tended to *underestimate* its population value in samples and did so nonnegligibly in samples less than 800. Then Bentler (1990) and McDonald and Marsh (1990) presented in the same issue of the *Psychological Bulletin* a modification of the NFI index that overcame the bias of the NFI. Instead of using chi-squares in the formula for the NFI, they replaced the chi-squares with estimates of the models' "noncentrality parameter" which is obtained from the chi-square by simply subtracting its degrees of freedom from the chi-square. Since in some cases sampling variation could produce a chi-square that is less than its degrees of freedom and thus a negative difference, Bentler further proposed forcing the resulting index, which he called FI, to equal 1 if it exceeded unity or to equal 0 if it became negative and called this the CFI.

$$\text{FI} = \frac{(\hat{\delta}^*_{\text{null}} - \hat{\delta}^*_k)}{\hat{\delta}^*_{\text{null}}} = \frac{[(\chi^2_{\text{null}} - \text{df}_{\text{null}}) - (\chi^2_k - \text{df}_k)]}{(\chi^2_{\text{null}} - \text{df}_{\text{null}})} \tag{15.36}$$

So,

$$\text{CFI} = \begin{cases} 0, & \text{if FI} < 0 \\ \text{FI} & \\ 1, & \text{if FI} > 1 \end{cases} \tag{15.37}$$

The CFI only shows nonnegligible bias in samples smaller than 50. But one should not perform confirmatory factor analysis or structural equation modeling with samples that small because other Monte Carlo studies show that below samples of 200, the results are highly variable.

Originally Bentler and Bonett (1980) recommended that NFI > .90 represented a "good approximation." However, subsequent experience has suggested that CFI>.95 is better for "adequate fit." However, considering the wide variety of models that might be fitted, and the wide variety of conditions by which the data, on which the empirical covariances are based, may be generated, the aim is not to establish how exactly the tested model is an approximation to the "real" model. There may not be a metric for such comparisons. And CFI>.95 is only a "rule of thumb," not an absolute "golden rule." The support for the hypothesized model is in how well the model-based reproduced covariance matrix fits the empirical covariance matrix. And this is only *prima facie* support, which may be overturned by producing a different model that fits as well or better—hopefully to within sampling error—with as many or more degrees of freedom.

The RMSEA index. The CFI index is a "goodness-of-fit index" that ranges between 0 and 1 to indicate degree of fit, with 0 horrible fit and 1 perfect fit. We will now look at an index that is a "badness-of-fit index," where 0 indicates perfect fit and larger values indicate lack of fit.

The estimate of the noncentrality parameter $\hat{\delta} = \chi^2_{\text{df}} - \text{df}$ estimates a lack-of-fit parameter that depends on the sample size. To indicate a degree of lack of fit for a model that is independent of sample size, McDonald (1989) suggested dividing the noncentrality parameter for a given sample size by $(N-1)$. This implied that its estimate would be given by $\chi^2_{\text{df}} - \text{df}/(N-1)$. Steiger and Lind (1980) recommended dividing this index by df, the degrees of freedom, and then taking the square root to produce an index that indicates the average lack-of-fit per degree of freedom, which allowed comparison of fit for models with different degrees of freedom. Browne and Cudeck (1993) further suggested setting the RMSEA to zero if the value under the radical is negative, to avoid an imaginary numerical value. Hence the RMSEA (which Steiger and Lind coined from "root mean square error of approximation") became

$$\text{RMSEA} = \sqrt{\text{Max.}\left\{\left(\frac{\chi^2_{\text{df}_k} - \text{df}_k}{(N-1)\text{df}_k}\right), 0\right\}} \tag{15.38}$$

Browne and Cudeck (1993) recommended that values less than or equal to .05 were "acceptable approximations."

Some have asserted that the RMSEA accounts for "parsimony" by dividing by the degrees of freedom. But this is not the case, because dividing by degrees of freedom makes it impossible to compare models on parsimony (i.e., relative economy in number of parameters or its obverse, the higher number of degrees of freedom relative to the potential number). Two models may differ in degrees of freedom but have the same value for RMSEA, because the additional degrees of freedom of the second may simply correspond to fixed parameters that add the same average noncentrality per degree of freedom to the base average noncentrality of the first model. But the model with more degrees of freedom is to be preferred because it subjects the model to more tests.

Jöreskog's GFI index. Jöreskog and Sörbom (1981) proposed a family of goodness-of-fit indices to be used with their LISREL program. These indices were inspired by Wright's index of determination

$$R^2 = 1 - \frac{\text{Error variance}}{\text{Total variance}}$$

Fisher (1925) further advocated this index as the intraclass correlation coefficient. The index ranges from 0 to 1, with 1 indicating "perfect fit."

The GFI index computes "error" as the sum of (weighted and possibly transformed) squared differences between the elements of the observed variance–covariance matrix **S** and those of the estimated model variance–covariance matrix $\hat{\mathbf{\Sigma}}_0$ and compares this sum to the total sum of squares of the elements in **S**. The matrix $(\mathbf{S} - \hat{\mathbf{\Sigma}}_0)$ is symmetric and produces the element-by-element differences between **S** and $\hat{\mathbf{\Sigma}}_0$. **W** is a transformation matrix that weights and combines the elements of these matrices, depending on the method of estimation. Thus we have

$$\text{GFI} = 1 - \frac{\text{tr}[\mathbf{W}^{-1/2}(\mathbf{S} - \hat{\mathbf{\Sigma}}_0)\mathbf{W}^{-1/2}][\mathbf{W}^{-1/2}(\mathbf{S} - \hat{\mathbf{\Sigma}}_0)\mathbf{W}^{-1/2}]}{\text{tr}\,[\mathbf{W}^{-1/2}(\mathbf{S})\mathbf{W}^{-1/2}][\mathbf{W}^{-1/2}(\mathbf{S})\mathbf{W}^{-1/2}]} \tag{15.39}$$

where

$\hat{\mathbf{\Sigma}}_0$ is the model variance–covariance matrix

S is the unrestricted, sample variance–covariance matrix

$$\mathbf{W} = \begin{cases} \mathbf{I} & \text{Unweighted least squares} \\ \mathbf{S} & \text{Weighted least squares} \\ \hat{\mathbf{\Sigma}}_0 & \text{Maximum likelihood} \end{cases}$$

Depending on the method of estimation, one modifies the GFI index accordingly.

Some of the criticisms of the GFI indices is that they tend to vary with sample size. Although Bollen (1989) notes that this is not due to the fact that N is involved explicitly in the formula for GFI, nevertheless Monte Carlo studies (Marsh et al., 1988) revealed that the average values of the GFI tend to increase with N. Some mathematical analysis by Steiger (1989, 1995) has revealed that the expected value of the GFI index tended to increase with sample size N. Steiger has also offered a way to construct a 95% confidence interval on a GFI index he has developed for use with models that are invariant under a constant scaling factor. However, Yuan (2005) and Yuan and Bentler (1998) have argued that the noncentral chi-square distribution is not correct for the chi-square that fails to fit, that a normal distribution is more appropriate. This will require reformulating the 95% confidence interval for the GFI.

Parsimony ratio. Although it is not a fit index, it is an index that is often used in association with fit indices: the parsimony ratio is the ratio of the number of degrees of freedom of the model to the total number of degrees of freedom possible as given by the number of nonredundant elements of the covariance matrix.

$$\mathrm{PR} = \frac{\mathrm{df}}{n(n+1)/2} \tag{15.40}$$

Because each degree of freedom represents a condition by which the model may fail to fit the data, the parsimony ratio indicates the degree to which the model was potentially disconfirmable. If the number of degrees of freedom of the model equals $n(n+1)/2$, there are no estimated parameters, and every asserted fixed parameter represents a condition by which the model could fail to fit. So when PR = 1.00, the model is maximally disconfirmable. This represents a maximum test of the model. When a model his a high goodness-of-fit index and a high parsimony ratio, it is a well-tested, good-fitting model. Mulaik et al. (1989) suggested one could multiply the parsimony ratio times the CFI, and Carlson and Mulaik (1993), recommended a value of $\mathrm{PR} \times \mathrm{CFI} \geq .85$ as indicating a well-fitting and well-tested model.

There are numerous other goodness-of-fit indices. We have shown the most popular here.

15.4 Designing Confirmatory Factor Analysis Models

15.4.1 Restricted versus Unrestricted Models

In Jöreskog (1969a, 1969b) Jöreskog was concerned to make a distinction between specifying a restricted solution and specifying an unrestricted solution when one fixes or frees certain parameters of the model. On the one hand, for a fixed value for the number of common factors, r, an unrestricted solution imposes no restrictions on the solution for $\mathbf{\Lambda\Phi}_{XX}\mathbf{\Lambda}'$ other than those equivalent to what is needed to obtain a principal axes solution for r common factors. In that case we require $\mathbf{\Phi}_{XX}=\mathbf{I}$, which fixes r diagonal elements to unity and $r(r-1)/2$ off-diagonal elements to zero, and requires further that $\mathbf{\Lambda}'\mathbf{\Psi}^{-2}\,\mathbf{\Lambda}$ is diagonal, placing $r(r-1)/2$ additional constraints on the columns of $\mathbf{\Psi}^{-1}\,\mathbf{\Lambda}$, to make them mutually orthogonal. The diagonal elements of $\mathbf{\Psi}^2$ are free parameters. In all, r^2 constraints are placed on the parameters of the model to yield a principal axes solution for r common factors. But there can be more than one unrestricted model. Suppose the principal axis solution is $\mathbf{\Sigma}_{YY} = \mathbf{\Lambda}_0\mathbf{\Lambda}_0' + \mathbf{\Psi}^2$. Then an equivalent model is

$$\mathbf{\Sigma}_{YY} = (\mathbf{\Lambda}_0\mathbf{T})(\mathbf{T}^{-1}\mathbf{T}'^{-1})(\mathbf{T}'\mathbf{\Lambda}'_0) + \mathbf{\Psi}^2 = \mathbf{\Lambda}_0\mathbf{\Lambda}'_0 + \mathbf{\Psi}^2$$

where $\mathbf{T}$ is any nonsingular $r \times r$ transformation matrix. Because $\mathbf{T}$ contains r^2 elements, this suggested to Jöreskog that a necessary but not sufficient condition that one has an unrestricted solution is that exactly r^2 constraints are distributed across the $\mathbf{\Lambda}$ and $\mathbf{\Phi}_{XX}$ matrices (Jöreskog, 1979). The significance of unrestricted solutions is that they correspond to exploratory factor analysis solutions where only minimal constraints are placed on the model to achieve a solution for r common factors. An unrestricted model for r common factors will yield the best fit of any model with r common factors. Jöreskog (1979; personal communication) suggested the following as one way to specify an unrestricted solution: (1) Fix the r diagonal elements of $\mathbf{\Phi}_{XX}$ to unity. This fixes the metric for the solution. The remaining off-diagonal elements of $\mathbf{\Phi}_{XX}$ are left free. (2) In each column of $\mathbf{\Lambda}$ free up one coefficient corresponding to an expected high loading; make sure that each freed parameter is in a different row. (3) In each row with a freed high loading, fix the remaining $r-1$ other coefficients to zero. (4) Free all other parameters in $\mathbf{\Lambda}$. There should now be $r-1$ zeros in each column.

In contrast, a restricted solution restricts the solution for $\mathbf{\Lambda\Phi}_{XX}\mathbf{\Lambda}'$ and in turn for $\mathbf{\Psi}^2$. This will occur whenever more than r^2 independent parameters of $\mathbf{\Lambda}$ and $\mathbf{\Phi}_{XX}$. are specified or when some of the diagonal elements of $\mathbf{\Phi}^2$ are specified. It will occur if the fixed values inserted into $\mathbf{\Lambda}$ and $\mathbf{\Phi}_{XX}$ cannot be obtained by a linear transformation of an unrestricted $\mathbf{\Lambda}$ with corresponding inverse transformations applied to $\mathbf{\Phi}_{XX}$. The fit of an unrestricted solution to a given sample variance–covariance matrix $\mathbf{S}$ will usually be better than a restricted solution.

Lower triangular specification. An alternative way of specifying an unrestricted model takes advantage of a theorem that one can always take a square matrix of full rank and by an orthogonal linear transformation convert it to a lower triangular matrix. That is, if $\mathbf{\Lambda}_0$ is a square matrix of full rank, then there exists an orthogonal transformation $\mathbf{T}$ such that $\mathbf{L} = \mathbf{\Lambda}_0\mathbf{T}$ is lower triangular. (We will not prove that here.) We can let the rows of $\mathbf{L}$ be any set of r linearly independent rows of $\mathbf{\Lambda}$. What we propose to do is to regard the rows for variables chosen to be the determiners of the factor-pattern matrix to be from a lower triangular matrix. The procedure for specifying this variant of the unrestricted model is similar to the previous one given by Jöreskog (1972): (1) Fix the r diagonal elements of $\mathbf{\Phi}_{XX}$ to unity. This fixes the metric for the solution. However, fix the remaining off-diagonal elements of $\mathbf{\Phi}_{XX}$ to zero. This forces the factors to be mutually orthogonal since $\mathbf{\Phi}_{XX}$ in this case is an identity matrix. (2) Pick r rows of $\mathbf{\Lambda}$ that one believes will be linearly independent. (3) In the first row of these r rows, free the first element, while fixing all remaining elements to 0. In the second of the r rows, free the first two elements and fix the remaining to zero. In the third of these r rows, free the first three elements and fix the remaining to zero. In general, in the ith row, free the first i elements, and fix the $r-i$ remaining elements to zero. In the rth row, free all r elements in the row. (4) Free all other parameters in $\mathbf{\Lambda}$. There should now be $r(r-1)/2$ zeros in $\mathbf{\Lambda}$. All remaining elements in $\mathbf{\Lambda}$ are free. There are $r(r-1)/2$ zeros in the off-diagonal of $\mathbf{\Phi}_{XX}$. In all, with r unities in the diagonal of $\mathbf{\Phi}_{XX}$ there are r^2 fixed parameters across $\mathbf{\Lambda}$ and $\mathbf{\Phi}_{XX}$. Jöreskog's solution is a linear transformation of this solution and yields the same chi-square. In theory the chi-square should be the same as obtained by a maximum-likelihood exploratory factor analysis for the same number r of common factors (Jöreskog, 1969a,b, 1972).

15.4.2 Use for Unrestricted Model

Jöreskog (1969a,b) notes that in exploratory factor analysis, no hypothesis is specified concerning [the nature of] the factors. Nevertheless the imposition of r^2 independent constraints on certain of the parameters of $\mathbf{\Lambda}$ and $\mathbf{\Phi}_{XX}$ is necessary to identify the exploratory common-factor model. In the maximum-likelihood case, the constraints are that $\mathbf{\Phi}_{XX}=\mathbf{I}$ and $\mathbf{\Lambda}'\mathbf{\Psi}^{-2}\mathbf{\Lambda}$ is diagonal. The unrestricted model, which also imposes r^2 independent constraints on these matrices, effectively is a rotation of an exploratory factor analysis model. Any of these variants of the unrestricted model or a maximum-likelihood exploratory factor analysis model with r fixed common factors may be used to test the following hypothesis: a common-factor model with r common factors (with the usual constraints on the relations between common and unique factors and among unique factors) is consistent with the data. The model may be rejected by having more common factors than hypothesized, correlations between common and unique factors, or correlations between pairs of unique factors. Thus an unrestricted model can be used in confirmatory factor analysis to test whether, at all, a common-factor model with r common factors is appropriate for the data, irrespective of any hypothesis one may have about how a certain set of r factors are related or unrelated to the observed variables. Effectively, the unrestricted model is said to be a model nested with the more constrained restricted confirmatory factor analysis model.

A series of models are parameter nested if they have the same structure and each successive model has the same constraints (e.g., fixed, constrained-equal, etc.) on parameters of the preceding model in the sequence to which are added new constraints on additional parameters. If one rejects any less constrained model, one necessarily will reject the more constrained model nested with it since it will contain the rejected constraints. A series of models are equivalence nested, if each model in the series can be placed in one-to-one correspondence with a model that has the same number of estimated parameters and covariance matrix for the observed variables in a parameter-nested series. This means one can substitute a model generating the same covariance matrix for a parameter-nested model in obtaining chi-square test values.

So, before performing a confirmatory factor analysis with hypotheses specified to test on the factors and their relations and/or nonrelations with the observed variables, you can first test the unrestricted model. If you pass the chi-square test, you can go on to test the confirmatory factor analysis model. But if you reject the unrestricted model with the same number of common factors as in the confirmatory factor analysis model, then you should stop and thoroughly reexamine whether you have enough factors, or whether there are correlated unique factors. If you can resolve the problem by freeing a few parameters, or adding more factors with corresponding free loadings, then you may go forward with a more constrained confirmatory factor analysis model, as long as you carry the freed parameters forward into later more constrained models. On the other hand, lack of fit may indicate that a quite different model is needed for the data, and one may decide, when that seems to be the case, to start over with a new study and a new model. Discussion of the unrestricted model and its use in structural equation modeling is given in Mulaik and Millsap (2000).

15.4.3 Measurement Model

We have already illustrated an ordinary confirmatory factor analysis study using the Carlson and Mulaik (1993) correlation matrix. No constraints were placed on the correlations between common factors, since they were all free parameters. This kind of model has been used in connection with certain structural equation models having at least 4 indicators per latent variable. Anderson and Gerbing (1988) noted that nested and embedded within such structural equation models was an implicit confirmatory factor analysis model with saturated (all free) correlations among the latent variables. They called this confirmatory factor analysis model the "measurement model." The reason they used this terminology was because they regarded the latent variables to be common factors of certain indicator variables. The confirmatory factor analysis model specified constraints on the loadings of the common factors which specified whether the latent was or was not a cause of the indicators in question. Implicit in this is the idea that a measurement is a variable that is caused by a quantity measured. Measurement, in this sense, is heavily theoretical, in that the researcher needs to explicitly state reasons why the indicators are effects of the latent indicated. A test of the measurement model is thus a test of the theory justifying the indicators and their relations to the latent variables. However, because the measurement model may be incompletely specified due to lack of prior knowledge, the test is only partial. Nevertheless in structural equation modeling Anderson and Gerbing (1988) argued that one should perform testing of the structural equation model in two steps: step 1 would be a test of the measurement model, and step 2 would be a test of the structural equation model. If the measurement model was rejected, that would imply that the structural equation model in which it was embedded would be rejected also. It also made problematic the assumptions made in the structural equation model that the latent variables were what they were supposed to be, and so any causal relations between latents would be problematic in their interpretation. In other words, the measurement model needed to be cleared up if it failed to fit, before going on to test the structural equation model.

15.4.4 Four-Step Procedure for Evaluating a Model

Mulaik and Millsap (2000) described a four-step nested series of models to test in structural equation modeling when the model has implicitly embedded in it a factor analysis model because it has four or more indicators of each latent: Step 1 tested the unrestricted model, step 2 the measurement model, step 3 the structural equation model (or in confirmatory factor analysis a model with constraints on correlations among the common factors). In step 4 one tested hypotheses about parameters that up to that point in the series of models had been free. For example, is the nonzero value of an estimated value significantly different from zero? If any of the earlier tests in the series rejected the model, the failure to fit needed to be resolved before going on to test subsequent models in the series.

15.5 Some Other Applications

15.5.1 Faceted Classification Designs

In psychology, factor analysis has made some of its most substantial contributions to the study of mental abilities as measured by tests of intellectual performance. Using exploratory factor analysis, psychologists have discovered that intellectual performance is highly complex and dependent upon many factors. Guilford (1967), for example, claimed to have established 82 factors of the intellect but postulated the existence of at least 120 factors. Guttman (1965) went farther than Guilford in conjecturing that the potential number of factors of the intellect is almost unlimited. Paradoxically, then, the technique of factor analysis, which originally was developed by psychologists such as Thurstone to help simplify the conceptualization of intellectual processes, had by the late 1960s achieved quite the opposite effect of inducing psychologists to consider intellectual processes as even more complex than they originally believed. As a result, faced with the growing plethora of factors coming from psychological laboratories using factor analysis, many psychologists began to doubt the value of factor analysis as a technique, with the question: What kind of coherent theory of mental processes can we construct with so many factors?

In response to such criticism, both Guilford (1967), with his structure-of-the-intellect model, and Guttman (1965), with his faceted definition of intelligence, attempted to bring order out of chaos by organizing mental tests (and their associated factors) into classification schemes which have some explanatory, theoretical significance. In doing so, Guilford (implicitly) and Guttman (explicitly) took the position that a coherent theory of mental processes will not come automatically from the methods of exploratory factor analysis but will have to be constructed on the basis of other, theoretical grounds. For example, Guttman (1965) recommended that researchers seek to identify common external features of intellectual-performance tests which may be used to classify these tests in the expectation that tests having the most classes in common will be intercorrelated the most, and those having the least in common will be intercorrelated the least [this is Guttman's (1959) contiguity metahypothesis]. Guilford (1967) appeared to use a kind of information-processing model of intellectual processes to guide him in specifying which tests elicit certain hypothetical intellectual processes. Guilford categorized tests according to their content (figural, symbolic, semantic, or behavioral), the mental operation involved (evaluation, convergent production, divergent production, memory, or cognition), and the kind of product elicited (units, classes, relations, systems, transformations, or implications).

In effect, drawing upon computer terminology, we might regard Guilford's system as classifying tests according to the kind of input involved, the kind of operations carried out, and the kind of output produced. In more elaborate forms, such an input-operation-output model may have some value in suggesting further research. In analogy with Guilford, who identified a factor of the intellect with each three-way combination of content, operation, and product, the input-operation-output model may identify a factor with each independent pathway through the system.

Interactions thus represent levels of functioning on pathways linking process centers, while main effects represent the levels of function of the centers themselves. The general level effect is a measure of the level of functioning of the whole system.

At this point the reader may see an analogy between a classification design and an analysis of variance design. In the present case variables are classified, while in analysis of variance, observations are classified. Nevertheless, the analogy suggests that we may anticipate common factors to correspond to the effects of the comparable analysis of variance design: there will be a general factor corresponding to the grand mean common to all observations; there will be main effect factors corresponding to main effects, and "interaction" factors corresponding to interaction effects in analysis of variance.

However, as in analysis of variance, not every effect anticipated by the design of the experiment, may be present, and so not every common factor corresponding to an "effect" of the design may be present either. Thus the researcher will need to justify including a common factor suggested by the classification design, using substantive theory.

For further discussion of classification designs see Mulaik (1975) and Mellenbergh et al. (1979).

15.5.2 Multirater–Multioccasion Studies

Researchers may be concerned to establish objective evidence for the stability of personality characteristics over time. To add further to establishing objectivity, they will often use more than one rater on a given occasion to describe the subjects. And beyond this, each rater may make several ratings on scales designed to measure the same trait, which also contributes to the objectivity of the final result.

The author collaborated with his dissertation professor, who was working with a group of nursing educators in the western United States (Ingmire et al., 1967) to evaluate a continuing education program they were conducting in several states. They had conceived a situational exercise in which their workshop participants were to interview a patient, played by a nurse serving as a patient-actor. After performing in the exercise, the patient-actor and an external observer made ratings of the participant on several five-point rating scales designed to rate the nurse's supportiveness of the patient. The situational exercise was given to the participants both at the beginning of the series of workshops (which were conducted several times during a year) and at the end of the year. The researchers then decided they also needed a control group, and obtained 352 otherwise eligible nurses who had not participated in the workshop to undergo evaluation in the same situational exercise with the same patient-actors and external observers. A possible source of uncontrolled variation and possible lack of fit was that in each state a different patient actor and a different external observer were used. Like the workshop participants, the control-group subjects were given the same exercise twice, a year apart. The research question was whether there is an enduring trait of "supportiveness" by which nurses may vary, yet be somewhat stable over a year's time within a given nurse. The control group would establish the stability. The experimental group participating in the workshops were to evaluate whether there was change produced by the workshop. The table of correlations among the 22 variables used in this study are in Table 15.4.

Several years later (Mulaik, 1988) I used data from the control group subjects to illustrate a confirmatory factor analysis involving a multirater-multioccasion model. I will now perform a new analysis of that data with an even more constrained model, with more degrees of freedom, that upholds the same results as in Mulaik (1988).

A path diagram of the multirater-multioccasion model applied to the correlation matrix in Table 15.4 is shown in Figure 15.1. The diagram shows that on each of two occasions, year 1 and year 2, the rated subject's behavior causes the patient/actor and the external observer to form judgments of the subject's supportiveness, which in turn caused each of the raters to make their ratings on the respective rating scales. Three invariants were hypothesized: (1) Stimulus to judgment invariance: subject's supportive behavior in year 1 would cause the same proportional degree of supportiveness judgment in the rater in year 2. Other aspects of the rater's judgment were idiosyncratic with the rater. Consequently the path coefficient from X_5 (subject's behavior in year 1) to X_1 (patient-actor's judgment in year 1) was constrained to equal the path coefficient from X_6 (subject's behavior in year 2) to X_2 (patient actor's judgment in year 2). Likewise, the path coefficient from X_5 (subject's behavior

TABLE 15.4

Correlations among 22 Rating Variables of 352 Nurses' Supportive Behavior in a Situational Exercise Conducted on Two Occasions a Year Apart[a]

Patient/Actor Ratings Year 1

1 100 Communicating understanding
2 64100 Friendliness
3 73 65100 Supportive
4 64 57 72100 Personal involvement
5 71 61 71 62100 Security as patient

Patient/Actor Ratings Year 2

6 32 27 34 27 29100 Communicating understanding
7 24 24 31 24 23 67100 Friendliness
8 24 24 31 22 25 72 70100 Supportive
9 19 24 26 23 21 64 62 71100 Personal involvement
10 31 28 34 29 31 73 72 79 66100 Security as patient

External Observer Ratings Year 1

11 44 36 46 39 39 32 31 31 25 34100 Respect for individual
12 48 40 47 39 44 36 30 31 23 31 68 100 Encourages patient to talk
13 45 39 51 40 42 31 25 26 20 28 71 70100 Recognizes need for security
14 42 39 47 36 40 36 24 28 31 30 61 67 65100 Information seeking
15 47 48 51 43 47 37 28 32 26 33 72 73 72 69100 Supportive
16 49 43 51 43 49 33 27 29 23 33 73 76 75 69 88100 Understanding

External Observer Ratings Year 2

17 20 24 21 10 20 45 50 47 40 49 34 33 25 25 32 29100 Respect for individual
18 21 19 24 17 21 50 47 50 42 52 31 39 32 31 34 34 68100 Encourages patient to talk
19 24 27 26 16 25 45 41 46 38 52 29 33 33 27 31 31 66 71100 Recognizes need for security
20 19 22 26 12 18 49 38 52 44 47 31 32 31 31 32 33 65 70 64100 Information seeking
21 24 25 26 17 27 52 52 59 49 61 32 44 36 32 37 36 69 73 73 68100 Supportive
22 20 25 27 17 27 54 51 57 49 61 32 39 33 32 36 34 71 75 74 72 87100 Understanding

[a] Decimal points have been omitted.

in year 1) to X_3 (external observer judgment in year 1) was constrained to equal the path coefficient from X_6 (subject's behavior in year 2) to X_4 (external observer's judgment in year 2). (2) Semantic invariance principle: the relation between judgment and rating for a given rating scale would be the same on each occasion a year apart. Hence, for example, the factor loading for variable Y_2 (Friendliness) on factor X_1 was constrained to equal the factor loading for variable Y_7 (Friendliness) on factor X_2.

Comparable constraints of equality were made for all factor loadings for comparable pairs of corresponding rating variables. (3) *invariance of variances of disturbances* of corresponding variables a year apart. Hence, for example, the error variance of ε_2 on variable Y_2 was constrained to equal the error variance

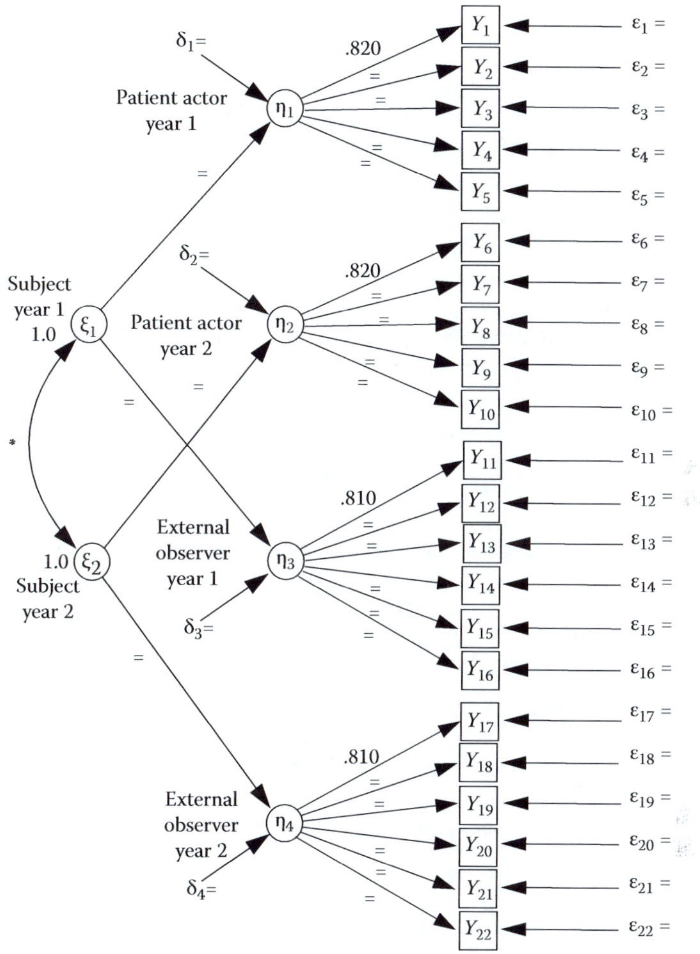

FIGURE 15.1
Path diagram for model of nurses' ratings by a patient actor and an external observer in a nurse-patient situational excise conducted twice a year apart. Numerical values for coefficients indicate fixed parameter values; = denotes parameter constrained equal to its counterpart a year apart; * denotes a free parameter.

of ε_7 on variable Y_7. Similarly the variance of the second-order disturbance δ_1 on X_1 was constrained to equal the comparable variance of the second-order disturbance δ_2 on X_2, and so on. The correlation between subject's supportiveness behavior in year 1 with that in year 2 was a free parameter.

The metric of the factor loadings was set by fixing the loading of Y_1 on X_1 and the loading of Y_6 on X_2 to .820. Similarly the loadings of Y_{10} on X_3 and of Y_{17} on X_4 were both fixed to .810. These values were determined by an exploratory factor analysis as being values consistent with an approximate variance of unity for the common factor in question. But the values were also set to be equal according to the semantic invariance principle. (Otherwise setting the metric is arbitrary.)

A confirmatory factor analysis of the model with these constraints was performed using the EQS program, Version 5.1, on an iMac computer. The results are shown in Table 15.5.

Although the chi-square goodness-of-fit statistic is significant with $p<.001$, indicating significant lack of fit, the model is still a very good approximation to the data, as indicated by the CFI and RMSEA indices. The model has plausibility in that it conforms well to a stimulus-response paradigm where the nurse-subject provides a common stimulus for two raters to rate, and their judgments as responses are then further expressed in ratings on several scales, which conform well to a single common factor among them. Lack of fit may be explained by small departures from causal homogeneity among the

TABLE 15.5

First- and Second-Order Factor-Pattern Matrices, First- and Second-Order Unique Variances, and Correlation between Second-Order Common Factors[a]

	First-Order Factors						Second-Order					Correlation	
	$\eta 1$	$\eta 2$	$\eta 3$	$\eta 4$	ψ^2		$\xi 1$	$\xi 2$		δ^2		$\xi 1$	$\xi 2$
1	**820**	**000**	**000**	**000**	306=	1	791=	**000**	1	407=	1	100	
2	762=	**000**	**000**	**000**	401=	2	**000**	791=	2	407=	2	58*	**100**
3	870=	**000**	**000**	**000**	219=	3	841=	**000**	3	246=			
4	763=	**000**	**000**	**000**	400=	4	**000**	841=	4	246=			
5	839=	**000**	**000**	**000**	274=								
6	**000**	**820**	**000**	**000**	306=								
7	**000**	762=	**000**	**000**	401=								
8	**000**	870=	**000**	**000**	219=		$\chi^2_{228}=409.636$		p<.001				
9	**000**	763=	**000**	**000**	400=								
10	**000**	839=	**000**	**000**	274=		CFI = .972						
11	**000**	**000**	**810**	**000**	374=								
12	**000**	**000**	847=	**000**	316=		RMSEA = .047						
13	**000**	**000**	834=	**000**	336=								
14	**000**	**000**	792=	**000**	402=		PR = .901						
15	**000**	**000**	935=	**000**	167=								
16	**000**	**000**	952=	**000**	135=								
17	**000**	**000**	**000**	**810**	374=								
18	**000**	**000**	**000**	847=	316=								
19	**000**	**000**	**000**	834=	336=								
20	**000**	**000**	**000**	792=	402=								
21	**000**	**000**	**000**	935=	167=								
22	**000**	**000**	**000**	952=	135=								

Note: Chi-square, CFI index, and parsimony ratio are given. Fixed parameters in boldface; = denotes constrained equal; * denotes free parameter.

[a] Decimal points have been omitted.

nurse subjects and also by a few replacements of the patient actor and external rater from one year to the next in a few state regions. Nevertheless, the model is highly tested with a parsimony ratio of .901. Thus the gem revealed by the study is an estimated correlation of .58 between the nurse-subject's objective supportiveness behavior measured twice over a year apart. This was also the finding in Mulaik (1988). The multirater-multioccasion paradigm seems to be an excellent research paradigm for studying trait stability. One may wonder why a multilevel analysis has not been done to deal with different regions and different actors and observers in the situational exercise. Unfortunately the original data that would make that possible is no longer in existence.

15.5.3 Multitrait–Multimethod Covariance Matrices

In seeking to formulate methods for establishing the objective validity of certain trait rating constructs Campbell and Fiske (1959) recommended that researchers obtain ratings on the same set of trait scales by more than one method. They called the correlation matrix among the trait rating scales, measured under the several methods, a multitrait-multimethod (MTMM) correlation matrix. They argued that the objective validity of the ratings would consist in the coherence of ratings on corresponding trait scales across the different methods. They described four conditions supporting the objectivity or *convergent validity* of the ratings: (1) Correlations between corresponding scales under different methods should be nonzero and statistically significant. (2) Correlations between corresponding scales under different methods should be higher than those between dissimilar scales compared across the different methods. (3) Correlations between corresponding scales measured under different methods should be higher than correlations measured within the same method between different scales. (4) The pattern of correlations between scales in different methods should be similar to the pattern of correlations between scales within the same method. The limitations of the Campbell and Fiske (1959) approach lay principally in the subjective way of determining convergent validity by merely inspecting the correlation matrix. Also, correlations were used, which would often force corresponding scale ratings into different metrics, by dividing covariances between the respective scales by different standard deviations within and between different methods. Corresponding variables also might have different reliabilities under different methods, varying then the pattern of relative magnitudes of correlations across different methods, even when there is convergent validity. Nevertheless various researchers sought to develop ways of analyzing multitrait-multimethod matrices to provide more objective ways of assessing convergent validity.

Jöreskog (1974) demonstrated how a multitrait-multimethod correlation matrix used by Campbell and Fiske (1959) to illustrate their method (Table 15.6) could be modeled by a confirmatory factor analysis model. Campbell and Fiske (1959) took data from a study conducted by Kelly and Fiske (1951)

TABLE 15.6

Multitrait-Multimethod Correlation Matrix Based on Ratings of Clinical Psychology Students on Five Traits by Three Kinds of Raters

		1P	2P	3P	4P	5P	1T	2T	3T	4T	5T	1S	2S	3S	4S	5S
Staff Ratings																
Assertive	1P	1.00														
Cheerful	2P	.37	1.00													
Serious	3P	−.24	−.14	1.00												
Unshakeable poise	4P	.25	.46	.08	1.00											
Broad interests	5P	.35	.19	.09	.31	1.00										
Teammate Ratings																
Assertive	1T	.71	.35	−.18	.26	.41	1.00									
Cheerful	2T	.39	.53	−.15	.38	.29	.37	1.00								
Serious	3T	−.27	−.31	.43	−.06	.03	−.15	−.19	1.00							
Unshakeable poise	4T	.03	−.05	.03	.20	.07	.11	.23	.19	1.00						
Broad interests	5T	.19	.05	.04	.29	.47	.33	.22	.19	.29	1.00					
Self-Ratings																
Assertive	1S	.48	.31	−.22	.19	.12	.46	.36	−.15	.12	.23	1.00				
Cheerful	2S	.17	.42	−.10	.10	−.03	.09	.24	−.25	−.11	−.03	.23	1.00			
Serious	3S	−.04	−.13	.22	−.13	−.05	−.04	−.11	.31	.06	.06	−.05	−.12	1.00		
Unshakeable poise	4S	.13	.27	−.03	.22	−.04	.10	.15	.00	.14	−.03	.16	.26	.11	1.00	
Broad interests	5S	.37	.15	−.22	.09	.26	.27	.12	−.07	.05	.35	.21	.15	.17	.31	1.00

Source: After Campbell, D.T. and Fiske, D.W., *Psychol. Bull.*, *56*, 81–105, 1959.

of the ratings of 124 clinical psychology students in a clinical setting on five different trait scales using three methods: staff ratings, peer ratings, and self-ratings. The correlation matrix among these scales is shown in Table 15.7. Jöreskog (1974) hypothesized five trait factors, T1,…, T5, one for each of the five traits rated. Corresponding scales were hypothesized to load on the same factor, so there were three indicators of each trait factor across the three methods of rating. Presumed nonzero loadings were treated as free parameters, no effort being made to prespecify a value to test for each of these loadings. Scales not corresponding to a factor in question were hypothesized to have zero loadings on the factor. Jöreskog also initially hypothesized three method factors, M1, M2, and M3, with scales rated by the same method presumed to load on the same method factor. Factor variances were fixed to unity to determine the metric of the solution. Correlations among trait factors were treated as free parameters, and

TABLE 15.7

Factor Loadings and Correlations among Factors in Multitrait-Multimethod Model (Jöreskog, 1974) for Correlations in Table 9.7

	Factor-Pattern Loadings							
	T1	**T2**	**T3**	**T4**	**T5**	**M1**	**M2**	u_{ii}^2
Staff Ratings								
Assertive	.871	**.000**	**.000**	**.000**	**.000**	.107	**.000**	.239
Cheerful	**.000**	.836	**.000**	**.000**	**.000**	.017	**.000**	.302
Serious	**.000**	**.000**	.573	**.000**	**.000**	−.296	**.000**	.583
Unshakeable poise	**.000**	**.000**	**.000**	.781	**.000**	−.253	**.000**	.318
Broad interests	**.000**	**.000**	**.000**	**.000**	.689	−.335	**.000**	.392
Teammate Ratings								
Assertive	.829	**.000**	**.000**	**.000**	**.000**	**.000**	.162	.291
Cheerful	**.000**	.697	**.000**	**.000**	**.000**	**.000**	.294	.468
Serious	**.000**	**.000**	.722	**.000**	**.000**	**.000**	.322	.349
Unshakeable poise	**.000**	**.000**	**.000**	.213	**.000**	**.000**	.533	.674
Broad interests	**.000**	**.000**	**.000**	**.000**	.599	**.000**	.439	.429
Self-Ratings								
Assertive	.552	**.000**	**.000**	**.000**	**.000**	.111	**.000**	.689
Cheerful	**.000**	.454	**.000**	**.000**	**.000**	.221	**.000**	.751
Serious	**.000**	**.000**	.428	**.000**	**.000**	.227	**.000**	.767
Unshakeable poise	**.000**	**.000**	**.000**	.429	**.000**	.380	**.000**	.678
Broad interests	**.000**	**.000**	**.000**	**.000**	.697	.622	**.000**	.168
Correlations among Factors								
T1	**1.000**							
T2	.559	**1.000**						
T3	−.371	−.438	**1.000**					
T4	.381	.662	−.082	**1.000**				
T5	.548	.292	.125	.430	**1.000**			
M1	**.000**	**.000**	**.000**	**.000**	**.000**	**1.000**		
M2	**.000**	**.000**	**.000**	**.000**	**.000**	−.208	**1.000**	

Note: Zeros and unities in boldface are fixed parameters.

correlations among method factors were also treated as free parameters. Correlations between trait factors and method factors were fixed to zero.

In his initial estimation of the free parameters of this model, Jöreskog obtained a solution in which method factors M1 and M3 were correlated 1.00 with each other. So, he reparameterized his model to postulate two method factors with scales under method 1 (staff ratings) and method 3 (self-ratings) loading on the same method factor M1 (in the new model) and scales rated by teammates loading on method factor M2.

I have reestimated the parameters of Jöreskog's model for the correlation matrix given in Table 15.6 using Bentler and Wu's EQS 5 program, and I have obtained essentially the identically same solution as Jöreskog (1974) (see Table 15.7). Initially when I postulated three method factors, they yielded estimated correlations of unity between methods 1 and 3. So, I reparameterized, as did Jöreskog (1974), to have only two method factors and the results are shown in Table 15.7. The chi square test of goodness of fit with 64 degrees of freedom was 61.512, which was not significant. Several goodness-of-fit indices indicated excellent fit as well: Jöreskog's GFI index was .941, whereas Bentler's CFI was 1.00. Some of our enthusiasm for the good fit of this model should be tempered by knowledge that the sample size was only 124, which, being less than 200, is currently regarded as inadequate for statistical inference purposes with chi square statistics in confirmatory factor analysis. The sample size of 124 may also lack power to reject the null hypothesis with other than large discrepancies. But parameter estimates should be relatively stable. Each of the trait factors had strong loadings on their respective factors. The method factors did not have strong influences on all ratings in all conditions presumed to be effected by them as evidenced by low loadings in many cases, and one near zero loading.

Jöreskog's demonstration of a confirmatory factor analysis model for the multitrait-multimethod correlation matrix has engendered numerous attempts to fit the model to other multitrait-multimethod matrices. These efforts have not always been successful. Frequently Jöreskog's (1974) model was empirically underidentified when applied to certain correlation matrices. Sometimes the algorithms for estimating parameters of this model did not converge. On other occasions it was discovered that freeing up correlations between the trait factors and method factors led to the model's being underidentified (Wohtke, 1984; Widaman, 1985). (This happens if one frees up the correlations between trait factors and method factors in Jöreskog's model (1974) applied to the Campbell and Fiske (1959) correlation matrix in Table 15.6.) The realization eventually emerged that Jöreskog's (1974) model was only one of several models that might be fit to this kind of data. For example, in some cases it was found that there were no trait factors, only method factors, and in other cases there were no method factors, only trait factors. On other occasions the number of trait factors was fewer than the number of trait scales. Wothke (1984) worked out mathematically necessary conditions for identification of the models and showed that in some cases the models could be empirically underidentified.

Widaman (1985) described sequences of "hierarchically nested" models that contain the multitrait-multimethod model of Jöreskog (1974) as a special case, that might be applied to a given multitrait-multimethod correlation matrix to find the best fitting model for that data. In developing these sequences he developed a way of classifying multitrait-multimethod models according to the properties of their trait factors and their method factors. For the trait factors there were four cases:

1, no trait factors

2, t trait factors, fixed unit intercorrelations (or *one* trait factor)

2′, t trait factors, fixed zero intercorrelations among trait factors

3, t trait factors, free intercorrelations among trait factors

Cases 2 and 2′ estimate the same number of parameters and are not strictly nested within one another. Widaman noted that it is not possible to perform a statistical test of the difference in fit between the two models, although one might prefer one to the other if its chi-square value were smaller. He also pointed out that cases designated by larger integers are more inclusive (or less restricted) than cases with smaller integers, so cases with smaller numbers are nested (special cases) of cases with larger case numbers.

Widaman (1985) also noted that a parallel set of cases could be formulated for the method factors of the multitrait-multimethod models:

A, no method factors

B, m method factors fixed unit intercorrelations (*one* method factor)

B′, m method factors, fixed zero intercorrelations among method factors

C, m method factors, free intercorrelations among method factors

Again cases B and B′ are not nested within one another as are cases 2 and 2′ above. Cases designated by letters that come earlier in the alphabet are nested within cases designated by letters that come later in the alphabet.

Widaman (1985) suggested that a taxonomy for multitrait-multimethod models could be generated by cross-classifying the four trait cases with the four method cases. A model could then be designed by its method and trait factor structures, respectively, e.g., 3B′ designates a model that has t freely correlated trait factors and m orthogonal method factors. Furthermore one could determine that one model is nested within another if both the trait and method cases respectively were nested within the trait and method cases respectively of the other model. For example, model 2A is a more restricted model nested within 3A, 2C, and 3C. However, model 2A is not nested within 1C, since model 2A has a higher number than 1C. The implication of one model's being nested within the other is that one can construct and test a nested sequence of models, beginning with the least restricted model. If the least restricted model does not fit the data, then one knows that the more restricted models will not either. On the other hand, if the least restricted model fits, then one can proceed to test more restricted models in the sequence until one reaches a model that does not fit.

In formulating his classification of models, Widaman (1985) excluded models in which correlations between method factors and trait factors were free parameters. Those cases led to underidentified models. He required trait factors to be uncorrelated with method factors. He also noted that some pairings of trait cases with method cases were problematic. A model 2B is a model with a single common factor in which method and trait factors are indistinguishable. A model 2′ B′ would have a single trait factor and a single method factor, each orthogonal to the other, but loading on all variables. Such a model is not identified.

In many cases, researchers have had difficulties in finding and fitting method factors to their data. Marsh (1989) proposed not including common method factors in the model but instead freeing the correlations between unique factors of trait scales in the same method while retaining fixed zero correlations between unique factors of variables under different methods. This is a less restricted way to introduce "method" effects that usually has better fit than models with a single common method factor for all scales in a given method. On the other hand, it abandons formulating and testing a hypothesis about the structure of the method effect within a given method. Still models with correlated uniquenesses within methods may be seen as less restricted models to compare with models that do impose method factor structure, using chi square difference tests, to test the nature of the structure. But Byrne and Goffin (1993) note several possible limitations to Marsh's correlated unique-factor model: The trait-factor loadings and correlations among traits in Marsh's correlated uniquenesses models are higher than they typically are in Jöreskog's common-factor model when applied to the same data set. (This may not really be a limitation but a difference in representation.) But Byrne and Goffin believe that as a result Marsh's correlated unique-factor model for handling method effects is biased toward providing stronger evidence for convergent validity and weaker evidence for discriminant validity. Perhaps a clearer limitation of the correlated unique-factor models is Byrne and Goffin's (1993) observation that because unique factors of variables under different methods cannot be freed to correlate (otherwise the model may become underidentified), this introduces an untestable and unrealistic assumption that method effects are uncorrelated—which Jöreskog's (1974) multitrait-multimethod factor analysis model can test, but only with well-defined common method factors. Testing the orthogonality assumption would require comparing the model in which these correlations are zero against an identified model in which they are free to be any permissible value. Jöreskog's (1974) model, on the other hand, has its own untestable and unrealistic assumption, that trait factors and method factors are uncorrelated, for, again, to test this orthogonality assumption would require estimating a model in which these correlations are free parameters, to yield a chi-square difference test when the chi square of the less restricted model is subtracted from the chi square of the more restricted model, and the difference chi square has degrees of freedom equal to the difference in degrees of freedom of the original chi squares.

Numerous excellent reviews have been written about multitrait-multimethod models within the confirmatory factor analysis framework. Readers may wish to consult Wothke (1984, 1996), Browne (1984), Marsh

(1988, 1989), Graham and Collins (1991), Marsh, Byrne and Craven (1992), and Byrne and Goffin (1993). There have also been explorations of other kinds of models besides the additive model of factor analysis. Browne (1989) formulated a multiplicative model whereby traits and methods multiply rather than add together:

$$\mathbf{\Sigma} = \mathbf{D}(\mathbf{P}_M \otimes \mathbf{P}_T + \mathbf{E}^2)\mathbf{D}$$

where

$\mathbf{\Sigma}$ is a $t \times m$ covariance matrix of t trait variables measured under m methods

$\mathbf{D}$ a $t \times m$ diagonal matrix of true-score variances for the $t \times m$ variables

$\mathbf{P}_M$ is an $m \times m$ matrix of correlations among m method variables

$\mathbf{P}_T$ a $t \times t$ matrix of correlations among t trait variables, with $\mathbf{E}^2$ a $tm \times tm$ diagonal matrix of unique variances

We will not pursue this model further, but refer the reader to Browne's paper and to papers by Cudeck (1988) and Byrne and Goffin (1993).

To close this discussion of multitrait-multimethod models, I would like to point out that the multitrait-multimethod factor analysis models described so far can be seen as special cases of the faceted design model. If the faceted design models we described earlier are seen to be built on an analogy with factorial analysis of variance designs with repeated observations per cell, the multitrait-multimethod model described by Jöreskog (1974) has a certain analogy with the randomized blocks design with one observation per cell. The weaknesses of the randomized blocks model rests among other things, on there being only one observation per cell of the design, which prevents consideration of interaction effects distinct from error. The same applies in the corresponding faceted design for multitrait-multimethod factor analysis with one indicator per trait-method combination. Many of the identification problems with this model arise because too few indicators are trying to do too much work. Having multiple indicators (at least 4) per trait-method combination would allow one to include not only main-effect factors for traits and methods, but factors corresponding to their interaction. An illustration of just such a model is shown in Figure 15.2.

15.6 Conclusion

When Jöreskog (1969a,b) brought forth confirmatory factor analysis, it represented a sea-change in correlational statistics. Instead of exploring, researchers could now test hypotheses about the structural composition of sets of correlated variables. However, Jöreskog (1974) soon eclipsed confirmatory

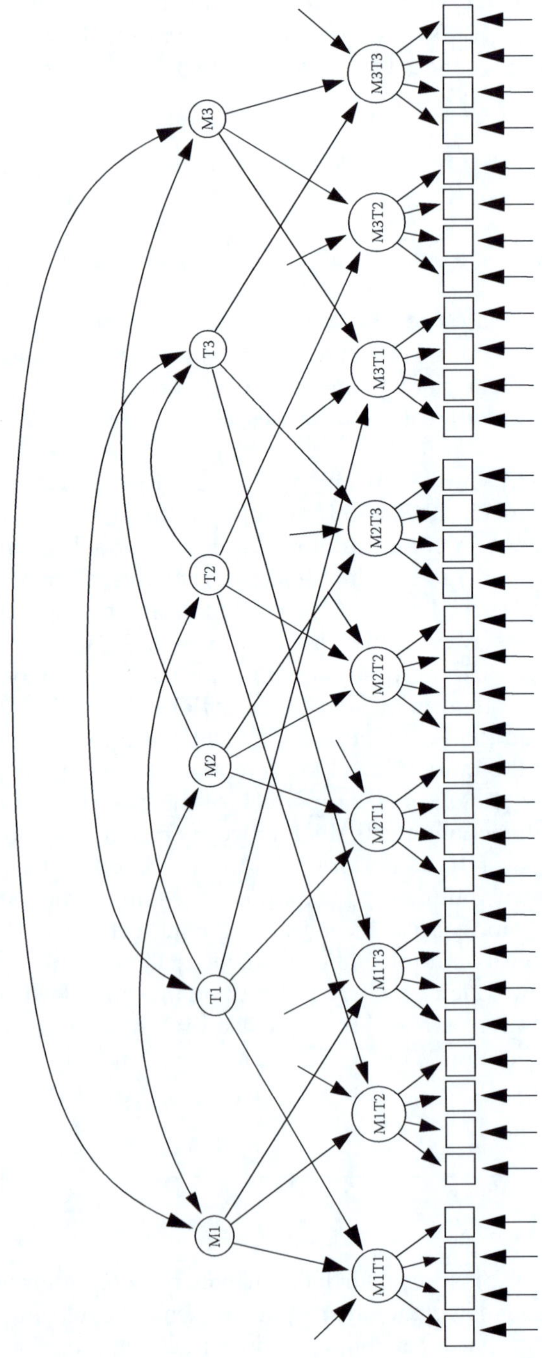

FIGURE 15.2
A facet design for a 3×3 MMMT model with 4 indicators per latent first-order factor representing a method/treatment combination.

factor analysis with analysis of covariance structures, and later with structural equation modeling with latent variables (Jöreskog. 1974), which subsumed confirmatory factor analysis as a submodel, and one that was limited in the forms of causality that it could represent to cases of common causes. Still the model remains a useful tool in the behavioral and social statistician's toolbox. Many situations conform to a common-factor framework. The fact that frequently a confirmatory factor analysis model is embedded within a structural equation model because each latent variable has four or more indicators of it allows for sequential testing of a series of nested models, from the unrestricted model, to the measurement model, and then the structural equation model to help disambiguate any lack of fit that might be found in the structural equation model.

For a much more thorough survey of various applications for confirmatory factor analysis than given here, consult Brown (2006).

References

Ahmavaara, Y. (1954). The mathematical theory of factorial invariance under selection. *Psychometrika, 19*, 27–38.

Aitken, A. C. (1934). Note on selection from a multivariate normal population. *Proceedings of the Edinburgh Mathematical Society, 4*, 106–110.

Anderson, T. W. (1984). *Introduction to Multivariate Statistical Analysis*, 2nd edn. New York: Wiley.

Anderson, J. C. and Gerbing, D. W. (1988). Structural equation modeling in practice: A review and recommended two-step approach. *Psychological Bulletin, 103*, 411–423.

Anderson, T. W. (1958). *An Introduction to Multivariate Statistical Analysis*. New York: Wiley.

Bargmann, R. E. (1955). A demonstration study on the effectiveness of factor-analytic methods. Hochschule f. Internationale Paed. Forschung, Frankfurt/Main, Forschungsbericht, June 1995.

Bartholomew, D. (1981). Posterior analysis of the factor model. *British Journal of Mathematical and Statistical Psychology, 34*, 93–99.

Bartholomew, D. (1996a). Comments on: Metaphor taken as math: Indeterminacy in the factor model. *Multivariate Behavioral Research, 31*, 551–554.

Bartholomew, D. (1996b). Response to Dr. Maraun's first reply to discussion of his paper. *Multivariate Behavioral Research, 31*, 631–636.

Bartlett, M. S. (1937). The statistical conception of mental factors. *British Journal of Psychology, 28*, 97–104.

Bartlett, M. S. (1950). Tests of significance in factor analysis. *British Journal of Psychology, 3*, 77–85.

Bartlett, M. S. (1951). A further note on tests of significance in factor analysis. *British Journal of Psychology, Statistical Section, 4*, 1–2.

Bentler, P. M. (1990). Comparative fit indexes in structural models. *Psychological Bulletin, 107*, 238–246.

Bentler, P. M. and Bonett, D. G. (1980). Significance tests and goodness of fit in the analysis of covariances structures. *Psychological Bulletin, 88*, 588–606.

Bernaards. C. A. and Jennrich, R. I. (2005). Gradient projection algorithms and software for arbitrary rotation criteria in factor analysis. *Educational and Psychological Measurement, 65*, 676–696.

Bock, R. D. and Bargmann, R. E. (1966). Analysis of covariance structures. *Psychometrika, 31*, 507–534.

Bollen, K. A. (1989). *Structural Equations with Latent Variables*. New York: Wiley.

Brown, R. H. (1961). A comparison of the maximum determinant solution in factor analysis with various approximations. Unpublished research report supported by a grant of the McDermott Foundation, Texas Instruments, Inc., Rolf E. Bargmann, principal investigator.

Browne, M. W. (1967). On oblique procrustes rotation. *Psychometrika, 32*, 125–132.

Browne, M. W. (1972). Oblique rotation to a partially specified target. *British Journal of Mathematical and Statistical Psychology, 25*, 207–212.

Browne, M. W. (1974). Generalized least-squares estimators in the analysis of covariance structures. *South African Statistical Journal, 8*, 1–24.

Browne, M. W. (1977). Generalized least-squares estimators in the analysis of covariance structures. In D. J. Aigner and A. S. Goldberger (Eds.), *Latent Variables in Socio-Economic Models*. Amsterdam: North-Holland, pp. 205–226.

Browne, M. W. (1982). Covariance structures. In D. M. Hawkins (Ed.), *Topics in Multivariate Analyses*. Cambridge, U.K.: Cambridge University Press, pp. 72–141.

Browne, M. W. (1984). Asymptotically distribution-free methods for the analysis of covariance structures. *British Journal of Mathematical and Statistical Psychology, 37*, 62–83.

Browne, M. W. and Cudeck, R. (1989). Single sample cross-validation indicies for covariance structures. *Multivariate Behavioral Research, 24*, 445–455.

Browne, M. W. and Cudeck, R. (1993). Alternative ways of assessing model fit. In K. H. Bollen and J. S. Long (Eds.), *Testing Structural Equation Models*. Newburg Park, CA: Sage Publications.

Butler, J. M. (1968). Descriptive factor analysis. *Multivariate Behavioral Research, 3*, 355–370.

Butler, J. M. and Hook, L. H. (1966). Multiple factor analysis in terms of weighted multiple regression. *Educational and Psychological Measurement, 26*, 545–565.

Byrne, B. (2006). *Structural Equation Modeling with EQS*, 2nd edn. Mahwah, NJ: Lawrence Erlbaum Associates.

Byrne, B. M. and Goffin, R. D. (1993). Modeling MTMM data from additive and multiplicative covariance structures: An audit of construct validity concordance. *Multivariate Behavioral Research, 28*, 67–96.

Campbell, D. T. and Fiske, D. W. (1959). Convergent and discriminant validation by the multitrait-multimethod matrix. *Psychological Bulletin, 56*, 81–105.

Carlson, M. and Mulaik, S. A. (1993). Trait ratings from descriptions of behavior as mediated by components of meaning. *Multivariate Behavioral Research, 28*, 111–159.

Carroll, J. B. (1953). An analytic solution for approximating simple structure in factor analysis. *Psychometrika, 18*, 23–38.

Carroll, J. B. (1957). Biquartimin criterion for rotation to oblique simple structure in factor analysis. *Science, 126*, 1114–1115.

Cattell, R. B. (1966a). The scree test for the number of factors. *Multivariate Behavioral Research, 1*, 245–276.

Cattell, R. B. (1966b). *Handbook of Multivariate Experimental Psychology*. Chicago, IL: Rand McNally.

Cooley, W. W. and Lohnes, P. R. (1962). *Multivariate Procedures for the Behavioral Sciences*. New York: Wiley.

Corbato, F. J. (1963). On the coding of Jacobi's method for computing the eigenvalues and eigenvectors of real symmetric matrices. *Journal of the Association of Computer Machinery, 10*, 123–125.

Crawford, C. (1967). A general method of rotation for factor analysis. In *Paper Read at Spring Meeting of the Psychometric Society*, Madison, WI, April 1, 1967.

Cronbach, L. J. (1951). Coefficient alpha and the internal structure of tests. *Psychometrika, 16*, 297–334.

Cronbach, L. J., Rajaratnam, N., and Gleser, G. C. (1963). Theory of generalizability: A liberalization of reliability theory. *British Journal of Statistical Psychology, 16*, 137–163.

Cureton, E. E. and Mulaik, S. A. (1975). The weighted varimax and the promax rotation. *Pschometrika, 40,* 183–195.

Dwyer, P. S. (1939). The contribution of an orthogonal multiple factor solution to multiple correlation. *Psychometrika, 4,* 163–171.

Dwyer, P. S. (1944). A matrix presentation of least-squares and correlation theory with matrix justification of improved methods of solution. *The Annals of Mathematical Statistics, 15,* 82–89.

Eber, H. W. (1966). Toward oblique simple structure maxplane. *Multivariate Behavioral Research, 1,* 112–125.

Eckart, C. and Young, G. (1936). The approximation of one matrix by another of lower rank. *Psychometrika, 1,* 211–218.

Embretson, S. E. and Reise, S. P. (2000). *Item Response Theory for Psychologists.* Mahwah, NJ: Lawrence Erlbaum Associates.

Fabrigar, L. R., Wegener, Dt. T., MacCallum, R. C., and Strahan, E. J. (1999). Evaluating the use of exploratory factor analysis in psychological research. *Psychological Methods, 4,* 272–299.

Ferguson, G. A. (1954). The concept of parsimony in factor analysis. *Psychometrika, 19,* 281–290.

Fisher, R. A. (1918). The correlation between relatives on the supposition of Mendelian inheritance. *Transactions of the Royal Society of Edinburgh, 52,* 399–433.

Fisher, R. A. (1925). *Statistical Methods for Research Workers.* London, U.K.: Oliver and Boyd.

Fletcher, R. and Powell, M. J. D. (1963). A rapidly convergent descent method for minimization. *The Computer Journal, 2,* 163–168.

Forster, P. (2001). Scientific inquiry as a self-correcting process. *Digital Encyclopedia of Charles S. Peirce.* http://www.digitalpeirce.fee.unicamp.br/home.htm

Galton, F. (1869). *Hereditary Genius, an Inquiry into Its Laws and Consequences* (2nd edn 1892). London, U.K.: Macmillan.

Galton, F. (1889). *Natural Inheritance.* London, U.K.: Macmillan.

Garrison, J. W. (1986). Some principles of postpositivistic philosophy of science. *Educational Researcher, 15,* 12–18.

Ghiselli, E. (1964). *Theory of Psychological Measurement.* New York: McGraw-Hill.

Gibbs, J. W. and Wilson, E. B. (1901). *Vector Analysis.* New Haven, CT: Yale University Press.

Glorfeld, L. W. (1995). An improvement on Horn's parallel analysis methodology for selecting the correct number of factors to retain. *Educational and Psychological Measurement, 55,* 377–393.

Goldstine, H. H., Murray, F. J., and von Neuman, J. (1959). The Jacobi method for real symmetric matrices. *Journal of the Association for Computing Machinery, 6,* 59–96.

Gorsuch, R. L. (1983). *Factor Analysis,* 2nd edn. Hillsdale, NJ: Lawrence Erlbaum Associates.

Green, B. F. (1952). The orthogonal approximation of an oblique structure in factor analysis. *Psychometrika, 17,* 429–440.

Guilford, J. P. (1941). The difficulty of a test and its factor composition. *Psychometrika, 6,* 67–78.

Guilford, J. P. (1956a). *Fundamental Statistics in Psychology and Education.* New York: McGraw-Hill.

Guilford, J. P. (1956b). The structure of intellect. *Psychological Bulletin, 53,* 267–293.

Guilford, J. P. (1959). Three faces of intellect. *American Psychologist, 14,* 469–479.

Guilford, J. P. (1967). *The Nature of Human Intelligence*. New York: McGraw-Hill.
Gulliksen, H. (1950). *Theory of Mental Tests*. New York: Wiley.
Guttman, L. (1940). Multiple rectilinear prediction and the resolution into components. *Psychometrika, 5*, 75–99.
Guttman, L. (1944). General theory and methods of matric factoring. *Psychometrika, 9*, 1–16.
Guttman, L. (1953). Image theory for the structure of quantitative variates. *Psychometrika, 18*, 277–296.
Guttman, L. (1954a). Some necessary conditions for common-factor analysis. *Psychometrika, 19*, 149–161.
Guttman, L. (1954b). A new approach to factor analysis: The radex. In P. F. Lazarsfeld (Ed.), *Mathematical Thinking in the Social Sciences*. Glencoe, IL: The Free Press.
Guttman, L. (1955). The determining of factor score matrices with implications for five other basic problems of common-factor theory. *British Journal of Statistical Psychology, 8*(2), 65–81.
Guttman, L. (1956). Best possible systematic estimates of communalities. *Psychometrika, 21*, 273–285.
Guttman, L. (1958). An estimate of communalities that is imagewise consistent and structure-free. Research Report 20, May 22, 1958, University of California, Berbeley, CA, Contract AF 41(656)-676.
Guttman, L. (1959). A structural theory for intergroup beliefs and attitudes. *American Sociological Review, 24*, 318–328.
Guttman, L. (1965). A faceted definition of intelligence. In R. Eiferman (Ed.), *Studies in Psychology, Scripta Hierosolymitana*, Vol. 14. Jerusalem, Israel: The Hebrew University.
Harman, H. H. (1960). *Modern Factor Analysis*. Chicago, IL: University of Chicago Press.
Harman, H. H. (1967). *Modern Factor Analysis*, 2nd edn. Chicago, IL: University of Chicago Press.
Harman, H. H. and Jones, W. H. (1966). Factor analysis by minimizing residuals (minres). *Psychometrika, 31*, 351–368.
Harris, C. W. (1962). Some Rao–Guttman relationships. *Psychometrika, 27*, 247–263.
Harris, C. W. (1967). On factors and factor scores. *Psychometrika, 32*, 363–379.
Harris, C. W. and Kaiser, H. F. (1964). Oblique factor analytic solutions by orthogonal transformations. *Psychometrika, 29*, 347–362.
Hayes, W. L. (1988). *Statistics*, 4th edn. New York: Holt, Rinehart and Winston.
Hayton, J. C., Allen, D. G., and Scarpello, V. (2004). Factor retention decisions in exploratory factor analysis: A tutorial on parallel analysis. *Organizational Research Methods, 7*, 191–205.
Heerman, E. F. (1963). Univocal or orthogonal estimators of orthogonal factors. *Psychometrika, 28*, 161–172.
Hendrickson, A. E. and White, P. O. (1964). PROMAX: A quick method for rotation to oblique simple structure. *British Journal of Statistical Psychology, 17*, 65–70.
Heywood, H. B. (1931). On finite sequences of real numbers. *Proceedings of the Royal Society of London, 134*, 486–501.
Holzinger, K. J. and Harman, H. H. (1941). *Factor Analysis*. Chicago, IL: University of Chicago Press.
Holzinger, K. J. and Swineford, F. A. (1939). A study in factor analysis: The stability of a bi-Jactor solution. *Supplementary Educational Monographs, No. 48*, University of Chicago, Chicago, IL.

Horn, J. L. (1965). A rationale and test for the number of factors in factor analysis. *Psychometrika, 30*, 179–185.

Horst, P. (1965). *Factor Analysis of Data Matrices*. New York: Holt, Rinehart & Winston.

Horst, P. (1969). *Generalized Factor Analysis*, Part l. April, University of Washington, Contract NONR-477(33), Office of Naval Research.

Hotelling, H. (1933). Analysis of a complex of statistical variables into principal components. *Journal of Educational Psychology, 24*, 417–441, 498–520.

Hotelling, H. (1936). Relations between two sets of variates. *Biometrika, 28*, 321–377.

Howe, W. G. (1955). *Some contributions to factor analysis*. Report No. ONRL-1919. Oak Ridge, TN: Oak Ridge National Laboratory.

Humphreys, L. G. and Montanelli, R. G. (1975). An investigation of the parallel analysis criterion for determining the number of common factors. *Multivariate Behavioral Research, 10*, 193–206.

Humphreys, L. G., Tucker, L. R., and Dachler, P. (1970). Evaluating the importance of factors in any given order of factoring. *Multivariate Behavioral Research, 5*, 209–215.

Hurley, J. R. and Cattell, R. B. (1962). The procrustes program: Producing direct rotation to test a hypothesized factor structure. *Behavioral Science, 7*, 258–262.

Jacobi, C. G. J. (1846). Ueber ein leichtes Verfahren die in der Theorie der Saecularstoerungen vorkommen-den Gleichungen numerisch aufzuloesen. *Journal fur die Reine und Angewandte Mathematik, 30*, 51–94.

Jastrow, J. (1900). *Fact and Fable in Psychology*. Boston, MA: Houghton Mifflin.

Jaynes, J. (1990). *The Origin of Consciousness in the Breakdown of the Bicameral Mind*, 2nd edn with an afterward. Boston, MA: Houghton-Mifflin & Co. First published in 1976.

Jennrich, R. I. (1979). Admissible values of γ in direct oblimin rotation. *Psychometrika*, 44, 173–177.

Jennrich, R. I. (2001). A simple general procedure for orthogonal rotation. *Psychometrika, 31*, 313–323.

Jennrich, R. I. (2002). A simple general method for oblique rotation. *Psychometrika, 67*, 7–20.

Jennrich, R. I. (2004). Derivative free gradient projection algorithms for rotation. *Psychometrika, 69*, 475–480.

Jennrich, R. I. (2006). Rotation to simple loadings using component loss functions: The oblique case. *Psychometrika, 71*, 173–191.

Jennrich, R. I. and Robinson, S. M. (1969). A Newton–Raphson algorithm for maximum likelihood factor analysis. *Psychometrika, 34*, 111–123.

Jennrich, R. I. and Sampson, P. F. (1966). Rotation for simple loadings. *Psychometrika, 31*, 313–323.

Johnson, R. A. and Wichern, D. W. (1998). *Applied Multivariate Statistical Analysis*, 4th edn. Upper Saddle River, NJ: Prentice-Hall.

Johnson, R. M. (1963). On a theorem stated by Eckart and Young. *Psychometrika, 28*, 259–263.

Jöreskog, K. G. (1962). On the statistical treatment of residuals in factor analysis. *Psychometrika, 27*, 335–354.

Jöreskog, K. G. (1963). *Statistical Estimation in Factor Analysis*. Stockholm, Sweden: Almqvist & Wiksell.

Jöreskog, K. G. (1965). Image factor analysis. Research Bulletin 65-05. Princeton, NJ: Educational Testing Service.

Jöreskog, K. G. (1966). Testing a simple structure hypothesis in factor analysis. *Psychometrika, 31,* 165–178.

Jöreskog, K. G. (1967). Some contributions to maximum likelihood factor analysis. *Psychometrika, 32,* 443–482.

Jöreskog, K. G. (1969). Efficient estimation in image factor analysis. *Psychometrika, 34*(1), 51–75.

Jöreskog, K. G. (1969a). A general approach to confirmatory maximum likelihood factor analysis. *Psychometrika, 34,* 183–202.

Jöreskog, K. G. (1969b). Factoring the multitest-multioccasion correlation matrix. Research Bulletin 69–62. Princeton, NJ: Educational Testing Service.

Jöreskog, K. G. (1979). Structural equation models in the social sciences: Specification, estimation and testing. In J. Magidson (Ed.), *Advances in Factor Analysis and Structural Equation Models.* Cambridge, MA: Abt Books.

Jöreskog, K. G. and Lawley, D. N. (1968). New methods in maximum likelihood factor analysis. *British Journal of Mathematical and Statistical Psychology, 21,* 85–96.

Jöreskog, K. G. and Sörbom, D. (1981). *LISREL V: Analysis of Linear Structural Relationships by the Method of Maximum Likelihood.* Chicago, IL: National Educational Resources.

Kaiser, H. F. (1958). The varimax criterion for analytic rotation in factor analysis. *Psychometrika, 23,* 187–200.

Kaiser, H. F. (1960). The application of electronic computers to factor analysis. *Educational and Psychological Measurement, 20,* 141–151.

Kaiser, H. F. (1961). A note on Guttman's lower bound for the number of common factors. *British Journal of Statistical Psychology, 14*(1), l.

Kaiser, H. F. (1963). Image analysis. In C. W. Harris (Ed.), *Problems in Measuring Change.* Madison, WI: University of Wisconsin Press, pp. 156–166.

Kaiser, H. F. (1970). A second-generation little jiffy. *Psychometrika, 35,* 401–415.

Kaiser, H. F. (1974). An index of factorial simplicity. *Psychometrika, 39,* 31–36.

Kaiser, H. F. and Caffrey, J. (1965). Alpha factor analysis. *Psychometrika, 30,* 1–14.

Katz, J. O. and Rohlf, F. J. (1974). Functionplane—A new approach to simple structure rotation. *Psychometrika, 39,* 37–51.

Kiers, H. A. L. (1994). SIMPLIMAX: Oblique rotation to an optimal target with simple structure. *Psychometrika, 59,* 567–579.

LaForge, R. (1965). Components of reliability. *Psychometrika, 30,* 187–195.

Lakoff, G. (1987). *Women, Fire and Dangerous Things: What Categories Reveal about the Mind.* Chicago, IL: University of Chicago Press.

Lakoff, G. (1993). The contemporary theory of metaphor. In A. Ortony (Ed.), *Metaphor and Thought,* 2nd edn. Cambridge, U.K.: Cambridge University Press, pp. 202–251.

Lakoff, G. and Johnson, M. (1980). *Metaphors We Live by.* Chicago, IL: University of Chicago Press.

Lakoff, G. and Johnson, M. (1999). *Philosophy in the Flesh.* New York: Basic Gooks.

Lakoff, G. and Nuñez (2000). *Where Mathematics Comes from. How the Embodied Mind Brings Mathematics into Being.* New York: Basic Books.

Lakoff, G. and Turner, M. (1989). *More Than Cool Reason: A Field Guide to Poetic Metaphor.* Chicago, IL: University of Chicago Press.

Landahl, H. D. (1938). Centroid orthogonal transformations. *Psychometrika, 3,* 219–223.

Lautenschlager, G. J. (1989). A comparison of alternatives to conducting Monte Carlo Analyses for determining parallel analysis criteria. *Multivariate Behavioral Research, 24,* 365–395.

Lawley, D. N. (1940). The estimation of factor loadings by the method of maximum likelihood. *Proceedings of the Royal Society of Edinburgh, 60*, 64–82.
Lawley, D. N. (1941). Further investigations in factor estimation. *Proceedings of the Royal Society of Edinburgh, 61*, 176–185.
Lawley, D. N. (1942). Further investigations in factor estimation. *Proceedings of the Royal Society of Edinburgh, Series A, 61*, 176–185.
Lawley, D. N. (1943, 1944). A note on Karl Pearson's selection formulae. *Proceedings of the Royal Society of Edinburgh, Series A, 62*, 28–30.
Lawley, D. N. (1949). Problems in factor analysis. *Proceedings of the Royal Society of Edinburgh, Series A, 62*, 394–399.
Lawley, D. N. (1958). Estimation in factor analysis under various initial assumptions. *British Journal of Statistical Psychology, 11*, 1–12.
Lawley, D. N. and Maxwell, A. E. (1963). *Factor Analysis as a Statistical Method*. London, U.K.: Butterworth & Co.
Lederman, W. (1937). On the rank of the reduced correlation matrix in multiple-factor analysis. *Psychometrika, 2*, 85–93.
Lord, F. M. (1953). The relation of test score to the trait underlying the test. *Educational and Psychological Measurement, 15*, 517–548.
Lord, F. M. (1958). Some relations between Guttman's principal components of scale analysis and other psychometric theory. *Psychometrika, 23*, 291–296.
Lorenzo-Seva, U. (1999). Promin: A method for oblique factor rotation. *Multivariate Behavioral Research, 34*, 347–365.
Magnus, J. R. and Neudecker, H. (1988). *Matrix Differential Calculus*. New York: Wiley.
Marshall, A. W. and Olkin, I. (1979). *Inequalities: Theory of Majorization and Its Applications*. New York: Academic Press.
Maraun, M. (1996a). Metaphor taken as math: Indeterminacy in the factor analysis model. *Multivariate Behavioral Research, 31*, 517–538.
Maraun, M. (1996b). Meaning and mythology in the factor analysis model. *Multivariate Behavioral Research, 31*, 603–616.
Maraun, M. (1996c). The claims of factor analysis. *Multivariate Behavioral Research, 31*, 673–689.
Mardia, K. V. (1970). Measures of multivariate skewness and kurtosis with applications. *Biometrika, 57*, 519–530.
Mardia, K. V. (1974). Applications of some measures of multivariate skewness and kurtosis in testing normality and robustness studies. *Sankhya, B36*, 115–128.
Marsh, H. W., Balla, J. R., and McDonald, R. P. (1988). Goodness-of-fit indices in confirmatory factor analysis: The effect of sample size. *Psychological Bulletin, 103*, 391–411.
McDonald, R. P. (1974). The measurement of factor indeterminacy. *Psychometrika, 39*, 203–222.
McDonald, R. P. (1996a). Latent traits and the possibility of motion. *Multivariate Behavioral Research, 31*, 593–601.
McDonald, R. P. (1996b). Consensus emerges: A matter of interpretation. *Multivariate Behavioral Research, 31*, 663–672.
McDonald, R. P. and Burr, E. J. (1967). A comparison of four methods of constructing factor scores. *Psychometrika, 32*, 381–401.
McDonald, R. P. and Marsh, H. W. (1990). Choosing a multivariate model: Noncentrality and goodness-of-fit. *Psychological Bulletin, 107*, 247–255.
McDonald, R. P. and Swaminathan, H. (1973). A simple matrix calculus with applications to multivariate analysis. *General Systems, 18*, 37–54.

Mendel, G. J. (1865). Versuche (Über Pflanzen-Hybriden). *Verh. Naturf. Ver. in Brunn, 4*, 3–47. (Appeared in 1866.)

Meredith, W. (1964a). Notes on factorial invariance. *Psychometrika, 29*, 177–185.

Meredith, W. (1964b). Rotation to achieve factorial invariance. *Psychometrika, 29*, 187–206.

Meredith, W. (1993). Measurement invariance, factor analysis, and factorial invariance. *Psychometrika, 58*, 525–543.

Miller, G. A. (1964). *Mathematics and Psychology*. New York: Wiley.

Mischel, W. (1968). *Personality and Assessment*. New York: Wiley.

Morrison, D. F. (1967). *Multivariate Statistical Methods*. New York: McGraw-Hill.

Mosier, C. I. (1939). Determining a simple structure when loadings for certain tests are known. *Psychometrika, 4*, 149–162.

Muirhead, R. J. (1982). *Aspects of Multivariate Statistical Theory*. New York: Wiley.

Mulaik, S. A. (1964). Are personality factors raters' conceptual factors? *Journal of Consulting Psychology, 28*, 506–511.

Mulaik, S. A. (1965). Reliability as the upper limit of a test's communality. *Perceptual and Motor Skills, 20*, 646–648.

Mulaik, S. A. (1966). Inferring the communality of a variable in a universe of variables. *Psychological Bulletin, 66*, 119–124.

Mulaik, S. A. (1972). *The Foundations of Factor Analysis*. New York: McGraw-Hill.

Mulaik, S. A. (1976a, April). Comments on the measurement of factorial indeterminacy. In *Paper Presented at the Joint Meeting of the Psychometric Society and the Mathematical Psychology Group*, Bell Laboratories, Murray Hill, NJ.

Mulaik, S. A. (1976b). Comments on "The measurement of factorial indeterminacy." *Psychometrika, 41*, 249–262.

Mulaik, S. A. (1985). Exploratory statistics and empiricism. *Philosophy of Science, 52*, 410–430.

Mulaik, S. A. (1987). A brief history of the philosophical foundations of exploratory factor analysis. *Multivariate Behavioral Research, 22*, 267–305.

Mulaik, S. A. (1993). Objectivity and multivariate statistics. *Multivariate Behavioral Research, 28*, 171–203.

Mulaik, S. A. (1996a). On Maraun's deconstructing of factor indeterminacy with constructed factors. *Multivariate Behavioral Research, 31*, 579–592.

Mulaik, S. A. (1996b). Factor analysis is not just a model in pure-mathematics. *Multivariate Behavioral Research, 31*, 655–661.

Mulaik, S. A. (2001). The curve-fitting problem: an objectivist view. *Philosophy of Science, 68*, 218–241.

Mulaik, S. A. (2004). Objectivity in science and structural equation modeling. In D. Kaplan (Ed.), *The Sage Handbook of Quantitative Methodology for the Social Sciences*. Thousand Oaks, CA: Sage, pp. 425–446.

Mulaik, S. A. (2005). Looking back on the factor indeterminacy controversies in factor analysis. In A. Maydeu-Olivares and J. J. McArdle (Eds.), *Contemporary Psychometrics: A Festschrift for Roderick P. McDonald*. Mahwah, NJ: Lawrence Erlbaum Associates, pp. 173–206.

Mulaik, S. A. and Cureton, E. E. (1975).The weighted varimax rotation and the promax rotation. *Psychometrika, 40*, 183–195.

Mulaik, S. A. and McDonald, R. P. (1978). The effect of additional variables on factor indeterminacy in models with a single common factor. *Psychometrika, 43*, 177–192.

Mulaik, S. A. and Millsap, R. E. (2000). Doing the four-step right. *Structural Equation Modeling, 7*, 36–73.

Neuhaus, J. O. and Wrigley, C. (1954). The quartimax method: An analytical approach to orthogonal simple structure. *British Journal of Statistical Psychology, 7*, 81–91.

Newton, I. (1704). *Optiks*. London.

Ortega, J. M. and Kaiser, H. F. (1963). The LLT and QR methods for symmetric tridiagonal matrices. *The Computer Journal, 6*, 99–101.

Osgood, C. E. (1952). The nature and measurement of meaning. *Psychological Bulletin, 49*, 197–237.

Pearson, K. (1895). Contributions to the mathematical theory of evolution, II. *Philosophical Transactions of the Royal Society of London, 186*, 343.

Pearson, K. (1901). Mathematical contributions to the theory of evolution, VII: On the correlation of characters not quantitatively measurable. *Philosophical Transactions of the Royal Society of London, Series A, 195*, 1–47.

Pearson, K. (1909). On a new method for determining the correlation between a measured character A and a character B. *Biometrika, 7*, 96.

Pearson, K. and Lee, A. (1903). On the laws of inheritance in man. *Biometrika, 2*, 357–462.

Peirce, C. S. (1931–1958). *Collected papers of Charles Sanders Peirce* (Vols. 1–8); C. Hartshorne and P. Weiss, Eds. (Vols. 1–6), and A. W. Burks, Ed. (Vols. 7–8). Cambridge, MA: Harvard University Press.

Piaggio, H. T. H. (1931). The general factor in Spearman's theory of intelligence. *Nature, 127*, 56–57.

Piaggio, H. T. H. (1933). Three sets of conditions necessary for a g that is real and unique except in sign. *British Journal of Psychology, 24*, 88–105.

Piaggio, H. T. H. (1935). Approximate general and specific factors without indeterminate parts. *British Journal of Psychology, 25*, 485–489.

Pope, D. A. and Tompkins, C. (1957). Maximizing functions of rotations—Experiments concerning speed of diagonalization of symmetric matrices using Jacobi's method. *Journal of the Association of Computer Machinery, 4*, 459–466.

Ralston, A. (1965). *A First Course in Numerical Analysis*. New York: McGraw-Hill.

Rao, C. R. (1955). Estimation and tests of significance in factor analysis. *Psychometrika, 20*, 93–111.

Rencher, A. C. (2002). *Methods of Multivariate Analysis*, 2nd edn. New York: Wiley.

Rippe, D. D. (1953). Application of a large sampling criterion to some sampling problems in factor analysis. *Psychometrika, 18*, 191–205.

Roff, M. (1936). Some properties of the communality in multiple factor theory. *Psychometrika, 1*(2), 1–6.

Rogers, G. S. (1980). *Matrix Derivatives*. New York: Marcel Dekker, Inc.

Rozeboom, W. W. (1991). Hyball: A method for subspace constrained factor rotation. *Multivariate Behavioral Research, 26*, 163–177.

Rozeboom, W. W. (1991). Theory and practice of hyperplane optimization. *Multivariate Behavioral Research, 26*, 179–197.

Rozeboom, W. (1996a). What might common factors be? *Multivariate Behavioral Research, 31*, 555–570.

Rozeboom, W. (1996b). Factor indeterminacy issues are not linguistic confusions. *Multivariate Behavioral Research, 31*, 637–650.

Saunders, D. R. (1953). An analytic method for rotation to orthogonal simple structure. *Research Bulletin 53-10*. Princeton, NJ: Educational Testing Service.

Saunders, D. R. (1962). Trans-varimax: Some properties of the Ratiomax and Equamax criteria for blind orthogonal rotation. In *Paper Presented at the Meeting of the American Psychological Association*, St. Louis, September 1962.

Schlosberg, H. (1954). Three dimensions of emotion. *Psychological Review, 61*, 81–88.

Schmid, J. and Leiman, J. M. (1957). The development of hierarchical factor solutions. *Psychometrika, 22*, 53–61.

Schönemann, P. H. (1965). On the formal differentiation of traces and determinants. *Research Memorandum No. 27*. Chapel Hill, NC: The Psychometric Laboratory.

Schönemann, P. H. (1971). The minimum average correlation between equivalent sets of uncorrelated factors. *Psychometrika, 36*, 21–30.

Schönemann, P. H. (1996a). The psychopathology of factor indeterminacy. *Multivariate Behavioral Research, 31*, 571–577.

Schönemann, P. H. (1996b). Syllogisms of factor indeterminacy. *Multivariate Behavioral Research, 31*, 651–654.

Schönemann, P. H. and Wang, M.-M. (1972). Some results on factor indeterminacy. *Psychometrika, 37*, 61–91.

Sherin, R. J. (1966). A matrix formulation of Kaiser's varimax criterion. *Psychometrika, 31*, 535–538.

Singh, J. (1959). *Great Ideas of Modern Mathematics: Their Nature and Use*. New York: Dover Publications.

Snow, W. S. (1966) *A Factor Analysis of Verbal Comprehension Tests*. Unpublished PhD dissertation, University of Utah.

Spearman, C. (1904). General intelligence, objectively determined and measured. *American Journal of Psychology, 15*, 201–293.

Spearman, C. (1927). *The Abilities of Man*. New York: Macmillan.

Spearman, C. (1929). The uniqueness of "g." *Journal of Educational Psychology, 20*, 212–216.

Spearman, C. (1933). The uniqueness and exactness of *g*. *British Journal of Psychology, 24*, 106–108.

Steiger, J. H. (1979). Factor indeterminacy in the 1930's and the 1970's: Some interesting parallels. *Psychometrika, 44*, 157–167.

Steiger, J. H. (1989). *EZPATH: A Supplementary Module for SYSTAT and SYGRAPH*. Evanston, IL: SYSTAT.

Steiger, J. H. (1994). Factor analysis in the 1980's and the 1990's: Some old debates and some new developments. In I. Borg and P. P. Mohler (Eds.), *Trends and Perspectives in Empirical Social Research*. Berlin: Walter de Grulyter, pp. 201–224.

Steiger, J. H. (1995). Technical aspects of SEPATH. *Statistica*. Tulsa, OK: StatSoft, pp. 3651–3684.

Steiger, J. H. (1996a). Dispelling some myths about factor indeterminacy. *Multivariate Behavioral Research, 31*, 539–550.

Steiger, J. H. (1996b). Coming full circle in the history of factor indeterminacy. *Multivariate Behavioral Research, 31*, 617–630.

Steiger, J. H. and Lind, J. C. (1980, May). Statistically-based tests for the number of common factors. In *Paper Presented at the Annual Meeting of the Psychometric Society*, Iowa City, IO.

Thomson, G. H. (1935). The meaning of 'i' in the measurement of 'g' (general intelligence). *Journal of Educational Psychology, 26*, 241–262.

Thomson, G. H. (1956). *The Factorial Analysis of Human Ability*. Boston, MA: Houghton Muffin.

Thomson, G. H. and Lederman, W. (1939). The influence of multivariate selection on the factorial analysis of ability. *British Journal of Psychology, 29*, 288–305.

Thurstone, L. L. (1935). *The Vectors of the Mind*. Chicago, IL: University of Chicago Press.

Thurstone, L. L. (1937). Current misuse of the factorial methods. *Psychometrika, 2*, 73–76.

Thurstone, L. L. (1947). *Multiple Factor Analysis*. Chicago, IL: University of Chicago Press.

Thurstone, L. L. (1951). *An Analysis of Mechanical Aptitude*. Psychometric Laboratory Report No. 62. Chicago, IL: University of Chicago.

Torgerson, W. S. (1958). *Theory and Methods of Scaling*. New York: Wiley.

Trendafilov, N. T. (1994). A simple method for procrustean rotation in factor analysis using majorization theory. *Multivariate Behavioral Research, 29*, 385–408.

Tryon, R. C. (1957). Communality of a variable: Formulation by cluster analysis. *Psychometrika, 22*, 241–260.

Tryon, R. C. (1959). Domain sampling formulation of cluster and factor analysis. *Psychometrika*, 24, 113–135.L. R. (1955). The objective definition of simple structure in linear factor analysis. *Psychometrika, 20*, 209–225.

Turner, N. E. (1998). The effect of common variance and structure pattern on random data eigenvalues: Implications for the accuracy of parallel analysis. *Educational and Psychological Measurement, 58*, 541–568.

Velicer, W. F., Eaton, C. A., and Fava, J. L. (2000). Construct explication through factor or component analysis: A review and evaluation of alternative procedures for determining the number of factors or components. In R. D. Goffin and E. Helmes (Eds.), *Problems and Solutions in Human Assessment: Honoring Douglas N. Jackson at Seventy*. Norwell, MA: Kluwer Academic.

Vernon, P. E. (1961). *The Structure of Human Abilities*, 2nd edn. London, U.K.: Methuen.

Vittadini, G. (1989). Indeterminacy problems in the LISREL model. *Multivariate Behavioral Research, 24*, 397–414.

Wilkinson, J. H. (1960). Householder's method for the solution of the algebraic eigenproblem. *The Computer Journal, 3*, 23–27.

Wilkinson, J. H. (1965). *The Algebraic Eigenvalue Problem*. London, U.K.: Oxford University Press.

Williams, J. (1978). A definition for the common factor analysis model and the elimination of problems of factor score indeterminacy. *Psychometrika, 43, 293–306*.

Wilson, E. B. (1928a). On hierarchical correlation systems. *Proceedings of the National Academy of Sciences, 14*, 283–291.

Wilson, E. B. (1928b). Review of "The abilities of man, their nature and measurement" by C. Spearman. *Science, 67*, 244–248.

Wilson, E. B. (1929). Comments on Professor Spearman's note. *Journal of Educational Psychology, 20*, 217–223.

Wishart, J. (1928). Sampling errors in the theory of two factors. *British Journal of Psychology, 19*, 180–187.

Wittgenstein, L. (1953). *Philosophical Investigations*. New York: MacMillan.

Wohtke, W. (1984). The estimation of trait and method components in multitrait-multimethod measurement. Unpublished doctoral dissertation, Department of Behavioral Science, University of Chicago, Chicago.

Wohtke, W. (1996). Models for multitrait-multimethod matrix analysis. In G. A. Marcoulides and R. E. Schumacker (Eds.), *Advanced Structural Equation Modeling*. Mahwah, NJ: Lawrence Erlbaum Associates.

Wright, S. (1921). Correlation and causation. *Journal of Agricultural Research, 20*, 557–585.
Yuan, K. H. (2005). Fit indices versus test statistics. *Multivariate Behavioral Research, 40*, 115–148.
Yuan, K. H. and Bentler P. M. (1998). Normal theory based test statistics in structural modeling. *British Journal of Mathematical and Statistical Psychology, 51*, 289–309.
Yule, G. U. (1897). On the theory of correlation. *Journal of the Royal Statistical Society, 60*, 812–821.
Yule, G. U. (1907). On the theory of correlation for any number of variables treated by a new system of notation. *Proceedings of the Royal Society, Series A, 79*, 182–193.
Zwick, W. R. and Velicer, W. F. (1986). Factors influencing five rules for determining the number of components to retain. *Psychological Bulletin, 99*, 432–442.

Author Index

A

B

C

D

E

F

G

H

Subject Index

D

I

J

P

T

U

V